APPLICATIONS
OF ION BEAMS
TO METALS

APPLICATIONS OF ION BEAMS TO METALS

Edited by

S.T. Picraux, E.P. EerNisse, and F. L. Vook

Sandia Laboratories
Albuquerque, New Mexico

PLENUM PRESS · NEW YORK AND LONDON

Library of Congress Cataloging in Publication Data

International Conference on Applications of Ion Beams to Metals, Albuquerque, N.M., 1973.
    Applications of ion beams to metals.

    Includes bibliographical references.
    1. Physical metallurgy — Congresses. 2. Ion implantation — Congresses. I. Picraux, S. T., 1943 —   ed. II. EerNisse, E. P., 1940 —   ed. III. Vook, Frederick Ludwig, 1931 —   ed. IV. Title.
TN689.2.I537   1973   669'.9   74-4395

ISBN-13: 978-1-4684-2081-4   e-ISBN-13: 978-1-4684-2079-1
DOI: 10.1007/978-1-4684-2079-1

International Conference on Applications of Ion Beams to Metals
held in Albuquerque, New Mexico, October 2-4, 1973

© 1974 Plenum Press, New York
Softcover reprint of the hardcover 1st edition 1974

A Division of Plenum Publishing Corporation
227 West 17th Street, New York, N.Y. 10011

United Kingdom edition published by Plenum Press, London
A Division of Plenum Publishing Company, Ltd.
4a Lower John Street, London W1R 3PD, England

FOREWORD

Conferences have been held in the past on atomic collision
phenomena and on the applications of ion beams to semiconductors.
However, within the past year it became apparent that there is a
growing new area of active research involving the use of ion
beams to modify and study the basic properties of metals. As a
result a topical conference was organized to bring together for
the first time scientists with a wide range of backgrounds and
interests related to this field. This book contains the proceedings of the International Conference on Applications of Ion Beams
to Metals which was held in Albuquerque, New Mexico, October 2-4,
1973. Much of the work presented herein represents ideas and
concepts which have had little or no previous exposure in the open
literature. The application of ion beams to superconducting properties for example is quite new, as is the chapter on ion induced
surface reactions, which includes primarily oxidation and corrosion
studies of implanted materials. These areas, as well as the
chapter on implantation alloy formation, indicate important future
areas of the application of ion beams to metals.

A reading of the chapters on superconductivity and on oxidation and corrosion can serve to bring one up to date on nearly
all the existing information in these areas of the ion beam modification of metals. A broad perspective of the oxidation area
is given in the invited paper by G. Dearnaley. In the chapter on
alloy formation by ion implantation an excellent series of examples,
as well as a valuable guide to important considerations for investigations in this area, are given in the paper by R. S. Nelson.

The application of ion beams to study thin films has now become
well established and the chapter on backscattering studies of thin
films is in no way intended to be comprehensive of that area.
However, it might be noted that there has been relatively little
ion backscattering work on metal-metal thin film interactions and
the work included here on diffusion and compound formation represents an important extension of the existing data. The introductory preview by J. W. Mayer gives a good overview of the current
status, as well as some important limitations, of the ion backscattering technique for thin film studies.

The chapter on lattice location of impurities in metals represents a field which has been actively pursued for several years. This area has profited immensely from the marriage of two solid state techniques, hyperfine interaction studies and ion channeling lattice location studies of implanted impurities. The introductory review paper by H. de Waard and L. C. Feldman not only provides a detailed guide to past work but also gives some important insights to recent extensions of these techniques. Many of the leading experts in this field were present at this conference and the papers in that section represent the most recent results. The papers indicate how an increasingly detailed microscopic picture of the implanted impurity and its surroundings is being obtained by these experimental techniques.

Finally, the book concludes with chapters on radiation damage, helium blister formation, and void formation studies. Radiation damage studies of metals, of which ion damage is one type, is an old established field. However, in recent years it has gained renewed importance, particularly with respect to how nuclear reactor materials will behave under the intense radiation environments which will be encountered. Ion beam simulations have become an important aspect of those studies. There has been no attempt in the radiation damage chapters to make a complete survey of that area with respect to ion beams, since there have been numerous review articles in the past and the field is diverse. However, the reader will find valuable perspectives offered in the review papers by G. L. Kulcinski, which focuses on the void problem in metals, and by M. Wilkens, which discusses the limits of electron microscopy techniques in obtaining a microscopic understanding of defect properties of metals.

This book brings together first results in an important new topical area of rapidly growing interest. The new results together with the carefully selected review papers should give researchers a good understanding of the current status of this area as well as some important perspectives on future directions. Also, the book is well suited towards reading selected sections of interest and it will be valuable to the reader wishing to keep up in some of the latest developments in solid state physics and technology.

S. T. Picraux
E. P. EerNisse
F. L. Vook
Albuquerque, New Mexico
November 1973

ACKNOWLEDGMENTS

The sponsorship of Sandia Laboratories and the financial support of the Office of Naval Research was vital to the success of the conference. We particularly wish to acknowledge help and advice from the International Advisory Committee: J. W. Corbett (SUNY), J. A. Davies (Chalk River), G. Dearnaley (AERE), J. W. Mayer (Cal Tech), R. S. Nelson (AERE), and A. Seeger (Max Planck), and from L. Cooper (ONR), W. Bauer (Sandia), L. C. Feldman (BTL), and D. K. Brice (Sandia).

The operation and success of the meeting is due in large measure to the expert organization of the local committee, including J. A. Borders, W. Beezhold, G. J. Thomas and R. S. Blewer, to the excellent skill of the conference secretary, Delores McKinley, and to the valuable support of Sharon Husa and Barbara McHaffie.

CONTENTS

CHAPTER I

IMPLANTATION MODIFICATION OF SUPERCONDUCTIVITY

SUPERCONDUCTIVITY OF PALLADIUM AND PD-ALLOYS CHARGED

WITH H OR D BY ION IMPLANTATION AT HELIUM TEMPERATURES

W. Buckel and B. Stritzker

Institut für Festkörperforschung der

Kernforschungsanlage, 517 Jülich, Germany

In this paper we would like to consider the production of metastable systems by means of ion implantation into a substrate held at liquid helium temperature. This technique has the great advantage that one can fabricate nearly all types of alloy systems. In most cases the thermal energy at 4 K is sufficiently low to prevent the diffusion and the clustering of the implanted atoms. The investigation of such metastable alloys is very interesting with respect to their superconducting properties. As long as the electron-phonon interaction is the only known interaction producing superconductivity, it is believed that the highest transition temperatures can only be achieved in such "highly sophisticated" metastable alloys.

We succeeded in preparing superconducting Pd-H alloys by means of H implantation into Pd at liquid He-temperature[1]. In contrast to the dilute alloys normally achieved by ion implantation, we have produced metastable alloys with very high concentrations of the implanted atoms.

If one looks for superconductivity of the transition metals, it will immediately be understood why the superconducting properties of such Pd-H alloys are quite astonishing. The first Figure shows a portion of the periodic table. The transition metals which become superconducting are illustrated by the crosshatching. The following can be seen. There is a large block of superconducting ele-

3

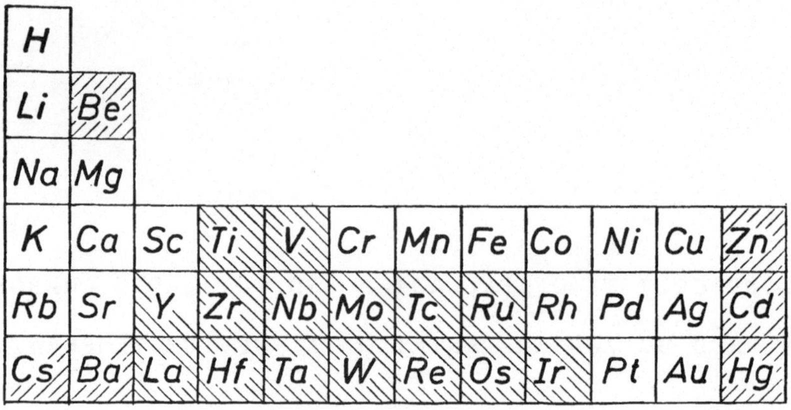

Fig. 1:   Portion of the periodic table illustrating the
          transition metals. The crosshatching shows the
          metals which become superconducting.

ments. At the higher number of valence electrons, there
exists a region of non superconducting metals, which con-
sits of the ferromagnets Cr, Mn and so on, and of the
nearly magnetic elements like Rh and Pd. These metals,
which do not become superconducting due to their magnetic properties, are followed by the noble metals. These
are not superconducting because of their weak electron-
phonon interaction.

     As can be noticed, Pd is located at a point in the
periodic table, which is thought to be unfavorable for
superconductivity. Furthermore, Pd is known as a good
example of an exchange enhanced paramagnet.

     This is the reason why it was very surprising when
about 1 1/2 years ago Skoskiewicz[2] found superconductivity in Pd after dissolving H. The next Figure shows
his measurements for the resistivity versus temperature
for different Pd-H alloys. The H-concentration is indicated by the ratio of H-atoms to Pd-atoms. One can see
the strong concentration dependence of the superconducting transition temperature, T_c. An alloy with a H-concentration of 0.73 is not yet superconducting above 1.2 K,
while an alloy with a H/Pd-ratio of 0.87 has a T_c of

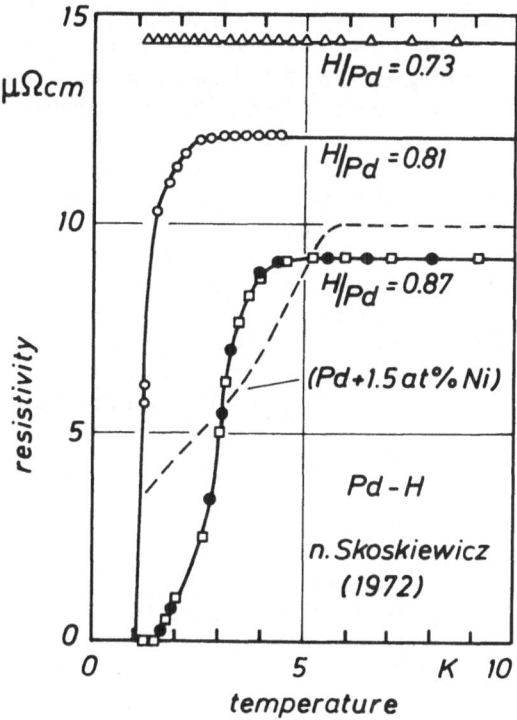

Fig. 2:   Resistivity versus temperature for different
          Pd-H-alloys after reference (2).

nearly 4 K. Skoskiewicz charged his Pd-samples by means
of electrolysis with H. With this method he was not able
to increase the H-concentration above 0.87. Another common method used to produce alloys of metals with H is
the charging of the metal under high H_2-gas pressure at
elevated temperatures. In the case of Pd, this method
also does not allow large H-concentrations without great
difficulties.

The strong increase of T_c with increasing H-concentration led us to attempt to enlarge the H-concentration
by means of our method of implantation at liquid helium
temperature. The H-implantation must be done at low temperatures because the diffusion of the H-atoms in the Pd
is too large at higher temperatures. Thus the implanted
atoms would leave the Pd-foil, and it would not be possible to raise the H-concentration. As is well-known, it

is not very difficult to make a hydrogen beam. One difficulty of our method consists of the long implantation
times necessary to get Pd-H alloys with concentrations of
more than one H-atom per Pd-atom. In order to shorten the
implantation time, we homogeneously precharged the 10 μm
thick Pd-foils in a H_2-gas pressure apparatus. In this
way one can easily obtain H-concentrations of about 0.7.
These foils had to be cooled down very rapidly in a cryo-
stat in order to prevent a large H-loss by diffusion at
room-temperature. This H-concentration is not yet sufficient to produce superconductivity. In order to further
save implantation time, we did not homogeneously raise
the H-concentration throughout the entire foil thickness.
We usually implanted the H_2^+-ions with one constant energy
of 130 KV.

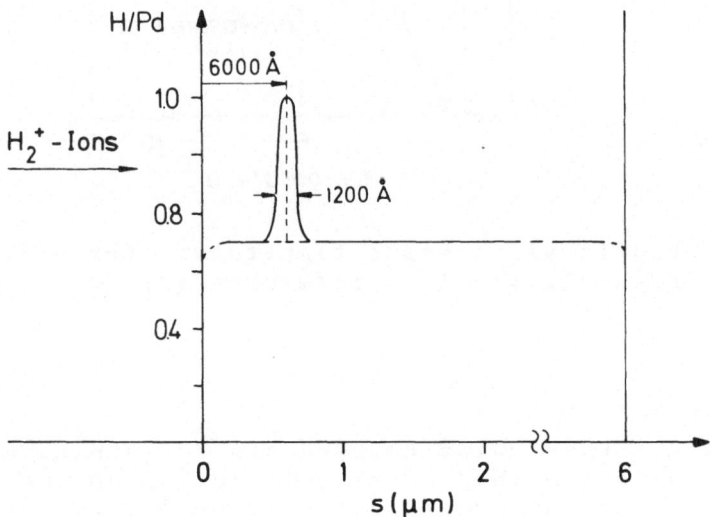

Fig. 3:   The profile of the H-concentration versus the
          foil thicknesses.

     Fig. 3 shows the resulting H-concentration profile
with respect to the foil thickness s. One can see the
homogeneous H-concentration from the precharging of the
whole foil. In the simplest model, the distribution of
the implanted atoms would be Gaussian and would have an
approximate 1200 Å width at a penetration depth of about
6000 Å [3]. In this way the H-concentration is raised in
only ≈ 1/30 of the foil thickness. This does not restrict

the measurement of the superconducting transition via
the electrical resistance. For if the thin part with the
high H-content becomes superconducting, it will short-
circuit the whole foil.

The hydrogen beam was scanned both in the vertical
and the horizontal directions to guarantee a homogeneous
irradiation. The maximum beam current density, about
$5 \mu A/cm^2$, was limited by the cooling power of our cryo-
stat. During this implantation, the foil temperature was
between 12 and 15 K while the substrate did not exceed
5 K.

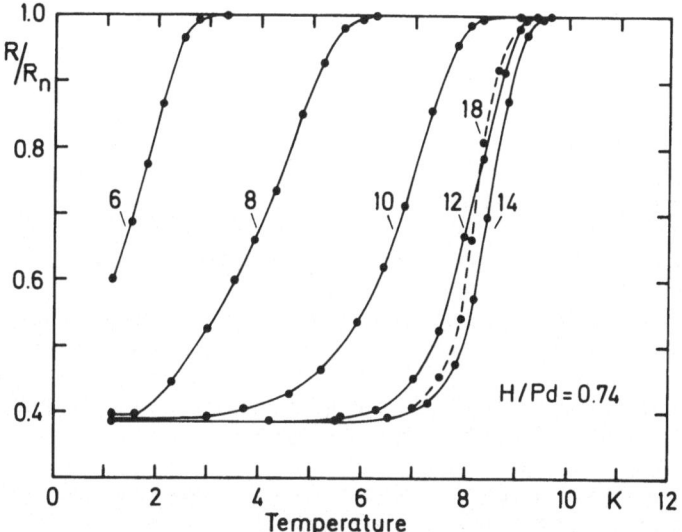

Fig. 4: Normalized resistance of a Pd-foil versus tem-
         perature for different doses, in mAs, of implan-
         ted H_2^+-ions with a starting concentration of
         $H/Pd \cong 0.74$.

The fourth Figure shows the normalized resistance
of a Pd-foil versus temperature with increasing dose of
implanted H. This foil was precharged with a H-concentration of 0.74. For our chosen sample geometry, 1 mAs
corresponds to 5×10^{16} atoms/cm². In order to always
have the same well-defined geometry, a mask shielded 1/3
of the foil length as well as the current and potential
contacts for the resistance measurement. The dose of
implanted H-atoms was determined by the charge reaching

the foil. In this way the H-dose for different foils
could be accurately compared.

The first drop of the resistance can be seen after
the implantation of 6 mAs. This decrease becomes sharper
and shifts to higher temperatures with increasing H-concentration. The superconducting transition reaches a maximum temperature of 9 K[4] and then slowly decreases
with additional implantation. With our technique we cannot exactly measure the absolute H-concentration because
we know too little about the distribution of the implanted H-atoms. By estimating the H-concentration which is
optimal for superconductivity, we get H/Pd-ratios between
1.0 and 1.2.

Since the implantation method is very flexible, we
could easily produce Pd-D alloys in order to measure an
isotope effect, and to learn more about the superconducting properties of the Pd-H system. We found a higher
T_c, of almost 11 K, for the heavier isotope.

Encouraged by this increase of T_c with greater mass,
we naturally wished to implant even heavier masses. Unfortunately, we could not use tritium because of its radioactivity. It was not expected that the implantation
of He could cause superconductivity. For we believe that
it is necessary to fill the d-band of the Pd with the H-
electrons, thus reducing the enhanced paramagnetism[5],
which is destructive for superconductivity. The inert He-
atoms will not give their electrons to the Pd. In spite
of this consideration, we implanted He^+-ions into a Pd-
foil precharged with H. With this experiment we could
exclude that distortions of the lattice, due to radiation damage, are responsible for the high T_c-values. The
implanted He-atoms did not give rise to superconductivity, but lowered T_c of an already superconducting Pd-H-
foil. In a following experiment, we implanted Li into
Pd-foils precharged with H. There was no superconductivity. We think that the Li-atoms are too large to be
built in as interstitials like the H. In a very recent
experiment, we tried to implant B which is known, to be
interstitially soluable in Pd[6]. This Pd-(H-B) system
becomes superconducting, but at lower temperatures than
we expected.

These experiments show that it is necessary for obtaining superconductivity in the Pd-H system, to destroy
the strong paramagnetism of the Pd by filling its d-band
with the electrons of the H. But the paramagnetism of the

Pd also vanishes in alloys of Pd with other metals which give their electrons to the d-band of the Pd[7], and which do not become superconducting. Well-known examples are Pd-Ag and Pd-Au alloys[8]. Why then is Pd-H superconducting and Pd-Ag not?

To answer the similar question: to what extent is the H responsible for the supercondictivity, we changed the Pd-matrix by alloying with Ag and Au. Surprisingly, we found very high transition temperatures of ≈ 16 K[9]. The implantation method for producing H-metal alloys has the important advantage that the implantation process is nearly independent of the host material. In contrast to this the electrolytic and the gas pressure methods are very sensitive to the metal you wish to charge with H. The result of our experiments in the Pd-noble metal-H system is shown in Fig. 5. The maximum of T_c is plotted versus Ag or Au concentration. The filled circles and the curve indicate the T_c values of the Pd-Ag-D system. T_c increases with increasing Ag-content until 15 at%. Then it remains nearly constant, and above 30 at%, T_c drops rapidly. The open circles represent the T_c values in the Pd-Ag-H system and are not much different from the D-

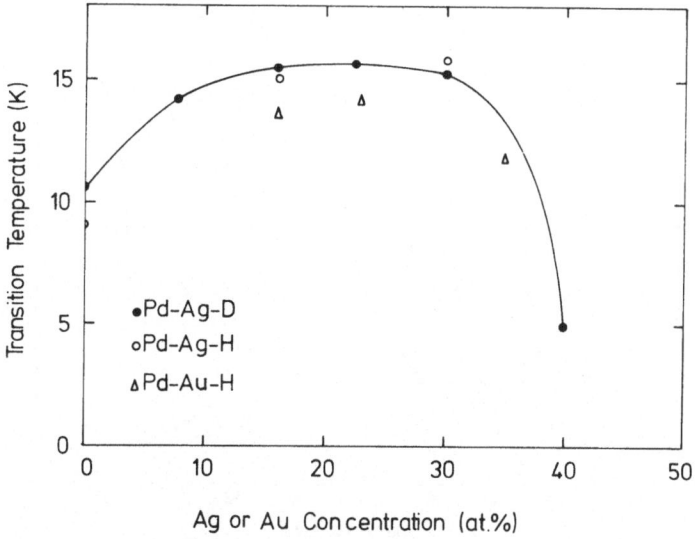

Fig. 5: Maximum T_c-values versus Ag or Au concentration for Pd-Ag or Pd-Au alloys charged optimally with hydrogen or deuterium.

points. The triangles show the maximum T_c-values for Pd-Au-H alloys which are about 2 K lower than the values of the Pd-Ag-D system.

These high T_c-values of 16 K, only 6 K below the highest known T_c [10], are quite astonishing in a part of the periodic table thought to be unfavorable for superconductivity. The strong dependence of T_c on the matrix indicates that not only the filling of the d-band of the Pd is responsible for superconductivity. There are some hints in the literature that the addition of H[11] as well as the substitution of Ag[12] or Au[13] slightly weaken the crystal-bonding of the Pd. We have found an interesting instability in the Pd-Ag-D system supporting the assumption of a weakened phonon spectrum which would be favorable for superconductivity. This instability, which is shown in the final figure, occurs at Ag-concentration in the region of the highest T_c-values. The resistance of a Pd-alloy with 16 at% Ag is plotted versus temperature. T_c increases with increasing dose of D-ions. But shortly before reaching the maximum value something very strange happens. The superconductivity has vanished below 1 K while the resistance has

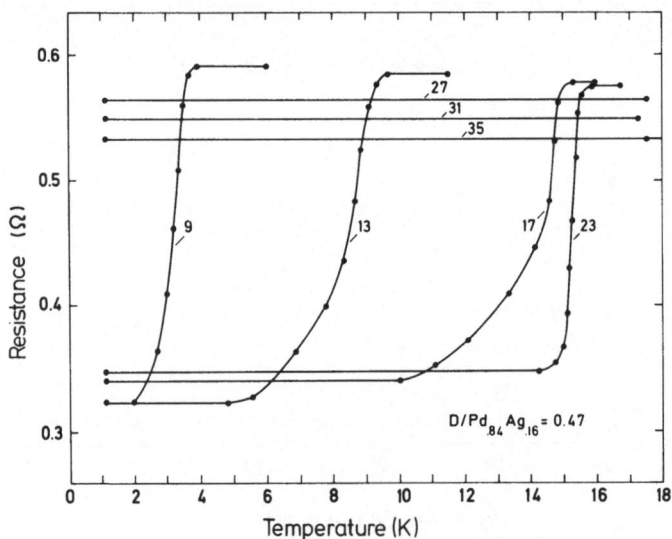

Fig. 6:   Resistance of a $Pd_{.84}Ag_{.16}$-foil versus temperature for different doses, in mAs, of implanted D_2^+-ions with a starting concentration of $D/(Pd+Ag) \simeq 0.47$.

dropped by about 5%. The residual resistance decreases more and more with further implantation. This behavior is even more astonishing if one keeps in mind that the D-concentration is raised in only 1/30 of the foil thickness.

This instability supports the assumption that the crystal bonding has become so weak that a new nonsuperconducting phase is formed. Such instabilities are well-known in superconductors with high transition temperatures. Since a strong electron-phonon interaction, i.e. weak phonon-modes, are necessary for high T_c-values, lattice instabilities can often occur which destroy the superconductivity. Therefore we think that the high T_c values in the Pd-Ag-D systems are achieved by the weakening of the phonon-spectrum which is caused by the D as well as the Ag.

These instabilities are the reason why it is so important to investigate metastable alloys, which are favorable for high temperature superconductivity. In this paper we have tried to show that ion implantation at liquid helium temperature is a valuable tool for preparing such metastable alloys. We strongly hope that this method will succeed in producing additional high temperature superconductors.

References

1. B. Stritzker and W. Buckel, Z. Phys. 257 (1972) 1.
2. T. Skoskiewicz, Phys. Stat. Sol. (a) 11 (1972) K 123.
3. W. Whaling, Encyclopedia of Physics, Vol. XXXIV, 193.
4. Theoretical models for the high T_c-values of Pd-H have been proposed by
   K.H. Bennemann and J.W. Garland, Z. Phys. 260 (1973) 367.
   B.N. Ganguly, Z. Phys. to be published.
5. H.C. Jamieson and F.D. Manchester, J. Phys. F 2 (1972) 323.
6. R. Mehlmann, H. Husemann and H. Brodowsky, Ber. Bunsenges. phys. Chem. 77 (1973) 36.
7. J. Wucher, Ann. Physique 7 (1952) 317.
8. F. Heininger, E. Bucher and J. Muller, Phys. kond. Materie 5 (1966) 243.
9. W. Buckel and B. Stritzker, Phys. Lett. A 43 (1973) 403.
10. J.R. Gavaler, Symposium on Superconductivity and Lattice Instabilities, Gatlinburg, Sept. 1973.

11. F.A. Lewis, The Palladium Hydrogen System, Academic
    Press London (1967), p. 44.
12. E. Walker, J. Ortelli and M. Peter, Phys. Lett. <u>31A</u>
    (1970) 240.
13. H. Masumoto and S. Sawaya, Trans. J.I.M. <u>11</u> (1970)
    51.

DISCUSSION

Q: (W. Bauer) Have you considered annealing your H implanted PdH
samples after implantation to temperatures ($\lesssim$ 100 K) where most
point defects anneal out but most of the implanted H will not
diffuse?

A: Yes, we annealed our samples at 80 K for about 15 minutes, but
we only found H diffusing out of the region of high H-content.
Annealing of a PdH sample with less than the optimal H concentration led to a decrease of T_c, while the annealing of a sample with
more than the optimal H concentration caused an increase of T_c.
Annealing of point defects would alter T_c only in one direction.

Q: (J. E. Schirber) Would you please describe how you determined
your H (or D) to metal ratios and estimate your uncertainty in
this quantity?

A: Assuming a reasonable width of the distribution of the implanted atoms (between 20-30% of the penetration depth) we
compared our T_c values at low H dose with those obtained from
Skoskiewicz (Ref. 2) in homogeneous PdH alloys. A rough extrapo-
lation to higher H doses led to an optimal H/metal ratio between
1.0 and 1.2.

Q: (H. deWaard) How is T_c defined in cases where the transition
to the superconducting state is rather gradual? Is this definition commensurate with the theoretical definitions, e.g. $T_c \sim \theta_D$
$\exp(-1/NV)$, etc.?

A: At low H doses, where the transitions are rather broad, we
defined T_c as that temperature where the residual resistance R_n
had dropped 10%. Thus, we could obtain a better fit to the T_c
values of Skoskiewicz (Ref. 2). The maximum T_c values of Fig. 5
are defined at 0.5 R_n. The corresponding transition widths are
usually smaller than 1 K. Theories like the BCS theory do not
give any information about the shape of the transition curve.
They assume an infinitely sharp transition. A broadening of the
transition can be explained by inhomogeneities or fluctuation
effects.

Q: (K. L. Merkle) What is the optimum concentration, i.e. the concentration that gives the highest T_c value, in the case of D in Pd? Also, how is the optimum H concentration related to the Ag and Au concentration in the case of the PdAg and PdAu alloys?

A: The optimum concentration of D is the same as that of H in Pd. We must always implant the same amount of H_2^+ ions independent of the Ag or Au content of the Pd alloys. Since the H concentration one can obtain by precharging the Pd-noble metal-alloys under H_2 gas pressure decreases linearly with Ag or Au concentration, the optimum H concentration decreases linearly with increasing noble metal content.

Q: (R. W. Bower) Did you try coating the Pd films to prevent outdiffusion of H_2?

A: No. Coating of the Pd would only be interesting if we could achieve the high required H concentrations throughout the whole foil thickness. To obtain this concentration in only $\approx 1/30$ of the foil thickness we had to implant more than 10^{18} H-atoms/cm^2. Thus a homogeneous implantation of these foils would take unrealistically long implantation times.

Q: (H. C. Freyhardt) Can you characterize in more detail the type of lattice instabilities you are talking about?

A: Until now we measured only the change of the electrical resistance during the transition (Fig. 6). There is an experiment in progress to investigate lattice instability by means of x-ray analysis.

ION IMPLANTATION IN SUPERCONDUCTING THIN FILMS

O. Meyer, H. Mann and E. Phrilingos

Institut für Angewandte Kernphysik

Kernforschungszentrum Karlsruhe

ABSTRACT

Selected ions were implanted at room temperature in thin films of the transition metal superconductors Ti, Zr, V, Nb, Ta, Mo, W and Re, the A-15 compound Nb_3Sn and the interstitial compounds NbC and NbN with NaCl structure. Both chemically active and inert ions were used in order to distinguish between radiation damage and other effects influencing the superconducting transition temperature T_c. Successively larger decreases in T_c, due to radiation damage, were obtained for NbC, NbN, Mo, Ta, V, Nb and Nb_3Sn in that order. Chemically active ions were found to produce the same reduction in T_c as inert ions in Ta, V, Nb and Nb_3Sn, whereas in those materials in which radiation damage is less influential, namely NbC, NbN, Mo, W and Re, T_c enhancement was found to occur and was dose-dependent.

INTRODUCTION

The influence of radiation damage on superconducting properties of material reported so far was mainly concerned with critical current and critical field measurements[1]. Light ions were usually used (up to mass 4) and ion doses were kept below 10^{13} ions/cm^2. The enhancement of the critical field due to the formation of pinning centers produced during irradiation was found by several investigators to have a maximum within this dose range. Only a slight effect on the transition temperature, T_c was observed in these earlier studies.

A far higher degree of disorder than that produced by light ion irradiation can be obtained in metal films made by vapour quenching[2,3] (evaporation on a cold (4^OK) subtrate). Such films usually show a fine grained structure with crystallite sizes smaller than about 50 $\overset{o}{A}$ and in some cases a transformation into a liquid-like amorphous phase appeared to take place. In general it was observed that for vapour-quenched non transition metal films T_c is substantially higher than that of bulk material. The reason for this T_c enhancement can probably be found in the decrease of the average phonon frequency, a phenomenon that has been discussed in several theoretical treatments [6,7]. Lattice disorder also influences the electronic band structure which seems to be important for the superconducting properties in transition metals, where the bulk T_c correlates well with the electronic density of states at the Fermi edge, $N(O)$. In general it was found that for vapour quenched films of group V-B (V, Nb, Ta with large $N(O)$) T_c's are lower than the bulk T_c's and in vapour-quenched films of Mo, W and Re (low $N(O)$) T_c's are higher than bulk T_c's [8]. The situation in sputtered films[4] and in films produced by electron beam evaporation in a vacuum of $10^{-5} - 10^{-6}$ Torr [5] is further complicated, as it has been shown that metastable high T_c phases can be stabilized by impurities, structural defects and intrinsic stresses.

The purpose of the present work was to investigate the influence of disorder and impurities produced by heavy ion bombardment on the superconducting properties. Ion implantation has been used to introduce damage and impurities in a controlled manner into solids. The effect of radiation damage on T_c can be separated from the effects of the incorporated impurities by either using inert gas ions or by implanting through the metal layer so that the ions come to rest in the substrate. Chemical effects on T_c have been studied by implanting ions of different solubilities in the host lattice. Additional information has been obtained by studying radiation damage and solubility levels in ion implanted vanadium single crystals using transmission-electron-microscopy and the backscattering and channelling technique[9].

EXPERIMENTAL AND ANALYSIS

Films of the transition metals Ti, Zr, V, Nb, Ta, Mo, W and Re together with Nb_3Sn with thicknesses from 100 to 4000 $\overset{o}{A}$, were prepared by electron beam evaporation and by co-deposition onto quartz substrates in an ultra-high vacuum system where a pressure of 1 to 5 × 10^{-8} Torr could be maintained during evaporation. The bakeable UHV-system consists of a turbomolecular pump, an ion getter pump and a titanium evaporation pump. NbN layers have been produced by reactive sputtering in an argon-nitrogen plasma. During the production of NbC layers in an argon-plasma the Nb-target has been covered by a perforated carbon plate[10]

These films were implanted with selected ions from every group of the periodic system. Ion energies used were in the range 50 to 400 keV, and doses were from 10^{14} to 10^{17} ions/cm^2. Dose rates were maintained between 0.1 and 10 µA/cm^2. Layer thickness, sputtering during implantation, and the depth distribution of implanted ions were recorded by use of the nuclear backscattering technique. The usefulness of this technique is demonstrated in Fig. 1 where the backscattering spectra, obtained with 2 MeV α-particles from a Mo layer are shown prior to and after implantation of 1×10^{17} S^+/cm^2 at 130 keV. By analysis of such spectra, which is described elsewhere[11], one can obtain the sputter yield (here about 3). By additional thickness measurements of the Mo-layer and the step between implanted and unimplanted Mo-layer, as indicated in Fig. 1, density changes in the implanted layer can be determined.

The superconducting transitions were measured resistively, with T_C being defined as the temperature at which the resistance decreased to half of its normal value. During implantation the layers were partially covered in order to restrict the implanted area to (5×10) mm^2. For the T_c measurements the voltage contacts (Fig. 2) were placed at equal distances from the boundary between the implanted and unimplanted areas. With this contact arrangement the resistance drop reveals a step as it is shown in Fig. 2 for V and Nb layers implanted with N^+ ions at various doses. For comparison the resistance drop of the unimplanted layer is included. The resistance curves of the implanted layers show two steps. The first step coincides with the resistance drop of the unimplanted layer, which can be used as a check to show that the shielded part of the layer is not influenced during the implantation procedure. The difference δT_c between the onset of the transition in the unimplanted layer and the temperature where the implanted layer was found to be completely superconducting was used as a measure of the damage influence on T_c. In the case where T_c enhancement is observed the results were confirmed by placing the voltage contacts directly onto the implanted layer. Diffraction patterns have been taken with a thin film X-ray camera in order to see if ion implantation has any effect on grain size and film structure.

RESULTS

Layers with the following T_c's have been routinely produced: Ti, Zr, Mo and W with $T_c < 1.2°K$, V $(4.5 - 5.3°K)$, Nb $(8.3 - 9.3°K)$ Ta $(4 - 4.45°K)$, Re $(1.5 - 1.7°K)$, Nb_3Sn $(18.2°K)$, NbN $(15 - 16°K)$ NbC $(9 - 11°K)$. The width of the transitions were usually $<0.1°K$.

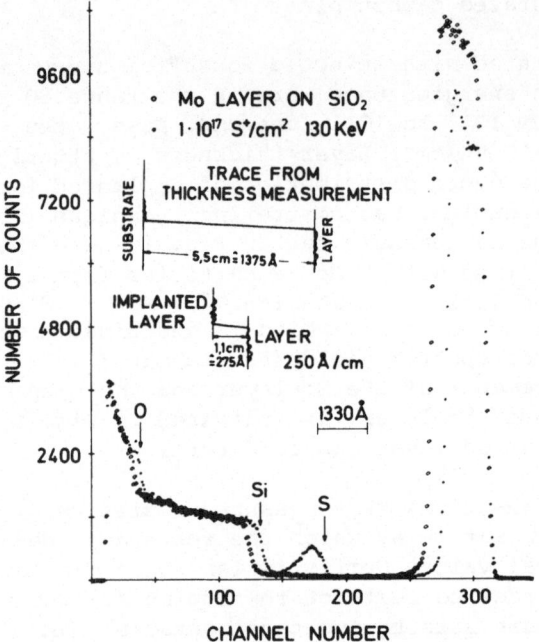

Fig. 1 Backscattering spectra obtained from an implanted and un-implanted part of a Mo layer. Together with thickness measurements, shown in the inset, the sputter yield and the density change in the implanted layer is determined.

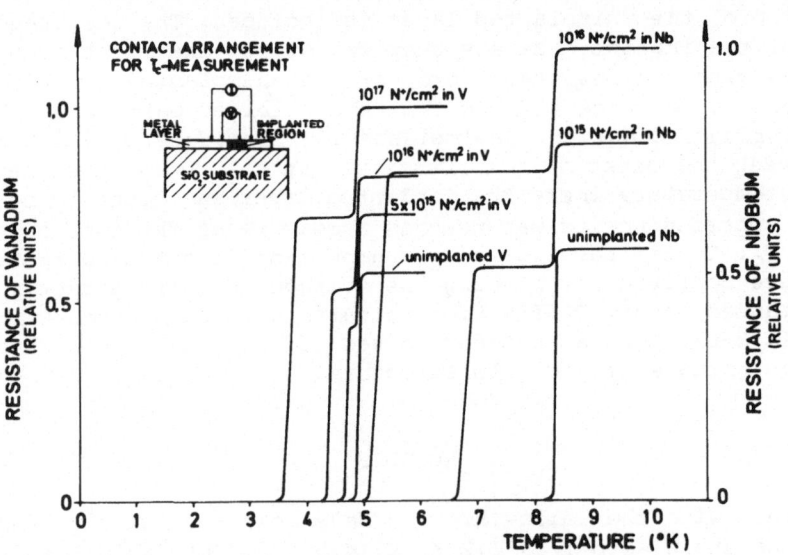

Fig. 2 Resistance versus temperature for Nb and V thin films, implanted with low energy N^+ ions at different doses. The inset shows the contact arrangement.

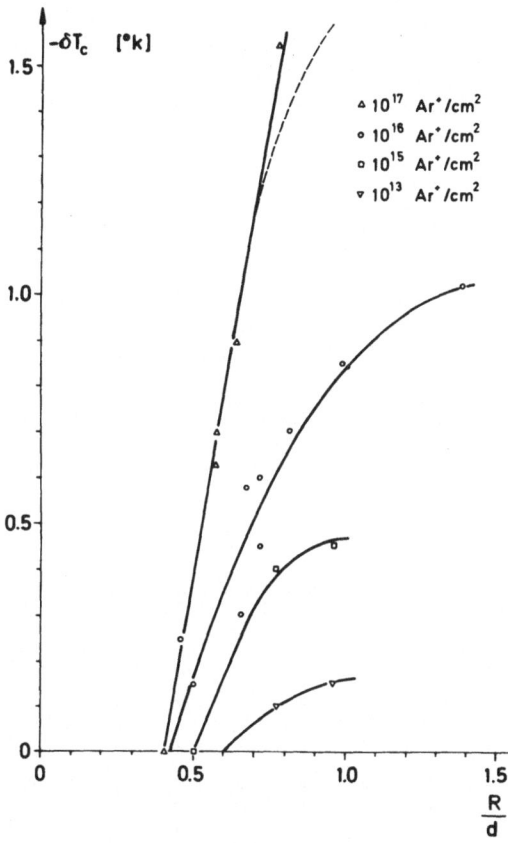

Fig. 3.   Decrease in T_c for V versus the ratio R/d with the ion dose as parameter.

a)   Influence of Radiation Damage on T_c in V.

        In order to study the influence of radiation damage on T_c, the metal films were implanted with the chemically inert Ar^+ and Kr^+ ions. The magnitude of the decrease in T_c, δT_c, was found to depend on the range of the implanted ions. This dependence is shown in Fig. 3 for Ar^+ ions implanted in V at four different doses. The observed δT_c is plotted as a function of the ratio R/d where quantity d is the layer thickness and $R = R_p + 1.17 \Delta R_p$. Here R_p and ΔR_p are the mean projected range and the range straggling, respectively, as determined from the LSS theory.[12] It can be seen that δT_c has a saturation level for R/d $\geq$ 1. In further experiments

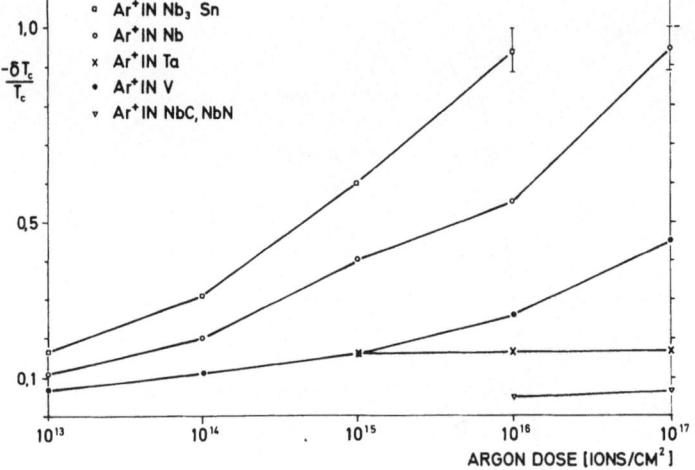

Fig. 4

Decrease in T_C, normalized to the bulk T_C, versus ion dose for different superconducting materials.

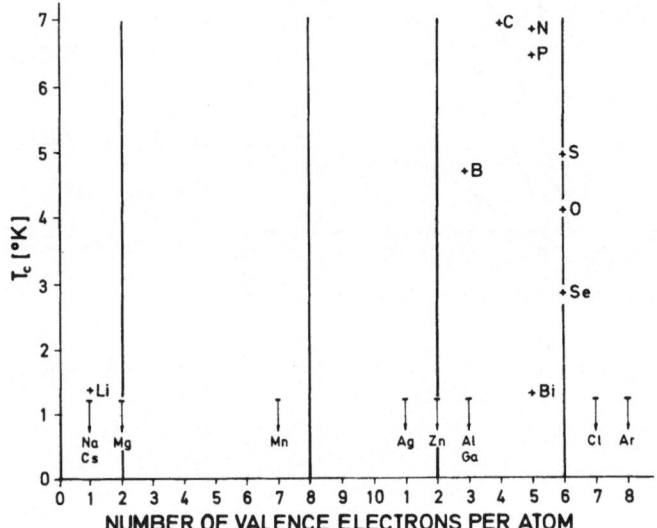

Fig. 5

T_C enhancement in Mo after implanting ions with various numbers of valence electrons per atom.

where R exceeds d by a factor 3 to 4 it has been verified that the decrease in T_C is due to the damage produced and not due to incorporated noble gas ions. The observed decrease in T_C for R/d < 1 is probably due to enhanced defect diffusion during implantation. If this were not the case the undamaged region of the metal layer would provide an effective short circuit. Defects deeper than R_p have also been found in ion implanted V single crystals by use of channelling technique and transmission electron microscopy[9]. In all further damage experiments R/d was kept about 1 by suitable choice of ion energy.

The influence of damage on T_C in different materials is summarized in Fig. 4 where δT_c, normalized to the corresponding T_C of the bulk material is presented as a function of argon ion dose. The effect of damage on the T_c of Nb_3Sn layers was found to be stronger than that observed in V and Nb. For example by the implantation of 10^{16} Ar^+/cm^2 the initial T_c of 17.8^oK was decreased to 2.0^oK. For a constant dose it can be seen that $\delta T_c/T_c$ increases for material with increasing electronic density of states at the Fermi edge. Vanadium is an exception as $N(0)_V > N(0)_{Nb}$. NbC and NbN are found to be highly radiation resistant. No effects on T_C were observed by bombardment with protons and He^+ ions at doses up to 10^{16} ions/cm^2 which is in agreement with results presented in [1]. A possible impurity effect due to the implanted ions has been further investigated by implanting Ga^+ ions in V and Nb. By comparison of δT_c values for Kr^+ implantation with those for Ga^+ implantation, no additional effect could be detected (within the limits of our measurement errors) with up to about 20 at % of Ga atoms incorporated in V and Nb films.

Since grain size is not affected by ion implantation as was verified by thin film X-ray diffraction and since resistivity measurements show the mean free path of electrons at $T > T_C$ is not strongly reduced, it is assumed that the smearing out of structure in the electronic density of states is not the only factor in the decrease of T_c by radiation damage. If this were true there should also occur an increase in T_c for argon implanted Mo due to damage, but this is not observed.

b) Chemical Effects of Implanted Ions

Possible chemical effects were studied by implanting ions from nearly all groups of the periodic system in those transition metals where the damage effect is presumably low, i.e. where N(E) has no sharp peaks at the Fermi edge. In Fig. 5 the T_C enhancement observed by implanting ions with various values of valence electrons per atom in Mo layers is summarized. T_C enhancement up to 7^oK was observed for implantation of elements from groups III(B), IV(C), V (N, P, Bi) and VI (S,O,Se).

For elements which are completely insoluble in Mo no T_C enhancement is observed. Those elements which form stable intermetallic compounds, either superconducting or non-superconducting, give T_C enhancement. It is possible that the elements that tend to form intermetallic compounds stabilize the disorder in Mo on a microscopic scale, thereby producing the T_C enhancement. Preliminary results also show enhancement of T_c by implantation of N^+ and S^+ ions in W and Re. A summary of the results obtained by ion implantation, T_c^{II}, in transition metals is given in TABLE I and are compared with the bulk value T_c^{Bulk}, theoretical values for amorphous transition metals [14] T_c^{Theory}, T_c^{VQ} for vapor-quenched films [8] and T_c^{SP} for "reactively"

Metal	T_c^{Bulk}	$T_c^{Theory}(VQ)$	T_c^{VQ}	T_c^{SP}	T_c^{II}
Ti	0.39	0.5	–	–	$\lesssim 1.2$
Zr	0.5	0.1-0.5	3.5	0	$\lesssim 1.2$
V	5.3	9	–	≤ 5.3	2-3
Nb	9.2	7.4-8.2	6	≤ 9.2	3-5
Ta	4.5	2.7-3.5	2	≤ 4.3	3-4
Mo	0.9	9.6	8.5	7-8	~ 7
W	0.01	4 -4.3	3.5	~ 4	~ 3.5
Re	1.7	6.2-7.5	7.5	~ 7	~ 4.5

Table I
Experimental (ion implantation, vapour-quenching, sputtering) and theoretical results for T_c of some transition metals.

sputtered films ([5],[17]). As the values of T_c^{II} are strongly dose dependent, maximum values were given for T_c enhancement, whereas for the T_c decrease the values obtained for 10^{16} Ar^+/cm^2 have been used.

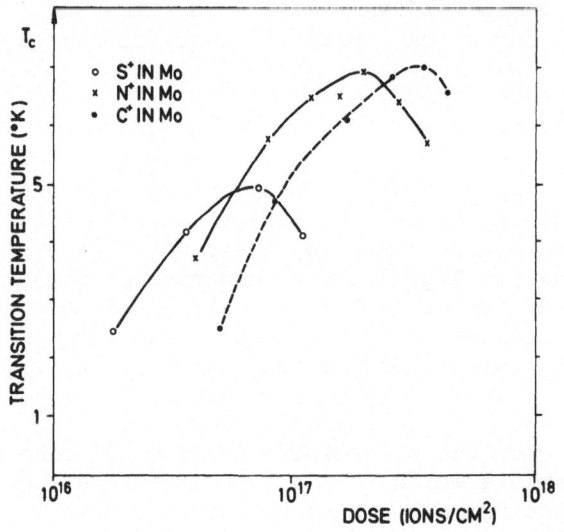

Fig. 6

Dose dependence of T_c enhancement after implanting S^+, N^+ and C^+ ions in Mo.

No increase in T_c was found after implanting ions with low solubilities in Ti and Zr, whereas after implanting Fe^+ ions in Ti a T_c of about $3^\circ K$ has been observed. This increase in T_c is probably due to the increase of the number of electrons per atom by implanting soluble elements, an effect which will be studied further in more detail. Preliminary results on non transition metals (Al, Sn) show that implanting B^+ and S^+ increased the T_C of Al to 2.5°K and of Sn to 4.4°K. Only a slight increase of about 0.1°K has been observed by implanting Ar^+ ions in Al under similar conditions. The dose dependence of the T_C enhancement is presented in Fig. 6 for S^+, C^+ and N^+ ions implanted in Mo. A maximum value in T_C is observed. T_C^{max} increases with decreasing mass of implanted ions and shifts to higher doses. These maxima were found to occur for lower ion doses in Re and for higher doses in W as compared with the results presented for Mo. More detailed studies especially on the density changes in the implanted layers

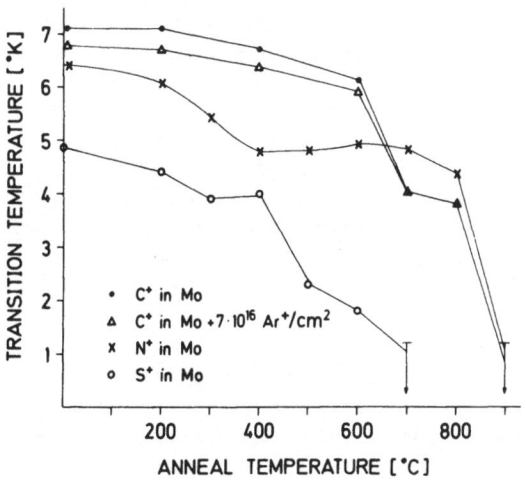

Fig. 7
Influence of an isochronal anneal process on T_C of implanted Mo layers.

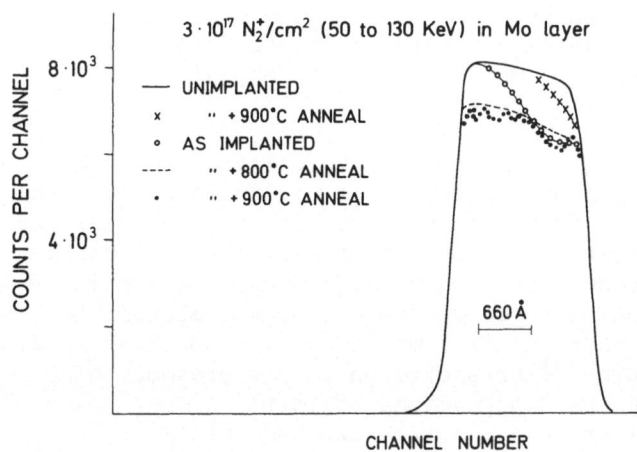

Fig. 8
Backscattering spectra from implanted and unimplanted Mo layers at room temperature and at 800°C and 900°C.

are necessary and will be done in the future.

c) Annealing Effects

Isochronal annealing processes in a vacuum of 10^{-7} Torr have been performed on implanted Mo layers and the dependence of T_c on annealing temperature is shown in Fig. 7. The T_c enhancement is found to be stable for annealing temperatures up to 400°C in S^+ implanted layers and up to 800°C in C^+ and N^+ implanted layers. The influence of lattice disorder was studied by additional implanting 7×10^{16} Ar^+/cm^2 in a C^+ implanted Mo layer with a T_c of 7.2°K. The observed T_c reduction of 0.5°K as shown in Fig. 7 is small compared to results presented in Fig. 4. From the backscattering spectra taken from implanted Mo layers and unimplanted layers it can be seen (Fig. 8) that at 800°C the nitrogen has moved throughout the metal layer. At 900°C additional oxygen has been incorporated as can be seen in the spectra of the unimplanted layer and the inhomogeneities visible in the spectra of the implanted layer seem to indicate the formation of precipitations. The increase of T_c due to V_3Si formation[13] has prevented study of the effect of anneal temperature on implanted V layers. In the case of the Nb_3Sn films however, the influence on T_c due to radiation damage has been found to be removed on annealing for several hours at 800°C.

DISCUSSION

To summarize, we found a strong decrease of the transition temperature of Ta, V, Nb, and Nb_3Sn films after bombarding with low energy high mass ions. The magnitude of the decrease in T_c was found to depend on mass, dose and range of implanted ions, and is therefore attributed to the radiation damage produced during implantation and not to chemical effects. It is assumed that the smearing out of structures in the electronic state density distribution is only one reason for the observed decrease in T_c. Otherwise, argon implantation in W, Mo and Re should produce a T_c increase for the same reason, but this is not observed. The disturbance of the long range order in Nb_3Sn may explain the strong decrease of T_c found for that material.

T_c enhancement was observed by implanting selected ions able to form intermetallic compounds but otherwise of low solubility in Mo, W, and Re films. It is assumed that these ions do not stabilize a high T_c phase after implantation but can stabilize a dislocation of the lattice atoms. The suggestion that implantation of impurities will stabilize high T_c phases largely composed of the element in question, as was observed in transition metal films evaporated in a vacuum of 10^{-5} to 10^{-6} Torr [4] or sputtered in the presence of gaseous impurities [5], can be discussed as follows:
In contrast to results given in [4] our layers had bulk T_c's in

the limit of measurement ($T_c \geq 1.2^\circ K$). X-ray patterns of unimplanted Nb and Mo films did only show the normal bcc phase. The implanted regions of these layers did not show any difference in the X-ray pattern. There was no evidence for the presence of other phases and there was no spread in the line width indicating small grain sizes. In addition the implanted Mo layers with high T_c were stable against irradiation with argon ions and stable against temperature treatment. In forming an impurity stabilized high T_c-phase one would also expect a saturation in the dose dependence of T_c rather than a maximum.

There are several arguments against the suggestion that the high T_c phases of MoN and MoC have formed during implantation. Firstly, the dose for maximum T_c enhancement was lower than necessary for the formation of stoichiometric MoN and MoC, secondly N^+ in W increased T_c but no high T_c WN-phase is known. N^+ in Ti and Zr did not increase T_c but TiN as well as ZrN are known to be superconducting.

The stabilized dislocation mentioned above may change the atomic volume[14] and therefore the electron density which causes a shift in the density of states at the Fermi edge. A further possible reason for this shift may be found in the high electronegativity of the implanted ions causing a T_c increase. The observed maxima in the dose dependence of T_c and the shift in the maxima for different material may be explained with the assumption that the Fermi energy is shifted across a peak in $N(E)$ resulting in a maximum in T_c. The observed increase or decrease in T_c may also be discussed in terms of the Eliasberg function $\alpha^2(w) F(w)$ [15] which describes the interaction between phonons and electrons causing the effective attraction between the electrons in a superconductor. It has been shown for nontransition metals[16] that the softening of the phonon distribution and therefore of $\alpha^2 F(w)$, in highly disordered superconductors can result in either a positive or a negative change in the transition temperature. In this regard the observed decrease in T_c for the group Vb elements may be due to a decrease in α or due to a shift of $F(w)$ to higher frequencies.

Further, by implanting ions of higher solubility in the host lattice it is possible to change the number of electrons per atom and to increase T_c in suitable ion - target systems.

ACKNOWLEDGEMENT

Thanks are due to M. Kraatz and R. Smithey for careful layer preparation and ion implantation and to F. Ratzel for performing T_c measurements. We further want to thank W. Buckel, J. Geerk, M. Gettings and G. Linker for many helpful discussions.

REFERENCES

(1)   G.W. CULLEN in BNL-50155 (1968) p. 437.

(2)   W. BUCKEL and R.J. HILSCH, Z. Phys. 132, 420 (1952);
       138, 109 (1954).

(3)   Review of M. STRONGIN Physica Vol. 55, 155 (1971).

(4)   W.L. BOND et al., Phys. Rev. Lett. 15, 260 (1965).

(5)   K.L. CHOPRA, M.R. RANDLETT, and R.H. DUFF, Phil. mag. 16,
       261 (1971).

(6)   R.C. DYNES, Solid State Communications, 10, 615 (1972).

(7)   J.W. GARLAND, K.H. BENNEMANN and F.M. MUELLER, Phys. Rev.
       Letters 21, 1315 (1968).

(8)   M.M. COLLVER, R.H. HAMMOND, Phys. Rev. Letters 30,92 (1973).

(9)   M. GETTINGS, G. LINKER and O. MEYER, III. Int. Conf. on Ion
       Implantation, New York, Dec. 11-14, 1972.

(10)   O. MEYER, G. LINKER and B. KRAEFT, Int. Conf. on Ion Beam
       Surface Layer Analysis, Yorktown Heights, New York, June
       18-20, (1973).

(11)   G. LINKER, O. MEYER and M. GETTINGS, Int. Conf. on Ion Beam
       Surface Layer Analysis. Yorktown Heights, New York, June
       18-20, (1973).

(12)   J. LINDHARD, H.E. SCHIØTT, and M. SCHARFF, Mat. Phys. Med. 33
       2 (1963).

(13)   F.J. CADIEU, in AIT Conf. Proc. No. 4, 213 (1972), ed. by
       D.H. Douglass.

(14)   K.H. BENNEMANN, Winter Colloquium in Schleching 19.2-23.2.
       (1973).

(15)   J.R. SCHRIEFFER, Theory of Superconductivity, Benjamin Inc.
       Publ. N.Y. 1964.

(16)   G. BERGMANN and D. RAINER, Z. Physik 263, 59 (1973).

(17)   D. GERSTENBERG, P.M. HALL, J. Electrochem. Soc. 111,
       936 (1964).

ION IMPLANTATION EFFECTS IN SUPERCONDUCTING NIOBIUM THIN FILMS

P. Crozat, R. Adde, J. Chaumont and H. Bernas

Institut d'Electronique Fondamentale, Centre de

Spectrométrie de Masse (CNRS), and Institut de Physique

Nucléaire, Université Paris Sud, 91405 Orsay - France

and D. Zenatti, Centre d'Etudes Nucléaires, DINR,

38041 Grenoble - France

Abstract - Critical temperatures and normal resistances were measured on thin niobium films after implantation of N, O, Nb and Er ions at doses ranging from 10^{13} to 10^{17} ions/cm^2. Radiation damage and magnetic impurity effects are presented.

Implantation effects in thin superconducting Nb films were studied by measuring the variation of the critical temperature T_c and of the normal resistance near T_c as a function of the implanted ion's nature and concentration.

Thin Nb films[1] were deposited on silicon substrates. To obtain high quality superconducting films, Nb was evaporated with an electron gun on the heated substrates (400°C) in a clean high vacuum (< 10^{-8} Torr). Special care was taken to eliminate impurities (especially oxygen)as Nb is an excellent getter. After evaporation and cooling down of the substrates, a gold film (2000 Å) is immediately deposited over the Nb film before opening the bell jar,both to protect the film against oxidation during storage of the samples and to make good electrical contacts. Nb films of thickness 500-600 Å are thus obtained with $T_c \sim 8.5$°K and a resistivity ratio $\rho_{300°K}/\rho_{77°K} \sim 2.8$. This high ratio is indicative of the Nb films' good quality. The sample shape shown in Fig. 1 is obtained from these films using standard photomasking and etching techniques.

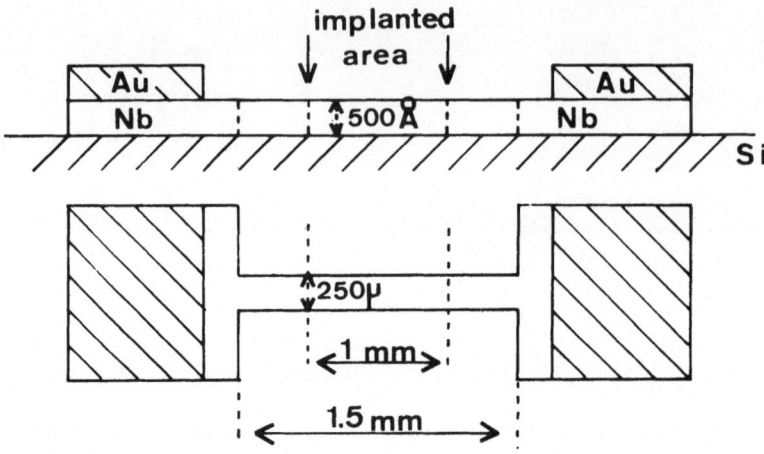

Fig. 1 : Niobium thin film (500-600 Å) deposited on
         Si substrate used for ion implantation.

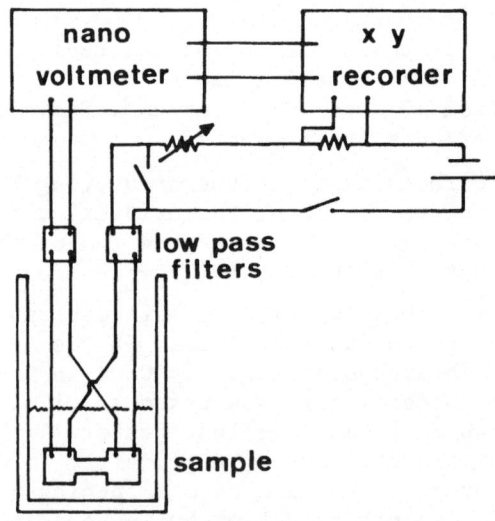

Fig. 2 : Experimental apparatus used for electrical measure-
         ments of ion implanted Nb films.

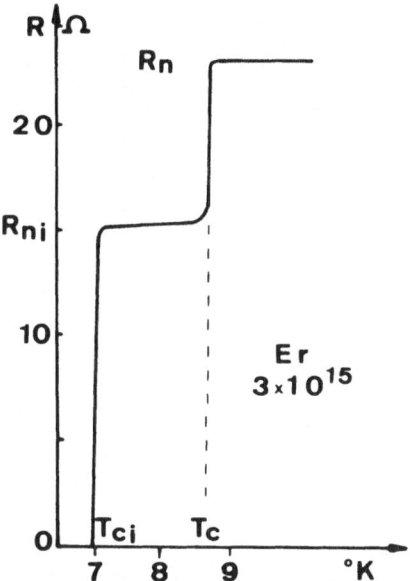

Fig. 3. Thin film resistance recorded versus temperature showing
$\Delta T_c = T_c - T_{ci}$ and normal resistance R_{Ni} of a Nb film
implanted with 3×10^{14} Er ions.

This method produces a constant Nb thickness all over the film.

Implantations were performed on the Orsay ion implantor [2]:
the implantation geometry is shown in Fig. 1. Nb^{++} implantations
were carried out at a single energy of 130 keV. In order to obtain
more uniform doping of the films, multiple implantations were carried out for O (15 and 40 keV) and Er (50 -, 110-, and 260 keV).
The doses indicated in the figures are sums over the various implantations.

Fig. 2 presents the experimental set-up for (I,V) characteristic measurements. Sample temperatures are studied from 1.8°K to
15°K : variations in T_c relative to the unimplanted film are measured to within < 0.1°K. Normal resistances, deduced from the (I,V)
curves are accurate to ∿ 5%.

Fig. 3 presents the film resistance measured as a function of
temperature : a small constant current (10 μA) is fed through the
sample and the dc voltage is simultaneously detected while slowly
varying the sample temperature. At the lowest temperatures, the entire film is superconducting and a zero resistance is measured. At
T_{ci} the transition temperature of the implanted region, the

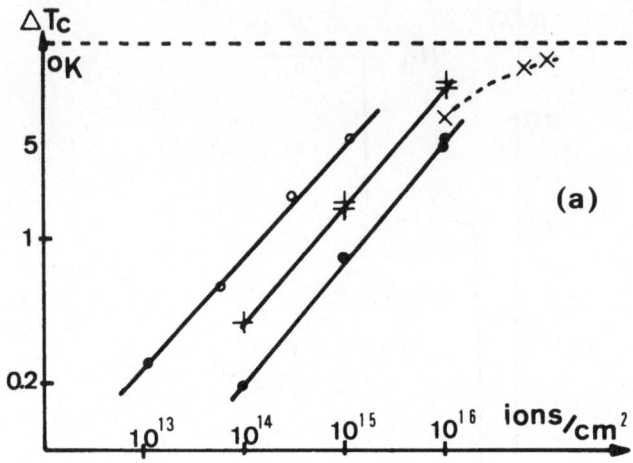

Fig. 4a : Measured $\Delta T_c = T_c - T_{ci}$ plotted versus total number of
implanted ions. 3×10^{15} implanted ions correspond
to 1% of the total number of Nb atoms in the implanted volume. As the Nb film T_c is $8.5°K$, the dotted
line indicates the maximum ΔT_c at which superconductivity is completely destroyed in the implanted area.

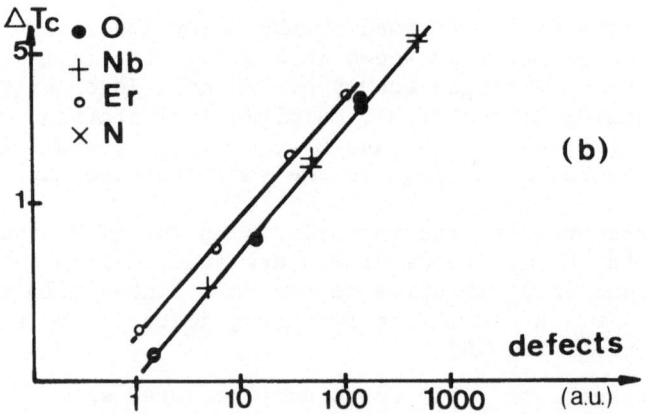

Fig. 4b : Measured $\Delta T_c = T_c - T_{ci}$ plotted versus total number of
created defects in arbitrary units.

resistance rises sharply up to R_{Ni} the normal resistance R_{Ni} of
the implanted region. If the temperature is further increased, R_{Ni}
is found to increase (generally very slowly) up to T_c the temperature at which the whole film becomes normal. In some cases, a rather strong variation of R_{Ni} was observed between T_c and T_{ci} , a
result which remains unexplained.

Fig. 4a shows the measured $\Delta T_c = T_c - T_{ci}$ versus the total number of implanted ions per unit area, while Fig. 4b presents the
same ΔT_c versus the total number of implantation induced defects,
measured in arbitrary units (we assume that the number of defects
is proportional to the ion energy). Fig. 4a shows that the critical
temperature is substantially reduced in all cases studied and varies approximately as the square root of the total number of implanted ions at least for two decades. When ΔT_c is plotted versus
the number of created defects (Fig. 4b) the curves for O and Nb
overlap exactly indicating that in these cases ΔT_c arises only
from the defects. Our values of ΔT_c are obviously much larger than
those observed when defects are introduced by less drastic procedures. This can probably be explained by lattice disordering
due to the implantation.

The ΔT_c due to Er implantation is larger than that produced by
O and Nb : it is particularly noticeable for low implantation doses. This further depression of T_c is possibly due to the pair -
breaking effect of the Er highly localized moment.

We found that the quantitative reproducibility of the results
depends rather critically on film preparation and protection. For
example Nb films exposed to air for long times ($\sim$ 3 months) prior
to implantation displayed a larger ΔT_c than that shown in Fig. 4
after Nb-implantation while a smaller ΔT_c was observed after O-
implantation. This could be related to trapping of light impurities
(especially oxygen) at the film surface.

REFERENCES

1. R. Adde , P. Crozat , S. Gourrier , G. Vernet , M. Bernheim ,
   D. Zenatti , Rev. Phys. Appliquée, Oct. 1973 (in press).

2. Chaumont J. et al., 3^d Int. Conf. Appl. Vac. Techn. Suppl.
   Le Vide, 52, 105 (1971).

DISCUSSION

Q: (P. Jung) Have you compared your results on the influence of
implanted N in Nb on superconducting properties to the influence
of thermally dissolved nitrogen?

A: No measurements and no comparisons were done.

Q: (H. C. Freyhardt) How does the critical temperature T_c of the Nb films change due to Fe implantation? You did not mention the results of these experiments. We are particularly interested in this question because we found a slight increase in T_c after high temperature Ni^+ irradiation of Nb.

A: We have only partial results on Fe in Nb. Its behavior seems close to that of O- and Nb-implants, but this remains to be checked.

Q: (P. Pronko) Your implants were performed at room temperature. Would you expect radical differences in the results if your implants were done at cryogenic temperatures?

A: As seen here and in the previous paper, radiation damage is an essential factor in reducing T_c. Hence, if the nature of the stable damage is changed by changing the implantation temperature, the results could well be different. However, since we don't know how the radiation damage produces a ΔT_c, it is safer to do the experiment before answering, and we are working in that direction.

Q: (J. W. Miller) In addition to changes in T_c due to effects of impurity concentration on the Debye frequency and density of states $N(0)$, is it not also likely that such changes are sensitive to impurity effects on the electron phonon interaction itself?

A: Yes, it is quite possible. Our main point is that lattice disorder is extremely high: for doses higher than $\sim 10^{14}$ at/cm^2, essentially all the atoms in the film have been displaced.

Q: (J. M. Poate) Did you worry about interaction between your Nb films and the Si substrate since the substrates during deposition are rather hot?

A: Yes, and we still do. Rutherford backscattering experiments are scheduled to study the interface. However, this problem does not affect the critical temperature results, since the highest T_c layer shorts the possible lower T_c layers. Also, note that the transition on the (R,T) curve is sharp, indicating reasonable homogeneity.

Q: (K. L. Merkle) What is the power law for the ΔT_c vs. dose
curve? Do you have any explanation for the fact that the observed
behavior extends to such high damage doses?

A: ΔT_c is approximately proportional to the square root of the
dose. We have no explanation at the moment for the observed power
law. The reduction in T_c is probably due to the reduction in the
Debye temperature and/or the change in the density of states produced by lattice disorder. Apparently, this process continues
until essentially all the Nb atoms in the film have been displaced.

SUPERCONDUCTING PROPERTIES OF THE DILUTE MAGNETIC ALLOYS

Pb-Mn, Sn-Mn AND Hg-Mn OBTAINED BY ION IMPLANTATION

W. Buckel and G. Heim*

University of Karlsruhe

*Present Address: IBM San Jose, California

1. INTRODUCTION

The study of magnetic impurities in non-transition metals is one of the most challenging topics to solid state theoreticians since the work of Kondo [1] (1964). Many fascinating results have been predicted (for a review see [2]), especially for superconducting alloys [3-5], but there has been a lack of techniques for preparing interesting alloys which often have a very low solubility limit (as many 3d-elements in non-transition metals).

So far either bulk alloys [6,7] or films evaporated onto cold substrates (quenched condensation, [8]) have been used in the field of superconductors containing magnetic impurities. These techniques suffer from characteristic disadvantages [9]:

- No solid solution was obtained because of a very low solubility limit (bulk alloys, e.g., Sn-Mn [7]).

- Heavy lattice disorder changes the properties of the host metal (quenched condensed films, [10]).

- Very low concentrations, which are often necessary to avoid interaction between the impurities, are hard to prepare and to determine.

In this situation ion implantation at very low temperatures offered the following possibilities [11,12]:

- Metastable alloys are obtained beyond the solubility limit.

- Lattice disorder, as introduced by radiation damage, is smaller by an order of magnitude as compared to quenched condensed films.

- Very low concentrations of several ppm are easily obtained and can be determined with the same relative accuracy as higher concentrations of several atomic percent.

This paper will discuss the use of the implantation technique for homogeneously doping film-targets. Results are presented for superconducting films of Pb, Sn and Hg, concerning the change of the transition temperature and the residual resistivity after implantation of Mn-ions. These results are presented to clarify the advantages and limitations of ion implantation as a technique to prepare dilute magnetic alloys.

2.   HOMOGENEOUS ALLOYING BY ION IMPLANTATION?

A homogeneous distribution of the dopant over the host sample is an essential condition of the alloying problem discussed here. As to implantation, the question arises, how to obtain a constant ion distribution over the target-depth. (It is no problem to reach homogeneity in a plane perpendicular to the ion beam. For this purpose the beam is raster-scanned across the target surface.)

It is proposed to approach this problem by using various ion energies and superposition of the resulting ion distribution functions [13,14]. Following Lindhard et al. (LSS-theory, 1963, [13]) a Gaussian distribution function $P(x_i)$ is assumed

$$P(x_i) = \frac{1}{(2\pi)^{\frac{1}{2}}} \exp(-x_i^2/2) \quad ,$$

where

$$x_i = (R_p(E_i) - \langle R_p(E_i) \rangle)/\langle \Delta R_p(E_i) \rangle$$

leading to an ion distribution function

$$F(x_i) = \frac{N_\square(E_i)}{\langle \Delta R_p(E_i) \rangle} P(x_i) \quad .$$

(Here $\langle R_p \rangle$ is the mean penetration depth, $\langle \Delta R_p \rangle$ is the standard deviation of the penetration depth and $N_\square(E_i)$ is the implantation rate, equal to all ions per unit area implanted at an energy E_i.) Finally the total ion distribution is obtained by summation over all energies used for implantation (Fig. 1):

$$F(x) = \sum_i F(x_i(E_i)) \stackrel{!}{=} \text{constant} \quad .$$

This function has to be made constant by choosing the energies E_i and the implantation rate $N_\square(E_i)$ in an appropriate way. For a simple calculation further approximations are made (for details, see [13]);

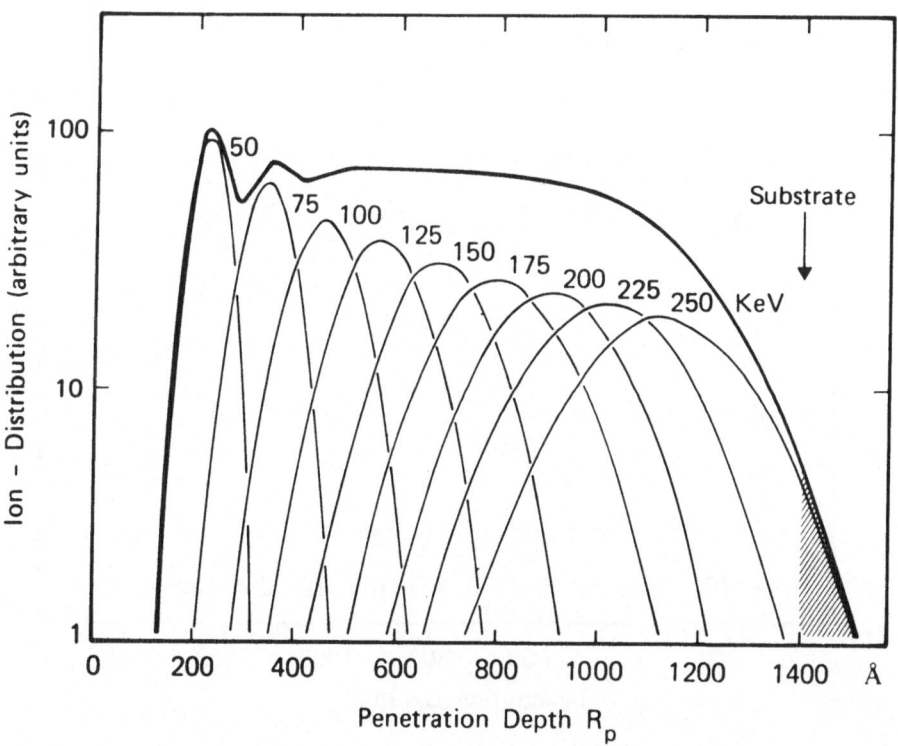

Figure 1. Gaussian distribution functions for different ion energies. ($\langle \Delta R_p \rangle / \langle R_p \rangle = 0.14$)

$$\langle R_p(E_1) \rangle \propto E_1$$

$$\langle \Delta R_p(E_1) \rangle \propto \langle R_p(E_1) \rangle \quad .$$

Then the implantation energies are taken at constant intervals ΔE_1 and a power-law is tried for the implantation rate:

$$N_\square(E_1) \propto E_1{}^n \quad .$$

This gives a total ion distribution function as shown in Fig. 2

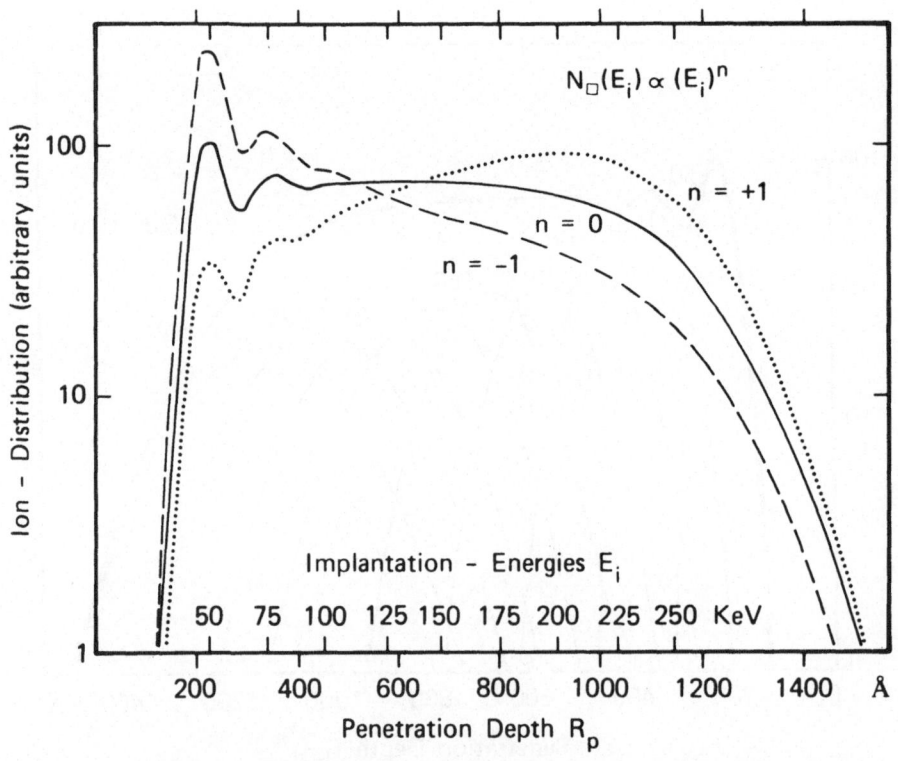

Figure 2. Total ion distribution for different implantation rates: $N_\square(E_1) \propto E_1{}^n$, the exponent n being varied as indicated.

for different values of n. It is concluded that a constant
implantation rate

$$N_{\square}(E_i) = \text{constant}$$

leads to a constant ion distribution over part of the
target-depth. The gradual decrease at high ranges is due to the
increasing width of the Gaussian curves. This deviation from a
desired step-function-like distribution remains true even if more
sophisticated models for an ion-distribution function are chosen
[15]. Here a major weakness of the implantation technique is
encountered:

- If the target thickness is chosen to coincide with the end
  of the constant distribution function, ions penetrate into
  the substrate, although they have been counted within the
  total ion dose. In this case the actual impurity concentration is overestimated. However, this approach might
  give the concentrations with the same relative accuracy at
  least for one sample if the same implantation technique is
  repeatedly used.

- If the target thickness is chosen large enough, all ions
  will come to rest within the target, but the condition of
  homogeneity is no longer satisfied.

In this work the maximum ion energy was kept sufficiently
low to deposit most impurities within the target film.
Inhomogeneity over dimensions smaller than the superconducting
coherence length (about 500 Å for the highly doped films) is
expected to be less effective on the superconducting properties.
Implantations were performed near 4K.

3. SUPERCONDUCTING PROPERTIES

a. Transition Temperature

Magnetic impurities in superconductors cause a strong
depression of the transition temperature T_c [3,4,5]. This was
also found for implanted films. In Fig. 3 the change of T_c with
manganese concentration is shown for the systems Sn-Mn, Pb-Mn
and Hg-Mn [16] in the low concentration regime (<400 ppm). The
slope of the curves is approximately constant at concentrations
above 100 ppm in agreement with theoretical results [3,4].
However, below 100 ppm there is a curvature in the $T_c(c)$-curves,
being positive for Sn-Mn, and negative for Pb-Mn and Hg-Mn,

respectively. It seems reasonable to assume, that scaling with impurity-concentration holds for very low concentrations. Therefore a different mechanism for the change of T_c has to be found.

The lattice defects generated by the implantation may provide the answer. The influence of lattice defects on the transition temperature of non-transition element superconductors is well understood today [17]. The change of T_c due to lattice disorder is found to be positive for weak-coupling superconductors (e.g., Sn) and zero or negative for strong-coupling superconductors (Pb and Hg, respectively). This can be proved by irradiating superconducting films with nonmagnetic ions, as has been done in the case of Sn (with Cu) and Pb (with Zn). Ions which are close to manganese in the periodic table are expected to generate a similar amount of defects in the lattice of the target. The

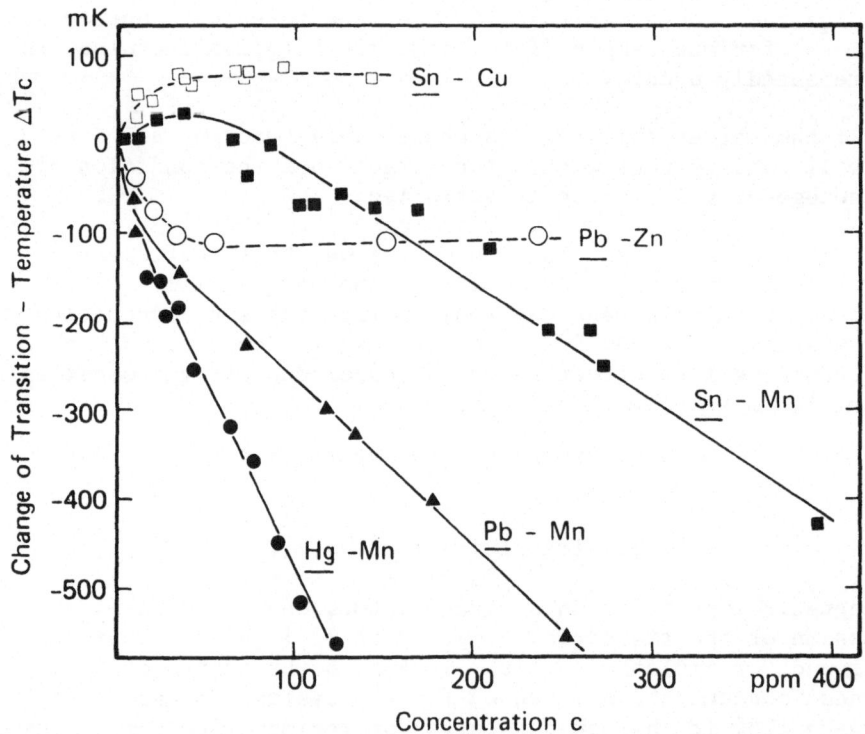

Figure 3. Change of transition temperature ΔT_c for superconducting films of Pb, Sn, and Hg, as caused by implantation of indicated impurities at $\sim$ 4K.

resulting change of T_c saturates at about 50 ppm of
impurity-concentration, indicating an equilibrium state of
radiation damage production and radiation annealing (see next
chapter).

So we conclude that the total change of T_c consists of two
contributions:

$$(\Delta T_c)_{total} = (\Delta T_c)_{magnetic} + (\Delta T_c)_{defect}$$

which may be separated in the way shown above. (The influence
of lattice defects on $(\Delta T_c)_{magnetic}$ which has been investigated
elsewhere [18-20], can be neglected for small defect
concentrations.)

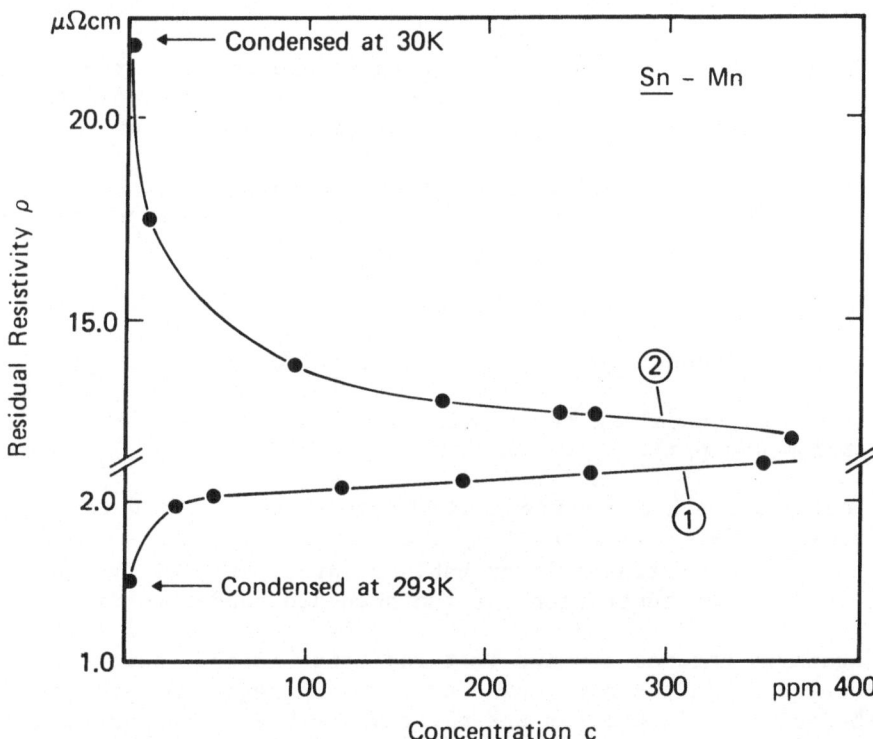

Figure 4. Residual resistivity of an annealed (1) and a disordered
(2) Sn film after repeated manganese implantation at $\sim$ 4K.

However, all results obtained in the higher concentration
regime (>50 ppm) have to be ascribed to a host superconductor
with slightly changed electronic properties, caused by a reduction
of the mean free path of the conduction electrons. This has to
be taken into account if a comparison is made to bulk alloys.

b. Residual Resistivity

The change of residual resistivity is obviously again
determined by the influence of two effects: radiation damage
and magnetic impurities, which are assumed to be additive:

$$(\Delta\rho)_{total} = (\Delta\rho)_{magnetic} + (\Delta\rho)_{defect}$$

The scattering of conduction electrons on localized impurity
spins is beyond the scope of this paper (see [11] for more
detail).

The contribution of defect centers to the residual resistivity
is clearly seen in Fig. 4, where two examples for Mn-implantation
into Sn-films are shown. Film (1) has been condensed at room
temperature, resulting in a minor impurity- and
defect-concentration. Film (2) has been condensed onto a cold
substrate (30K), where a heavy lattice disorder is frozen in
[19].

Implantation of Mn-ions produces now two competitive effects.

- A defect production (dominant in film (1) below 50 ppm).

- A defect annealing by irradiation (dominant in film (2)
  below 400 ppm).

Both effects reach a dynamic equilibrium state, leading to a
stationary defect concentration, which is not the same for films
(1) and (2). One reason is probably a larger initial (residual
gas) impurity concentration for the quenched condensed film.

Again it becomes obvious that radiation damage effects are
dominant only in the very low concentration regime and the effects
of the magnetic impurities are clearly visible at concentrations
above.

c. Kondo-Effect

Several dilute magnetic alloys have been found to show anomalies in the temperature dependence of some electronic properties, e.g., a resistance minimum at low temperatures. Since the theoretical work of Kondo [1] these anomalies are understood as a consequence of a negative exchange interaction between the spins of the conduction electrons and the localized spins of the magnetic impurities.

Resistance minima have been found in implanted Sn-Mn alloys slightly above the transition to the superconducting state (see Fig. 5). The logarithmic increase of the resistivity towards lower temperatures indicates that the Kondo-effect is present in these alloys. The Kondo temperature may be estimated to lie well

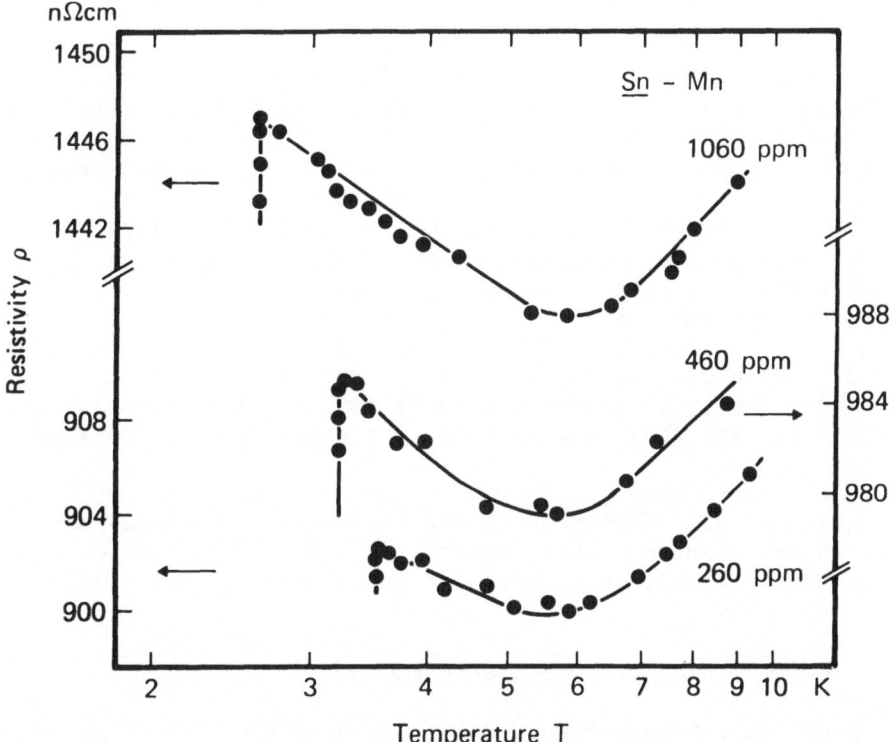

Figure 5. Kondo-effect in the resistivity of Sn-Mn films. (Notice the strong magnification of the resistivity scale.)

below the transition temperature of the host superconductor, but cannot be calculated from the T_c-c-dependence alone, because the value of the localized impurity spin is unknown.

More information is expected from critical current measurements, where another 'pair-breaking' mechanism [21] is introduced into the superconductor, thus lowering the transition temperature. In this method the Kondo temperature may be found from the temperature dependence of the pair-breaking parameter.

Insight in the Kondo-effect may be obtained from tunneling measurements, as well [22]. Here ion implantation offers new possibilities, if electron-tunneling into implanted films proves to be successful. Preliminary results obtained in our laboratory indicate, that the implantation of ions does not destroy evaporated tunneling devices. However, the problem of which implantation profile is appropriate for this case still exists.

4. SUMMARY

To summarize, ion implantation has proved to be a useful tool for the preparation of homogeneous dilute alloys. The ability to obtain low concentrations of few ppm is the major advantage compared to other techniques. However, medium concentrations are also accessible, if it is desired to investigate interaction effects between the impurities. The absolute accuracy of the impurity-concentration is still a problem, being limited to about 30% in the present state.

Further investigations are necessary on the subject of impurity-oxides, either introduced by implantation or built later in the film, and their influence on the properties of the studied alloys.

Once the basic features of this technique are known, many other so far unknown dilute magnetic alloys may be studied, e.g., mercury base alloys, which are hard to prepare by other methods.

5. REFERENCES

1. Kondo, J., Progr. Theor. Phys. <u>32</u>, 37 (1964).
2. Fischer, K., Springer Tracts mod. Physics <u>54</u>, 1 (1970).
3. Abrikosov, A. A. and Gorkov, L. P., Soviet Phys. JETP <u>12</u>, 1243 (1961).
4. Müller-Hartmann, E. and Zittartz, J., Phys. Rev. Letters <u>26</u>, 428 (1971).

5. Ludwig, A. and Zuckermann, M. J., J. Phys. F(GB), $\underline{2}$, L21
    (1972).
6. Suhl, H. and Matthias, B. T., Phys. Rev. $\underline{114}$, 977 (1959).
7. Boato, G., Gallinaro, G., and Rizzuto, C., Rev. Mod. Phys.
    $\underline{36}$, 162 (1964).
8. For a review see: Hilsch, R., Minnigerode, G. V., and
    Schwidtal, K., Proc. 8th Int. Conf. on Low Temp. Phys.,
    London 1962.
9. Wassermann, E., Z. Physik $\underline{220}$, 6 (1968).
10. Buckel, W. and Hilsch, R., Z. Physik $\underline{138}$, 109 (1954).
11. Geerk, J., Heim, G., and Kessler, J., Z. Physik $\underline{242}$,
    86 (1971).
12. Buckel, W., Dietrich, M., Heim, G., and Kessler, J.,
    Z. Physik $\underline{245}$, 283 (1971).
13. Lindhard, J., Scharff, M., and Schiøtt, H. E., II Mat.
    Fys. Medd. $\underline{33}$, 14 (1963).
14. Gibbons, J. F., Proc. IEEE $\underline{56}$, 295 (1968).
15. Furukawa, S. and Jshiwara, H., J. Appl. Phys. $\underline{43}$, 1268 (1972).
16. Krauß, G., Diplomarbeit Universität Karlsruhe, 1973.
17. Buckel, W., "Supraleitung," p. 190, Physik Verlag 1972.
18. Falke, H., Jablonski, H. P., Kästner, J., and Wassermann,
    E. F., Z. Physik $\underline{259}$, 135 (1973).
19. Fallenbüchel, P., Diplomarbeit Universität Karlsruhe, 1973.
20. Hauser, J. J., Solid State Comm. $\underline{11}$, 507 (1972).
21. Maki, K., in Superconductivity, p. 1035, ed. R. Parks,
    Marcel Dekker, New York, 1969.
22. Chaba, A. N. and Singh Naghi, A. D., Lettre Al Nuovo
    Cimento $\underline{4}$, 794 (1972).

DISCUSSION

Q: (H. Bernas) The Mn concentrations are quite high in your
SnMn alloy. Aren't you worried about interaction effects?

A: Yes. Interaction effects may be present at least in the concentrated SnMn alloy (1060 ppm). This is supported by the fact
that the slope of the logarithmic part of the resistivity curve
is smaller than in the case of the less concentrated alloy.

ION IRRADIATION AND FLUX PINNING IN TYPE II SUPERCONDUCTORS

Herbert C. Freyhardt, Anthony Taylor and Benny A. Loomis

Argonne National Laboratory, Argonne, Illinois 60439

ABSTRACT

Niobium foils containing 10 and 900 wt ppm oxygen were irradiated with 3.5 MeV $^{58}Ni^+$ to between 50 and 100 displacements per atom. The irradiation at 900°C produced a bimodal void distribution with a large number of small voids and with large diameter voids that act as strong pinning centers for fluxoids in type II superconductors. Transverse critical current measurements showed a substantial increase in both the upper critical field H_{c2} and the critical current density J_c. J_c was anisotropic and exhibited a maximum for fluxoids parallel to the irradiated surface. From the void size and number densities, determined by transmission electron microscopy, interaction forces between fluxoids and voids as well as volume pinning forces are estimated and compared with the experimental results.

INTRODUCTION

Defects and defect clusters produced during particle irradiation of superconductors change the electronic and thermal properties and, therefore, influence the fundamental superconducting parameters. Whereas homogeneously distributed defects or agglomerates of defects, which are much smaller than the diameter D_{FL} of a fluxoid (which is determined by the coherence length ξ of the superconductor), have a negligible interaction with the fluxoids, inhomogeneous distributions of defects or defect clusters, which are of sufficient size relative to D_{FL} can cause considerable flux pinning and may lead to high critical current densities in

type II superconductors(1). In this investigation we are particularly interested in radiation-induced flux pinning due to voids.
Finely dispersed voids can form and grow in metals during neutron,
heavy particle or electron irradiation in a temperature region
close to or below half the melting temperature(2,3), where both
the interstitials and the vacancies are mobile. Choosing appropriate irradiation conditions, the size, shape and distribution
of voids can be changed and regular arrays can form. In the type
II superconductors, voids with their simple geometry provide a
useful tool for flux-pinning measurements. This report is a first
attempt to investigate, by transverse critical current measurements, the interaction between fluxoids and voids in heavy-ion
irradiated superconductors. The voids are generated by $^{58}Ni^{+}$ bombardment of Nb and of Nb containing oxygen at temperatures up to
900°C. Although neutron irradiation would result in a homogeneous
distribution of the damage throughout the specimen, ion irradiation,
affecting only a thin layer near the surface, was chosen because
of the high displacement rates obtainable.

ION IRRADIATION AND LOW-TEMPERATURE MEASUREMENTS

In the investigation, 100-μm-thick Nb foils with low Ta content were used. The heat treatment(4), with a final annealing at
2350°C in a vacuum of 3×10^{-9} Torr, reduced the level of impurity
content to 4 ppm nitrogen, 14 ppm carbon and 10 ppm oxygen (wt
ppm) prior to a controlled doping of the foils with oxygen.

After the foils were electrochemically polished in an $HF-HNO_3$
solution, they were irradiated with 3.5 MeV $^{58}Ni^{+}$ at 800 or 900°C
to between 50 (3×10^{16} ions/cm^{2}) and 100 displacements per atom
(DPA). During the irradiation a vacuum of 10^{-7} Torr was maintained.

For the low-temperature measurements, the specimens, typically
1 mm wide, 5 to 9 mm long and 100 μm thick, were immersed in
liquid helium, and the transverse critical currents I_c were
measured versus an externally applied magnetic field H produced
by a superconducting solenoid. In these four-probe measurements,
I_c was usually determined as the current at which a voltage drop
of 0.01 to 0.05 μV appeared across the gauge length. The complete
transition curves were studied in detail. Using a similar four-
probe arrangement, the transition temperature T_c of the irradiated
foils was determined by monitoring the resistivity versus the
specimen temperature. Because of the small size of the specimens
available, magnetization measurements at 4.2K were extremely difficult. However, an integration technique combined with a
sensitive pick-up coil system allowed reproducible results to
be obtained(5).

RESULTS

Under the irradiation conditions chosen, the Ni ions penetrate to an average depth of 9400 Å(6), whereas the defect production rate shows a maximum around 7500 Å. The damage is thus
confined to a surface layer that represents only a small fraction
of the foil volume. Both primary defects (vacancies and interstitials) produced during the ion irradiation are mobile at the
irradiation temperature chosen. Because of the slightly larger
interaction between interstitials and dislocations compared with
the interaction between vacancies and dislocations, the interstitials precipitate preferentially at dislocations, whereas the
vacancies agglomerate into voids (Fig. 1). The large voids show
a broad maximum in the void size distribution at a diameter of
280 Å for a 900°C irradiation of pure niobium with 10 ppm oxygen
(Fig. 1a), whereas the small voids have a sharp peak around 30 Å
with a density of 2×10^{16} cm^{-3}, which is a factor of 40 greater
than for the large voids. The small voids appear nearly spherical,
and the large voids of this bimodal distribution show a characteristic polyhedral shape, described in detail by Loomis et al.(7).
Under the image condition of Fig. 1, the simultaneously present dislocation loops cannot be seen. A larger oxygen content resulted
in smaller void sizes and reduced void spacings. As an example,
Fig. 1b shows the void arrangement in a niobium specimen containing
900 ppm oxygen irradiated at 800°C to 50 DPA.

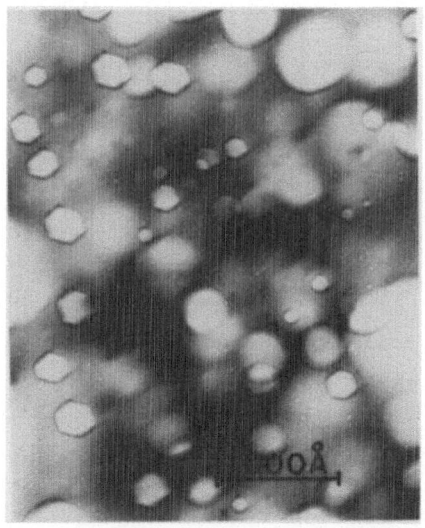 

Fig. 1a   Fig. 1b
Voids formed during irradiation of niobium with 3.5 Mev $^{58}Ni^{+}$ ions.
Fig. 1a, Nb with 10 ppm oxygen irradiated to 50 DPA at 900°C.
Fig. 1b, Nb with 900 ppm oxygen irradiated to 50 DPA at 800°C.

   Although the damage in the niobium foils due to the Ni-ion
irradiation is confined to a layer below the surface, the bombarded specimens show a pronounced increase in the transverse
critical current density J_c (Fig. 2) relative to the unirradiated
sample and a considerable hysteresis of their magnetization curves
(Fig. 3). The increase of J_c is several orders of magnitude
greater. However, we have to keep in mind that during the high-
temperature irradiation at a vacuum of 10^{-7} torr the foils pick up
a considerable amount of gas atoms, which also can produce an
increase in $J_c(H)$. This was confirmed by investigating specimens
that had been exposed to the same vacuum and heating conditions
but had not been irradiated. A further proof will be reported

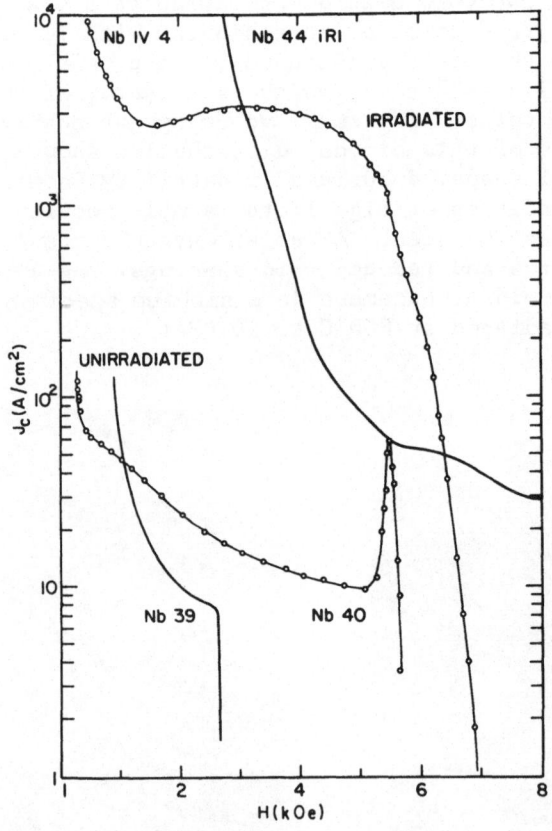

Fig. 2
Critical current density J_c versus applied magnetic field H at
4.2K for niobium foils with 10 (Nb 39, Nb 44) and 900 (Nb 40, Nb
IV 4) ppm oxygen before and after $^{58}Ni^+$ ion irradiation. Nb 44,
irradiated to 50 DPA at 900°C; Nb IV 4, irradiated to 100 DPA at
800°C.

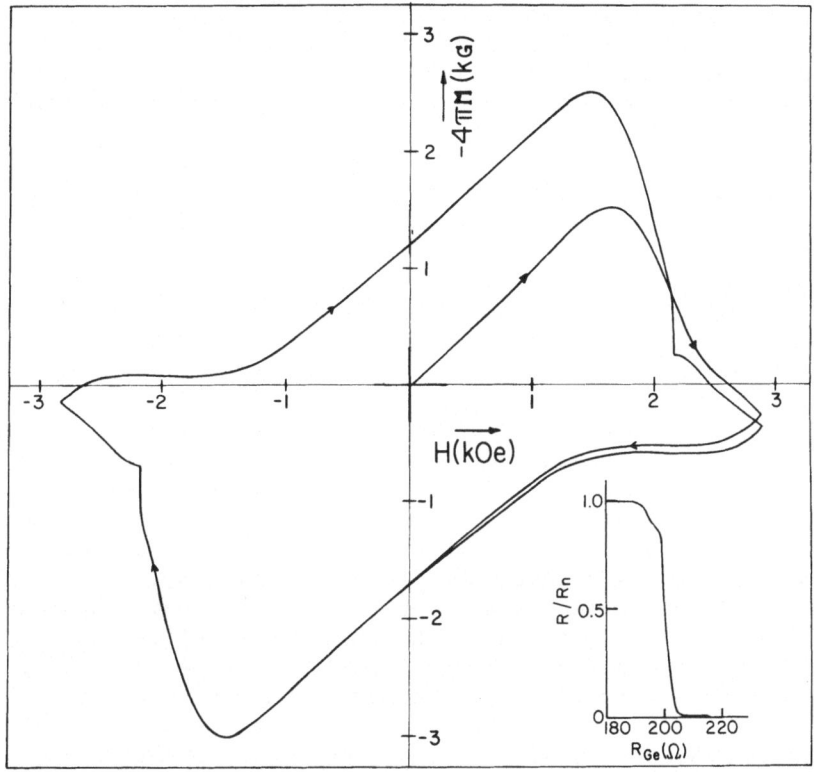

Fig. 3

Magnetization $-4\pi M$ versus external magnetic field H at 4.2K for the
Ni-ion-irradiated niobium sample Nb 44 IR 1 (the irradiation conditions are the same as in Fig. 2). The inset gives the ratio
of the resistivity R at a given specimen temperature to the
normal state resistivity R_n (directly above the transition) versus
the temperature (measured as the resistivity of a germanium
resistor) for the specimen Nb IV 4.

later. The increase in J_c due to the contamination of the specimen
during the irradiation is particularly pronounced for the pure
niobium in which small critical current values could be obtained
in the annealed state. In the oxygen-doped specimen (Nb 40, Fig.
2), an extremely sharp peak below H_{c2} is observed in the unirradiated state, which is subject to further investigations. This
peak can be seen in the irradiated state also, although barely
indicated, because of the broad maximum and the sharp drop in
$J_c(H)$ near the peak field.

After irradiation the surface layer is enriched with Ni.
The Ni ions exhibit a Gaussian range distribution around the
average penetration depth of 9400 Å, which is broadened because
of radiation induced diffusion that occurs at the irradiation temperature. This Ni enrichment and, to a smaller extent, the pick-
up of gaseous impurity atoms during the irradiation are responsible for the increase in the upper critical field H_{c2}. The
increase is again more pronounced for the pure niobium because
the relative increase in the residual resistivity ρ_n during the
bombardment is larger than for the oxygen-doped specimen and,
therefore, according to the Goodman relation (cf. reference (1)):
$H_{c2}/(\sqrt{2} H_c) = \kappa_o + $ const. x ρ_n (H_c is not altered significantly)
the relative H_{c2} increase should also be larger.

From earlier low temperature ac susceptibility measurements
as a function of temperature, we knew that no drastic change in
the transition temperature T_c of the Ni enriched layer was to be
expected. However, the resolution of this former arrangement was
insufficient and, therefore, the T_c determination was repeated
using a resistivity measurement (inset of Fig. 3). T_c of the
major transition, which we ascribe to the bulk, unirradiated
portion of the foil (niobium with 900 ppm oxygen) is 9.1K. This
is roughly 0.1K smaller than the transition temperature of pure
Nb. The suppression of T_c is accompanied by an enhanced residual
resistivity ρ_n up to 2.6 $\mu\Omega$ cm. Thus a smearing of the sharp
density of states in niobium due to the interstitial oxygen
atoms, acting as scattering centers for the electrons, might
be responsible for this effect(8). Recent experiments on ion
implantation in Nb films show similar results(9). A smaller
fraction of the specimen, however, exhibits a slightly higher
transition temperature, obviously caused by the Ni ions implanted
in the surface layer.

Several experiments have been performed to demonstrate that
the increased pinning is introduced by defects produced in the
surface layer during the heavy-ion bombardment. One of these
experiments is shown in Fig. 4, in which the irradiated surface
layer has been removed step by step by electrochemically polishing
the sample in an $HF-HNO_3$ solution. Particular attention was paid
to the effect of the polishing procedure on sample quality (compare curves I and II). The material was removed at a rate of
roughly 700 Å/sec. In the first step, only the edges were polished, and then the edges and the unirradiated surface were polished. The long tail in the $J_c(H)$ curve of specimen Nb 44 IR,
extending to fields above 15 kOe, almost disappeared (Fig. 4 and
2). This tail may have been brought about by a rearrangement of
the Ni and gaseous impurity atoms near the edges during the spark
erosion of this specimen after the irradiation (only this particular specimen was treated in this manner). After the first

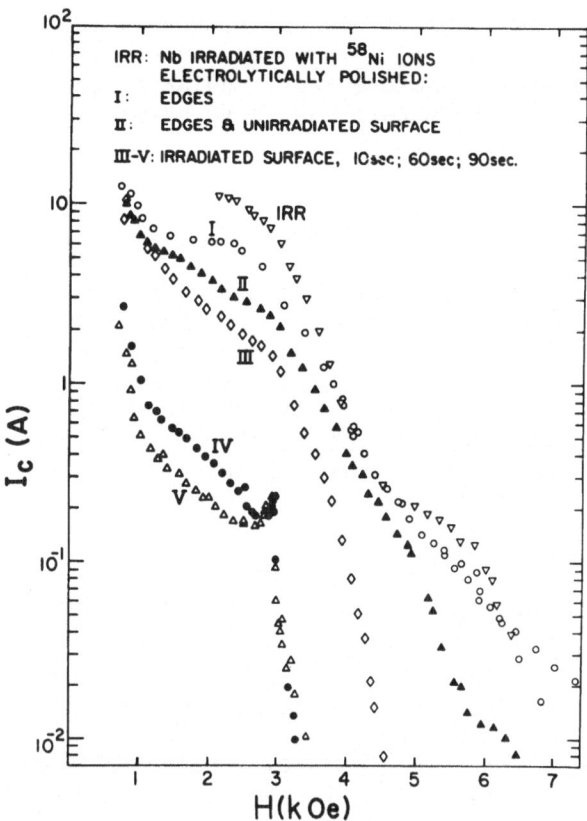

IRR: Nb IRRADIATED WITH ^{58}Ni IONS ELECTROLYTICALLY POLISHED:
I: EDGES
II: EDGES & UNIRRADIATED SURFACE
III-V: IRRADIATED SURFACE, 10sec; 60sec; 90sec.

Fig. 4

Critical current I_c versus external field H at 4.2K for pure Nb irradiated to 50 DPA at 900°C (Nb 44 IR) after various electrochemical polishing treatments. During the polishing, the cross-sectional area is reduced from 9.1×10^{-4} cm^2 (IRR) to 8.3×10^{-4} cm^2 (IV).

polishing of the bombarded side, approximately half of the irradiated layer was removed, and J_c(H) is seen to be only slightly changed (Fig. 4). It is only after the subsequent polishing treatment, during which most of the layer is removed, that the critical current densities drop drastically. In addition, a sharp peak below H_{c2} now appears, which we found to be characteristic of Nb foils deliberately doped with oxygen. From the position of this peak and from the value of H_{c2}, we know from former results that during the irradiation the bulk of the sample picked up between 100 and 140 ppm impurity atoms. Additional prolonged polishing left the upper critical field unchanged and decreased the level of $J_c(H)$ only slightly. Under the available irradiation conditions, a reduction of the critical currents to their original values before the irradiation cannot be expected, indicating that pinning centers must have been created in the bulk of the foil during the irradiation.

DISCUSSION

A quantitative interpretation of the critical current measurements reported above is rather difficult, because we do not have sufficient knowledge of the superconducting parameters of the irradiated layers with their spatially varying Ni concentration. However, we can qualitatively discuss the results, following along the lines of a statistical theory developed by Labusch[10], who connected the volume pinning force p_V acting on the unit volume of the fluxoid lattice in a type II superconductor due to a statistical array of obstacles with the elementary interaction force p_{max} between one fluxoid and one obstacle

$$p_V = N \, p_{max}^2 \, (B/\phi_o)^{3/2} \, \alpha/<c_{FLL}>,$$

where N is the area density of obstacles, α is a constant, $\phi_o = 2 \times 10^{-7}$ G cm^2, the unit flux quantum, and $<c_{FLL}>$ is some mean elastic constant of the fluxoid lattice that can be calculated from the reversible properties of the material and is dependent on the flux density B. This volume pinning force is (in high κ superconductors) directly proportional to $J_c B$. The equation describes quite adequately the observed behavior of the critical current. Above H_{c1}, the fluxoids penetrating into the superconductor form a regular lattice that, due to the mutual electromagnetic repulsion among the fluxoids, becomes rapidly more rigid as the fluxoids approach each other. Consequently, the elastic constants rapidly increase above H_{c1} with field, and the volume force p_V decreases, because the obstacles are prevented from acting with their maximum pinning force p_{max}. Below H_{c2}, p_{max} depends on the flux density causing a less pronounced field dependence of J_c, even though the fluxoid lattice becomes softer again; the shear modulus in a plane perpendicular to the fluxoids vanishes[11]. This behavior is well developed in the unirradiated specimens Nb 39 and 40 (Fig. 2). The higher pinning in the oxygen-doped, compared with the pure niobium, sample is due to a nonrandom distribution of impurity atoms, which also cause the sharp peak below H_{c2}. Quantitative evaluations of similar curves have been reported elsewhere[12].

The voids produced in the surface layer now represent strong obstacles for the fluxoids, because energy is required to drive

normal electrons in the core of a fluxoid after it is moved from
the void ("core interaction" between fluxoids and voids). The
interaction energy δE for this case is simply the condensation
energy of a superconductor times the volume of the void: $\delta E =$
$(\mu_o H_c^2/2)$ $(\pi D^3/6)$, where H_c is the thermodynamic critical field of
the superconductor, and D is the void diameter, which is, in our
case, smaller than the fluxoid diameter $D_{FL} \sim 2 \xi$ (ξ: coherence
length). Using H_c of pure niobium at 4.2K[12], δE for a 200 Å
large void is estimated to be about 0.4 eV. The interaction force
of this void with a fluxoid is then given by $p_{max} \sim \delta E/\xi$ ($2\xi > D$).
If stress fields occur around voids (caused by a segregation of
foreign atoms at the voids[3,13], additional interaction mechanisms
will contribute to the pinning. However, during recent transmission
electron microscopy observations, no such stress field could be
detected in our specimen. A thorough study of the possible pinning mechanisms is in progress.

Using previous results for the field dependence of the
elastic constants of Nb[12], we can estimate the magnitude of the
volume pinning force and the critical current density J_c. For
the pure Nb specimen (Nb 44 IR) irradiated at 900°C to 50 DPA, a
critical current density of 10^5 A/cm^2 flowing in the damaged
layer can be estimated using the size and number densities of the
large (D = 280 Å) voids. This value is calculated for magnetic
fields between 2 and 3 kOe. Because of our actual foil dimensions,
this large value of the current density J_c allows no direct comparison of the estimated J_c and the experimental results obtained
from measurements in which the fluxoids are perpendicular to the
foil. Therefore, transverse critical current measurements have
been performed[14] with the fluxoids parallel to the specimen surface (Fig. 5). These investigations reveal a strong dependence
of J_c on the angle between specimen surface and external field
direction. J_c is maximum for fluxoids parallel to the surface and
minimal if they are perpendicular to the surface. In the first
case, the total current through the foil can be obtained from
the high current density carried in the surface layer and the lower
current density of the bulk of the specimen. The magnitude of
the latter can be found from the current flowing through the foils
after the surface layer had been removed (Fig. 4). Consequently,
the average current density of the foil (field parallel to the
surface) is of the order of 7 to 10 x 10^3 A/cm^2 at fields around
2 to 3 kOe, which is in agreement with our experimental data. The
small 30 Å voids, simultaneously present in the irradiated specimen
Nb 44, can be shown to give a negligible contribution to J_c.
Their pinning forces with the fluxoids are much too small. However,
in a quantitative evaluation, one must also include the interactions with dislocation loops present in the material, although
their influence is not dominating the effect due to the voids.

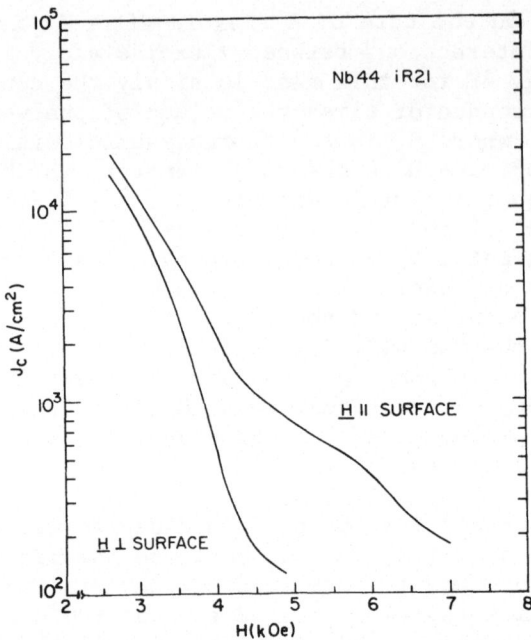

Fig. 5
Critical current J$_c$ versus external field H at 4.2K for different
angles between the field and the specimen surface. Voltage
criterion, 1 μV.

     Specimen Nb IV 4 after the irradiation exhibits a broad
maximum around H = 3 kOe superimposed upon the monotonically
decreasing J$_c$(H). In addition, the sharp peak near H$_{c2}$, characteristic of the unirradiated specimen, is still indicated in the
irradiated state. The broad maximum is believed to be caused by
a superposition of several maxima due to a matching between the
spacing of the fluxoids (varying with the magnetic field) and
given spacings of the regular void array, which was shown to be
developed in this particular foil(15).

                           CONCLUSIONS

     In the present investigation, we have (for the first time)
performed flux-pinning experiments on niobium foils containing
voids that are produced by heavy-ion irradiation. It is demonstrated that large interaction forces between voids and fluxoids
can occur which result in critical current densities of up to
10^5 A/cm^2. To achieve these high forces, voids with sizes comparable to the fluxoid diameter have been chosen. The high

current density in the irradiated surface layer should result in considerably smaller ac losses in the superconducting state. The known geometry of voids and the possibility of generating regular arrangements of pinning obstacles make them especially useful for fundamental investigation of flux-pinning problems.

ACKNOWLEDGMENTS

We would like to thank S. B. Gerber for valuable help during the specimen preparation. W. Schlump and L. Schultz for their cooperation and T. D. Ryan for the computer calculations.

REFERENCES

1) A. M. Campbell and J. E. Evetts, Adv. in Physics 21 (1972) 199.

2) D. I. R. Norris, Radiation Effects, 14 (1972) 1; 15 (1972) 1.

3) "Radiation – Induced Voids in Metals", USAEC Technical Information Center, Oak Ridge, Tennessee, CONF-710601 (1972) and "Voids Formed by Irradiation of Reactor Materials", Proc. Brit. Nucl. Energy Soc., Reading, U.K. (1971).

4) B. A. Loomis and S. B. Gerber, Acta Met., 21 (1973) 165.

5) These measurements have been performed in the experimental set-up of M. A. Kirk, ANL, whom we would like to thank for the help.

6) Calculated by T. D. Ryan using D. K. Brice, RASE3 and DAMG2: Routines for Ion Implantation Calculations, Sandia Laboratories Report SLA-73-0416 (April 1973). Diffusion effects at the irradiation temperature have not been taken into account.

7) B. A. Loomis, A. Taylor, T. E. Klippert and S. B. Gerber, Proc. Int. Conf. on "Defects and Defect Clusters in B.C.C. Metals and their Alloys", R. J. Arsenault (editor), Nuclear Metallurgy, vol. 18 (1973), p.332.

8) J. E. Crow, M. Strongin, R. S. Thompson and O. F. Kammerer, Phys. Letters, 30 A (1969) 161; W. De Sorbo, Phys. Rev., 132 (1963) 107.

9) H. C. Freyhardt and K. L. Merkle, unpublished.

10) R. Labusch, Crystal Lattice Defects, 1 (1969) 1.

11) R. Labusch, Phys. Stat. Sol., 32 (1969) 439.

12) H. C. Freyhardt, Phil. Mag. 23 (1971) 345 and 369.

13) F. V. Nolfi, Jr., Scripta Metallurgia, in print.

14) Measurements, performed at the Institut für Metallphysik, Göttingen, Germany.

15) H. C. Freyhardt, B. A. Loomis, P. Okamoto and A. Taylor, to be published.

DISCUSSION

Q: (G. L. Kulcinski) (1) Have you looked at lower irradiation temperatures where voids are not formed to determine the effectiveness of dislocation loops for fluxoid pinning? If lattice strain is important in pinning fluxoids, then I would expect that 200 Å loops would be more effective than 200 Å voids. (2) You mentioned the use of ordered void arrays as pinning sites. Did you see such ordered arrays and are they similar to those observed at Pacific Northwest Labs (PNL) at 800 C and 900 C?

A: (1) It is unlikely that the superconducting properties measured in this study were influenced significantly by stress fields from dislocations. In general the dislocation density in ordered void structures is quite small. Experiments to characterize the dislocation structure are in progress, however. We have not yet investigated flux pinning due to dislocation loops (in Nb) that are produced at lower irradiation temperatures. However, we have studied in detail the interaction between fluxoids and dislocations in plastically deformed niobium. These results and the very large core interaction of the fluxoids with the voids seem to indicate that voids (which can have stress fields) are more effective pinning centers than dislocation loops of comparable size. However, this has yet to be shown by quantitative calculations. From TEM investigations of our specimen we know that the dislocation loops present in our foils do not dominate the effect due to the voids. (2) Yes, in the niobium specimen NbIV4 (irradiated with 3.5 MeV Ni ions at 800 C to 100 DPA) which showed a broad maximum in $J_c(H)$ at a magnetic field of 3 kOe we found well developed regular void arrangements. The void size and number densities of these regular arrays are comparable to those observed at PNL. At a lower dose of 50 DPA, void ordering was also found in Nb alloys containing Zr. These specimens exhibit a similar broad maximum in the $J_c(H)$ curves, which could be resolved as a superposition of several maxima caused by a matching of the fluxoid spacings (which vary with the magnetic field) with given void spacings.

Q: (J. E. Schirber) Do you intend to make a.c. loss measurements on these materials?

A. Yes, we intend to perform a.c. loss measurements in the future. Because of the high increase in the critical current density J_c of the irradiated layer we expect a reduction of the losses, which are (in the Bean model) proportional to J_c^{-1}.

Q: (P. Jung) What was your quenching rate from irradiation to measuring temperature? Was it not fast enough to prevent the picked-up gases in the bulk from clustering or precipitating?

A: The initial quenching rate was about 100°C/sec, but despite this, a small amount of pinning in the bulk occurred.

CHAPTER II

ION INDUCED SURFACE REACTIONS

THE USE OF ION BEAMS IN CORROSION SCIENCE

G. Dearnaley

AERE, Harwell, England

INTRODUCTION

Two major preoccupations of corrosion science are to find ways of improving the corrosion resistance of metals by alloying and, secondly, to achieve a better understanding of the often complex mechanisms of corrosion. I wish to demonstrate how ion implantation may be used to carry out both these tasks, particularly when employed in conjunction with ion beam and other, more conventional, techniques of surface analysis and inspection.

The usual problem is one of preparing sufficient samples of metal alloyed with different constituents to different levels of concentration. A good example of the kind of work involved is provided by the development of the zircaloys under the general direction of the Naval Reactors Branch of the USAEC [1] . One difficulty in the comparison of specimens is that of achieving a consistent grain structure, since this in itself may influence the corrosion. Another problem is where to start in choosing alloying additions: some rule of thumb is required. The only scientific guide is that provided by the so-called Wagner-Hauffe rules [2] which apply to parabolic corrosion (film thickness proportional to $(time)^{\frac{1}{2}}$) controlled by the rate of migration of charged species across the corrosion film. It is usual to divide oxides into the following categories:-

1. <u>Oxides with anion defects</u>

    (a) oxygen-deficient with anion vacancies (n-type semiconductor) e.g. Nb_2O_5, ZrO_2

(b) Oxygen-excess, with interstitial anions
     (p-type semiconductor) e.g. UO_2

2. Oxides with cation defects

   (a) metal deficient, with cation vacancies
        (p-type semiconductor) e.g. NiO, Cr_2O_3, FeO

   (b) metal-excess, with interstitial cations
        (n-type semiconductor) e.g. ZnO.

In addition one must know whether the rate-determining process is
the migration of anions or cations, or the counterflow of electrons.
Thus addition of trivalent ions such as Cr^{3+} to a bivalent metal-
deficient p-type oxide such as NiO will increase the concentration
of cation vacancies (due to the maintenance of electrical
neutrality) but will decrease the hole concentration. Since in
this instance the transport of cations determines the oxidation
rate, there is an enhancement of the oxidation. Additions of
monovalent ions, e.g. Li^+, have the opposite effect and reduce the
oxidation rate. Similar rules, based upon valency arguments,
apply to the other cases [2] . Notice that the rules reverse
depending upon whether it is ionic or electronic transport which
is rate-determining.

      However, there are so many exceptions observed in practice
[3,4] that these rules are of no great value, except to students.
We shall see that a study, made by ion implantation, suggests that
a reformulation of the rules would be more appropriate. Even so,
it must be realized that complex factors such as grain boundary
diffusion, and layered scale formation are bound to lead to
departures from rules of such a kind. In other cases it is the
ionic size and the way in which this governs the mechanical stress
within the oxide that determines the influence of a constituent.

      Ion implantation clearly offers us the means of introducing
essentially any doping species into a metal in reasonably well-
known concentrations. Subsequent corrosion tests allow one to
determine the beneficial species and devote more attention to
these. Moreover, as I shall demonstrate, the grain size is not
modified and the effect of the implantation can be studied within
individual grains, oriented at different angles to the surface.
Unlike the case of alloyed constituents, implanted material is
not likely to be bound up in intermetallic precipitates, the
presence of which may have a complicating and often deleterious
effect on corrosion. There are cases when, due to widely differing melting points, it is extremely difficult to prepare an alloy,
e.g. of lithium in nickel; ion implantation easily overcomes
this problem.

Next, we need to combine ion implantation with an accurate
means of determining the oxygen uptake in the implanted compared
with the unimplanted region of metal. In this way each receives
exactly the same surface preparation and corrosion treatment. The
conventional method is to measure the weight gain as a function of
time; we have used this, but in general it is too cumbersome and
requires larger implanted areas. For initial surveys something
more sensitive is needed, and so we have developed an ion backscattering technique which makes use of the very large elastic
scattering cross-section of O^{16} for protons above 4 MeV [5]
(Figure 1). Over an adequately wide range of scattering angle
and energy the differential scattering cross-section remains
constant and of the order 80 times the Rutherford value, due to a
nuclear resonance. This overcomes the otherwise low sensitivity
for oxygen compared with that for common metals. Thus the experiment consists of a series of implantations, oxidations, and
backscattering analyses.

Helium ion backscattering, at more modest energies, is
valuable in establishing the stoichiometry of an oxide as a
function of depth below the surface. It may also serve to reveal
the distribution of implanted material after the oxidation; it is
rarely left undisturbed by the ionic migration during corrosion.
We shall hear later of some experiments of this kind in anodic
films on aluminium [6] .

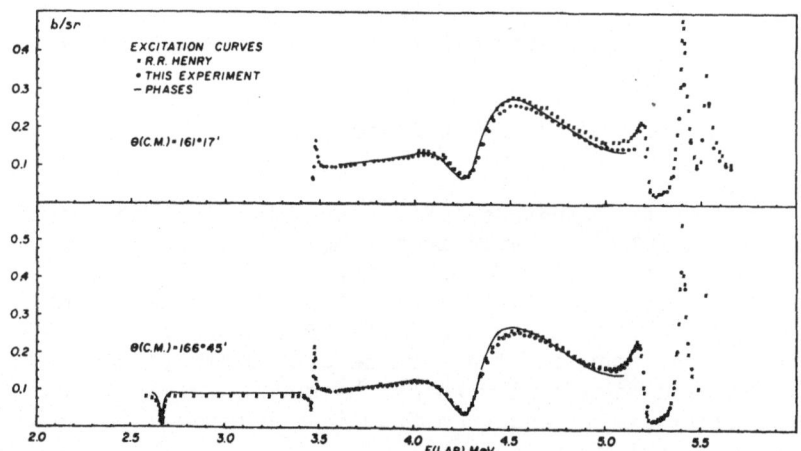

Figure 1.
Differential elastic scattering cross section, in barns per unit
solid angle, as a function of proton energy for the O^{16} (p,p)
process at two scattering angles.
(from Harris et al. [5]).

This type of study may complement the use of the ion microprobe analyser as a means of determining the depth distribution of
the components of a surface film; it has the advantages of being
more quantitative and more trustworthy as regards the depth scale.

In some work on corrosion, the species to be incorporated into
a metal or alloy may be well known from earlier studies, but one
may choose to introduce it by ion implantation rather than by
alloying throughout. I shall give an example from the nuclear
reactor field [7] . The use of ion implantation here is an
example of a way of meeting a surface requirement of a material
without sacrificing its bulk properties. The environment is one
of several in which a conventional plating or coating technique
would not be adequate on grounds of safety. An ion implanted
layer is, however, an integral part of the material and free from
any interfacial weaknesses.

In this review I shall provide examples of various ways in
which ion beam techniques have been brought to bear on corrosion
problems in metals and alloys. Most of them necessarily stem from
our own laboratory, since there has been little or no other work
reported prior to this conference. It is heartening to see other
groups now under way in this field.

One important aspect which I shall not review is that of
aqueous corrosion. Experiments are in progress here, [8,9] but
it is difficult to be sure that the results are not significantly
influenced by traces of hydrocarbon deposited during ion bombardment of a metal surface. Extreme caution is necessary, whereas
in high temperature oxidation experiments any such layer would
be quickly burnt away.

TITANIUM AND STAINLESS STEEL

A survey, of the type described above, was carried out by
Dearnaley et al. [10] to investigate the effects of various metal
species upon the thermal oxidation of polycrystalline titanium and
18/8/1 stainless steel. These are both technologically important
materials often employed because of their combination of strength
and corrosion resistance. The purpose was to study the initial
stage of the oxidation during which the mechanism may be expected
to be relatively simple and free from the mechanical complexities
which accompany thick scale formation.

Finely polished specimens as well as 5 μm thick foil specimens
of the two metals were implanted with ten different ion species at
energies of a few hundred keV, chosen to give an ion range
approaching 1000Å. A 500 keV Cockcroft-Walton accelerator and
sputtering ion source proved a versatile means of achieving the
required implantations. The specimens were then oxidized in dry

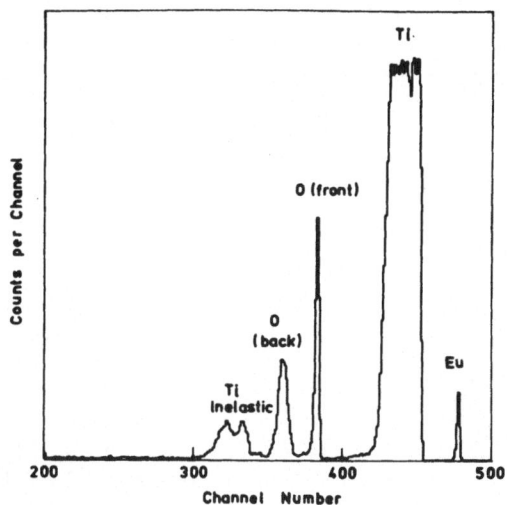

Figure 2.

Energy spectrum of 4 MeV protons scattered at 164° from a 5μm thick foil of titanium, implanted with europium ions and subsequently oxidized (from ref. /‾10‾/).

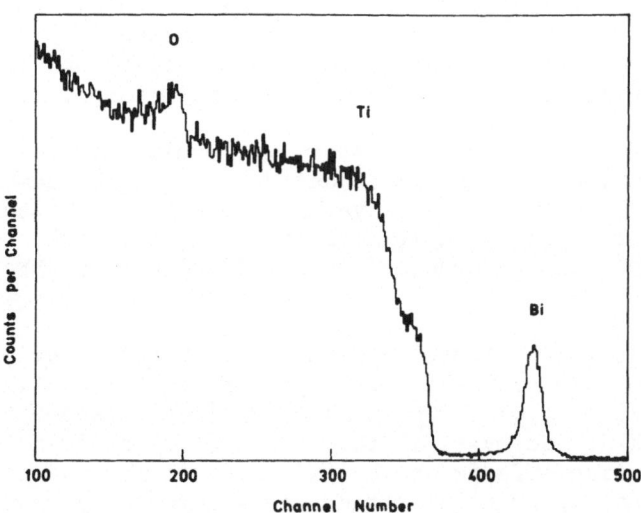

Figure 3.

Energy spectrum of 1.5 MeV He^4 ions scattered at 164° from a titanium specimen, implanted with bismuth ions and oxidized. (from ref. /‾10‾/)

O_2 at $600^{\circ}C$ (for titanium) or $800^{\circ}C$ (for stainless steel) for times up to about 30 minutes. The optical interference colour gave an immediate qualitative indication of the oxide thickness.

The oxygen take-up was determined quantitatively by proton elastic scattering at 164°, in this case at 4.0 MeV. It was soon found preferable to use this method with the thin foil specimens, as $Ti^{48}(p,p)$ shows a strongly resonant behaviour above about 3.5 MeV. The peaks in the scattered proton spectrum from a foil, due to oxygen and to metal are then completely separated (figure 2). Measurements in thick specimens showed that if all the visible oxide was removed by vibratory polishing there was still an appreciable amount of oxygen dissolved in the metallic-looking surface.

Helium ion backscattering at 2 MeV (figure 3) shows, both in the oxygen and titanium scattering profiles, that there is a variation in composition with depth, the oxide in this case being about 2000 Å thick. The ratio of Ti to O increases with depth and some of the oxygen is probably dissolved in a non-transparent layer of metal at the interface.

Calcium ions were the most effective in inhibiting the oxidation of titanium, their effect increasing with fluence up to a value of 5×10^{16} per cm^2. Europium ions similarly led to a reduction in oxidation. All other ions investigated (Zn^+, In^+, Ce^+, Al^+, Y^+, Ni^+ and Bi^+) enhanced oxidation, again to an extent which increased with fluence. Argon ions produced only a minor effect, however, suggesting that under the conditions of these experiments radiation damage is of minor importance.

In stainless steel the interesting result emerged that those ions (Ca^+, Eu^+) which inhibit oxidation of Ti were the ones that enhanced oxidation while, vice versa, those ions that enhance oxidation in Ti were found to inhibit it in stainless steel. Moreover, the relative order of their effectiveness appeared to be the same.

It is clear that these results do not agree at all with the Wagner-Hauffe valency rules, although certainly there is some trend from divalent Ca to pentavalent Bi (note however the position of Ni). However, if the ions are arranged in the order of their relative effects on the oxidation of the two metals (Table 1) it is apparent that there is a remarkably good correlation with the electronegativity [11] of each ion. In Pauling's terms this is a measure of the power of an ion to attract electrons, and hence it is reasonable that it should be a more significant and sensitive parameter than the valency in determining the distribution of charged defects in an oxide lattice.

TABLE 1

Relative effectiveness of ion	Electronegativity	Position in Electrochemical Series	Ionisation Potential 1st	2nd	3rd	Ionic crystal Radius Å
Calcium	1.0	Ca^{2+} - 2e⁻ -2.76V	6.1	11.87	51.2	0.99
Europium	1.1	Not quoted	5.67	11.24	N-q	0.95
Cerium	1.1	Ce^{3+} - 3e⁻ -2.34V	5.6	12.3	20.0	1.03
Yttrium	1.3	Y^{3+} - 3e⁻ -2.37V	6.38	12.23	20.5	0.89
Zinc	1.5	Zn^{2+} - 2e⁻ -0.76V	9.39	17.96	37.7	0.74
Aluminium	1.5	Al^{3+} - 3e⁻ -1.706V	5.98	18.82	28.4	0.51
Indium	1.7	In^{3+} - 3e⁻ -0.34V	5.78	18.86	28.03	0.81
Nickel	1.7	Ni^{2+} - 2e⁻ -0.23V	7.63	18.15	35.16	0.69
Bismuth	1.9	Not quoted	7.29	16.68	25.6	0.96

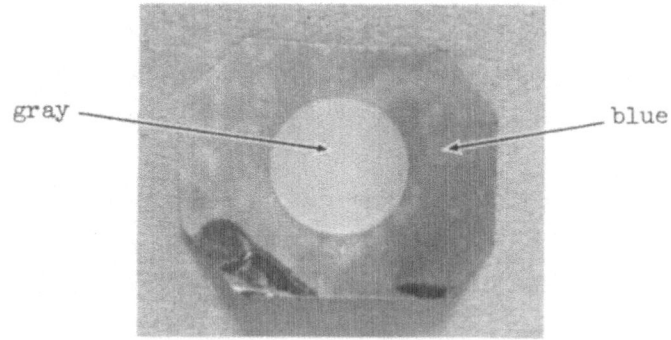

Figure 4.
Photograph of an oxidized specimen of bismuth-implanted titanium foil.

Dearnaley et al. have revised their opinion of the interpretation of these results, originally attributed [10] to trapping of oxygen ions. Since the parabolic oxidation law is obeyed for titanium at 600°C [12] it is permissible to adopt the basic approach of Wagner and Hauffe and regard TiO_2 as an oxygen-deficient n-type semiconductor with electron transport as the rate-determining process. Then substitutional species with low electronegativity, such as Ca^{2+}, can accept electrons from Ti^{3+} donors and so reduce the electronic conductivity and the oxidation rate. Calcium is well known to act as a good electron trap in TiO_2 [13]. These conclusions are entirely consistent with measurements made by Suffield and Dearnaley on the electronic conductivity

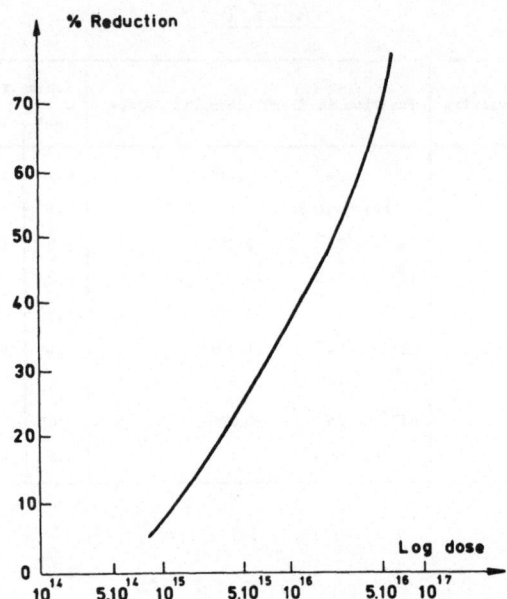

Figure 5.
The effect of calcium implantation on the oxidation of titanium in
O_2 at 1 atm. at 600°C, as a function of ion fluence.

Table II

Spec. No.	Ion type and dose ions/sq cm		Oxidation Effect	
S1	Ca	10^{15}	enhanced	20%
S2	Ca	10^{16}	enhanced	40%
S3	Ca	10^{16}	enhanced	51%
S4	Ca	5×10^{16}	enhanced	100%
S5	Y	10^{16}	reduced	30%
S6	Y	5×10^{16}	reduced	50%
S7	Eu	5×10^{16}	enhanced	50%
S8	Eu	5×10^{16}	enhanced	62%
S9	Bi	10^{16}	reduced	28%
S10	Bi	10^{16}	reduced	31%
S11	Bi	2×10^{16}	reduced	50%
S12	In	2×10^{16}	reduced	32%
S13*	Al	10^{17}	reduced	52%

The effect of ion implantation on the oxidation of 18/8/1
stainless steel.

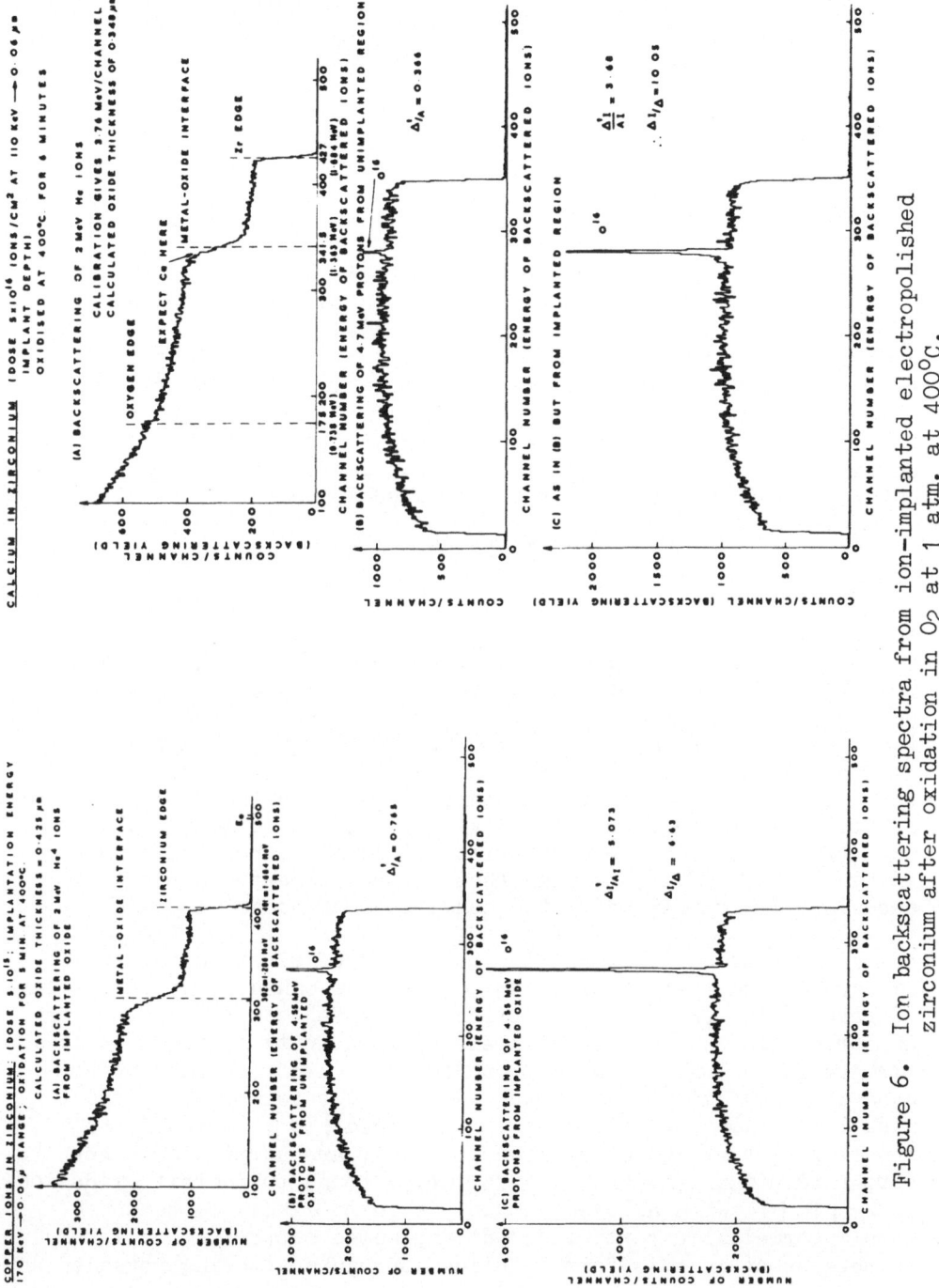

Figure 6.   Ion backscattering spectra from ion-implanted electropolished
zirconium after oxidation in O_2 at 1 atm. at 400°C.

of anodic TiO_2 grown on ion-implanted titanium [14]. Different
impurities stabilized different impedance states of the oxide in
a manner correlated with their electronegativity.

In 18/8/1 stainless steel, in which exactly opposite effects
occur, (Table 2) it is proposed that the early stage of oxidation
results in formation of a metal-deficient p-type semiconductor
film in which compensation is achieved by substitutional additions
of high electronegativity. This is consistent with the known
semiconducting behaviour of Cr_2O_3, and the oxide on stainless steel
is known to be very rich in chromium oxide [3] .

It must be emphasized that these ideas are bound to break
down, just as do the earlier rules, when the scale thickness
increases and inhomogeneities such as those discussed by Wood [3]
appear. The study of implantation effects on the oxidation of
titanium has been extended recently to higher temperatures and
longer times by Lucas et al. [15] and it is clear that the
phenomena are quite different. Strong electric fields and space
charge effects of the type discussed by Cabrera and Mott [16] may
not be absent in the thin oxides studied by Dearnaley et al. [10] .

ZIRCONIUM

A similar survey to that made in Ti has been carried out by
Weidman, Goode and Dearnaley [17] but using an even wider variety
of ions. Polycrystalline discs of Zr were prepared by mechanical
polishing and electropolishing before implantation. Oxidation was
performed in dry O_2 at 1 atm. for times of about 5 minutes at
temperatures between 380° and 400°C. The oxygen taken up was
determined by proton scattering at 164° at an energy of 4.55 MeV
(the peak of the O^{16} (p,p) resonance [5]). In this case, since
Zr itself shows no resonant behaviour, thick specimens could be
used (figure 6). Bombardment with Zr^+ ions produced only an
insignificant effect, proving that damage processes are of
secondary importance under the conditions of these tests.

It was soon clear that the behaviour of zirconium is totally
different from that of titanium: most ions enhance the oxidation,
(figure 7) the principal exceptions being Fe^+ and Ni^+. There is
no correlation whatsoever with electronegativity (Table 3). It is
already known that the Wagner-Hauffe rules are violated in the
case of zirconium [18] and the beneficial effect of iron and
nickel additions has also been established [18]. Some of the
results, such as those observed with Cu^+, are surprising since
copper is often (though not invariably) found to inhibit oxidation
in Zr [18] . When alloyed, copper is quite likely to be bound
into intermetallic particles, but implanted copper is not likely
to be present in this form. Helium ion backscattering revealed an

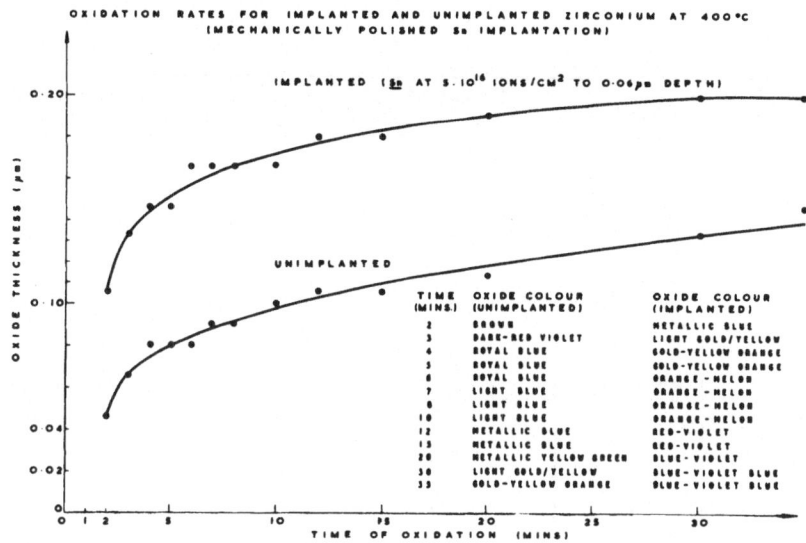

Figure 7.
The growth of oxide on finely-abraded zirconium in O_2 at 1 atm.
at 400°C, as determined from the optical interference colour.
(from ref. $\boxed{17}$).

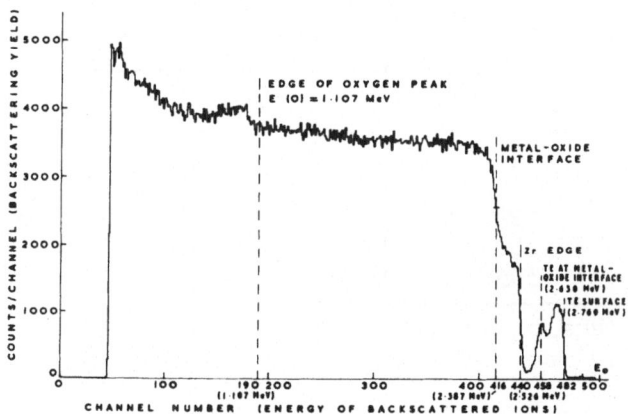

Figure 8.
3 MeV He^4 ions back-
scattered from oxidized
thallium-implanted Zr,
showing an anomalous
impurity distribution
(from ref. $\boxed{17}$).

interesting effect in Tl^+ -implanted zirconium: after oxidation
the thallium was distributed in two layers, one near the surface
and one near the metal-oxide interface (figure 8). The reason for
this is not understood, but unusual distributions of implanted
species in anodic oxides have been observed by Mackintosh and
Brown [6] and ascribed to electrochemical differences between metal

Table III

Ion species	Fluence	Oxidation temperature $^{\circ}C$	Oxidation time, mins.	Oxygen ratio
Electropolished Specimens				
Ni	5×10^{15}	400	5	0.53
Nb	5×10^{15}	400	5	0.84
Fe	5×10^{15}	400	5	0.86
Fe	10^{15}	400	5	0.93
Cr	5×10^{15}	400	5	0.95
Ni	10^{15}	400	5	0.98
Cr	5×10^{16}	400	5	1.03
Zr	10^{16}	400	5	1.04
P	5×10^{15}	400	5	1.06
Ca	5×10^{15}	400	5	1.10
Y	5×10^{16}	390	5	1.13
Si	5×10^{15}	400	5	1.18
Fe	10^{16}	400	5	1.47
B	5×10^{16}	405	45	1.48
Nb	5×10^{16}	380	7	1.56
B	5×10^{16}	380	6	1.63
Fe	2×10^{16}	400	5	1.80
Sn	5×10^{15}	400	5	1.93
Fe	5×10^{16}	400	5	2.18
Ca	5×10^{16}	400	5	2.36
Eu	5×10^{16}	410	3.5	5.75
Cu	5×10^{15}	400	5	6.63
Mechanically polished specimens				
Ni	5×10^{16}	380	5	2.1
Tl	5×10^{16}	395	7	2.5
Cs	5×10^{16}	385	5	4.0
Ca	5×10^{16}	400	6	10.0
Al	5×10^{16}	395	8	13.7

The ratio of oxygen up-take in implanted and unimplanted Zr.

and impurity. Similar shifts may occur due to the electric fields in thin tarnishing films [16] .

Since the grain size of the electropolished specimens used by Weidman et al. [17] was about 50 μm it was possible to see visible effects of implantation within an individual grain (figure 9). Inspection of the optical interference colours shows that different

grain orientations oxidize to different extents, as was already
known [19] but also that ion implantation of a given dopant
affects the oxidation to a degree dependent on the grain orientation. The proton and He^4 ion beams used for backscattering were
some 500 μm in diameter, and hence probed average effects in about
100 grains. Clearly there is much scope for investigations in
large-grained specimens in which channeling techniques would be
feasible. Experiments on these lines are now under way.

The effect of implanting different amounts of iron and
nickel into electropolished Zr is shown in figure 10. There
appears to be an optimum fluence of about 3×10^{15} ions per cm^2, and
it is interesting to observe that this corresponds to a mean
concentration of transition metal of about the same as that found
to give optimum oxidation resistance in zircaloy 2 [1] . It is
very simple to establish this trend by a series of implanted
specimens.

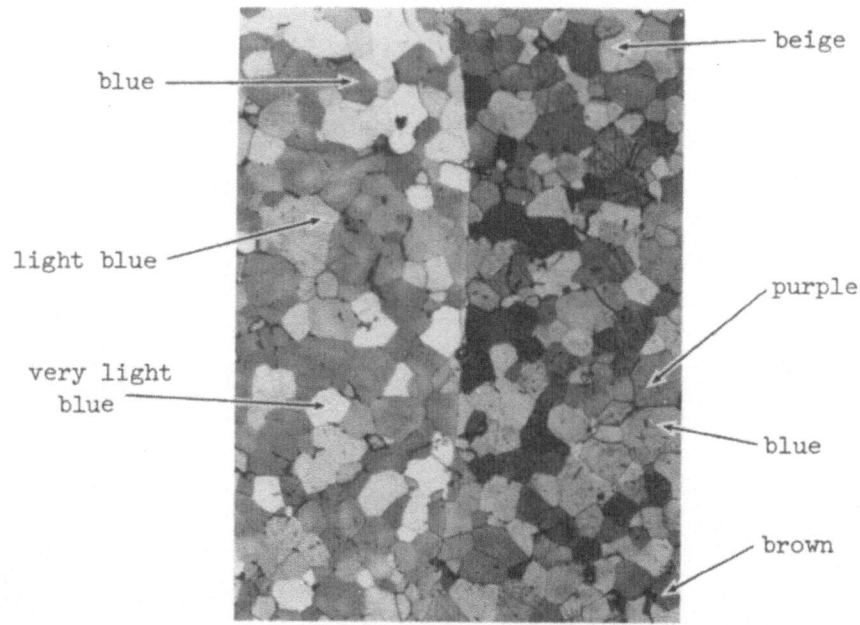

Figure 9.
Photograph of oxidized ion-implanted specimen of electro-polished
zirconium. The left half had been implanted with 5×10^{16} calcium
ions per cm^2.

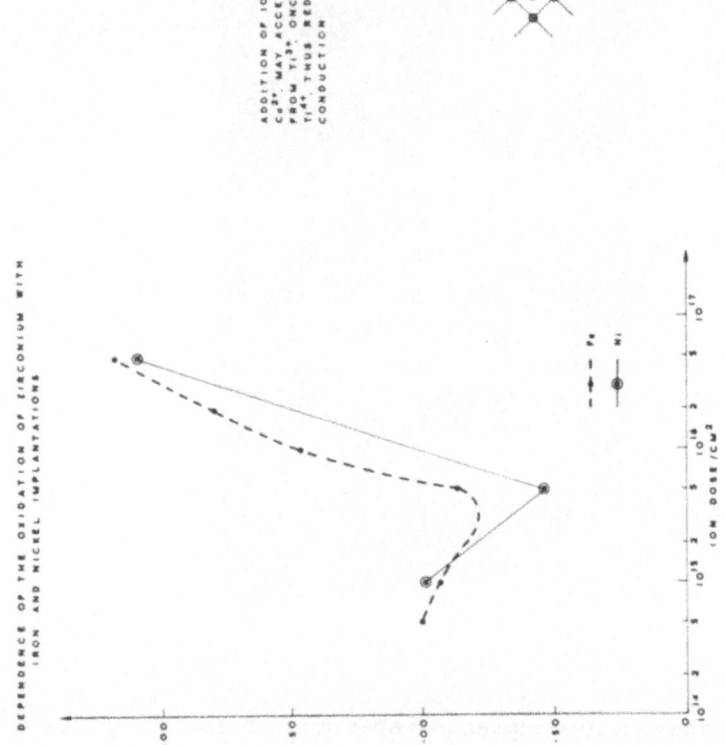

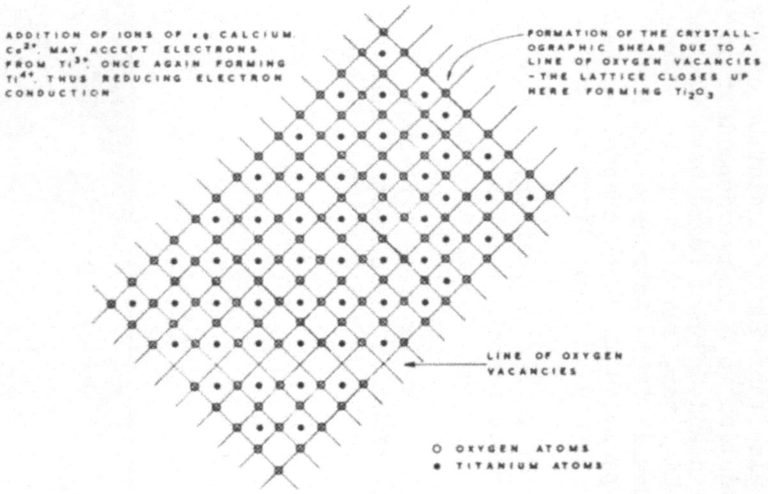

Figure 11.
Crystallographic shear in anion-deficient TiO_2

Figure 10.
Effect of Fe^+ and Ni^+ implantation
upon the oxidation of electropolished zirconium, as a
function of ion fluence.

COMPARISON OF TITANIUM AND ZIRCONIUM

The results of ion implantation favour the idea, discussed by Kubaschewski and Hopkins [20] and Wanklyn [18] that it is the mechanical stress within zirconium oxide which determines the formation of cracks and pores, and so influences the rate of further oxide growth.   Cox [21] has obtained good evidence of pore formation in even thin oxide films as zirconium.   Under these conditions it is the ionic size, coupled with the lattice site occupancy and solubility of dopant ions within the oxide that will govern the degree of stress.   Transitions from tetragonal to mono-clinic forms of ZrO_2 are believed to occur during growth and some ions, such as Ca^+, are known to stabilize the high temperature cubic form.

This situation is far too complex to lend itself to generalized rules, whether based upon valency, electronegativity or ionic size. It is possible for low concentrations of impurity to reduce lattice parameters, while higher concentrations exceed the substitutional solubility limit and lead to a lattice expansion, so accounting for the behaviour shown in figure 10.

Shirvington and Cox [22] have shown that in the oxidation of pure zirconium (but not of zircaloy) electron transport is the rate-determining process.   This may well be so within a barrier layer of impermeable oxide near the metal interface.   Above this there seems likely to exist a highly disordered, porous structure con-sisting of several crystal forms.

It may at first seem strange that the two Group IV B metals, Ti and Zr, should behave so differently after implantation with the same dopant species.   The answer probably lies in their different crystal structure and the remarkable ability of titanium oxides to eliminate oxygen vacancies by the mechanism of crystallographic shear [23] .   Anion vacancies become ordered into planes to form so-called 'Wadsley defects': the lattice closes up and a lower stoichiometry exists locally (figure 11).   It is therefore erroneous to discuss the oxidation of titanium, as has often been done, in terms of the migration of oxygen ions via a vacancy-assisted diffusion.   It is not yet established whether the diffusion is interstitial or by progressive shrinkage of Wadsley and other related defects.   What is clear, however, is that the crystal structure and, in particular, the defect behaviour of an oxide plays a very important part in determining the oxidation of a metal and the influence of impurities upon the oxidation rate.

Zirconium dioxide, by contrast, does not exhibit crystallo-graphic shear.   It therefore remains rich in anion vacancies and hence is one of the best ionic conductors, while this conductivity

is enhanced by additions of calcium and yttrium [24] . For this
reason, we can understand the uniform stoichiometry of the oxide
apparent in figure 6, compared with the non-uniformity of titanium
oxide (figure 3). The latter corresponds to an increasing
concentration of Wadsley defects nearer to the metal, and these
allow a continuous variation in stoichiometry.

COPPER

Perhaps the first reported observation of the effect of ion
implantation upon the subsequent corrosion of a metal was that by
Crowder and Tan [25] following boron implantation of copper. Other
workers have noticed a reduction in the atmospheric tarnishing of
copper plates exposed to ion bombardment [26] but the first systematic study has been initiated by Rickards [27] in carefully-prepared
single crystals, and is the subject of a paper at the present
conference.

LONG-TERM OXIDATION OF STAINLESS STEEL IN CO_2

Alloyed additions of yttrium or rare earths are known to be
effective in inhibiting the high temperature oxidation of iron and
nickel based alloys, some of which are of considerable technological
importance. The effects of small concentrations of yttrium on the
oxidation, in carbon dioxide, of an austenitic stainless steel
containing 20% Cr, 25% Ni and stabilized with 1% Nb had previously
been studied, within the temperature range $800^{\circ}C$ - $1000^{\circ}C$. A few
parts per thousand of yttrium reduced the overall attack and
improved the oxide adherence considerably. However, there are
drawbacks to the use of yttrium-bearing alloy, the chief of these
being the reduction in ductility and tensile strength of the
material, due to grain boundary segregation of yttrium oxide.

Implanted yttrium, in concentrations of about 0.2 per cent
through a depth of 0.2 μm (obtained with a fluence of 3.5×10^{15} Y+
ions/cm^2 in multi-energy implantations) was found to be just as
effective as alloyed yttrium in reducing oxidation [7] (figure 12).
It also lessens the percentage of oxide that spalls. The implantation treatment is effective over a period of at least 6000 hours
(8 months) at $850^{\circ}C$ during which time an amount of steel equivalent
to over 20 times the implanted depth has been converted to oxide.

Ion microprobe analysis of a bevelled section of implanted
steel, after oxidation, has shown the accumulation of yttrium at
the metal-oxide interface, where it is believed to form an
impermeable layer rich in the perovskite $YCrO_3$. The improved
adherence of the oxide may be due to a modification of the vacancy

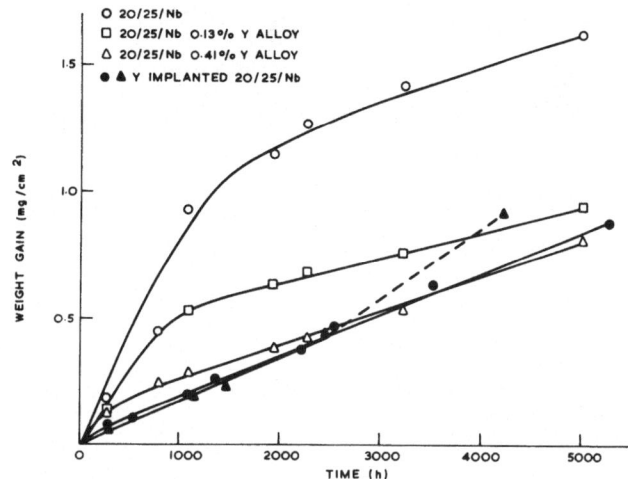

Figure 12.
Effect of yttrium implantation on the oxidation of a stainless
steel in CO_2 at 800°C compared with the behaviour of alloyed
yttrium (from ref. [7]).

condensation mechanism at the interface where voids are formed that
lead to spalling. The effectiveness of an implanted layer rules
out some other explanations of improved oxide adherence, such as
one involving keying at grain boundaries. Thus an initially
shallow implantation may provide a long-lasting protection, and
the evidence is that relatively little of the implanted material
is lost by spalling.

ION MICROBEAM ANALYSIS

In the previous section reference was made to the use of the
ion microprobe analyser [28] for analysis of a corrosion film.
Figure 13 shows the layout of the CAMECA analyser: a focussed ion
beam of argon or oxygen ions at energies around 20 keV sputters the
surface of the specimen.   Secondary ions are mass-analysed and
recorded.

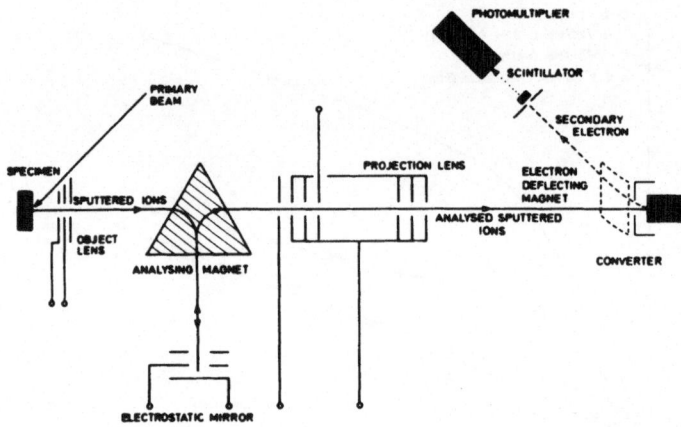

Figure 13.
Layout of the CAMECA ion microprobe analyser.

There are two modes of operation.   The ion beam may be
scanned over a small area of target so as to erode it progressively.
The time dependence of the yield of a given ion species provides
information regarding its depth distribution.   An example of this
technique in the examination of oxide scale on Nimonic alloys has
been given by Stott, Lin and Wood [29] .   Unfortunately it is not
easy to be sure of the sputtering rate, particularly in an
inhomogeneous layered film, and so there is some uncertainty in the
depth scale.

An alternative, as mentioned above, is to bevel the specimen
at a known angle and to scan the ion beam across the bevel.   The
image recorded in the detection plane for a selected ion species
can then reveal its true depth, though without careful calibration
the method is not quantitative.

The ion microprobe analyser has poor sensitivity for some
species, and its competitor, the electron probe microanalyser, is
not suitable for identifying light atoms or even heavier constituents present in very small concentration.   Then analysis by

means of an ion microbeam at MeV energies has advantages.
Figure 14 shows the 4-element magnetic quadrupole attached to the
Harwell 3.5 MeV Van de Graaff by Cookson et al. [30] and which
provides a beam spot typically 5 μm square.   Backscattering,
nuclear reactions, X-ray and γ-ray yields can be measured as the
beam is scanned electrostatically over a specimen.   Examples of
its use are given in the paper at this conference by Hartley [31].

Figure 14.
Photograph of the ion microbeam facility on the Harwell 3.5 MV
Van de Graaff.

CONCLUSIONS

Ion implantation, when carried out with a versatile ion source
and combined with ion backscattering and ion microbeam analysis, has
greatly extended the scope for exploratory studies in corrosion
science, as a result of which a better understanding of certain
oxidation mechanisms is beginning to emerge.   Thus the electronegativity of additive ions may well prove a more significant parameter than the valency in determining corrosion rates.   The importance of the defect structure in the corrosion layer has been
demonstrated by a comparison of titanium and zirconium.

There may be differences between the behaviour of an alloyed
constituent and its action when implanted to the same concentration,
owing to differences in the degree of formation of intermetallic
compounds.   Both chemical doping and lattice damage effects may be
important in some cases; in others damage appears to be insignificant.

Ion implantation appears to be a viable means of providing a
long-lasting inhibition of the corrosion of metal components which

are important parts of costly systems, such as occur in the nuclear
energy, aerospace and military fields. The use of small amounts
of additives introduced by ion implantation into the surface of a
metal rather than alloyed throughout is in keeping with the trend
towards conservation of material resources and optimum control of
material properties.

There is a wide field of investigation open for the exploita-
tion of ion beams in corrosion research. Both ion backscattering
and ion microprobe analysis can reveal the depth variation of
stoichiometry and the distribution of components in a layered
scale. These experiments can be particularly fruitful in single-
crystal or large-grain polycrystalline specimens, in which channelling techniques can be employed. The degree of epitaxy of a
corrosion film can thereby be studied, and the effect of implantation or different crystal orientations may be compared. Such
experiments are only just commencing. The application of ion beams
to corrosion science may be expected to develop rapidly. It
promises to be a fruitful area of interdisciplinary research.

ACKNOWLEDGMENTS

The author wishes to thank Dr. J.E. Antill, Dr. M.J. Bennett,
Dr. R.J. Brook, Dr. A.M. Stoneham, Mr. J.F. Turner and Mr. Lewis
Weidman for valuable discussions and comments during the prepara-
tion of this review.

REFERENCES

1.   S. Kass, 'Corrosion of Zirconium Alloys', ASTM Publication
     no. 368, p.3 (1964).
2.   K. Hauffe, 'Oxidation of Metals' 1965 (Plenum Press, New York).
3.   G.C. Wood, Oxidation of Metals, $\underline{2}$ no. 1, 11 (1970).
4.   U.R. Evans, 'The Corrosion & Oxidation of Metals', Supplement,
     1968 (Arnold, London).
5.   R.W. Harris, G.C. Phillips and C. Miller Jones, Nucl. Phys. $\underline{38}$
     259 (1962).
6.   W.D. Mackintosh and F. Brown (this conference).
7.   J.E. Antill et al., Proc. Conf. on Ion Implantation, Yorktown
     Heights, 1972 (Plenum Press, New York: to be published).
8.   A.D. Street, W.A. Grant and G. Carter, ibid.
9.   W.A. Grant (priv. comm.).
10. G. Dearnaley, P.D. Goode, W.S. Miller & J.F. Turner, Proc.
     Conf. on Ion Implantation, Yorktown Heights, 1972 (Plenum
     Press, New York: to be published).
11. L. Pauling, 'The Nature of the Chemical Bond', p.58
     (Cornell Univ. Press) 1945.
12. P. Kofstad, K. Hauffe & H. Kjøllesdal, Acta. Chem. Scand. $\underline{12}$,
     239 (1958).

13. L.E.J. Roberts (ed.) 'Solid State Chemistry' (Butterworths, London). (1972).
14. N.W. Suffield & G. Dearnaley, Proc. Conf. on Ion Implantation, Yorktown Heights, 1972 (Plenum Press, New York: to be published).
15. H. Bernas (priv. comm.)
16. N. Cabrera and N.F. Mott, Rep. Prog. in Phys. 12, 163 (1949).
17. L. Weidman, P.D. Goode and G. Dearnaley (to be published).
18. J. Wanklyn, 'Corrosion of Zirconium Alloys' ASTM Publication No. 368, p.58 (1964).
19. J.P. Pemsler, J. Electrochem. Soc. 105, 315 (1958).
20. O. Kubaschewski and B.E. Hopkins, 'Oxidation of Metals and Alloys' 1953 (Butterworth, London).
21. B. Cox, J. Nucl. Mat. 29, 50 (1969).
22. P.J. Shirvington and B. Cox, J. Nucl. Mat. 35, 211 (1970).
23. S. Andersson and A.D. Wadsley, Nature 211, 581 (1966).
24. B.C.H. Steels, 'Solid State Chemistry' p.117 ed. L.E.J. Roberts (Butterworths, London, 1972).
25. B.L. Crowder and S.I. Tan, IBM Technical Disclosure Bulletin 14, 198 (1971).
26. J. den Boer (Priv. Comm.).
27. J. Rickards and G. Dearnaley, these conference proceedings.
28. R. Castaing and G. Slodzian, J. de Microscopie 1, 395 (1962)
29. F.H. Stott, D.S. Lin and G.C. Wood, Corrosion Sci. 13, 449 (1973).
30. J.A. Cookson, A.T.G. Ferguson and F.D. Pilling, J. Radio-analytical Chem. 12, 39 (1972).
31. N.E.W. Hartley, this conference.

DISCUSSION

Q: (L. C. Feldman) Did you implant with a beam of the same species as the host material? If so, what was the effect?

A: Yes. As you will see in Table III, Zr^+ bombardment of electro-polished Zr samples gave rise to only an insignificant enhancement of the subsequent oxidation.

Comment: (J. P. Biersack and D. Fink) We observed an abundant oxidation of Nb to Nb_2O_5 after implantation of 220 keV Li^+ leading to an Li concentration of a few percent in the region of 0 to 0.8 μm depth. The oxidation occurred after 1 hr at 700C in air. The oxidation exceeded 1 mg/cm^2 and was much higher than the reported oxidation after Ca implantation which has the same electronegativity but differs considerably in the ionic radius.

A: I am most interested to hear of this observation. Nb_2O_5 is
held to be an oxygen-deficient n-type semiconductor, and implanted
Li^+ (which would probably form interstitial donors) would increase
the oxidation if electron transport is the rate-determining factor.
I believe the electronegativity of Ca to be somewhat higher than
that of Li, but because of the larger ionic radius (approximately
1 Å) it may need to substitute for Nb and so behave quite differently in the oxide lattice.

Comment: (S. T. Picraux) In addition to the modification of
oxidation/corrosion properties after implantation there is another
important effect which you have not mentioned. This is oxidation
during irradiation. S. M. Myers and I have observed in Be, for
example, several orders of magnitude increase in oxidation rate of
Be under Ne irradiation. In contrast to many of the studies you
reported on, the radiation damage would be expected to be a major
factor here.

Q: (W. Mackintosh) You noted movement of Y^+ implants in stainless
steel and Tl^+ in Zr. Did you observe movements of other implants
in other materials?

A: Yes. Ion backscattering experiments frequently revealed redistribution of (heavy) implanted atoms during thermal and anodic
oxidation of Ti and Zr. See Refs. 7 and 10.

Q: (H. J. Smith) I want to make a remark about the use of the ion
microprobe analyses: I do not know what the case will be for oxygen
atoms in metals, but we have found definite enhanced diffusion of
metal atoms in metals during ion bombardment at room temperature,
and there is also evidence of this for noble gas ions. This indicates that in some applications the validity of the ion microprobe
measurements could be questioned.

A: I agree. This is only one of several factors which make depth
distribution measurements by the ion microprobe analyses insuffi-
ciently quantitative for some applications.

Q: (E. Taglauer) You found the implanted Y being concentrated
close to the oxide layer by ion microprobe analysis. Could that
not be a typical effect of enhanced secondary ion yield due to the
oxide at the interface?

A: A small-angle bevelled section was made through both the oxide and substrate metal, and the ion microbeam was scanned along this section. The well-known enhanced secondary ion yield from oxidized surfaces would have been evident across the entire thickness of exposed oxide, but the yttrium was observed only near the metal-oxide interface. Therefore, we believe the yttrium to be concentrated (together with a certain amount of silicon) in this region. As a general point, however, I agree that this effect must be borne in mind in the application of the ion microprobe analyzer to the study of oxide films.

Comment: (H. Bernas) In work performed at Orsay, the oxidation properties of Ti at 800C were studied after implantation of C, O, Ti, Aℓ, and Ca at depths ranging from 400 Å to 800 Å and doses between 10^{14} and 5×10^{16} ions/cm^2. In all cases, the oxygen weight gain Δm depended on the oxidation time t in the well-known way = $(\Delta m/s)^2$ = kt; (s = sample surface area, k = proportionality constant). C, O, and Ti-implanted samples were found to have the same value as unimplanted Ti samples. Aℓ- and Ca-implantation produced differences. It is interesting that, contrary to your results at 600C, we find that at 800C Aℓ-implantation slows down oxidation while Ca-implantation enhances oxidation. In both cases, this is still true after 10 hours of oxidation (corresponding to oxide thicknesses of some 75,000 Å). The reversal between 600 and 800C could be due to different oxidation mechanisms. At 600C, oxygen diffuses into the sample, while at 800C it has been shown that Ti diffuses out of the bulk, into the oxide layer. This work was done by X. Lucas, E. Garcia, G. Béranger, J. Chaumont and H. Bernas.

THE EFFECTS OF ION BOMBARDMENT ON THE THIN FILM OXIDATION

BEHAVIOR OF ZIRCALOY-4 AND Zr-1.0 Nb

J. A. Spitznagel, L. R. Fleischer, W. J. Choyke

Westinghouse Research Laboratories

Pittsburgh, Pennsylvania   15235

ABSTRACT

The oxidation rates of Zircaloy-4 and Zr-1.0 Nb in oxygenated water at 360°C were suppressed by prior ion bombardment. O^+, Ar^+, and Xe^+ accelerated to energies in the 67 to 150 keV range to fluences of 5×10^{13} to 1×10^{16} ions/cm^2 provided lattice damage on the order of 10 to 100 dpa. Subsequent autoclave oxidation was used to explore the damage effects on reaction rate. Since annealing accompanied oxidation, several sequential bombardment/oxidation cycles were performed on each sample. Oxidation of bombarded surfaces was much more uniform than that observed on control samples.

INTRODUCTION

Oxidation of zirconium alloys, in the thin film region, by water at temperatures of approximately 300°C is normally controlled by anion diffusion through an increasingly protective oxide film. However, growth of the hypostoichiometric oxide does not necessarily follow a parabolic rate law.[1] The oxidation behavior of Zircaloy-4 (Zr-4) and Zr-Nb alloys under these conditions differs in two important ways. The Zr-Nb alloys generally oxidize more rapidly than Zr-4. Zr-Nb alloys are also sensitive to the concentration of oxygen in the corrodent while Zr-4 is not. It is thought that these differences may be due to different ionic transport mechanisms. Anion diffusion by a vacancy mechanism is presumed to dominate oxygen transport throughout the oxide film thickness on Zr-4. Conversely, it has been proposed[2] that, in the outer

87

two-thirds of the oxide on Zr-Nb alloys, oxidation is dependent on interstitial oxygen anions.

In the presence of a neutron flux and an oxidizing environment the oxidation rate of Zr-4 may be measurably enhanced at temperatures up to 400°C,[3] and the corrosion rate of the alloy is observed to be sensitive to the oxygen content of the environment. The corrosion rate of Zr-Nb alloys remains sensitive to the oxygen concentration in the corrodent but, in general, is not strongly enhanced by the neutron flux.[4]

It was anticipated that the oxidation of zirconium alloys might be influenced by the bombardment of the oxide layer in two ways. From the dynamic point of view, vacancies and interstitials produced by bombardment at moderate temperatures provide a vacancy concentration higher than the thermal equilibrium concentration. As a result, the coefficient for diffusion dependent on vacancy movements is increased. Under suitable conditions, processes controlled by diffusive mass transport, such as creep, are therefore enhanced in nuclear reactors.[5] Similar arguments have been applied to the transport of reactive species in oxidation. However, the increased diffusivity failed to account for the magnitude of the rate increase observed in-pile.[6]

Alternatively, it has been suggested that relatively stable regions of damage might provide high diffusivity paths for diffusion through the oxide.[7] These could range from particle tracks to finer oxide crystallites nucleated under bombardment. These hypotheses have not been pursued experimentally before.

EXPERIMENTAL

Specimens (19 mm square x 0.51 mm thick) of Zircaloy-4 (Zr-1.4 Sn-.2 Fe-.1 Cr-.004 Ni) and Zr-1.0 Nb were sheared from rolled strip. The Zircaloy-4 strip had been recrystallization annealed at 700°C for 1/2 h following a 50% reduction, while the Zr-1.0 Nb was used in the beta quenched and 50% cold rolled condition. The specimens were cleaned by pickling and carefully rinsed in boiling distilled water to minimize contamination.

Pre-bombardment (and post-bombardment) oxide films were formed by exposing the samples to demineralized neutral water containing approximately 5 ppm dissolved oxygen at 360°C in a nickel-lined stainless steel autoclave. Temperature control during oxidation was better than ± 5°C. The initial autoclave exposure time was selected to produce oxide films with average thicknesses from 75 to 200% of the calculated mean projected range of the ion to be implanted. This permitted some control of the location of the energy deposition resulting from the subsequent ion bombardment. Thus, the effect of concentrating the lattice damage partly in the metal substrate or entirely in the oxide could be studied. Average film thicknesses were generally in the range .05-.4 μm following the first oxidation step.

Implantation was accomplished with a 10 to 180 kv accelerator using post-acceleration magnetic separation and electrostatic horizontal and vertical scanning. During implantation, pressure at the sample was about 10^{-7} Torr.

The scanned beam was intercepted by a 3 cm^2 aperture so that the beam hitting the sample was extremely uniform in circular regions on the faces of the square specimens leaving only the extreme edges unirradiated. Both sides of at least three specimens were irradiated in every cycle under identical conditions. For high temperature implants the pre-oxidized specimens were affixed to a boron-nitride holder in a specially designed tantalum sheath furnace.[8]

Most of the specimens were bombarded at room temperature with 150 kev oxygen, argon or xenon ions to a fluence of 10^{16} ions/cm^2. This permitted a comparison of the effects on oxidation of lattice damage resulting from bombardment with chemically active or inert ions. The use of ions of different mass also permitted independent variation of energy deposition and ion fluence. Broadening of the energy deposition peak in several samples was accomplished by bombarding sequentially at several accelerating voltages, i.e., 150, 120, 100 and 67 kV. The incident ion species implanted in a given specimen remained the same if the specimen was re-bombarded during the course of multiple irradiation-oxidation cycles. Single specimens were implanted at 288° and 360°C and subsequently oxidized at these same temperatures.

Oxidation weight gain for each autoclave cycle was determined after cleaning the specimens in an alkaline potassium permanganate solution. Direct measurements of oxide film thickness using interferometry and stylus-profilometer measurements proved unreliable because of large variations in local film thicknesses.

Most samples were subjected to a sequence of bombardment-oxidation cycles. Control specimens which were not ion bombarded were also placed in the autoclave during the oxidation portion of each cycle.

Range statistics for the penetration of oxygen, argon and xenon ions in ZrO_2 and Zr were obtained from the code written by Johnson and Gibbons[9] based on the model of Lindhard et al.[10,11] Estimates of the total energy available to produce atomic displacements were obtained with the IONDOSE code.[12]

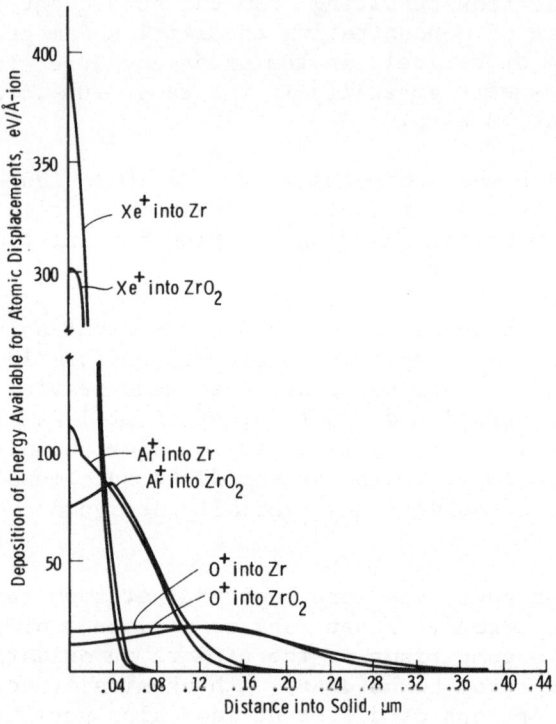

Figure 1.   Distributions of energy available for atomic displacements in zirconium metal (Zr) and zirconium oxide (ZrO$_2$)
resulting from bombardment with 150 keV ions.

RESULTS

    Figure 1 shows the curves for energy deposition resulting
from nuclear collision phenomena in ZrO$_2$ and Zr calculated by
IONDOSE for 150 kev O$^+$, Ar$^+$ and Xe$^+$ ions. The curves for energy
deposition in the case of zirconium metal do not take into account
the energy losses and straggling which must occur if the ions
first traverse an oxide film of finite thickness. They are included
only for comparison purposes. As pointed out by Garner[13] the
curves are necessarily approximate because of the limitations of
LSS theory at low ion energies. The distance from the surface and
broadening of the energy deposition peaks are seen to increase
with decreasing ion mass. These data were translated into atomic
displacement concentrations using an average binding energy of
33 eV for both zirconium and oxygen atoms, and the Kinchin and
Pease Model.[14] The number of displacements per atom in the vicinity

of the energy deposition peak in ZrO_2 is approximately 7 dpa for implantation of 10^{16} O^+ ions/cm^2 at 150 kV. Bombardment with 10^{16} Ar^+ or Xe^+ ions/cm^2 at the same energy moved the energy deposition peak toward the external surface of the oxide and increased the calculated number of displacements by factors of 4 and 15 respectively. Displacements resulting indirectly from ionization processes were not taken into account in the calculation.

Several interesting effects of the lattice damage on the oxidation processes could be ascertained visually. Delineation of the damaged region on all bombarded samples was difficult until the samples were oxidized further in the autoclave. Observation of the interference colors on the Zr-4 samples after one to four hour autoclave exposure invariably showed that the ion bombarded regions oxidized much more uniformly and at a lower rate than the unirradiated corners or the control samples. Prolonged oxidation of either alloy without re-irradiation, however, eliminated interference color differences.

On a number of specimens a white deposit was observed to have formed preferentially on the irradiated areas during autoclaving. The deposit was loosely adherent and could easily be wiped off with tissue paper. Examination of the deposit in the scanning electron microscope revealed numerous discrete particles. Energy dispersive x-ray analysis indicated only a nickel peak. It was not possible to determine whether the deposit was a nickel oxide, however, since the characteristic nickel and nickel oxide lines

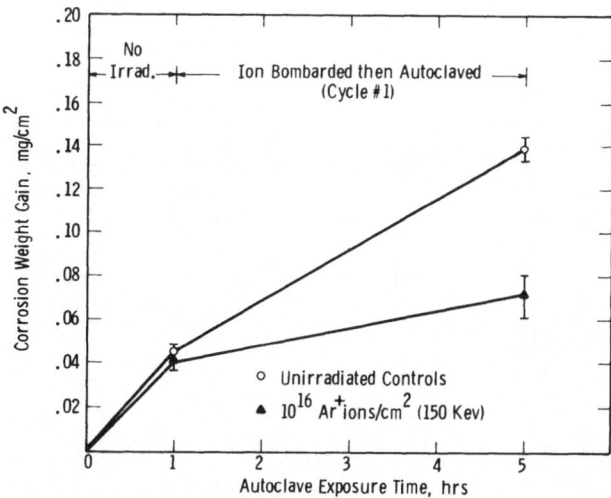

Figure 2.   Effect of Ar^+ bombardment on the oxidation of Zr-1.0 Nb in 360°C H_2O + 5 ppm O_2.

92 J. A. SPITZNAGEL, L. R. FLEISCHER, AND W. J. CHOYKE

overlap. The occurrence of the deposit necessitated a thorough cleaning of all specimens in alkaline potassium permanganate solution after autoclaving and prior to measuring the weight gains.

The effect of argon bombardment of the oxide layer on the subsequent oxidation rate of Zr-1.0 Nb is shown in Fig. 2. Localizing the damage in the oxide resulted in a suppression of the oxidation rate during subsequent exposure to 360°C water.

The same effect was observed by damaging the oxide layer on Zr-4 alloys using oxygen ions as shown in Fig. 3. The suppression occurred whether single or multiple energy implants were used. A slight effect could still be seen for oxidation times of 72 hours following re-bombardment, cycle #2, but the difference in average oxidation rate for the controls and ion bombarded samples decreased with increased oxidation times. Continued oxidation without additional ion bombardment (data not shown) eliminated the effect after exposures of 72-720 hours. The spread in corrosion weight gain was consistently less for groups of ion implanted samples than for the controls, Fig. 3. This was in agreement with the visual observations of more uniform oxidation of the ion bombarded samples.

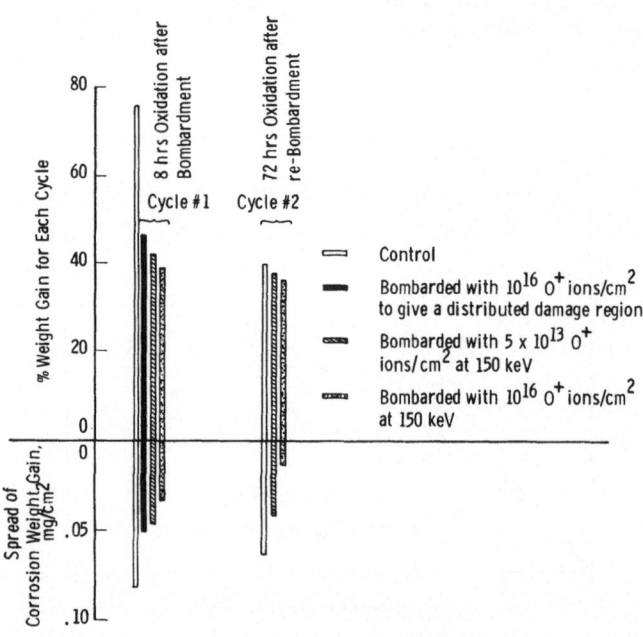

Figure 3. Effect of O^+ bombardment damage distribution and fluence on the oxidation of Zircaloy-4 in 360°C H_2O + 5 ppm O_2.

For all experiments on Zr-4, the decrease in spread paralleled the reduction in oxidation rates, and both effects disappeared with prolonged autoclave exposure.

Bombardment of lightly preoxidized samples with argon or oxygen ions, accelerated to several energies to broaden the damage deposition peak, was used to simultaneously damage the Zr-4 metal substrate and the oxide adjacent to the metal/oxide interface. To achieve the desired damage distribution, it was necessary to use samples with very thin (approximately 600 Å) oxide films with Ar^+ bombardment, and somewhat thicker films (approximately 1200 Å) with O^+ bombardment. The specimens in the batch with the thickest films (1800 to 4800 Å) were separated for use as controls. It would be expected, in the absence of lattice damage effects, that the thickest films would be the most protective and, therefore, that the samples with thinner oxide layers would oxidize faster than those with thicker films. However, the results of post-bombardment oxidation, shown as cycle #1 of Fig. 4, were a suppression in the oxidation weight gain of the bombarded samples relative to the unbombarded controls after one hour in the autoclave.

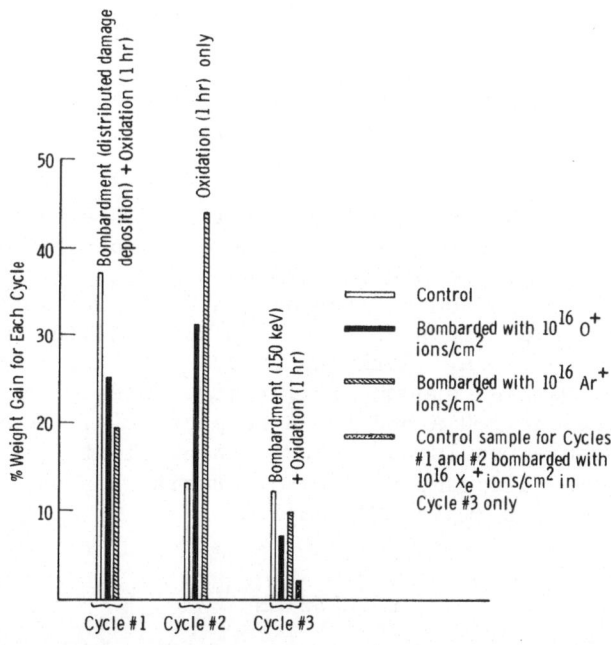

Figure 4. Effect of bombarding ion species on the oxidation of Zircaloy-4 in 360°C H_2O + 5 ppm O_2.

After a second hour of autoclaving (without further
bombardment), the relative weight gains of the bombarded samples
were larger than those of the control specimens. At this stage of
the experiment (end of cycle #2) the thicknesses of the films on
the bombarded samples had increased by a factor of two to three
and the damaged metal regions had oxidized.

The samples were rebombarded (with monoenergetic ions) at the
start of cycle #3, but the oxide thicknesses prevented the ions
from reaching the metal; all of the damage was deposited in the
oxide. Again, the oxidation rate of the bombarded samples was
less than that of the controls.

One of the unirradiated control samples which had the highest
oxidation rate at the end of cycle #2 was implanted with
10^{16} Xe^+ ions/cm^2 at 150 kev at the start of cycle #3. This showed
the lowest corrosion rate of all during the subsequent autoclave
exposure.

The range of weight gain values for the 1 hour oxidation
increments shown in Fig. 4 is quite large and, on a purely
statistical basis, one would have to conclude that the lattice
damage did not significantly perturb the oxidation rate. However,
observation of the changes in interference colors on the ion
bombarded areas following oxidation supports the trends indicated
in Fig. 4.

Implantation of preoxidized specimens with 10^{16} 0^+ ions/cm^2
at the oxidation temperature was observed to produce the same
effect as room temperature bombardment, i.e., a reduction in the
subsequent oxidation rate.

Thin films of the oxide suitable for 100 kV electron microscopy
were prepared according to the technique described by Airey and
Sabol.[15] Figure 5 is a centered-beam dark field electron micrograph of the oxide from one of the oxygen bombarded samples of
Fig. 4. The oxide consists of many fine crystallites with an
average diameter of 85 Å. Both the unirradiated controls and
argon implanted samples showed similar structures with average
crystallite sizes of 160 Å and 100 Å respectively. Many fine
pores or gas bubbles could be observed in a bright field micrograph
of the argon bombarded samples. Such porosity was not observed in
the control or oxygen implanted samples.

DISCUSSION

Diffusion measurements on thin zirconium oxide films by Cox
and Pemsler[16] have shown that oxygen anion diffusion along
crystallite boundaries is approximately 10^4 times faster than bulk

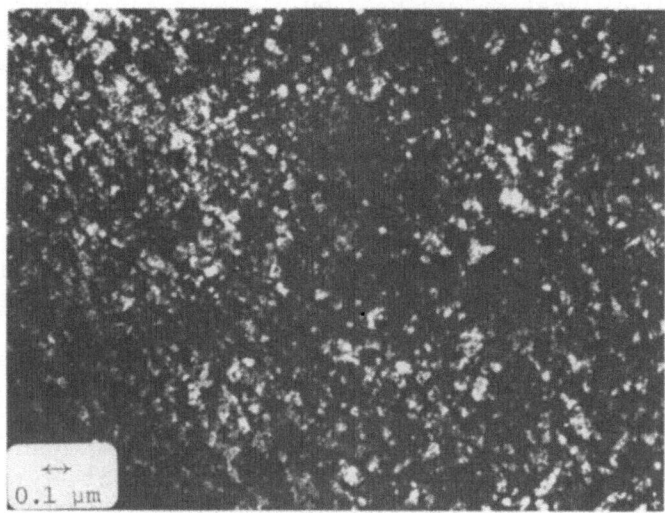

Figure 5.   Centered beam dark-field transmission electron
micrograph of the oxide layer grown on oxygen
bombarded Zircaloy-4.

diffusion at 400°C. It is likely that oxygen ion transport at lower
temperatures must be predominantly by boundary diffusion. Thus, it
might be expected that the increase in boundary cross section
accompanying a refinement of crystallite size would promote anion
diffusion and increase the oxidation rate. Conversely, it could
be argued that an increased number of boundaries only a few
hundred Angstrom units apart would amplify the scattering of
electrons, impeding the corrosion current, and retarding the
corrosion rate.[3]

Recently, Sabol, McDonald, and Airey[17] have shown that oxide
films produced on Zr-4 by exposure to 360°C water consist of many
fine crystallites. The average diameter of the crystallites found
adjacent to the oxide/metal interface increases from approximately
50 Å to 300 Å (in thin films) as the interface moves away from the
anion source. Preferential growth of the most favorably oriented
new oxide crystallites has been proposed to explain the observed
results. It can be inferred from this work that the highest oxidation rates accompany the finest crystallite sizes because of rapid
oxide diffusion down crystallite boundaries. Thus, it is likely
that the expected parabolic rate law is modified by the additional
dependence of diffusivity on crystallite size.

The average crystallite sizes reported by us suggest that the
ion bombardment develops somewhat smaller crystallites than those

found for the oxides on the control samples. However, the spread
in crystallite diameters measured on a single micrograph was larger
than the maximum difference in the average values for the bombarded
and control samples. Therefore, a reduction in crystallite size
due to ion bombardment cannot be demonstrated.

Several other mechanisms based on lattice damage in the oxide
or metal resulting from fast neutron bombardment have been proposed
to account for observed increases in oxidation rate. They are
equally applicable to ion bombardment and include: phase transformations in the oxide[18]; embrittlement of the oxide film[19];
enhanced electronic transport[20]; and perturbations in the relative
magnitudes of ionic and electronic transport processes.[3] Cox[3] has
evaluated the proposed mechanisms and concluded that existing data
best support a shift in the relative rates of the ionic and
electronic transport processes resulting from cumulative lattice
damage in the oxide film. While such damage could take the form
of the microstructural changes alluded to in the other proposed
mechanisms, there is little evidence to support the formation of
embrittlement cracks, pores, or second phases in thin, neutron
bombarded oxide films. Similarly, the transmission electron microscopic examination of bombarded and oxidized films in the present
work disclosed no cracks or second phases. Gas bubbles attributable
to ion implantation were observed only in Ar^+ bombarded samples.
However, the oxidation behavior of those samples was similar to
that of samples bombarded with other ions. Therefore, the presence
of the bubbles does not appear to influence the oxidation rate.

Bombardment with both chemically active, soluble ions (O^+) and
chemically inactive, relatively insoluble ions (Ar^+, Xe^+) produced
the same retardation of subsequent oxidation. An estimate of the
concentration of oxygen ions at the "concentration peak" (based on
the calculated peak width at half-height[21]) suggests that a maximum
of one oxide ion was implanted for every 100 oxide ions already
present. The total quantity of oxygen implanted in any cycle was
small compared to the amount required for the subsequent weight
gain during autoclaving.

The data suggest that the suppression in oxidation weight gain
following bombardment was slightly damage level dependent in that
larger reductions seemed to accompany higher fluences of the same
bombarding ion or equivalent fluences of heavier ions. However,
with the weight gain technique used, only the increased uniformity
and suppression in oxidation rate relative to the control samples
can be considered statistically significant, and not the effect
of relative damage levels.

The reductions in weight gain were apparently independent of
whether the damage was concentrated entirely in the oxide or partly
in the underlying metal. It was also independent of whether the

damage was "localized" by bombardment with monoenergetic ions or distributed over about three times the volume by multiple energy bombardments.

In a previous study, Harrop, Wilkins and Wanklyn[20] reported that the oxidation rates in 1 atm, 250°C steam of thinly oxidized zirconium and Zircaloy-2 samples could be increased by bombardment with 65 kev neon ions. Damage to the metal was deemed necessary to produce the enhancement. No effect was seen if the damage was localized in the oxide. The results of the current investigation are not inconsistent with these findings. However, the use of higher ion fluences, anodic oxide films and different oxidizing conditions in the earlier study makes a detailed comparison impossible.

The observed retardation of corrosion of ion bombarded Zr-4 and Zr-1.0 Nb samples cannot be explained on the basis of the results thus far available. The effect does not seem to be alloy sensitive; the fluence dependence is small; and the increase in retardation with increased damage concentration is also of minor extent. One might suggest that the suppression may be due to space charge effects. This would help to explain the preferential deposition in the autoclave of nickel-containing particles on the bombarded areas. Alternatively, ion bombardment may have affected the ionic mobility by creating charged defects which temporarily trap the diffusing ions or electrons.

Microstructural examination revealed no visible lattice damage capable of perturbing ionic or electronic transport. The crystallite boundaries probably represent the principal sinks for the defects created by bombardment. It appears that lattice damage on the order of 10 to 100 displacements per atom is accommodated in fine crystallite size oxide films.

These exploratory experiments have outlined a regime of conditions suitable for studying an unexpected lattice damage connected phenomenon. More systematic bombardment/oxidation experiments coupled with studies of conductivities and charge carriers in the oxide films should be made.

ACKNOWLEDGMENT

The authors wish to thank Dr. G. P. Sabol of the Westinghouse Research Laboratories for supplying the Zr-4 and Zr-1.0 Nb strip used in this investigation and for many helpful discussions. The assistance of N. J. Doyle in conducting the ion implantations and G. Economy for performing the autoclave exposures is gratefully acknowledged. Calculations of the energy available for atomic displacements were made by Dr. F. A. Garner of the Westinghouse Advanced Reactors Division.

REFERENCES

1. W.W. Smeltzer. R.R. Haering, and J.S. Kirkaldy, Acta Met. 9, 880 (1961).
2. C.S. Campbell and C. Tyzack, Br. Corros J. 5, 172 (1970).
3. B. Cox, J. Nucl. Mater. 28, 1-47 (1968).
4. R.C. Asher, D. Davies, T.B.A. Kirstein, P.A.J. McCullen, and J.F. White, Corros. Sci. 10, 695 (1970).
5. F.A. Nichols, J. Nucl. Mater. 30, 249 (1969).
6. P.J. Harrop, J.N. Wanklyn, J. Nucl. Mater. 22, 350 (1967).
7. B. Cox, K. Alcock, and F.W. Derrick, J. Electrochem. Soc. 106, 129 (1961).
8. N.J. Doyle, J.M. Bogdon, and W.J. Choyke, to be published.
9. W.S. Johnson, J.F. Gibbons, Projected Range Statistics in Semiconductors, Stanford University Press, Palo Alto (1969).
10. J. Lindhard, M. Scharff, Phys. Rev. 124, 128 (1961).
11. J. Lindhard, M. Scharff, H.E. Schiott, Mat. Fys. Medd. Dan. Vid. Selsk. 33 (1963).
12. F.A. Garner, to be published.
13. F.A. Garner, personal communication.
14. G.H. Kinchin, and R.S. Pease, Rept. Prog. Phys. 18, 1 (1955).
15. G.P. Airey and G.P. Sabol, J. Nucl. Mater. 45, 60 (1972).
16. B. Cox and J.P. Pemsler, J. Nucl. Mater. 28, 73 (1968).
17. G.P. Sabol, S.G. McDonald, and G.P. Airey, to be published in the Proceedings of the ASTM/AIME Symposium on Zirconium in Nuclear Applications, Portland, Oregon, August 1973.
18. J.R. Johnson, Trans. AIME 212, 13 (1958).
19. P.J. Harrop and J.N. Wanklyn, J. Nucl. Mater. 21, 310 (1967).
20. P.J. Harrop, N.J.M. Wilkins, and J.N. Wanklyn, Corros. Sci. 7, 289 (1967).
21. J.W. Mayer, L. Eriksson, and J.A. Davies, Ion Implantation in Semiconductors, Academic Press, New York, 19 (1970).

DISCUSSION

Comment: (E. P. EerNisse) Two phenomena concerned with ionization induced damage in oxides may be important here. First, some glass compositions crystallize at elevated temperatures if first exposed to ionizing radiation. Such crystallization would affect diffusion processes. Second, natural oxides grown on metal substrates always have considerable stress in them; recent measurements by myself indicate that these stresses can be relaxed considerably by ionizing radiation. Since your results are similar for ions depositing their atomic collisions in the metal, the only effect left is the ion energy deposited into ionization processes.

A: We agree that the observed suppression in oxidation rate is probably the result of ion energy deposited into ionization processes. Creation of charged defects which modify the electric field gradients or ambipolar diffusion which may accompany the growth of thin oxide films, as shown by Hauffe, would seem to be the most likely cause. There is good reason to believe that stress relaxation in the oxide film favors <u>higher</u> oxidation rates as evidenced by the shift to linear kinetics at oxide thicknesses greater than 2 μm. No evidence of a statistically significant amount of crystallization or recrystallization resulting from ion bombardment could be seen from examination of electron diffraction patterns or dark field electron micrographs.

ION IMPLANTATION AND BACKSCATTERING FROM OXIDIZED SINGLE-CRYSTAL COPPER

J. Rickards

Instituto de Fisica, University of Mexico

and G. Dearnaley

AERE, Harwell, England

It is well known that small amounts of impurities in metals strongly affect their corrosion properties. In the particular case of copper, alloys containing varying amounts of aluminum are quite common, and are denominated aluminum bronzes. These alloys have improved characteristics like corrosion resistance, high strength, hardness and wear resistance. Alloys containing up to 18.4% aluminum have been studied as well as their different metallurgical phases, and their resistance to corrosion (1).

One explanation given for the resistance to oxidation of copper-aluminum alloys is the presence of a thin film of alumina formed on the surface (2). When this film is free of copper, a very high resistance to oxidation was found; the constant in the parabolic growth law diminished 2×10^5 times.

The purpose of the present study is the implantation of Al^+ ions into single-crystal copper, the subsequent oxidation, and a backscattering study of the composition of the oxide layer both on the implanted area and on the unimplanted region. The channeling technique also provided a tool for studying the oxide.

*Supported by Instituto Nacional de Energia Nuclear.

The implantation technique differs from the usual methods in that one does not have to develop the metallurgy of the species in order to obtain the desired configuration. In the present case, for instance, the specimens implanted were the $\langle 110 \rangle$ faces of high purity (5N) single crystal copper, the final configuration obtained being essentially different from an alloy. New information regarding possible mechanisms for the oxidation resistance is therefore to be expected. Also in implantation, by knowing the energy and dose of ions, one can control the depth and concentration of the implant, leaving the bulk of the specimen essentially unchanged.

The first sample studied was implanted with $10^{16} cm^{-2}$, 220 keV Al^+ ions. From an LSS calculation of projected range and straggling, 70% of the ions are deposited in a region of depth 1150 ± 400 Å, giving an approximate concentration in this region of 1% in number of atoms.

It was then thermally oxidized in dry oxygen at 320°C for 20 min. During the oxidation the change in colors indicated that the implanted region resisted oxidation by about 70% with respect to the rest.

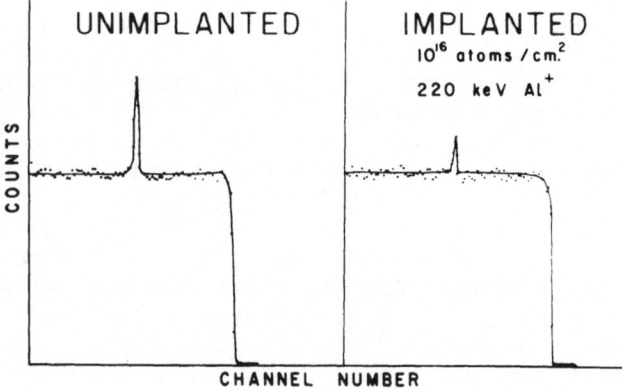

Fig. 1.- 4 MeV proton backscattering spectra from unimplanted and Al^+ implanted samples.

In order to obtain a quantitative result, backscattering spectra were run on the two areas. A proton energy of 4 MeV was chosen because at this energy there is a broad and large resonance in the oxygen scattering, so the cross section is many times Rutherford, and the oxygen peak stands out clearly from the solid copper background (3). A comparison of the two spectra is shown in figure 1. A 70% reduction in oxidation was obtained for the implanted region. From the size of the oxygen peak and the solid copper background, an estimate was made of the oxide thicknesses, giving 2800 Å for the unimplanted region and 850 Å for the implanted region. This calculation was based on the value 130 mb/sterad for proton scattering from oxygen at about 165° taken from reference 3; proton scattering from copper was assumed Rutherford, based on an experiment (4) up to 3.7 MeV where deviations from Rutherford were found to be less than 10%.

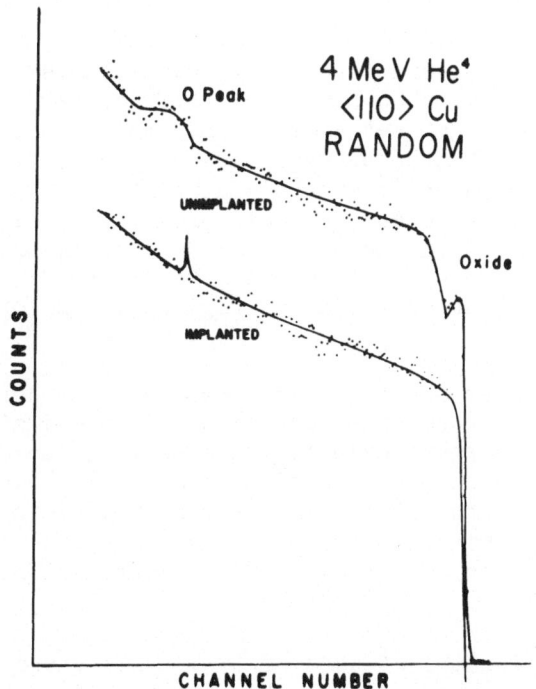

Fig. 2.- 4 MeV He$^+$ backscattering from the same samples as in figure 1.

To verify this, backscattering spectra were obtained for 4 MeV He[+] backscattering. See figure 2. In this case the oxygen peak is less well defined, but it was still possible to extract an 80% reduction in the oxidation, which is quite near the other value. The oxide thicknesses were obtained in the same manner as before. In this case there are two different experiments on He scattering from oxygen which give cross sections differing by 80%. When the value of Cameron (5) of 25 mb/sterad was used, the thickness for the unimplanted oxide was calculated to be 2900 Å, in good agreement with the thickness from proton backscattering. However with the more recent value of McDermott et al. (6), 45 mb/sterad, the thickness is 80% lower. Since there is no information on He scattering from copper, this was assumed to be Rutherford.

Two other samples were implanted and oxidized, and the resistance to oxidation observed from the difference in colors. One was implanted with the same dose but lower energy of 120 keV, and the oxidation was reduced by 30%. The other was implanted at the lower energy of 120 keV, but with a dose of 5×10^{16} cm^{-2}, in this case the oxidation was reduced by 70%. These numbers indicate that the reduction in oxidation goes with increasing energy and increasing dose.

In an attempt to learn more about the mechanism of oxide resistance, an electropolished sample of ⟨110⟩ copper was implanted 10^{16} cm^{-2} Cu$^+$ ions at 200 keV energy, and then oxidized under similar conditions. Proton backscattering spectra at 4.55 MeV bombarding energy were taken in order to compare the oxides of the implanted and unimplanted areas. See figure 3. In this case the oxidation was enhanced in the implanted region, by almost 300%. The two oxides looked quite different. On the implanted region it was smooth, while on the unimplanted region it was rough and flaky.

It was thought at the beginning that the disorder caused by implanting aluminum caused the reduction in the oxidation rate. However, after the copper implant, it seems that the crystal lattice is so damaged that the mechanism of oxidation was affected. The ⟨110⟩ face of copper is one that strongly resists oxidation, especially in this temperature range (7). Possible chemical effects were expected to be minimized by implanting the same species in the copper.

Besides studying the effects of implantation and comparing with the unimplanted regions, a backscattering study was made of copper oxide films grown under different conditions on the ⟨110⟩ faces of single crystal copper samples. Since there has been quite a bit of conflicting evidence regarding the composition of oxides grown under varying conditions (1), information from backscattering could help to clear up some of the doubts.

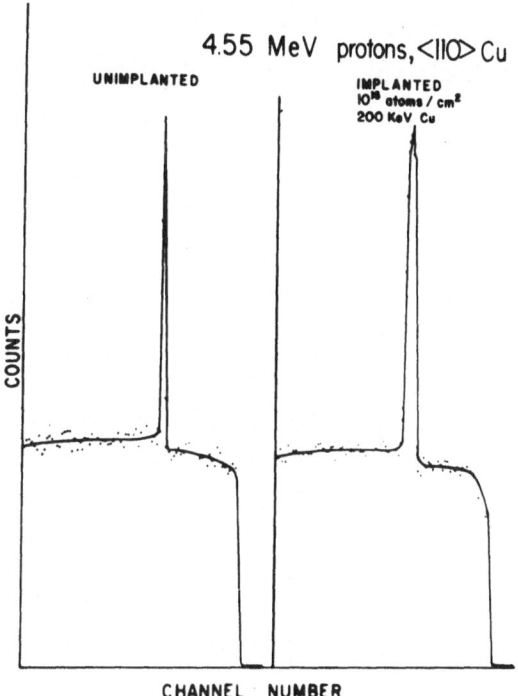

Fig. 3.- Oxidation enhancement by implanting Cu⁺ ions in copper.

As figure 2 shows, from the backscattering spectra one can obtain information about the depth distribution of the elements in the sample. One sample was oxidized up to 350°C for several minutes until the oxide was beginning to flake off. Then it was observed by backscattering 1.5 MeV He ions. The spectrum taken in the random direction shows a step corresponding to the copper oxide at the surface, as can be seen in figure 4. The copper concentration at the surface of the oxide of 60% is consistent with the stoichiometry of either Cu_2O or CuO. However, the concentration increases with depth to about 80% when the oxide ends and the pure copper begins. This of course indicates the non-stoichiometry of the oxide. A channeled spectrum was obtained by aligning the

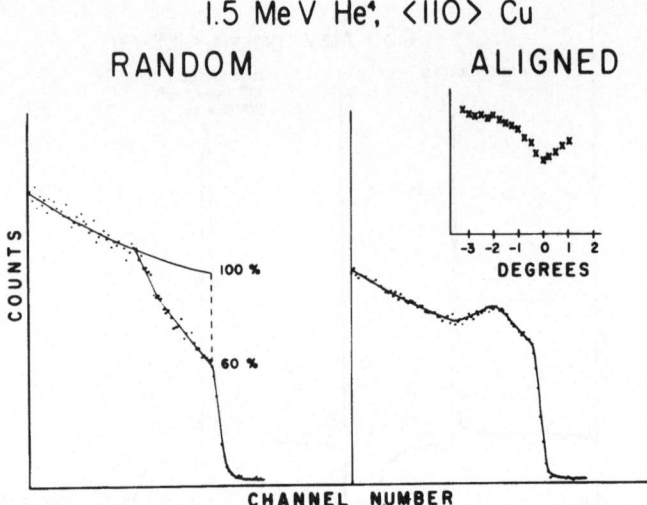

Fig. 4.- Variation of the stoichiometry of copper oxide.

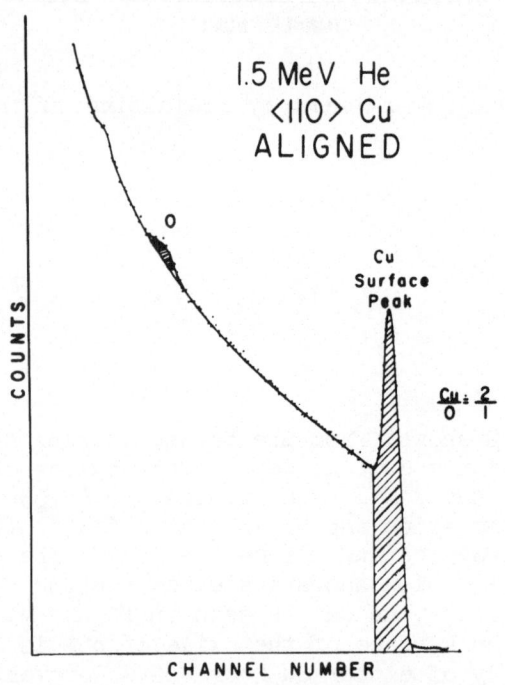

Fig. 5.- Cu/O ratio obtained from a channeled backscattering spectrum.

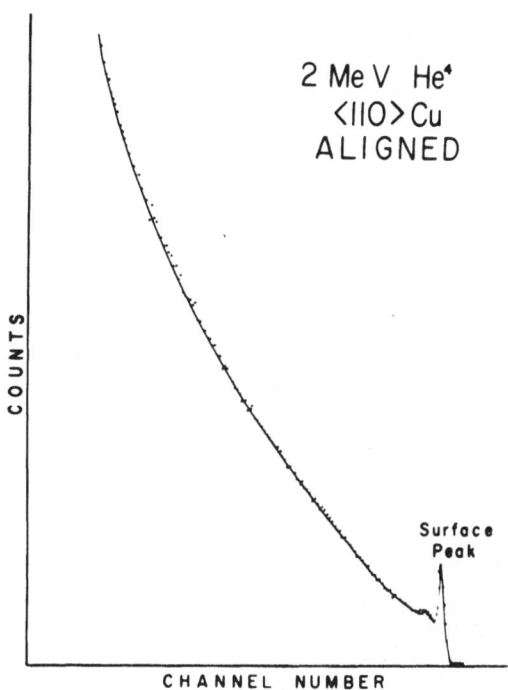

Fig. 6.- Oscillation just below the surface peak in channeling
in copper.

sample with the incoming beam, and the same figure shows the structure in the oxide, the oxide-copper interface not so well defined,
and the bulk counts fewer because of the channeling. Another backscattering spectrum, taken with 3.5 MeV He, yields a Cu/O ratio of
2, consistent with Cu_2O, if one assumes pure Rutherford mechanism.
This is so in spite of the structure of the oxide.

Another specimen was oxidized slightly until it was silvery
and then bombarded with 1.5 MeV He. The spectrum is in figure 5,
under aligned conditions. The Cu/O ratio 2, indicating Cu_2O, and
the thickness of the oxide obtained was about 400 Å.

Figure 6, shows the best channeled spectrum obtained for the copper crystals used. The sample was prepared by first annealing for 24 hours at 600°C and then removing about 0.5 mm from the surface by electropolishing for 120 minutes, and not oxidizing. The surface peak is clear, as is an oscillation just below the surface, due to the varying probability of nuclear encounter as the particle is channeled through the crystal (8).

References

1.- H. Leidheiser Jr., The Corrosion of Copper, Tin, and their Alloys, Wiley, 1971.
2.- L. E. Price and G. J. Thomas, J. Inst. Met. 63 (1938) 21, 29; 65 (1939) 247.
   U. R. Evans, An Introduction to Metallic Corrosion, E. Arnold Ltd., 1963, p. 23.
   U. R. Evans, The Corrosion and Oxidation of Metals, E. Arnold Ltd., 1971, p. 76.
3.- R. W. Harris, G. C. Phillips and C. Miller Jones, Nuclear Phys. 38 (1962) 259.
4.- V. Ya. Golovnya, A. P. Klyncharev, B. A. Shilyaev and N. A. Shlyakhov, Sov. J. Nucl. Phys. 4 (1967) 547.
5.- J. R. Cameron, Phys. Rev. 90 (1953) 839.
6.- L. C. McDermott, K. W. Jones, H. Smotrich and R. E. Benenson, Phys. Rev. 118 (1960) 175.
7.- F. W. Young, J. V. Cathcart and A. T. Gwathmey, Acta Metallurgica 4 (1956) 145.
8.- J. H. Barrett, Phys. Rev. B3 (1971) 1527.

DISCUSSION

Q: (J. E. Westmoreland) (1) Did you characterize the Cu single crystals in terms of dislocation count or mosaic spread? (2) What was the channeling minimum yield prior to implantation?

A: (1) No. (2) 3%.

Q: (R. S. Nelson) Did you study the position of the $A\ell$ in your samples after oxidation?

A: No. Ion backscattering was not applicable, but in principle a nuclear reaction such as $^{27}A\ell(p,\gamma)$ provides a means of studying this and we intend to carry out such experiments.

Q: (S. T. Picraux) In your second slide the Cu backscattering profile for the unimplanted sample indicates a nonuniform depth distribution of O to Cu ratio, with the maximum O concentration near the interface. Is this nonuniformity reasonable to expect for single crystal Cu oxidation?

A: The statistical accuracy of the data points do not allow the O:Cu ratio to be extracted from the high-energy edge of this profile. To that extent the solid line drawn through these points may be misleading. For accurate determination of stoichiometry as a function of depth we preferred to use 1.5 MeV He^4 (rather than 4 MeV) and such a profile is shown in Fig. 4.

MOVEMENT OF IONS DURING THE ANODIC OXIDATION OF ALUMINUM

W.D. Mackintosh and F. Brown

Atomic Energy of Canada Limited

Chalk River Nuclear Laboratories

Chalk River, Ontario, Canada

INTRODUCTION

In an extensive series of experiments, we have studied the movement of foreign atoms during the anodic oxidation of aluminum by introducing them into the surface layers of specimens by ion implantation at low energy (20-40 keV) and determining the depths of the implants after oxidation by means of Rutherford backscattering analysis. Two different situations were investigated; each species was implanted into specimens covered only by natural air-formed oxide ($\approx$ 0.5 $\mu g/cm^2$ thick) and into specimens covered by a pre-formed oxide 11 $\mu g/cm^2$ thick. In the former the greater part of the implanted atoms was located in the metal, in the latter they were entirely contained by the oxide layer; see Fig. 1. The depths of the implants after both the air-formed and pre-formed oxide films had been thickened anodically, were determined using a 2 MeV He beam in the manner described in detail in Ref. (1).

We reported in Ref. (1) that after anodic oxidation, implanted noble gases were found at a depth $\sim$ 2/5 of the oxide thickness. As the noble gases are immobile during oxidation (2) this observation confirmed the conclusion reached in earlier work (3) that both oxygen and aluminum moved during anodic oxidation and transport numbers

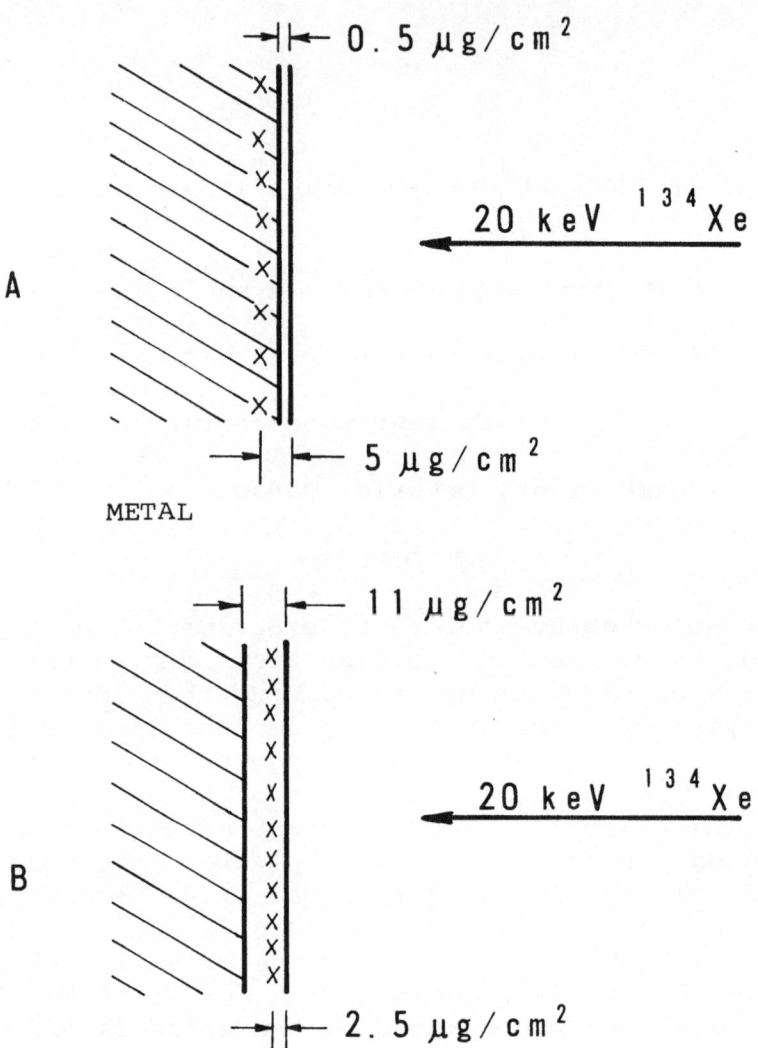

Fig. 1 The two methods of implanting. A, into metal
        beneath the air-formed oxide, B, into pre-formed
        oxide. X, Xenon implants.

for any particular electrolyte and current density are
obtainable. Halogens were found at greater depth in the
oxide than the noble gases, alkali metals at lesser
depths. From this we concluded that these species were
mobile and moved in the direction expected. Typical
positions are indicated in Figure 2.

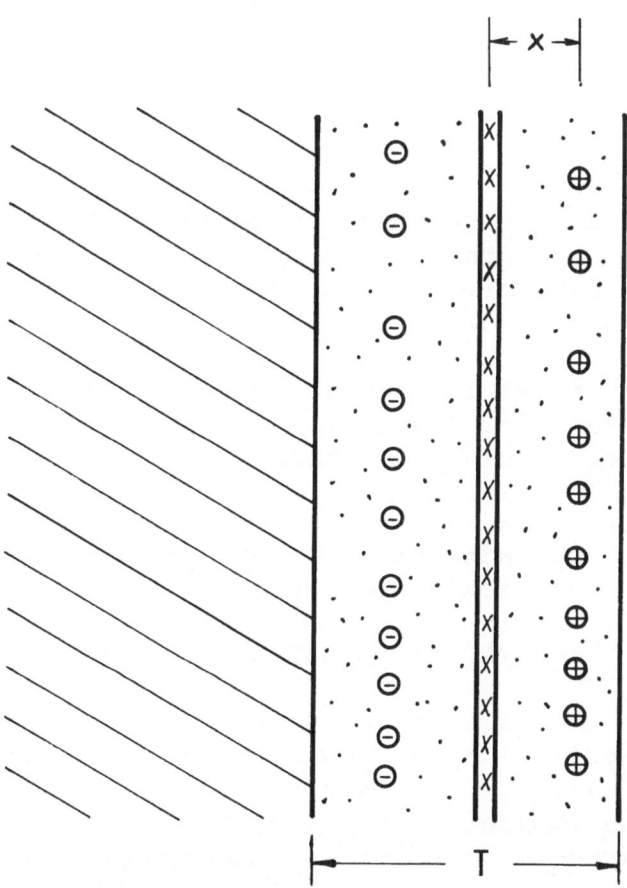

Fig. 2 The position of noble gases (X) and typical
        positions of alkali metals ⊕ and halides ⊖
        after anodic oxidation. Migration Number = X/T

We quantitatively described the mobility by obtaining
a migration number. As shown in Fig. 2, this is defined
as the ratio of the distance between the implant and the
original outer surface of the specimens (i.e. the position of the noble gas markers) X and the thickness of
the oxide grown T. The numbers obtained for the halogens were less than the transport numbers for oxygen;
those for the alkali metals less than aluminum. In
these experiments it was immaterial whether the atoms
were implanted in the metal beneath the air-formed oxide
or into a pre-formed oxide film.

Since publication of Ref. (1) we have found that
all metal species investigated move outwards when implanted into a pre-formed oxide film which is subsequently thickened. Most are more mobile than aluminum
and are lost to the electrolyte. However, we have found
that the same pattern does not emerge for many metal ions
if they are implanted into a specimen covered only with
the natural air-formed oxide. In most cases, after anodizing to form an oxide considerably thicker than the
initial depth of the implant, a substantial fraction is
located in the metal substrate under the oxide film
while the remaining fraction of the implant behaves as
if it had been implanted into pre-formed oxide (Fig. 3).

In this present paper we show which species exhibit
this effect and which do not. Our results lead to conclusions of practical importance.

OBSERVATIONS AND DISCUSSION

Fifteen metallic species in addition to those reported in Ref. (1) have been implanted into specimens of
aluminum covered only by air-formed oxide and the positions of the implants determined after anodizing in a
saturated aqueous solution of ammonium pentaborate at a
current density of 2 mA/cm^2. Of these fifteen only
three behaved like the alkali metals. Ba, Ca and Sr
moved outwards either remaining in the oxide film between original and final surfaces or passing into the
electrolyte. A fraction of each of the other twelve
species was found in the metal beneath the anodic oxide

films. Since the oxide film grows in part by inward
migration of oxygen, the positions originally held by
the implants have been overwhelmed. Thus the species
have been swept further into the metal by the advancing
oxide front in a direction opposite to that of same implants introduced into pre-formed oxide films, Fig. 3.
The fraction of each that evades capture in the metal
behaves is it would if it had been implanted in pre-
formed oxide; i.e. moves outwards. The twelve exhibiting this behaviour are, Ag, Co, Cu, Fe, Ga, Hg, In, Mn,
Ni,Tl, Sb and V. The fractions remaining varied from
95 to 50% and showed considerable variation for one
species from specimen to specimen.

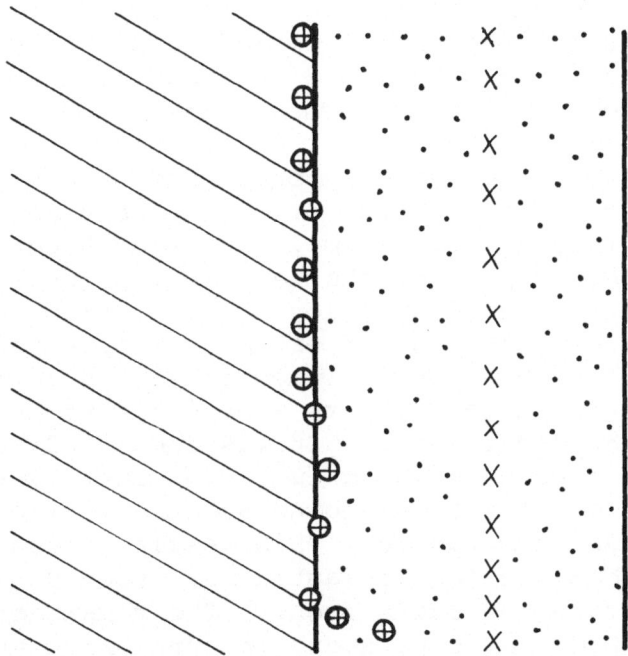

Fig. 3 Positions of implants into metal under natural
        air-formed oxide after anodic oxidation, ⊕ Tl,
        and X Xe.

All these species have lower oxidation potentials
than aluminum, the alkali metals and earths higher potentials. Thus the behaviour of the implant during anodic oxidation is correlated with its normal electrochemical properties. However, there is no apparent
correlation between amounts found in the substrates and
electrochemical properties. Using Tl implants, we
found that as the oxide front moves inwards forcing the
bulk of the implanted atoms ahead a trail is left behind
in the oxide. As Tl atoms move in the oxide at much the
same rate as Al atoms (illustrated in Fig. 3) this trail
can be seen in the backscattered spectra. We have further
found that the rate at which Tl atoms are left behind in
the oxide has a first order dependence on the concentration of these atoms in the metal beneath the interface.

In all the experiments leading to the above observations, the amount of foreign atoms implanted ranged
from 10^{14} to 10^{16} atoms/cm^2 depending on the mass (since
the sensitivity of Rutherford backscattering analysis is
mass dependent). The maximum average concentration over
the full width of the implant distribution was thus
$\sim$ 0.3 at.% with the maximum concentration (at the median
range) being about a factor of two larger. In order to
determine if behaviour of implants at such concentrations
could be used to predict the behaviour of bulk impurities, we investigated an alloy of Al containing 0.3 at.%
Ag. It was ascertained, after mechanically polishing a
specimen, that the concentration of Ag was constant
throughout the surface layers accessible to analysis by
Rutherford backscattering. It was then anodized to give
an oxide thickness of 15 μg/cm^2. On analysis an excessive amount of silver was found at the metal/oxide interface and none in the oxide. This result is consistent
with that obtained for implanted Ag; some 90% of silver
retreats before the oxide front. The remainder is lost
to the electrolyte, consistent in turn, with the behaviour of Ag when implanted into pre-formed oxide.
Further anodizing in steps of the same thickness resulted
in no increase in the concentrations of Ag in the substrate surface. The excess silver amounted to an average

of 2×10^{15} atoms/cm^2. As the thickness of oxide
approaches the maximum attainable (100 μg/cm^2), some
silver appeared in the oxide layer. The quantity of Ag
in the substrate surface remains unchanged if the oxide
is dissolved in a phosphoric-chromic acid mixture.

One further experiment was carried out with a similar specimen. Rather than anodize, we electropolished
the specimen in a perchloric acid-ethyl alcohol mixture
for successive short periods of time (30 sec) and analyzed after each. There was a build up of Ag on the
surface (in electropolishing the oxide is removed as
fast as formed); the amounts increased linearly from
1.4×10^{15} atoms/cm^2 for the initial step to 3.3×10^{15}
atoms/cm^2 for the fifth. The amount of surface Ag was
not reduced on immersion in phosphoric-chromic acid
solution, nor was it reduced on treatment with hot KCN
solution or dilute HNO$_3$ for an hour. Immersion of the
specimens in water through which H$_2$S was bubbling produced no blackening and subsequent dipping in hot HNO$_3$
(in which the sulphides of Ag are very soluble) left the
Ag concentration unchanged.

Swanson, Maury and Quenneville (4) also observed a
surface peak of Mn after electropolishing a single crystal of Al alloyed with 0.1 - 0.3 at.% Mn.

The depth resolution attainable in backscattering
analyses using a solid state detector ($\sim$ 3 μg/cm^2 in Al)
is not sufficient to establish whether the excess of
alloying atoms lies as a separate entity on the surface
of the substrate or whether these atoms are mixed at
higher concentrations in the surface layers. The resistance to attack by agents known to attack Ag, suggests that the latter alternative obtains. In addition,
we have established that Tl ions retreating before an
oxide front are, at least in part, dissolved in the matrix. A single crystal of Al was implanted with Tl and
an oxide film grown. Ninety percent of the Tl atoms
implanted were found in the metal beneath the oxide film.
The film was then dissolved and a channeling experiment
performed (see e.g. Ref. 5). Some 50% of the Tl atoms

were in a disordered surface layer; it was not possible
to distinguish it either as a discrete surface layer on
the surface or a mixture with disordered Al atoms.
However, the remaining 50% was clearly distributed in
substitutional sites in a thin layer of ordered substrate
Al atoms.

When the near metals, As, Se and Te are implanted
into the metal substrate and oxide film grown they are
found dispersed throughout the oxide with none in the
metal under the oxide film.

CONCLUSIONS

The foregoing observations form only a part of a
continuing research program. These were chosen to present here because a number of conclusions of practical
importance can be drawn from them.

(a) The use of the anodizing and stripping techniques
(6) for such purposes as measuring ion ranges and diffusion profiles is limited to non-metals, near metals,
and those metals which have higher oxidation potentials
than Al. Any species not falling into these categories
will move before the advancing oxide front, so that the
amount found in the oxide solvent will be less than the
amount contained in the thickness of matrix converted to
oxide. The amount remaining in the matrix will be enhanced by some fraction of foreign atoms originally contained in this thickness (Fig. 4). In view of our experience with Tl implants where the amount moving before
the oxide front depended on concentration it would seem
hazardous to attempt to apply correction factors. It
should be noted that our experiments do not, as yet,
encompass very low implant doses or very dilute alloys.

It would seem prudent to recognize that similar
phenomena may arise in materials other than aluminum.
An experimenter investigating the suitability of other
systems for measuring ion ranges or diffusion profiles
(or for that matter, any other kind of concentration
gradient) may well be able to make this simple check;
anodize the specimen to a thickness certain to over-

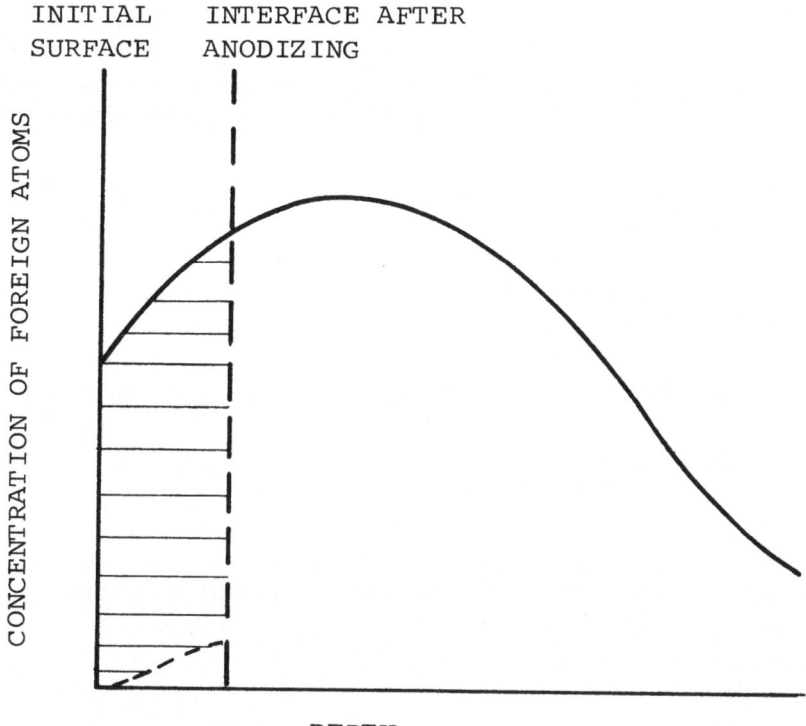

Fig. 4   Sectioning Al to determine the profile of a
         distribution of a metal with a lower oxidation
         potential than Al. The curved line represents
         a distribution, the shaded area the fraction of
         the distribution which will not be removed from
         the specimen on anodizing and stripping down
         to the vertical dashed line.

whelm the entire distribution of the embedded foreign
atoms, strip, and analyze the remaining matrix. The
presence of the foreign atoms in the matrix should be
viewed with suspicion.

    Some species (e.g. Ca and Sr) although they are not
trapped at all in the metal under the oxide, move outwards more rapidly than Al during oxidation and dissolve
in the electrolyte. In these cases analysis of the

stripping solutions will not reveal the amount contained
in the thickness of matrix oxidized. It is necessary to
either obtain the amount by the difference in residual
amounts in the matrix or analyze the electrolyte as well.

(b) A metal with a lower oxidation potential existing
as a bulk impurity will concentrate near the surface
during repeated anodizing and stripping or when electropolishing a specimen. This fact combined with the observation that the increased concentration is apparently
not accessible to attack by specific reagents produces
a dilemma in the preparation of the Al alloy surfaces
for such experiments as are described in Ref. (4).
Electropolishing produces the excess of the alloying
component at the surface. It can be removed by mechanical polishing but this produces damage that decreases
the suitability of the material for channeling experiments. Again we have not yet established whether or not
this rule will hold for other substrates than Al.

(c) In view of the complex behaviour of many foreign
atoms during oxidation it is unsafe to deduce the relative mobilities of oxygen and aluminum from the position of such ions in a specimen after oxidation. With
the exception of noble gases all of the twenty odd
species implanted into a thin pre-formed oxide film
moved during the anodic thickening of the film so that
none could be regarded as a marker for the original
film. Implantation into metal substrate covered only
with air-formed films can lead to quite contradictory
interpretations.

An example of the confusion that could arise if
any other atom than a noble gas were used as a marker
is given by chromium. If implanted into a pre-formed
oxide which is then thickened the implant disappears as
it is more mobile than aluminum. Thus there is no indication whether the oxide has grown below the original
film, above it, or on both sides of it. When implanted
in the metal substrate chromium is found under the oxide
after anodizing. This would lead to the conclusion that
the film had grown on top of the original surface atoms

i.e. by Al migration only. However if a specimen of Al
is immersed in a phosphoric-chromic acid solution a
small quantity of chromium is retained at the surface
(7). After anodizing this chromium is found on the surface. As the oxide is beneath the chromium this would
lead to the conclusion that oxidation had taken place by
oxygen migration only.

REFERENCES

(1)   F. Brown and W.D. Mackintosh, J. Electrochem. Soc.
      120, 1096 (1973).

(2)   J.P.S. Pringle, ibid, 120, 398 (1973).

(3)   J.A. Davies, J.P.S. Pringle, R.L. Graham and
      F. Brown, ibid, 109, 999 (1962).

(4)   M.L. Swanson, F. Maury and A.F. Quenneville,
      International Conference, Application of Ion Beams
      to Metals, Albuquerque, October, 1973.

(5)   J.W. Mayer, L. Eriksson and J.A. Davies, Ion
      Implantation in Semiconductors, Academic Press,
      New York (1970).

(6)   J.A. Davies, J. Friessen and J.D. McIntyre, Can. J.
      Chem. 38, 1526 (1960).

(7)   J.E. Lewis and R.C. Plumb, Intern. J. Appl.
      Radiation Isotopes 1, 36 (1956).

FRICTION AND WEAR OF ION IMPLANTED METALS

N.E.W. Hartley, G. Dearnaley, J.F. Turner and J. Saunders

A.E.R.E. Harwell, Didcot, Berkshire, U.K.

ABSTRACT

Recent work has demonstrated that large changes in friction can occur as the result of the implantation of various selected ions into steel $/\underline{\ 1\ }/$. The present contribution supplements these data by describing experiments in which equally striking reductions in wear are observed for the case of a pin under heavy load rubbing on an ion implanted disc. B^+, N^+ and Mo^+ implanted into steel to doses between 10^{16} and 10^{18} ions-cm^{-2} reduce wear on a test pin by more than a factor 10. Similar effects are observed for implantations into Cu. The observation that the beneficial effect of ion implantation persists for times very much greater than those sufficient for complete removal of the implanted layer during rubbing contact suggests that some initial rapid wear stage is substantially reduced. Alternatively, a large decrease in oxidation and subsequent wear particle nucleation may have occurred. Backscattering experiments with a 2.0 MeV He^+ ion microbeam reveal that for a steel specimen implanted with $2.8 \cdot 10^{16}$ ions-cm^{-2} of Ag^+ a substantial proportion of the implanted metal remains within the surface even after repeated severe deformation during friction testing.

INTRODUCTION

The development of heavy-ion accelerators has been accompanied by an upsurge of interest in the applications of ion beams to materials other than semiconductors. It has been demonstrated by a variety of centres of research activity throughout the world that physical, chemical, optical and now mechanical properties of

surfaces may be markedly influenced by ion bombardment or implantation.   It is perhaps surprising that despite our familiarity
with friction and wear phenomena the one field of surface interactions which has received least attention is the mechanical one.
Possibly the reason for this apparent neglect is the difficulty of
imagining that ion ranges could conceivably suffice to affect the
mechanical behaviour of any surface which is not atomically smooth.
Yet 18 years ago $\boxed{2}$ it was shown that even atomically smooth
surfaces can show very high friction, and we now know that for two
hard "rough" surfaces, contact exists only over intensely local
areas.   There is now a considerable volume of work where solid
state physics appears in the discussion of tribological phenomena
(see, for example, $\boxed{3}$).

   The classical theory of metallic friction considers that the
frictional force arises due to the force required to shear numerous
metallic junctions which are set up between sliding surfaces $\boxed{4}$.
A discussion of the relationship between contact area and expressions for friction coefficient and wear rate has been given by
Archard $\boxed{5}$.   When inter-surface adhesion occurs, the amount of
wear (i.e. the separation of small fragments of material from one
or other of the sliding surfaces) will depend on the strength and
location of the shearing junctions.   For example, if a junction
is weaker than the metals themselves, in the presence of an oxide
film, shearing will occur at the actual interface where the junction
is formed within the oxide, and the amount of material removed from
either surface will be small.   Similar principles can be applied
to account for the loss of material from the metal surfaces $\boxed{4}$.
It is therefore unlikely that any direct relation should exist
between the coefficient of friction and the amount of wear.
(Early experiments show that it is possible for metallic transfer
to be reduced by a factor 10^4 while the friction is reduced by a
factor 10 $\boxed{6}$).   However, each process involves similar basic
mechanisms and in practice friction and wear are physically
associated.   In certain situations, therefore, physical or
chemical changes brought about by ion implantation (e.g. the
incorporation of a "solid lubricant" within sliding surfaces) may
be expected to alter friction and wear characteristics.

   Frictional changes as a result of ion implantation in steel
have already been reported $\boxed{1}$.   For completeness some of these
results are reviewed below together with additional data and
discussion.   The wear experiments are the first reported for ion
implanted metals.   Experiments using a standard wear testing pin-
and-disc machine are described.   To date, the implanted ions have
been B^+, N^+ and Mo^+.   The substrates and sliding pins are various
combinations of steels, copper and graphite summarised in Table 1.

Table 1

Materials

Substrate	Ion	Slider
Mild Steel	B^+ N^+	440C Steel Pin
440C Steel	Mo^+	" " "
Copper	B^+	Graphite Pin

FRICTION

Friction tests were carried out on En 352 which is a case-hardening steel containing several alloying elements (Mn, Ni, Cr, S, ...) to a combined total of about 4 atomic % [7]. The specimens were spark-machined from large discs and had dimensions of 2 cm diameter x about 0.3 cm thickness before implantation. Each specimen was hand ground and polished to a final finish with 0.25 µm diamond paste in a suspension of dilute glycerol. The Harwell 500 kV Cockroft-Walton accelerator [8] was used to implant a variety of metallic and non-metallic ions (e.g. Pb, Ag, Sn, In, Mo, S, Kr) to doses ~ 10^{16} ions-cm^{-2}. The implantation energies ranged between 150 and 400 keV for singly charged ions. Implantation details have been given elsewhere [1]. Two MeV He$^+$ backscattering experiments revealed that the implantation depths correlated well with the predicted ranges of Lindhard, Scharff and Schiøtt [9] (i.e. ~ 0.02 to 0.04 µm).

Friction measurements were obtained using the apparatus shown in figure 1 (described in more detail in reference [1]), in which the specimen is driven slowly beneath a loaded WC - 6% Co ball. The horizontal friction force was recorded, after suitable amplification, on a pen recorder sweeping synchronously with the specimen.

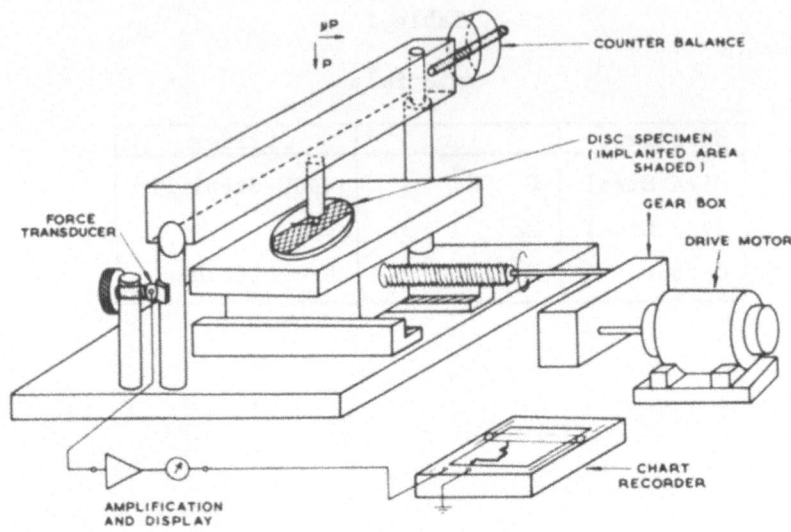

Figure 1.   Schematic diagram of friction apparatus.

Table 2 (from ref. $\lfloor 1 \rfloor$)

Friction Data for Ion Implanted En 352 Tested in Air

Ion	Dose (ions cm^{-2})	μ_{En352}	$\mu_{Impl.}$	Remarks
Kr	2.8×10^{16}	.24	.24	Friction Peak at Implantation Boundary
Sn	2.8×10^{16}	.24	.09	
In	2.8×10^{16}	.30	$.31 \pm .05$	Erratic; Transfer of Ions during wear
Ag	2.8×10^{16}	.22	.26	As In; Adhesion
Pb	6.3×10^{16}	.23	$.33 \pm .08$	Stick-slip; Adhesion Friction Peak (as Kr)
Mo	2.8×10^{16}	.26	.24	
S	6.1×10^{16}	.20	.19	
Mo+2S	2.8×10^{16} $+$ 5.6×10^{16}	.26	.20	

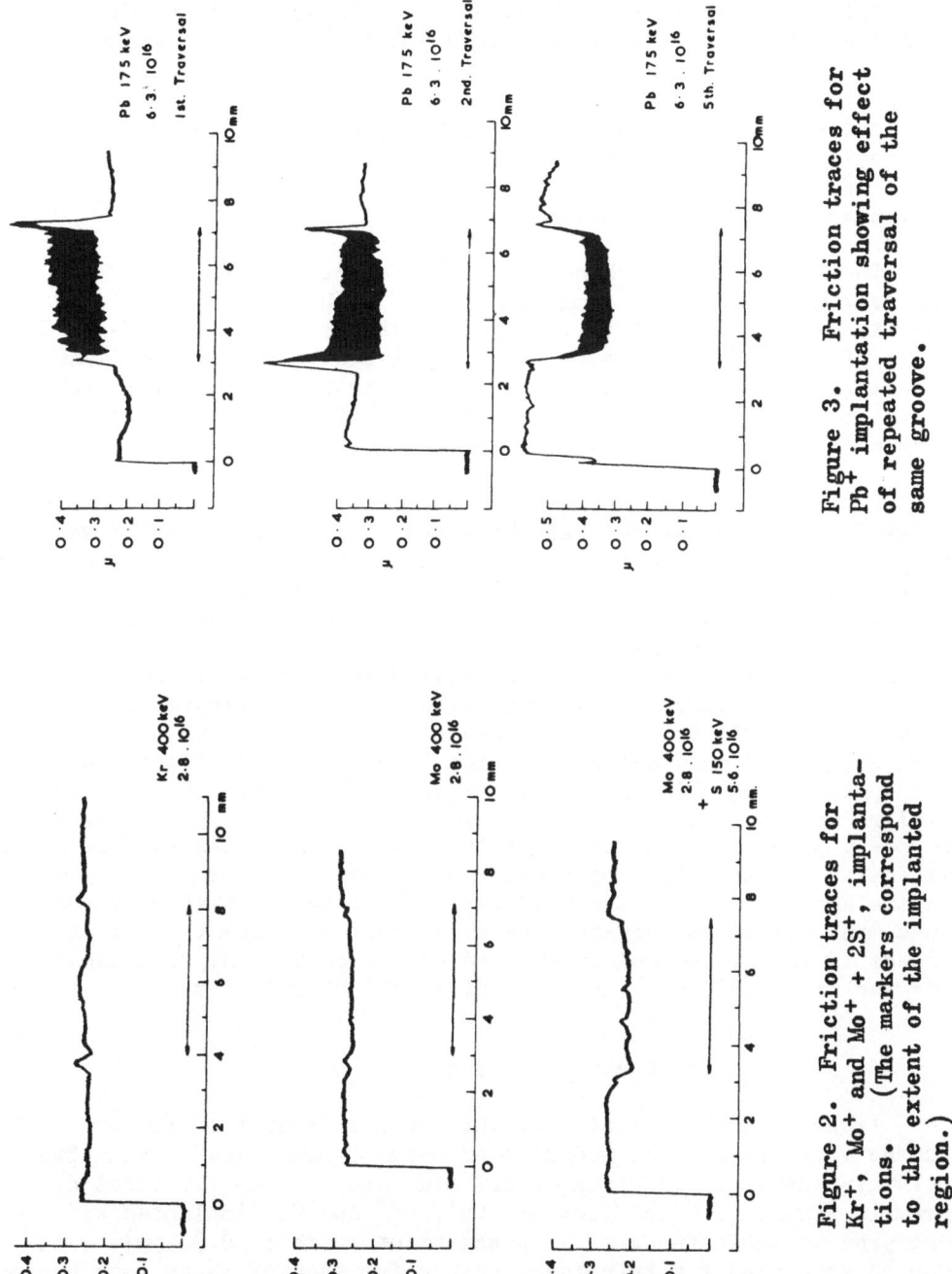

Figure 3. Friction traces for Pb$^+$ implantation showing effect of repeated traversal of the same groove.

Figure 2. Friction traces for Kr$^+$, Mo$^+$ and Mo$^+$ + 2S$^+$, implantations. (The markers correspond to the extent of the implanted region.)

Friction Results

All the implantations produced a change in friction with the exception of a sample implanted with 2.8×10^{16} ions-cm^{-2} of Kr$^+$ (400 keV) which only revealed a friction peak at the implantation boundary (figure 2, top trace). Large friction changes were observed with the metals used in the manufacture of bearings (e.g. Ag, Sn, In and Pb), the greatest reduction in friction being recorded for an implantation of 2.8×10^{16} ions-cm^{-2} of Sn$^+$, 380 keV (see summary Table 2 below).

In experiments in which 2.8×10^{16} ions-cm^{-2} of Mo$^+$ (400 keV) was followed by 5.6×10^{16} ions-cm^{-2} of S$^+$ (150 keV), it was found that the combination of Mo + 2S gave a larger effect than Mo or S alone (figure 2). There is no conclusive data yet from ESCA(XPS) experiments that MoS2 was formed in this case, although this may be due to the relatively weak penetration of the electron beam (see also ESCA experiments section, below).

Implantation did not invariably reduce friction. For example, 6.3×10^{16} ions-cm^{-2} of Pb$^+$ (175 keV) produced a succession of very large "stick-slip" events, characteristic of local adhesion followed by intermittent sliding. Rapid changes in friction appear as a broadened trace on the pen recording. Retracking the ball several times over the same friction groove reduced the amplitude of the adhesion oscillations and also inhibited the gradual increase in friction which occurs as the untreated steel is re-traversed (figure 3). The friction increase on steel may be due to the ploughing in of hard oxide particles ($Fe_2 O_3$, $Fe_3 O_4$) formed at hot spots during sliding. The presence of a softer Pb-oxide layer will reduce the friction over the implanted region if intra-oxide forces are weak. ESCA experiments on a Pb$^+$ implanted surface after extensive friction testing showed an increase of approximately 40% in the amount of Pb in the oxidized state compared with a standard specimen, and it is therefore possible that the formation of an excess of Pb_2O_3, PbO and other lead-oxygen containing complexes has the effect of reducing the coefficient of friction.

Dependence of Friction on Dose

Figure 4 presents data for the dependence of the relative change in friction as a function of dose of implanted ions. The relative change in friction is defined according to the inset diagram in figure 4. The ions are Pb$^+$, Se$^+$ and Mo$^+$ implanted at energies suitable for similar penetration range (~ 0.03 μm). It can be seen that the frictional change for each of these ions becomes more pronounced as the implantation dose is increased. For Pb$^+$ implantation a very rapid increase in friction is observed for doses higher than 3.0×10^{16} ions-cm^{-2}.

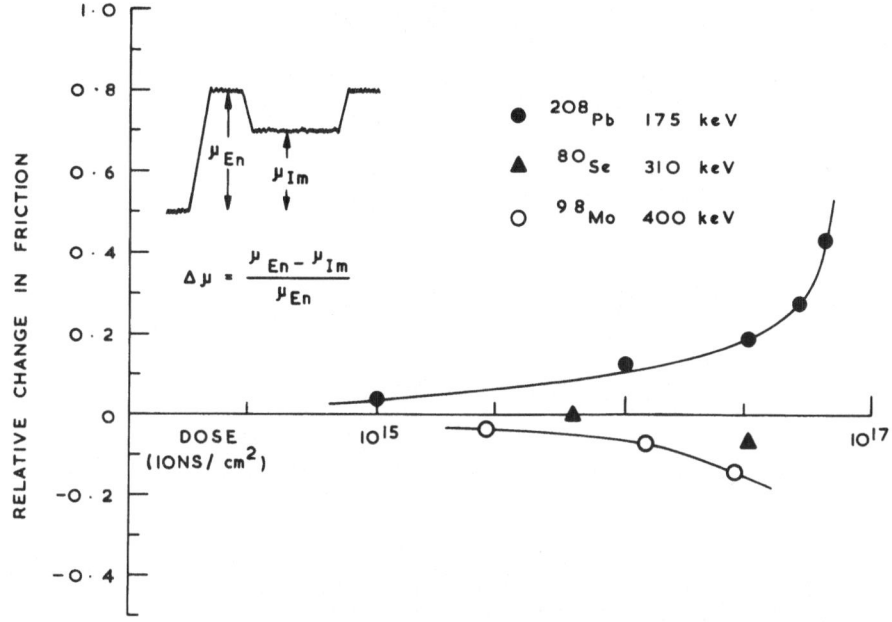

Figure 4. Effect of increasing ion dose on the relative change in friction for Pb^+, Se^+ and Mo^+ ions.

Backscattering analysis has shown that the surface concentration of Pb implanted into En 352 to a dose of 6.3×10^{16} ions-cm^{-2} is 4.2% [1]. Higher surface concentrations appear to be limited by sputtering. Sputtering may be responsible for the peak friction values for Pb (and some other ions) seen at the edge of the implanted region, figure 3, [10]. The sputtering yield for Se, Mo and Pb, calculated assuming hard sphere collisions ranges between 34 atoms-ion^{-1} (Se) and 92 (Pb). The greatest degree of sputtering therefore occurs for the ion which gives the largest increase in friction (Pb). However, if damage effects are contributing towards the frictional changes observed, it remains to explain why increases as well as decreases in μ can be augmented by increasing the dose. If the production of soft oxide is important then the thermodynamic preference of Mo to oxidise as compared with Pb will have to be considered. Additional physical parameters such as the hardness and crystal structure of the oxide as well as the melting point of the metal may also have to be considered when attempting to account for the frictional changes. It may be significant that a low melting point metal (Pb) shows increase in friction with dose (figures 2 and 4), whereas the opposite is observed for Mo (m.p. 2883K).

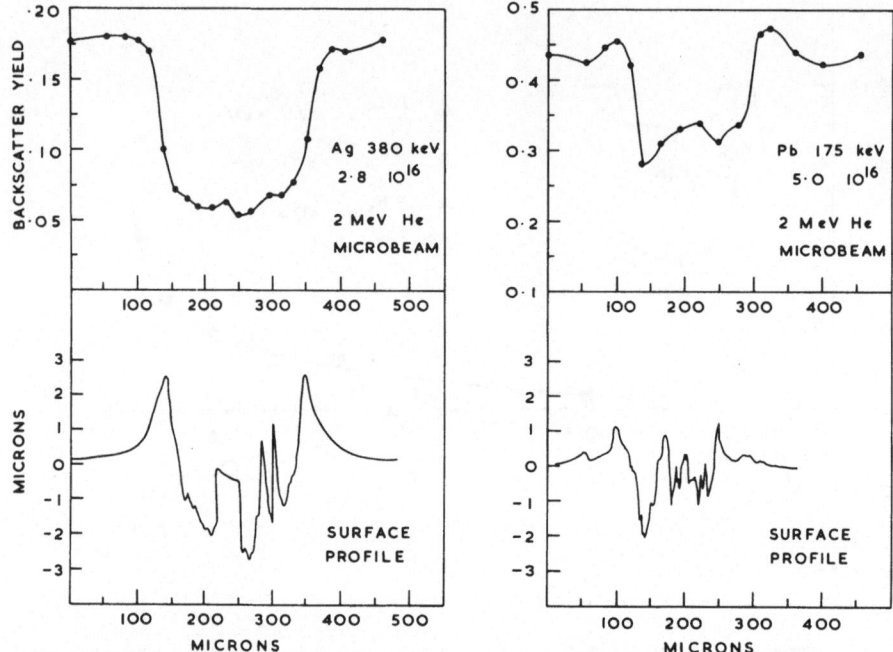

Figure 5 a) & b). 2.0 MeV He$^+$ ion microbeam backscattering data for friction grooves on Ag and Pb implanted surfaces.

MICROBEAM BACKSCATTERING

A beam spot of micron dimensions $\boxed{11}$ can be positioned accurately on selected areas of interest by employing a specially designed target chamber incorporating stepping motors. Back-scattering experiments with a 15 μm x 15 μm 2 MeV He$^+$ ion beam scanned through 100 μm were used to examine the distribution of implanted ions after friction testing. The data of figure 5 compare the backscattering yield of implanted Ag$^+$ and Pb$^+$ ions with their respective surface profile across a friction groove. The data show that for the case of an Ag$^+$ implanted friction track, after 5 traversals at 2 Kg applied load approximately 34% of the original concentration of Ag is still present. With lead, one traversal at 1 Kg applied load reduces the original Pb concentration to approximately 66%. The remaining fraction of the implanted

species is presumably transferred to the slider.   However, additional experiments have so far failed to reveal evidence of Pb or
Ag contamination on steel balls.   It is clear from the results of
figure 5 that even after very high contact pressures ($\sim$ 200 Kpsi)
a substantial proportion of the implanted metal remains within the
surface.

The roughness of the surface profile of a Pb^+ implanted
friction groove is associated with the intermittent adhesion
phenomena discussed in a previous section, and shown in figure 3.
Scanning electron microscope examination reveals tearing of the
surface in these regions $\diagup 1 \diagdown$.   The peak in Pb backscattering
yield in the centre of the groove (figure 5(b)) may be due to the
agglomeration of Pb-oxide complexes in regions where the shear
stress is highest.   It may be significant that Ag, which has a
very low free energy of oxide formation, does not exhibit the
concentration peaking after one friction traversal, characteristic
of Pb.

ESCA (XPS) ANALYSIS

X-ray photoelectron spectroscopy (XPS) is a particular method
of surface analysis from the range of techniques known as electron
spectroscopy for chemical analysis (ESCA).   All these techniques
have the creation of a hole in an inner atomic level as the first
step in the individual process $\diagup 12 \diagdown$.   Each of the variations in
electron spectroscopy is therefore concerned with surface composi-
tional analysis, since the binding of the initially ionised level
is reflected in the kinetic energy of the electron whose energy is
analysed in the subsequent emission stage.   A soft X-ray photon
of known wavelength impinges on the specimen surface.   If the
effective work function of the analyser is ϕ_{anal}, then the kinetic
energy E_{kin} of the re-emitted electron measured by the analyser is

$$E_{kin} = h\upsilon - E_B - e\,\phi_{anal} \qquad \dots (1)$$

where E_B is the binding energy of an electron in a core level of a
surface atom.   (Fuller details of the energy transitions and a
comprehensive reference list is given in a recent comparison of
ESCA techniques by Rivière $\diagup 12 \diagdown$).   Since E_B is usually only a
fraction of an eV wide (unless it is in an energy band), the only
term in equation(1) which governs the width of the observed peak
E_{kin} is $h\upsilon$, the exciting radiation.   For aluminium $K_{\alpha1}$ lines
(1.4866 keV) the inherent width is 1·12 eV.   The available energy
resolution is therefore sufficient to record shifts in E_B due to
differing chemical environment.   Recognising the obvious difference
in penetration depth, ESCA and backscattering may nevertheless be
regarded as complementary since each provides information about
atoms within the surface monolayers, chemical and "nuclear"
respectively.

XPS analyses on friction-tested samples implanted with S ions (either singly or in combination with metallic ions) showed that at the surface about 50% of the small S concentration was associated with sulphide configurations. The possibility of the formation of solid lubricants (e.g. MoS_2) in situ cannot therefore be ruled out. Qualitative information was obtained from a set of Pb^+ implanted specimens which revealed that after removing approximately 40 Å of steel by argon ion bombardment, 46% of the Pb present was in an oxidised state after friction testing.   This is consistent with the hypothesis that lead oxide is formed during sliding and may account for the effects observed in figures 3 and 5(b).

WEAR

Specimen Preparation and Testing

One of the most universally adopted methods of wear testing involves the "pin-and-disc" configuration, in which a disc is rotated beneath a loaded pin (see insert on figure 7).   Wear occurs on the disc, in the form of a groove or wear track, and on the pin whose contact area is worn flat.   It is generally easier to measure the wear on the pin for a given testing time, since a small volume of material removed may be simply related to a measurable change in the area of contact.   Various pin geometries may be used; in this study the 440C pins were finished to a right circular cone of apex angle $120^o \pm 10'$ and the graphite pins had hemispherical ends. An Avery-Denison T62 Tribotester was used for the wear experiments. The machine consists of a motor control unit for rotating the specimen disc up to 6000 r.p.m., a lubricant circulating pump and a load cell with amplifier for recording the frictional force experienced by the pin as the test disc is turned beneath it.

Figure 6. Photograph of wear testing machine.

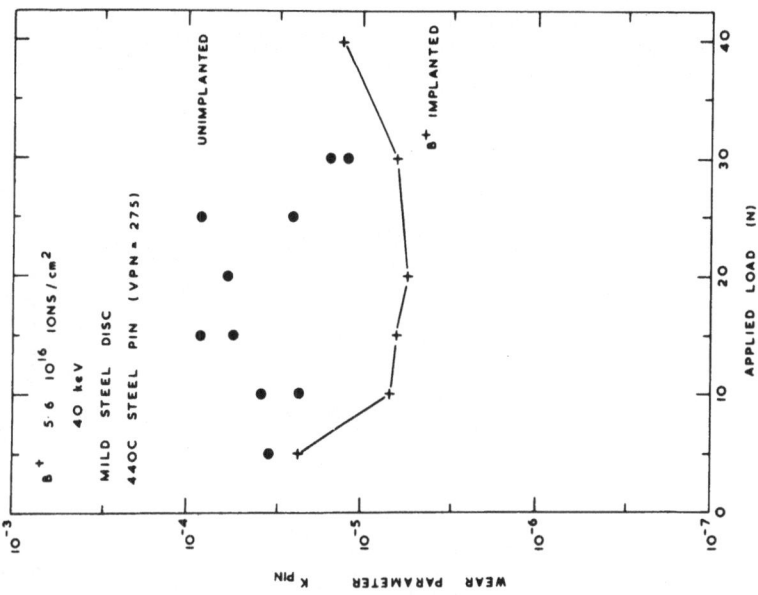

Figure 8. Wear parameter for B⁺ implanted mild steel disc.

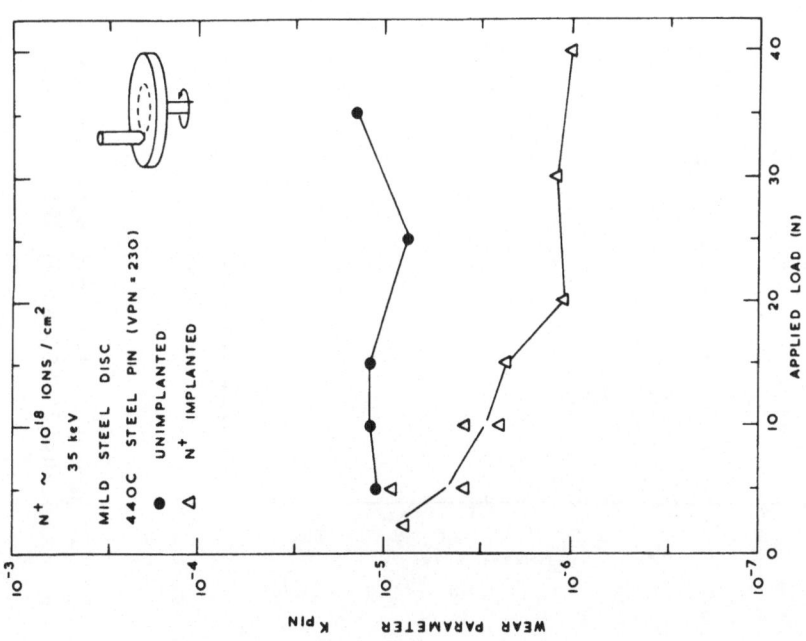

Figure 7. Wear parameter for N⁺ implanted mild steel disc as a function of applied load.

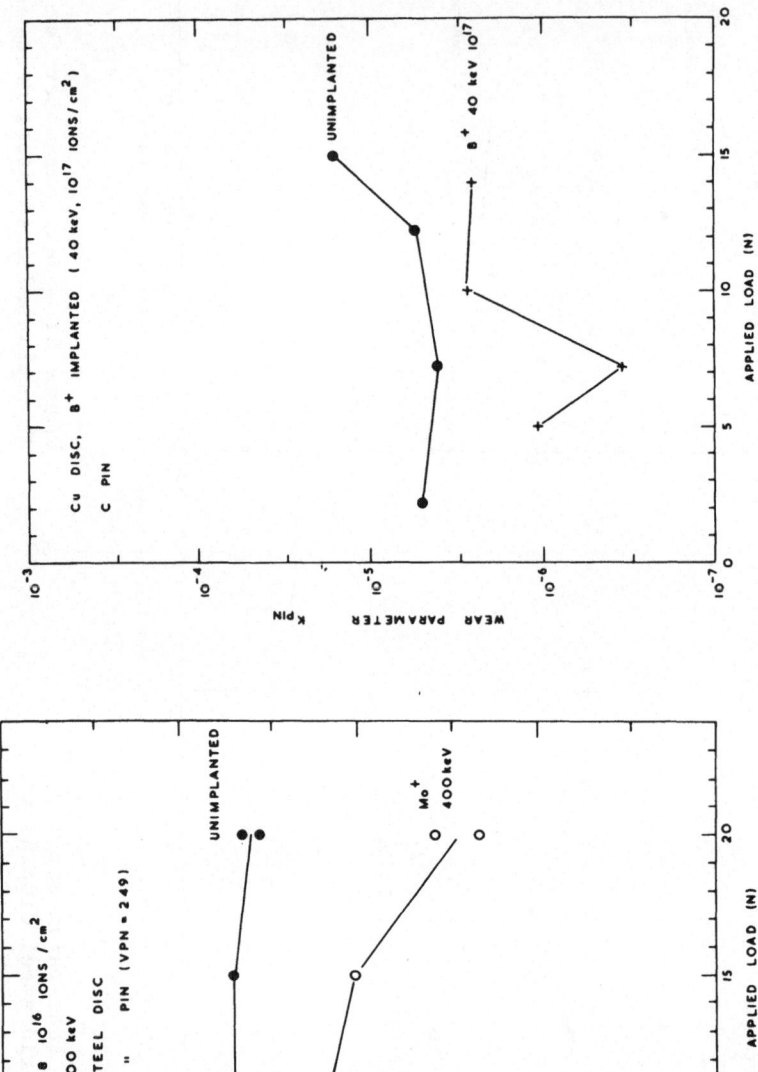

Figure 10. Wear parameter for B⁺ implanted Cu disc. graphite pin.

Figure 9. Wear parameter for Mo⁺ implanted 440C steel disc.

Estimations of wear performance are made by calculating the wear parameter, K, for a given set of conditions. A relationship between K and the total volume of material removed (ΔV) can be derived from the primary concept $\boxed{4}$ that friction is related to the true area of contact. Thus

$$K = \frac{3\ (\Delta V)\ H}{lx} \qquad \ldots (2)$$

where H, l and x are the hardness, applied load and sliding distance respectively. Note that K is independent of speed, and this is observed to be the case for many systems. Under severe wear conditions (where there is much welding and large metallic wear particles are produced) K has values typically of the order of 10^{-2} to 10^{-3}; under mild wear conditions (when oxide films prevent severe damage) K is typically between 10^{-5} and 10^{-6}. For such values of K, therefore, only about one in 10^6 encounters between asperities results in a wear particle $\boxed{5}$.

1" diameter test discs were ground flat and parallel to within 0.0001" which is the accepted tolerance for wear testing. The speed of revolution varied between 300 and 2000 r.p.m. White spirit lubricant was pumped at constant speed throughout the tests, with the exception of the Cu-graphite experiments (figure 10) which were run dry. Variations in the bulk temperature of the lubricant and the ambient temperature had no significant effect on the measurement of K.

Wear Results

Plots of the wear parameter for a pin sliding on implanted and unimplanted surfaces are given in figures 7 to 10 on the previous page. In each case the same pin was used to eliminate the effects of varying pin hardness. In figure 7, the data relating to an unimplanted surface was obtained from the reverse side of an implanted disc as a further precaution against specimen variation. Additional tests showed that the effects of heating of the specimen during implantation could account for a small decrease in K_{pin}; also that the temperature rise during N^+ implantation (about 150°C) was essentially uniform throughout.

The data show that after a high dose of N^+ (10^{18} ions-cm^{-2}, 35 keV) the wear parameter measured on the pin is substantially reduced, and the proportional reduction becomes more pronounced as the applied load is increased. Similar effects are observed for a high dose of boron (implanted using a Harwell-Lintott isotope separator), figures 8 and 10, and molybdenum, figure 9. For boron implantation the reduction in wear is not as marked as it is for the other two ion species, and the data for B^+ do not exhibit the rapid decrease in K_{pin} observed with N^+ and Mo^+ implantation. These effects are discussed below.

DISCUSSION

The friction experiments show that large changes in the coefficient of friction can be brought about by the implantation of specific ions, and the frictional properties introduced by this means appear to be due to the nature of the implanted ion.   It is diffi-cult to explain the large positive and negative variations in friction using a theoretical model based on damage or contamination phenomena alone.   From the limited amount of wear tests concluded to date, a reduction in wear parameter is measured for each of the combinations of test disc and wear pin.   If radiation damage is primarily responsible for surface hardening than it occurs suffi-ciently to reduce the wear parameter for substrates possessing a wide range of physical properties (Cu, mild steel, 440C steel) implanted with ions of widely differing mass (B, N, Mo).   It was concluded in earlier work that quantitative discussion on the nature of the friction effect is difficult since the wear and corrosion products are unknown.   The same is true for wear tests, although strong evidence for oxide formation under contact stress situations has now been established as a result of the XPS experiments.   This helps to reinforce the conclusion that metals which possess soft oxides may be implanted with advantageous results, provided the chemical environment is compatible.   For the case of wear, there is sufficient evidence to indicate that the surface films most effective in promoting corrosion resistance are also effective in preventing damage from sliding.   An important characteristic on which this improvement depends is the ductility of the film. Ductile films are most able to repair themselves to prevent rapid corrosion and wear damage $\boxed{3}$.   Molybdenum is frequently used in stainless steels to promote the stability of corrosion resistant surface films; boron and nitrogen may serve as wear inhibitors by enhancing the production of soft oxides (e.g. $B_2 O_3$).   It is clear that considerable overlap exists between tribological and corrosion studies $\boxed{13}$, and ion implantation may prove to be very useful in the identification of various common features because of its versatility $\boxed{14}$.

Initially, during the mechanism of mild wear, fragments of material are transferred due to the shearing of small metallic welds. These particles are smoothed out and oxidised (enhanced by frictional heating), and are subsequently rubbed off the surface $\boxed{6}$.   The important parameter is the rate of oxidation of the transferred layer, which may be reduced by compaction (implantation resulting in a high lateral surface compressive stress), metallurgical transformation (the N^+ - induced formation of iron nitrides) or passivation (Mo).   During wear tests on the unimplanted samples a large quantity of wear debris in the form of a fine black powder (probably $Fe_2 O_3$) was found to accumulate on the specimen and on the leading edge of the pin.   With the implanted samples, the lower wear rate was accompanied by much less wear debris.

Transfer of material between surfaces in relative motion is a frequent tribological event.   In the evaluation of solid lubricants by wear testing, therefore, it is emphasised that coatings should be applied to the moving surface (in this case the rotating disc), little advantage being gained in treating both surfaces $\underline{/}$ 15 $\underline{/}$. The study of transfer mechanisms on implanted surfaces which possess an interface-free "treated region" may have certain advantages over surfaces treated by conventional deposition techniques, from which a weakly bonded surface layer is often produced.

CONCLUSIONS

Ion implantation can induce frictional changes on steel surfaces and can substantially reduce the wear parameter measured under pin-and-disc conditions for implanted steel and copper. Oxidation occurs as a result of frictional contact.   Implanted Pb and Ag ions remain within the surface after severe deformation but may become redistributed across the regions of contact.

ACKNOWLEDGEMENTS

The X-ray photoelectron spectroscopy experiments of J.P. Coad and the assistance of D.K. Sood with the microbeam experiments of figure 5 are gratefully acknowledged.   A vacation studentship for J. Saunders was provided by the U.K.A.E.A.

REFERENCES

1. N.E.W. Hartley, G. Dearnaley and J.F. Turner, Proc. 3rd Internat. Conf. on Ion Implantation, Yorktown Heights, U.S.A., Dec. 11-14, 1972 (to be publ. by Plenum Press): available as AERE - R 7441 (1973).
2. A.I. Bailey and J.S. Courtney-Pratt, Proc.Roy.Soc.$\underline{A227}$(1955),500.
3. P.M. Ku (Ed.), Interdisciplinary Approach to Friction and Wear, NASA SP-181 (1968).
4. F.P. Bowden and D. Tabor, The Friction and Lubrication of Solids. Oxford University Press, Part I (1950), Part II (1964).
5. J.F. Archard, J.Appl. Phys. $\underline{24}$ (1953), 981.
6. See Discussion in Ref. $\underline{/}$ 2 $\underline{/}$.
7. J. Woolman and R.A. Mottram, The Mechanical and Physical Properties of the British Standard En Steels (B.S. 970-1955), Pergamon Press, (1969), $\underline{3}$, 482.
8. P.D. Goode, Nucl. Instr. and Methods $\underline{92}$ (1971), 447.
9. J. Lindhard, M. Scharff and H.E. Schiøtt, Mat. Fys. Medd. Dan. Vid. Selsk. $\underline{33}$ (1963), No. 14.

10. C.A. Maloney, Private Communication.
11. J.A. Cookson and F.D. Pilling, A 3 MeV Proton Beam of less than Four Microns Diameter, AERE – R 6300 (1970).
12. J.C. Rivière, Ionization (Characteristic Loss) Spectroscopy, and Appearance Potential Spectroscopy: Comparison with Auger Spectroscopy and X-ray Photoelectron Spectroscopy, AERE–R 7368 (1973).
13. G. Dearnaley, these Conference proceedings.
14. G. Dearnaley, New Trends in the Use of Accelerators in Solid State Physics, AERE–R 7455 (1973).
15. J.K. Lancaster in Tribology Handbook, M.J. Neale (Ed.), Butterworths, London, (1973).

DISCUSSION

Q: (R. L. Cohen) On the wear experiments with, e.g., Mo implantation, have you performed microprobe scans to see if there was Mo in the bottom of the wear groove?

A: The experiments you refer to are currently in progress. The contact pressure during wear testing was several orders of magnitude lower over that for friction testing, and it is therefore probable that a proportion of the implanted ions remains within the bottom of the wear groove also.

Q: (H. C. Freyhardt) How does differing surface roughness influence the results of your friction experiments?

A: Ground discs of En352 implanted with Mo showed similar reduction in friction to polished surfaces. It would therefore appear that, within the limits of our experiments, surface roughness does not influence the frictional change.

CHAPTER III

THIN FILMS AND INTERFACES

ION BEAM STUDIES OF METAL-METAL AND METAL-SEMICONDUCTOR REACTIONS

J. W. Mayer

California Institute of Technology

Pasadena, California   91109

ABSTRACT

Backscattering techniques allow determination of elemental composition as a function of depth with typical values of depth resolution of 100 to 300Å. The technique is ideally suited to measurements in thin film systems of thicknesses of several thousand Ångstroms and has provided a basis for examination of thin film interactions. This presentation is intended as an introduction to the papers presented in this session of the conference. These papers represent examples of the type of studies carried out in this field: interdiffusion and mixing between thin metal films, and reaction kinetics and silicide formation in systems composed of metal films deposited on Si or SiO_2. These studies involve solid-solid diffusion and reactions since the process temperatures are below the eutectic. Although backscattering techniques provide depth-microscopy, supplemental measurements are often required to establish lateral uniformity and the relative importance of grain boundary and bulk diffusion.

INTRODUCTION

Studies of interdiffusion in thin film systems require measurement of composition changes over depths of a hundred to a few thousand Ångstroms. Backscattering techniques have the depth resolution capability to permit mass-sensitive, depth microscopy over these dimensions. The five papers presented in this session give adequate evidence of the usefulness of backscattering techniques in thin film studies. In fact these papers provide a general introduction to the type of investigation carried out in metal-metal and metal-semiconductor interactions.

The use of ion beams for analysis of thin solid films has been discussed in detail in a recent conference[1] and in review articles.[2-4] To summarize: mass identification is determined by scattering kinematics (energy loss experienced in a large-angle scattering event) and depth information is determined by the energy loss of the particles in their inward and outward trajectory through the layer under analysis. For MeV ^{4}He ions, the depth resolution is about 200-300Å for depths up to 0.5μm.

In comparison[5] with other surface analysis techniques, such as Auger electron spectroscopy or secondary ion mass analysis, backscattering has the characteristic that depth distributions are based on energy loss of detected particles rather than on layer removal by use of ion sputtering. In this sense backscattering is nondestructive; however, radiation damage is produced during analysis and care must be taken to ensure that beam induced damage does not alter the composition of the sample.

Backscattering measurements provide information on the composition of samples as a function of depth. It is an ideal tool for studying diffusion profiles or the kinetics of compound formation. For example, the reaction kinetics of compound formation of W films on Si and the ratio of Si to W in the compound layer have been found in a straightforward fashion.[6] However, the composition of the compound layer is often not sufficient for an unambiguous identification of the phases that are present. The most direct method to identify phases is by means of x-ray diffraction. For thin films, it is necessary to use geometries where the x-rays are incident at glancing angles to ensure adequate sensitivity.[7] In systems where there are a number of possible phases, such as in the work of Kraütle et.al,[8] both backscattering and diffraction techniques must be employed.

It is generally found[9] that silicide formation occurs uniformly across the interface producing a well-defined layer of nearly constant thickness. Only a few exceptions have been noted.[10] The situation is reversed in metal-metal thin-film systems that have been studied by backscattering. The interdiffusion seems to be dominated by fast grain boundary migration which produces a non-uniform front.[11-13] Only in one case, the classical Au-Al couple, were the compound layers nearly uniform in thickness.[14] The beam spot of the MeV ^{4}He beam is typically 1-2 mm in diameter. Consequently, it is often a wise precaution to supplement backscattering measurements with techniques that provide lateral resolution, as scanning electron microscopy.

II. CHARACTERISTICS OF THIN FILMS

Interdiffusion in thin film systems can be a very complex process in that evaporated metal films are generally poly-crystalline and the effects of migration along grain boundaries must always be considered. The starting point should be the characterization of the as-deposited films. Backscattering techniques alone are not adequate. In fact, at all stages in the reaction process it is generally desirable to supplement backscattering measurements with other techniques such as x-ray and transmission-electron diffraction.

Some of the properties of thin films and the techniques used to study their properties are given in Table I. This is not intended as a complete listing, but primarily directed toward parameters evaluated in this session .

TABLE I. Thin Film Studies

	Property	Measurement Technique
1.	Film Thickness	Backscattering[4]
2.	Contamination (oxygen, argon)	Backscattering[8,10]
3.	Structure	Transmission Electron[11,15] Diffraction
4.	Interface Barrier (oxide layer)	Auger and Sputtering[16] Backscattering[4]
5.	Compound Formation Kinetics	Backscattering[4,6,8,9,14]
6.	Phase Identification	X-ray Diffraction[7-10]
7.	Thin Film Epitaxy	Reflection-Electron Diffraction[17] Channeling[18]
8.	Lateral Uniformity	Electron Microscopy[19]
9.	Stress	X-ray Diffraction[7]

Backscattering techniques are extremely powerful for determination of the thickness (in terms of atoms/cm^2) of deposited
films and the presence of contaminants. In the latter case it
is often desirable to deposit comparison films on carbon or
beryllium substrates to increase the sensitivity for detection
of low mass elements such as oxygen or nitrogen.

The determination of the structure and grain size of the
films requires other techniques, typically transmission electron
diffraction. This was carried out in one study presented in
this session.[11] In other work[15] it was found that the grain
size of deposited Pd films on Si was dependent on the crystal
orientation of Si substrate. If grain boundary diffusion is a
dominant factor, it will be necessary to characterize the
structure of the films.

The evaluation of data on reaction kinetics is critically
dependent on knowledge of the presence of diffusion barriers
at the interface. It is well-known, for example, that thin
oxide layers on silicon surfaces can significantly retard and
even block silicide formation.[9] In one paper in this session[8]
it was shown that different silicide phases are formed when the
same metal is deposited on Si as compared to SiO_2 substrates.
It is often extremely difficult to detect the presence of thin
interface barriers. One approach is to sputter clean the
surface; however, the substrate surface is heavily damaged and
the sputtering gas is retained in the surface layer.

The kinetics of interdiffusion can generally be measured
in a straightforward fashion by use of backscattering techniques.
As pointed out earlier, if compounds are formed, it is usually
required to use other techniques such as x-ray diffraction for
proper identification of the phases.[8]

When metal films are deposited on single crystal substrates,
the compound layers may show preferred orientation. This is
found for example with Pd_2Si[18] and $NiSi_2$.[17] For the palladium
silicide case it was noted[20] that the temperature for the transformation from Pd_2Si to PdSi depended on the degree of epitaxy.
Channeling measurements when combined with backscattering are
convenient when the crystallites are aligned within 1 to 2° of
each other.[18] In general, however, diffraction techniques are
more applicable.

Stress develops in thin-film structures either during deposition or subsequent thermal processing. One often observes

cracking or peeling of the films. This is an area that has not been adequately investigated.

III. GENERAL OBSERVATIONS

In the past few years there have been a marked increase in the number of studies of thin-film reactions. Backscattering measurements have played a major role in this development as evidenced by the papers in this session and at the conference on analysis of surface layers.[1]

Metallization and formation of silicide layers have played an important role in integrated circuit technology. From an operational standpoint it is of interest to determine the temperatures at which interdiffusion and silicide growth occurs, kinetics, identity and stability of any phases that are formed, and the influence of interface conditions. This information can be determined from combined use of diffraction and back-scattering techniques.

There are two aspects of silicide formation that have not been satisfactorily answered: the influence of migration along grain boundaries and the identity of the diffusing species (Si or metal atoms). Palladium reacts with Si to form Pd_2Si at temperatures as low as 200°C. This would suggest grain boundary growth; however, the same diffusion activation energy was found from 200 to 700°C indicating that the same diffusion mechanism was involved. The same growth rate is found for Pd_2Si on different oriented samples where the grain size and degree of epitaxy are different. These findings suggest[4,15] that bulk diffusion is dominant even at temperatures of 200°C. Consequently, it may be inferred that reaction temperature alone is not an adequate criterion to choose between grain boundary and bulk diffusion.

There have been only two cases, $HfSi$[10] and WSi_2[6], in which there has been an identification of the diffusing species. For both examples, the data indicated that Si was the diffusing specie. In order to gain further understanding of the mechanisms involved in silicide formation, identification in other systems must be made.

The papers in this session and a recent review[9] of silicide formation present strong evidence that the tools are now available to study reactions in thin-film systems. It is now necessary to continue investigations to uncover the basic mechanisms involved in interdiffusions and compound formation in these systems.

REFERENCES

1. Proceedings of the International Conference on Ion Beam
    Surface Layer, Yorktown Heights, June 1973.
2. M-A. Nicolet, J.W. Mayer and I.V. Mitchell, Science 177,
    841 (1972).
3. A. Turos and Z. Wilhelmi, Nukleonika 13, 975 (1968) and
    14, 320 (1969).
4. W.K. Chu, J.W. Mayer, M-A. Nicolet, T.M. Buck, G. Amsel and
    F. Eisen, Thin Solid Films 17, 1 (1973).
5. A. Turos and J.W. Mayer, Thin Solid Films 19, 1 (1973).
6. J.A. Borders and J.N. Sweet, these conference proceedings.
7. K.N. Tu and B.S. Berry, J. Appl. Phys. 43, 3283 (1972).
8. H. Krautle, W.K. Chu, M-A. Nicolet, J.W. Mayer and K.N. Tu,
    these conference proceedings.
9. J.W. Mayer and K.N. Tu, J. Vac. Sci. and Tech. (to be
    published).
10. J.F. Ziegler, J.W. Mayer, C.J. Kircher and K.N. Tu., J. Appl.
    Phys. 44, 3851 (1973).
11. W.J. DeBonte, J.M. Poate, C.M. Melliar-Smith and R.A.
    Levesque, these conference proceedings.
12. J.E.E. Baglin, V. Brusic, E. Alessandrini and J.F. Ziegler,
    these conference proceedings.
13. J.A. Borders, in Ref. 1.
14. S.U. Compisano, G. Foti, F. Grasso, J.W. Mayer and E. Rimini,
    these conference proceedings.
15. S.S. Lau and D. Sigurd, Phys. Stat Solidi (submitted).
16. T. Narusawa, S. Komiya and A. Hiraki, Appl. Phys. Lett.
    21, 272 (1972).
17. H. Krautle, W.K. Chu and K.N. Tu (unpublished data on $NiSi_2$).
18. D. Sigurd, R.W. Bower, W.F. van der Weg and J.W. Mayer in
    Ref. 1.
19. D. Sigurd, G. Ottaviani, V. Marrello, J.W. Mayer and J.O.
    McCaldin, J. Non-Cryst. Solids 12, 135 (1973).
20. G.A. Hutchins and A. Shepala, Thin Solid Films (to be
    published).

RUTHERFORD SCATTERING STUDIES OF DIFFUSION IN THIN MULTILAYER METAL
FILMS

W. J. De Bonte, J. M. Poate, C. M. Melliar-Smith and

R. A. Levesque

Bell Laboratories, Murray Hill, New Jersey   07974

ABSTRACT

Interdiffusion in Ti:Pd:Au and Ti:Rh:Au thin film metallization
schemes has been studied by Rutherford backscattering. Grain sizes,
as a function of annealing temperature, have been measured by transmission electron microscopy. The diffusion of Au into Rh is dominated by grain boundary effects. When Rh is the intermediate layer,
Au saturates the fine structured Rh grains in the as-deposited films
without any heat treatment. The diffusion behavior after annealing
is explained in terms of a crude model in which grain boundary
diffusion is assumed to keep the Rh grain boundaries filled with
Au, which subsequently diffuses into the grains by bulk diffusion.
This gives $Q = 1.34$ eV and $D_o = 5 \times 10^{-7}$ cm^2/sec for bulk diffusion
of Au into Rh. For equivalent annealing conditions considerably
greater diffusion is observed between Pd:Au than Rh:Au.

1.   INTRODUCTION

The study of interdiffusion between thin metal films has been a
neglected field apparently for the reason that such diffusion profiles are difficult to measure. This paucity of experimental
information is reflected in the shortage of theoretical treatments
of diffusion in thin films. There is now, however, considerable
impetus to understand such diffusion processes because of the role
that thin metal films play in integrated circuits. We will consider
in this paper multilayer Au-based metallization schemes and their
thermal stability. Gold has many attractions for use as a thin
film conductor on integrated circuits [1] because of its excellent
conductivity, lack of oxidation and resistance to electromigration
failures. The fact that Au is very unreactive also means that it

does not adhere well to SiO_2. To overcome this a metallic "glue" layer such as Ti is introduced. Unfortunately the Ti:Au couple is susceptible to corrosion and a barrier metal such as Pd or Pt has to be introduced [2]. Figure 1 shows schematically such a metallization system. Interdiffusion in such complex systems can result in loss of conductivity. It is important therefore to establish the degree of interdiffusion.

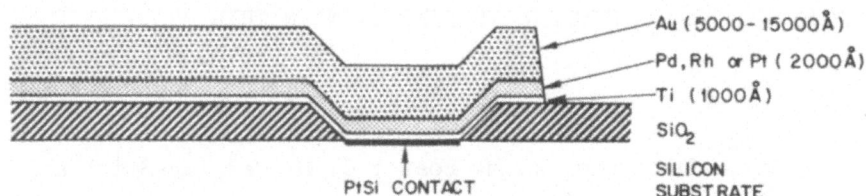

Fig. 1   Typical multilayer metallization scheme.

We have discussed elsewhere [3] interdiffusion in the more usual Ti:Pd:Au or Ti:Pt:Au metallization schemes. Here we will specifically compare the ternary systems of Ti:Pd:Au and Ti:Rh:Au. The latter structure was of interest because Rh serves the same corrosion protection purpose as Pd or Pt and has the advantage of a higher electrical conductivity. In addition its refractory nature was expected to reduce interdiffusion effects. Diffusion profiles have been measured by the Rutherford backscattering technique which is proving peculiarly suitable to such thin film analysis. (The equipment has been described elsewhere [4].) The changes in microstructure of the deposited films as a function of annealing temperature have been investigated by transmission electron microscopy.

2.   EXPERIMENTAL

The films used in this study were deposited onto sapphire substrates using an ion pumped, electron-beam evaporator. Sequential evaporations were made without breaking vacuum and pressures during the evaporation did not exceed 1×10^{-6} torr. The substrates were clamped to a large Cu block and the measured temperature rise of the substrates during evaporation was restricted to less than 15 C.

Annealing was carried out in two different ambients, air and high vacuum. Air anneals were performed in a standard tube furnace with constricted tube ends. Vacuum anneals were performed in a vacuum of $1-5 \times 10^{-7}$ torr produced by turbo-pumping. Thin film samples were radiantly heated from the back (substrate) side, while the temperature was sensed by a thermocouple attached to the front side. The rise in pressure to $\sim 5 \times 10^{-7}$ torr occurred at the beginning of each anneal and was associated with outgasing in the heated region.

Typically, the annealing temperature was reached in 1-2 min. Cooling at the end of the anneal was accomplished in a period of 1-2 minutes by back-filling with N_2 cooled to $\sim$88 K. The cold back-filling with N_2 not only brought the sample temperature down quickly but also assured that most of the residual gas in the vacuum chamber during annealing was N_2 rather than O_2 or H_2O.

Grain size measurements and electron diffraction patterns were obtained using a 100 keV transmission electron microscope. For these measurements 700 Å metal films were evaporated separately onto oxidized Si wafers, annealed in air and then stripped from the substrates for microscopy. A number of composite (Ti:Rh:Au) films were also stripped from the substrates after annealing and then selectively etched to allow the individual films to be viewed separately in the electron microscope.

3. RESULTS

3.1 Rutherford Backscattering

The backscattering spectra for the Ti:Pd:Au and Ti:Rh:Au systems before and after vacuum annealing are shown in Fig. 2. The spectra show pronounced interdiffusion in the Pd:Au case and much less in the case of Rh:Au. To aid in detailed interpretation of the Pd:Au spectra, the order of the Pd-Au evaporation is reversed so that backscattering spectra will give the diffusion profiles directly [3]. Figure 3 shows backscattering spectra from such reverse evaporations and the diffusion profiles are clearly seen after the anneals. Assuming the Au-Pd interdiffusion is the same in the Ti:Pd:Au and Ti:Au:Pd systems then the directly measured profiles can be used to unfold the Ti:Pd:Au data. The dashed curves in the Pd:Au spectrum of Fig. 2 are the unfolded profiles carried out in this manner. In Fig. 2 the measured Au diffusion profile of Fig. 3 was normalized to the Au at the point on the spectrum corresponding to the kinematic position of Pd on the surface of the Au. This Au profile was then subtracted from the total spectrum (containing Pd and Au counts) to give the Pd diffusion profile. It can be seen that the diffusion profiles for Pd in the Ti:Pd:Au and Ti:Au:Pd systems are very similar in shape even though somewhat different in magnitude. Comparison of the spectra for Ti:Rh:Au and Ti:Au:Rh show that the diffusion strongly depends on the sense of the evaporations.

3.2 Transmission Electron Microscopy

The grain size distribution as a function of annealing conditions is shown, for single films, in Figure 4. The grain size measurements from the composite films proved somewhat inconclusive due to the problems involved in chemically separating the films after annealing. This may have been due to the combined effects of

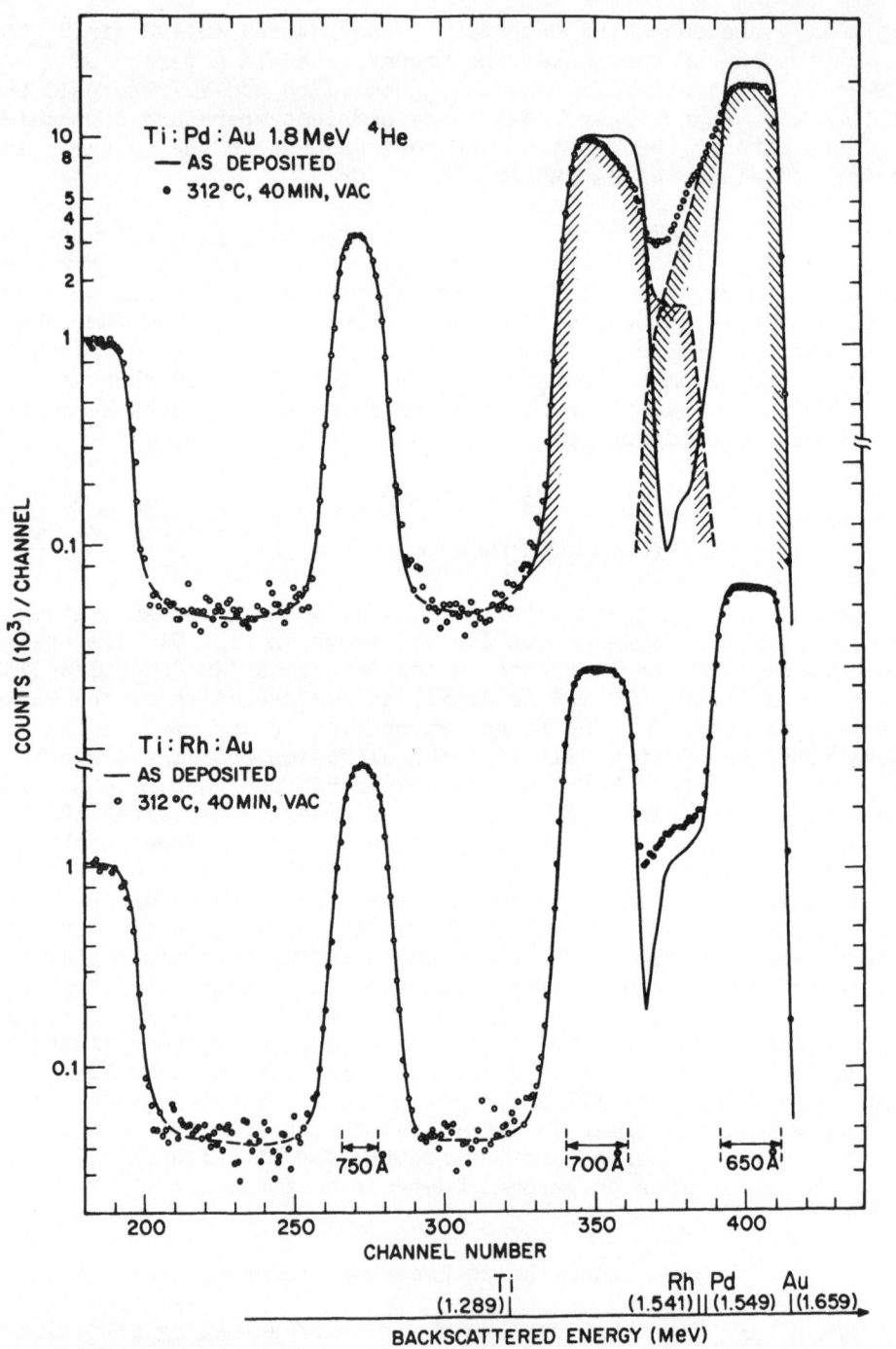

Fig. 2 Backscattering spectra of Ti:Pd:Au and Ti:Rh:Au films before and after annealing.

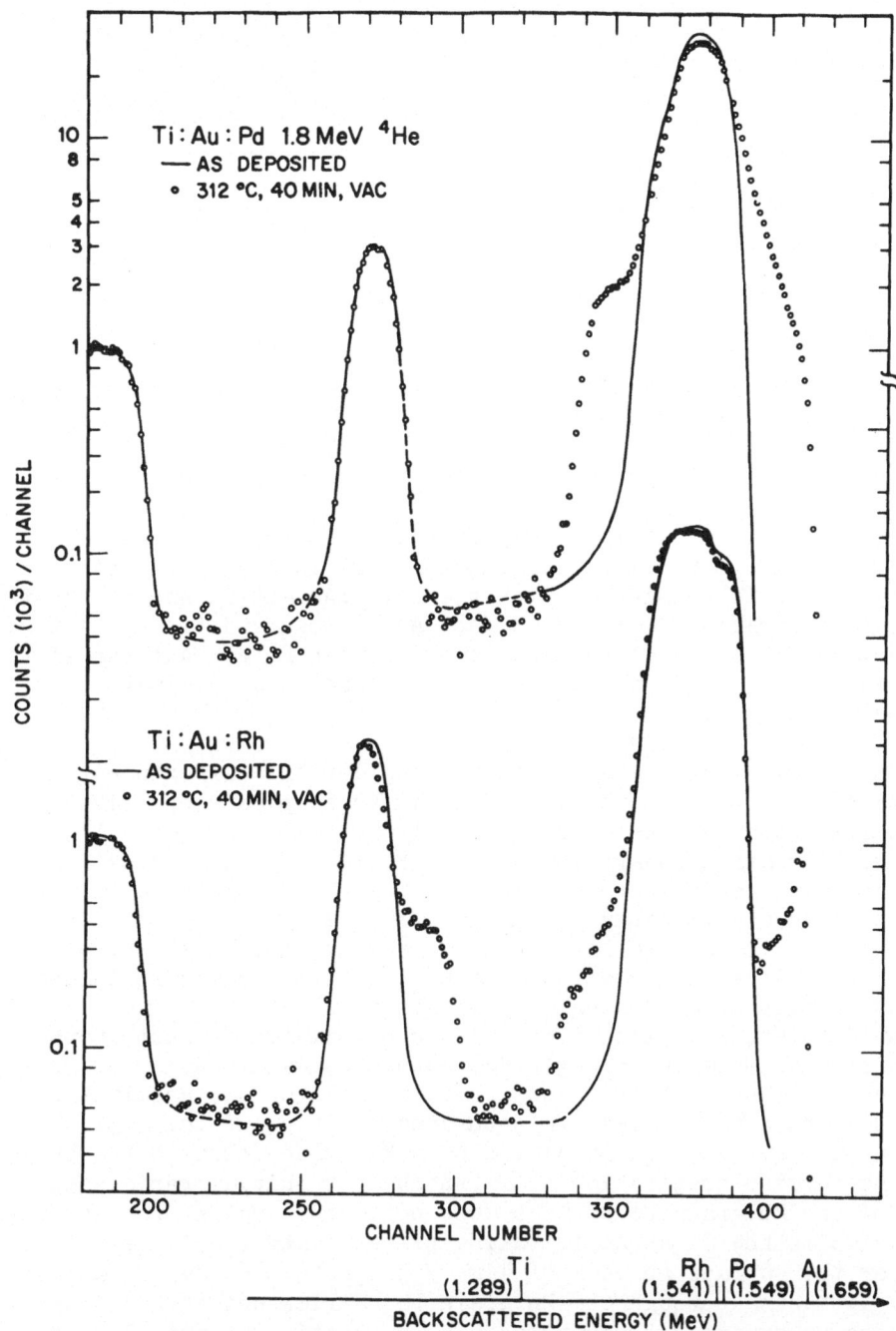

Fig. 3 Backscattering spectra of Ti:Au:Pd and Ti:Au:Rh films before and after annealing.

interdiffusion and oxidation. Initial data however indicate that
film interdiffusion does not have a major effect on the grain
growth during annealing.

Electron diffraction patterns show evidence for progressive oxidation of the Pd (to PdO) and Rh (to Rh_2O_3) films when annealed in
air above 312 C. Quantitative data are not available but the pattern intensities suggests oxidation considerably in excess of
monolayer amounts.

4. DISCUSSION

We will first discuss qualitatively the salient features of the
interdiffusion in these thin film systems. In the annealing range
of 200-389 C no Ti interdiffusion is observed into Pd or Au for
the Ti:Pd:Au or Ti:Au:Pd systems respectively. However for the
couples Ti:Pd and Ti:Au considerable Ti interdiffusion is observed
[3]. The explanation for this difference in Ti diffusion in the
binary or ternary system is that Pd and Au interdiffuse very
rapidly blocking the diffusion paths for Ti into Pd or Au respectively. This is consistent with the Ti:Rh:Au and Ti:Au:Rh results
where considerable Ti diffusion into Au is observed for the latter
system. Little interdiffusion occurs between Rh and Au; therefore
when Au is next to Ti the interdiffusion should approximate that
of the Ti:Au couple.

Inspection of Figures 2 and 3 shows considerable differences in
interdiffusion between the Pd:Au and Rh:Au couples respectively.
It is known that diffusion in such polycrystalline films will
consist of grain boundary and bulk diffusion. We will consider
the Rh:Au and Au:Rh systems first as they show unambiguously the
effects of grain boundary diffusion. In the as deposited Ti:Rh:Au
films the Au is seen to be interdiffused into the Rh at approximately the 5% level throughout the Rh. This extremely rapid interdiffusion must occur during evaporation or shortly afterward.
Whilst the detailed mechanisms of this phenomenon are not known
[5] diffusion must be taking place down the grain boundaries. We
have analyzed the Ti:Rh:Au isochronal anneals on the assumption
that the grain boundaries are saturated by this initial rapid diffusion and that subsequent diffusion takes place through the bulk.
For the reverse evaporation of Ti:Au:Rh this initial rapid saturation of the Rh grain boundaries does not occur suggesting that the
diffusion in the Ti:Rh:Au is taking place during the Au deposition.
However the surface Au peak of Fig. 3 suggests that during annealing
Au is diffusing through the Rh grain boundaries and then diffusing
over the Rh surface. This process as a function of annealing temperature is shown in Figure 5. The as-deposited profile may represent a low level of Au diffusion but the counting uncertainties due
to pileup make this interpretation tentative. The flat profile at
200 C is interpreted as Au diffusing through the Rh grain boundaries.

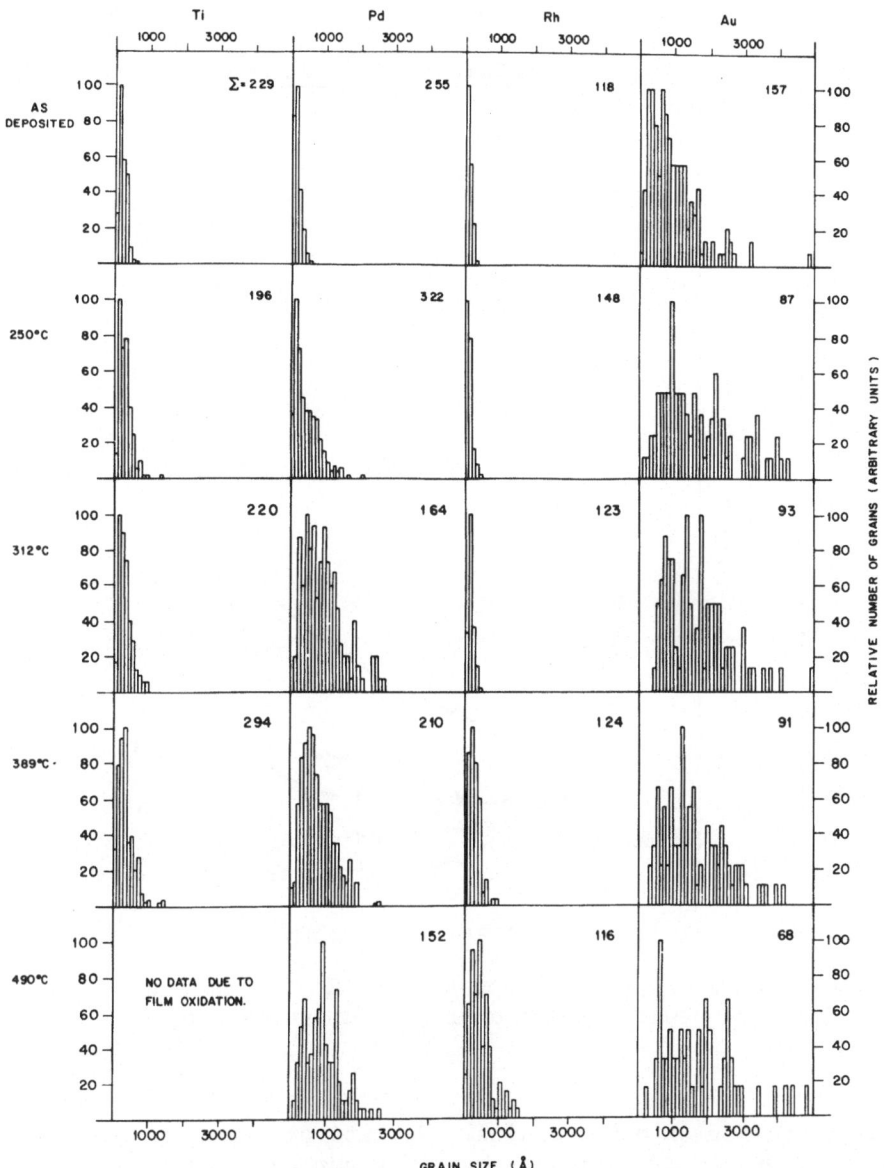

Fig. 4 Grain-size distribution in single-layer films as function
of annealing. Number in upper right corner of each histogram is
number of grains measured.

The fact that the Au grain boundary concentrations in Ti:Rh:Au
and Ti:Au:Rh are significantly different is explained by differences
in grain sizes. At 250 C the flat profile has risen indicating

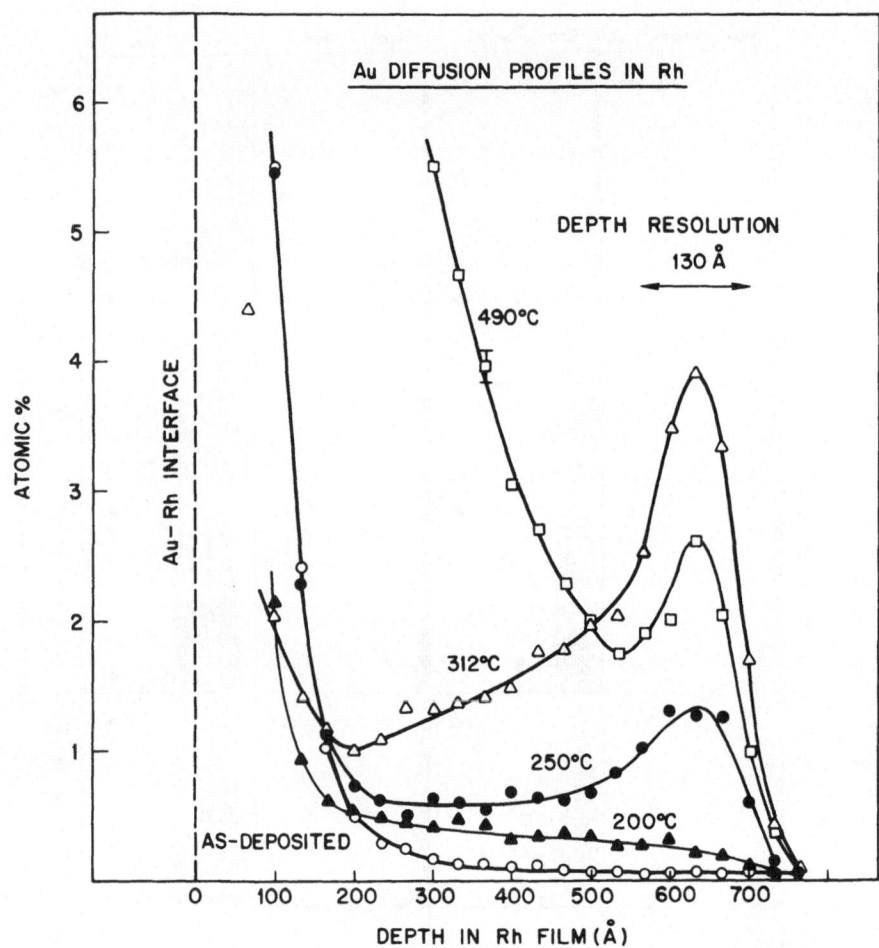

Fig. 5 Diffusion profiles for Au into Rh in Ti:Au:Rh films as
function of annealing temperatures. Top surface of Rh film is at
650 Å.

bulk diffusion into the grains. We interpret the surface peak
as being due to Au diffusing over the surface of the Rh on emer-
gence from the grain boundaries. The number of Au atoms/cm^2 in
this peak ($1.3 \times 10^{15}/cm^2$) is almost exactly a monolayer coverage.
At 490°C bulk diffusion is becoming the dominant process. The
reversed slope on the profile at 312 C appears anomalous and may be
due to such competing factors as grain growth and Ti diffusion into
the underlying Au film. If equivalent anneals are performed in
air, no surface peaking effects are observed. Oxidation of the Rh
surface presumably produces a barrier to surface diffusion.

The extraction of meaningful diffusion coefficients from back-scattering spectra is complicated in the Ti:Rh:Au system by the competition between bulk and grain boundary effects. The grain boundaries provide diffusion channels which at low temperatures are much more efficient in transporting material than are paths through the bulk. However, the boundaries become less important at high temperature, not only because the bulk diffusion coefficient is growing more rapidly with temperature than the grain boundary diffusion coefficient (since the activation energy for bulk diffusion is generally larger than that for grain boundary diffusion), but also because the density of grain boundaries is decreasing as grain growth occurs (see Fig. 4). Thus, we have three simultaneous temperature-dependent effects to consider.

We will analyze in detail the Au-Rh diffusion profiles such as shown in Fig. 2. The Au appears to saturate the Rh grain boundaries throughout the film and Au concentration through the film rises uniformly as a function of temperature below $\sim$400 C. Our data for the diffusion of Au into Rh in Ti:Rh:Au are analyzed subject to the following assumptions: (1) at high temperatures, (490 C), Au diffusion into Rh is dominated by bulk diffusion; (2) at low temperatures, (312 C and below), grain boundary diffusion is infinitely faster than bulk diffusion; (3) the average width of the Rh grain boundaries is $\sim$5 Å, so that by rough geometrical arguments a total grain boundary volume of 5% of the film volume corresponds to an average initial Rh grain size of $\sim$300 Å (within a factor of two of the average as-deposited Rh grain size in Fig. 4); (4) grain growth during annealing occurs at a rate which may be obtained from the grain-growth measurements on the initially smaller Rh grains analyzed in Fig. 4. From these assumptions on grain boundary behavior, one can formulate a crude model to treat the low-temperature data. Grain boundary diffusion is assumed to keep the Rh grain boundaries filled with Au, which subsequently diffuses into the cubic grains (of average edge length 2a) by bulk diffusion governed by a diffusion coefficient D_B. The average normalized Au concentration in the Rh film is then

$$c_{Au} = v_G + (1-v_G)F(D_B t/4a^2) \qquad (1)$$

where v_G is the relative volume of grain boundaries in the Rh film, t is the annealing time, and the function F is given by [6]

$$F(D_B t/4a^2) = 1 - \left\{ \frac{8}{\pi^2} \sum_{n=0}^{\infty} (2n+1)^{-2} \exp\left[- \pi^2(2n+1)^2 D_B t/4a^2 \right] \right\}^3 .$$
$$(2)$$

With our knowledge of v_G and 2a as a function of annealing temperature, we can fit our data for c_{Au} to Eq. (1) to obtain D_B for the
low annealing temperatures. These data, combined with the data
for bulk-dominated diffusion at 490 C (from an error function fit)
give $D_o = 5 \times 10^{-7}$ cm^2/sec for the pre-exponential factor and
Q = 1.34 eV for the activation energy for bulk diffusion of Au
into Rh.

Overlapping spectral peaks and competing Ti/Au interdiffusion make
analysis of the diffusion of Rh into Au difficult. However, if
bulk diffusion is assumed to be dominant above ~400 C, the activation energy and pre-exponential factor for this process may be
tentatively set at Q = 1.0 eV, $D_o = 9 \times 10^{-9}$ cm^2/sec.

The interdiffusion of Pd and Au in Ti:Pd:Au and Ti:Au:Pd manifests
anomalous behavior when correlated with the grain size measurements of Fig. 4. While the Pd grains are consistently smaller
than the Au grains, especially in the as-deposited films and in
films annealed at low temperatures, our spectra in Figs. 2-3 would
lead one to believe that Pd diffusion into Au is assisted by grain
boundary diffusion while Au diffusing into Pd is dominated by
bulk diffusion. These results will be presented elsewhere [3].

ACKNOWLEDGMENTS

We are indebted to Miss L. V. Haller for preparing the films, to
Miss S. E. Koonce and A. G. Cullis for electron microscopy work,
to P. A. Turner and W. L. Brown for many helpful discussions and
to P. J. Silverman for help in the initial stages of the experiment.

REFERENCES

1. M. P. Lepselter, Bell System Tech. J. 45, 233 (1966).

2. A. T. English and P. A. Turner, J. Electronic Materials 1, 1
    (1972).

3. J. M. Poate, P. A. Turner, W. J. De Bonte, J. Yahalom to be
    published.

4. T. M. Buck, J. M. Poate, K. A. Pickar, C-M. Hsieh, Surface
    Science 35, 362 (1973).

5. C. Weaver and L. C. Brown, Phil. Mag, 17, 881 (1968).

6. H. S. Carslaw and J. C. Jaeger, Conduction of Heat in Solids,
    2nd ed., Oxford Univ. Press, p. 185 (1959).

DISCUSSION

Q: (T. S. Noggle) In many cases, the buildup of evaporated metal films occurs by an "island" growth which tends to leave grooves at the junctures of grains. The filling of such grooves in the Rh layer by the deposited Au would lead to backscattering distributions similar to those observed. Have you studied your films to evaluate whether this possibility occurs?

A: We have carried out backscattering measurements on the Ti:Rh couple alone. If grooves in the Rh film were the cause of the anomalous diffusion, we should see a similar shoulder on the leading edge in the backscattering peak for Ti in the Ti:Rh couple, as was seen for Au in Ti:Rh:Au. We did not see this effect. However, the Rh film in the Ti:Rh couple was somewhat thicker ($\sim$ 1200 Å), than the Rh film ($\sim$ 700 Å) in the Ti:Rh:Au system.

Q: (A. Hiraki) What is the origin of the one-way diffusion of Au into Rh? Is it an atomic size effect?

A: We are not certain of the mechanism for the anomalous diffusion of Au into Rh, except that it appears to occur during deposition. However, the experimental evidence indicates that the diffusion is taking place down the grain boundaries and is therefore probably dominated by the grain structure of the film rather than by atomic size effects.

ANALYSIS OF COMPOUND FORMATION IN Au–Al THIN FILMS[+]

S.U.Campisano,G.Foti,F.Grasso,J.W.Mayer[x] and E.Rimini

Istituto di Struttura della Materia dell'Università

Corso Italia, 57 - I 95129 Catania, Italy

Backscattering techniques were used to evaluate intermixing in evaporated thin films of Al and Au. The film thicknesses ranged between 2000 Å and 7000 Å. In the analysis of 2.0 MeV He$^+$ backscattering spectra the gold yield in a virgin sample has been used to normalize the gold concentration in the reacted sample. The aluminum component in the spectra gives the complementary concentration, but counting statistics are poor and background corrections must be made. At temperatures around 100 C the backscattering spectra show that the compound Au_2Al forms at the interface between Au and Al. The growth of the compound is proportional to the square root of the annealing time. At temperatures around 200 C the backscattering spectra show that the compound $AuAl_2$ forms and grows proportional to the square root of the annealing time.

INTRODUCTION

Diffusion in thin vacuum deposited films has received attention only during recent years. To examine diffusion phenomena it is necessary to use experimental methods which detect composition changes over extremely short distances (of the order of several hundreds Å). Backscattering techniques are ideally suited to explore the behaviour of intermixing in these thin couples (1).

- - - - -

(+) Work supported in part by Gruppo Nazionale di Struttura della Materia del Consiglio Nazionale delle Ricerche and by Centro Siciliano di Fisica Nucleare e di Struttura della Materia.
(x) Permanent address: California Institute of Technology, Pasadena 91109,Ca,supported in part by N.S.F.(USA–Italy cooperative program).

Interdiffusion in thin metal systems has been studied by backscattering technique, and in almost all cases there is a fast migration
related to grain boundary diffusion effects, as for example in the
Au-Cr (2) and Au-Cu (3,4) systems. In the copper gold system at
temperatures of 150 C - 200 C diffusion of gold through copper and
of copper through gold was noted. There was no evidence of the
growth of a distinct phase at the interface between the two metals.
To date only the two metal system Al-Ti indicates the existence of
a phase (Al_3Ti) at the interface between the two metal layers (5).

We have chosen among the compound forming systems, the gold
aluminun case since it has been studied extensively by other techniques (6). The earliest results were obtained by measuring the
optical reflectivity of films evaporated onto glass microscope
slides. It was found that the reflectivity fells sharply after a
time which depends on the film thickness. This was interpreted
with the formation of layer of Au_2Al compound between the gold and
aluminum that grows by diffusion in the temperature range 70 C -
150 C. Diffusion in this system was also investigated by adhesion
and resistance measurements. Both adhesion and optical reflectivity
measurements indicated an activation energy for interdiffusion of
about 1 eV (23.5 kcal/mol). When an excess of aluminum was present
the distinct purple color of the $AuAl_2$ compound appeared at higher
annealing temperatures. The formation kinetics of the $AuAl_2$ compound was not investigated in detail but results indicated that
also this second compound grows by a phase boundary movement limited by diffusion.

The phase diagram of the Au-Al system, reported in Fig.1,
shows the existence of five intermetallic compounds. Electron
diffraction measurements indicate only the presence of Au_2Al and

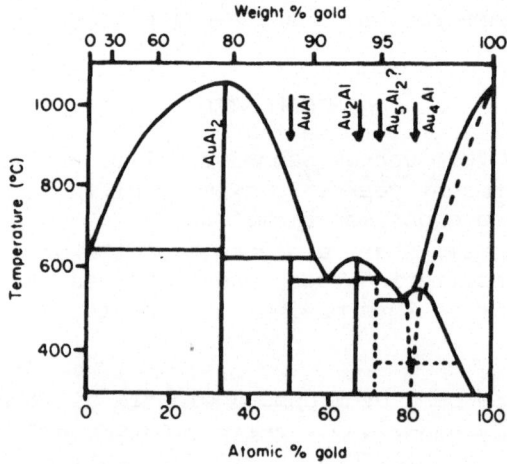

Fig.1 - Phase diagram of Au-Al system (taken from Ref.6)

of $AuAl_2$ in thin film specimens (6). It should be noted however
that in bulk samples of Au-Al diffusion couples all the compounds
are formed with the most rich gold compound near the gold boundary
and the aluminum rich compound near the aluminum boundary (7). The
purpose of this work was to use backscattering techniques to study
interdiffusion in evaporated films of gold and aluminum. The emphasis was placed on the analysis procedure and to determine if sharply defined phases are formed. The existence of sharply defined
phases of uniform thickness is an important condition in backscattering analysis since the beam diameter of 1 mm is about 10^4 times
greater than the film thickness. Any non uniformity in the thickness of the phases makes it difficult to determine diffusion kinetics and to identify the compound that is formed.

                              EXPERIMENTAL

     The gold-aluminum samples were prepared by evaporating gold
layers onto formvar covered aluminum and then without breaking vacuum by depositing an aluminum layer. The vacuum before evaporation
was approximately 10^{-6} torr. The thickness of aluminum films was
generally about two times that of the gold films. The thicknesses
of aluminum ranged between 3000 Å and 7000 Å, while the corresponding gold thicknesses ranged between 1500 Å and 3000 Å.

     Thermal processing was carried out in a tube furnace under
vacuum (10^{-6} torr) or in purified argon atmosphere. These two ambient conditions show no difference in the results. The temperature
range was between 70 C to 240 C, and the temperature was held $\pm 1°C$.
Backscattering measurements have been made using the 2.5 MeV Van
de Graaff. Helium particles backscattered through an angle of 150°
were energy detected by a surface barrier detector and conventional
electronic were used to amplify and to display the signals. Secondary electron suppression techniques were used and the dead time during analysis was less than 5%.

                          ANALYSIS   PROCEDURE

     Schematic backscattering spectra are shown in Fig.2 for a vir-
gin sample of evaporated aluminum and gold and for a sample in which
uniform thickness of Au_2Al has been formed at the interface between
the aluminum and gold. In this schematic spectrum the positions of
Au and Al on the surface are indicated by arrows. The Al yield has
been increased by a factor ten. In Fig.2a the front edge of the Al
coincides with the surface position and the width of the Al is a
direct indication of the number of aluminum atoms per cm^2. For convenience we have used the bulk density of aluminum (6.02×10^{22}
atoms/cm^3) to give film thickness in Å. We have also used a linear
energy-loss to thickness relation which introduces only a 5%

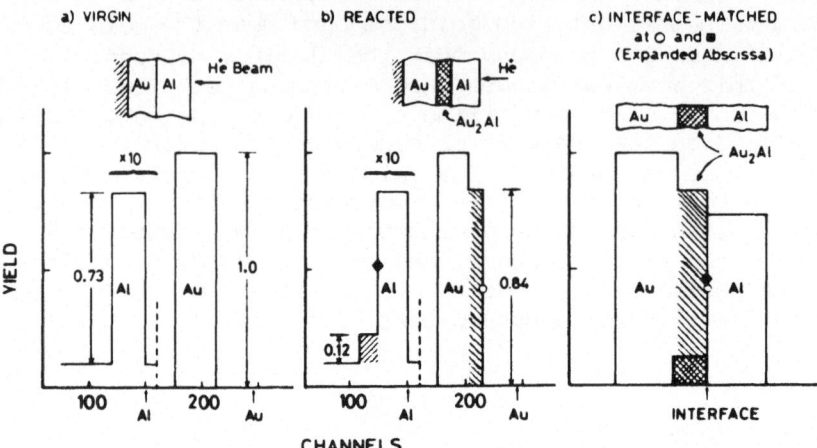

Fig.2 – Schematic diagram of backscattering spectra for evaporated Au–Al films; (a) virgin sample; (b) sample with Au$_2$Al between gold and aluminum. The ◆ and ○ indicate the interface between Au$_2$Al and Al; (c) the same spectrum shifted so that the interface positions are matched. (Note that the channel scale is doubled).

error for film thickness of 7000 Å. From the stopping power data of Chu and Ziegler (8) the energy to depth scale conversion for 2.0 MeV He$^+$ is 51 eV/Å for Al. The front edge of the gold signal is shifted below the surface peak because of the presence of the aluminum film on the top. The conversion of energy scale to depth scale for gold is 146 eV/Å, using the bulk density of Au (5.9X10^{22} atoms/cm^3). At the interface between gold and aluminum (high energy edge of the gold component and low energy edge of the aluminum component) the aluminum yield is 0.073 of the gold yield. The low energy tail extending below the gold component (~1% of the gold yield) produces a background of about 10% of aluminum yield.

The spectrum for a reacted sample with Au$_2$Al compound at the interface is shown in Fig.2b. The presence of this compound is indicated by a step in the high energy edge of the gold yield and a step at the low energy end of the aluminum. The height of the gold step is 0.84 of the gold yield in the untreated sample. (The height of the step would be 0.56 if AuAl$_2$ was formed). The leading edge of gold (○) is also shifted to high energy because some of the overlaying aluminum is consumed. The corresponding interface edge (◆) of the aluminum yield is also shifted to higher energies. The height of the aluminum step is 0.012 of the gold reference and is comparable with background. It is then difficult to analyse the composition of the compound from the aluminum component. Fig.2c shows on an expanded scale the aluminum and gold components shifted to match the interface. This interface matching procedure is also used in Figs. 4 and 6.

For the analysis of the backscattering yields we have assumed the additivity of the stopping cross section (Bragg's rule). The validity of this assumption has been tested recently (9) for the same system. The relative yield of the gold in a mixture increases (compared to the virgin sample) with gold concentration. The curve is not linear due to change in the stopping cross section with varying the gold concentration as shown in Fig.3a. The composition ratio can be also obtained by the ratio between the gold and aluminum yields measured at the same depth. This is nearly linear to the composition ratio because it depends on the ratio between the stopping cross section for beam particles scattered from gold and aluminum atoms as reported in Fig.3b. The stopping ratio is 0.905 for Au_2Al and 0.885 for $AuAl_2$. Since the Rutherford cross section is 38.5 times higher for gold than for aluminum counting statistics and background correction produce large uncertaines in the gold to aluminum concentration when composition is determined from Y_{Au}/Y_{Al} ratio. Both methods are described in detail in the appendix.

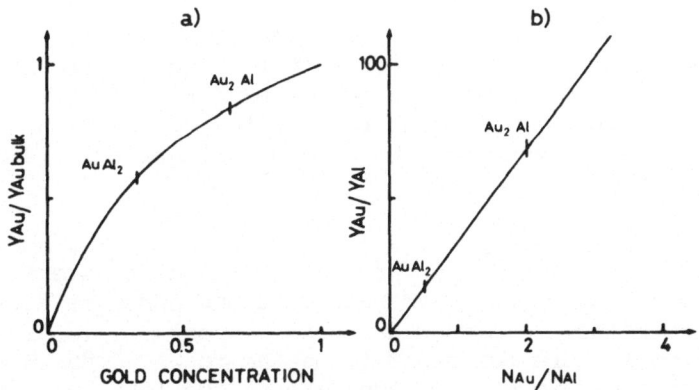

Fig.3 — (a) The ratio of the gold yield in Au-Al alloy relative to bulk gold yield versus the gold concentration in the alloy; (b) the ratio of the gold to the aluminum yield versus the composition ratio.

RESULTS

The results for the Au_2Al compound are shown in Fig.4 according to the scheme described in Fig.2. The arrows represent the calculated yield relative to the bulk gold $(Y_{Au\ bulk})$. The experimental spectra are in reasonable agreement for composition ratio corresponding to Au_2Al compound. In these spectra each channel corresponds to 7.75 KeV and for the gold signal one channel corresponds to 67.5 Å of Au_2Al, so that the width of the step is directly related to the thickness of the compound layer. The increase in the thickness is proportional to the square root of the annealing time as shown in Fig.5 for a sample annealed at 85 C.

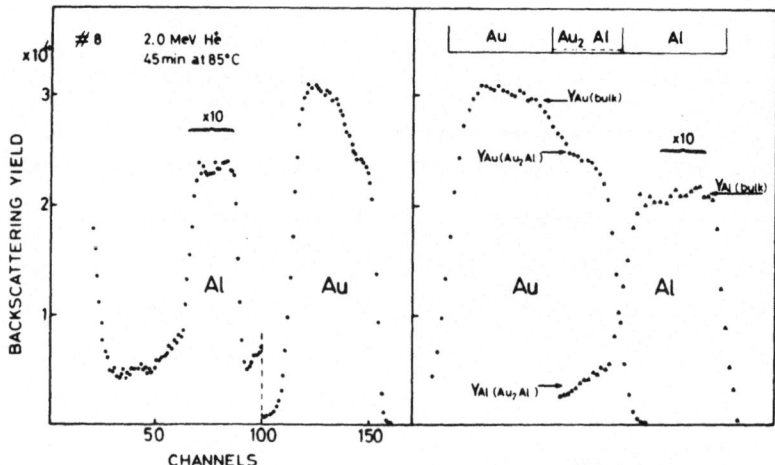

Fig.4 – Backscattering energy spectra for a 3900 Å film of Al deposited on 2100 Å film of Au reacted at 85 C for 45 min. The right hand portion shows the same spectrum shifted to match the interface with channel spacing doubled. The arrows indicate the yields relative to bulk gold.

If sufficient aluminum is present the gold is enterely converted to Au_2Al. At higher temperatures a new phase is formed at the interface between aluminum and Au_2Al layers. Fig.6 gives the backscattering spectra for a sample heated to 200 C for 5 min. The right hand side of Fig.6 shows the data of the left hand side after the interface matching. Here again the calculated yields shown by arrows agree with the measured yields. Also in this case the width of the gold step corresponding to $AuAl_2$ compound increases linearly with the square root of the time as shown in Fig.5 for a sample annealed at 200 C. In this case one channel (7.75 KeV) corresponds to 90 Å of $AuAl_2$.

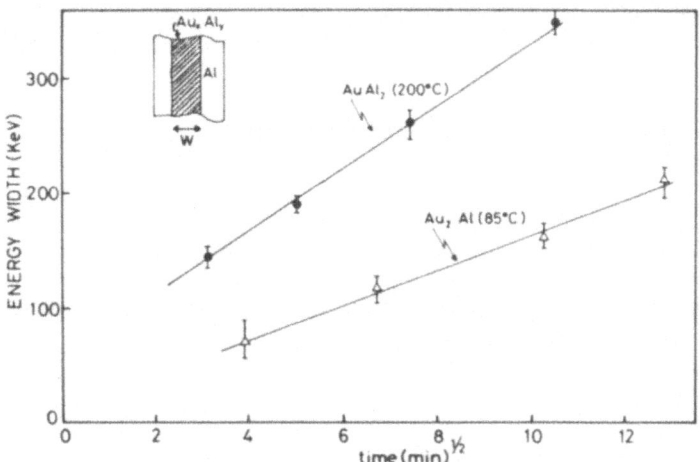

Fig.5 – Energy width of the step corresponding to the thickness of the compound versus square root of the annealing time.

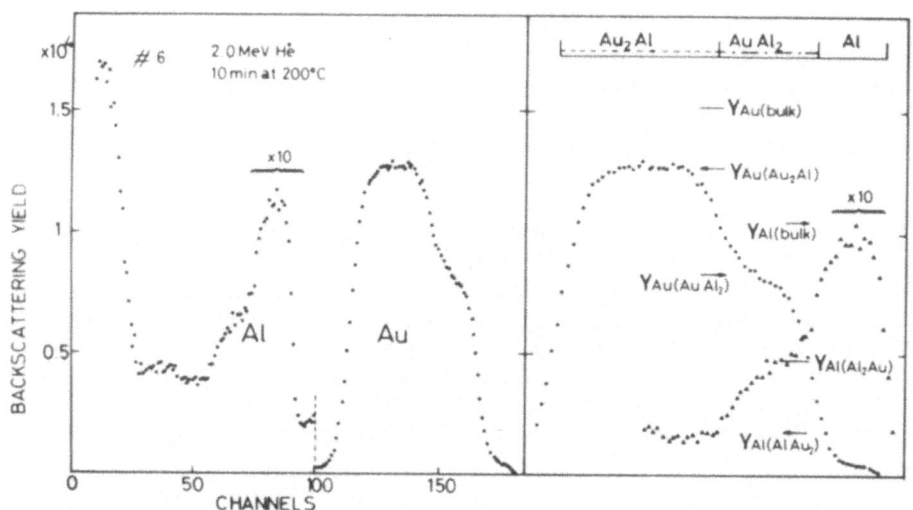

Fig.6 – Backscattering energy spectra for 3000 Å film of Al deposited on 1700 Å film of Au reacted at 200 C for 10 min. The right hand side shows the same spectrum shifted to match the interface between Au_2Al and Al with channel spacing doubled. The arrows indicate the yields relative to bulk gold.

CONCLUSION

Backscattering technique applied to the Au–Al thin film system indicates that the Au_2Al is formed as a uniform layer at the interface between Al and Au at temperatures around 100 C; at temperatures around 200 C the $AuAl_2$ compound is formed instead at the interface between Al and Au_2Al if sufficient free aluminum is present. Both phases grow proportional to the square root of the annealing time. At the present we are continuing this investigation to determine the activation energies associated with growth of these two compounds. The results so far obtained are in substantial agreement with previous work (6). In comparison with previous evaluation technique which relay on changes in optical reflectivity and film adhesion, backscattering technique offer a simple method for determining reaction kinetics in thin metal films.

Thanks are due to G.Caruso and V.Scuderi for the help in performing these measurements. One of the authors (J.W.M.) would like to thank A.Rimini for hospitality during the course of the experimental work.

REFERENCES

1) W.K.Chu, J.W.Mayer, M.A.Nicolet, T.M.Buck, G.Amsel and F.Eisen. Thin Solid Films, August 1973.
2) J.K.Hirvonen, W.H.Weisenberger, J.E.Wesmoreland and R.A.Maussner. Appl.Phys.Lett. 21, 37, (1972).
3) S.U.Campisano, G.Foti, F.Grasso and E.Rimini. Proc. of the Conference on Ion Beam Surface Layer Analysis. Thin Solid Films (to be published).
4) J.A.Borders. Proc. of the Conference on Ion Beam Surface Layer Analysis; Thin Solid Films (to be published).
5) R.W.Bowers. Solid State Electronics (to be published).
6) C.Weaver. Physics of Thin Films, vol.6, p.315, (1971).
7) E.Philofsky. Sol. St. Electr. 13, 1391, (1970).
8) W.K.Chu,and J.F.Ziegler. Nuclear Data Tables (to be published).
9) J.S.Y.Feng, W.K.Chu and M.A.Nicolet. Proc. of the Conference on Ion Beam Surface Layer Analysis. Thin Solid Films (to be published).

APPENDIX

The composition of a mixture of atomic density $N_t(c)$ containing c fraction of element A and (1–c) of element B can be obtained in backscattering spectra by measuring (a) the yields Y_A and $Y_A(c)$ of A atoms in a pure A sample and in the mixture respectively or by measuring (b) the yields $Y_A(c)$ and $Y_B(c)$ on the same sample and at the same depth.

a) Assuming the additivity of the stopping cross section, i.e.
$\varepsilon(c) = c\varepsilon_A + (1-c)\varepsilon_B$, being ε_A and ε_B the stopping cross section for pure A and B respectively, the yields Y_A and $Y_A(c)$ are given by

$$Y_A = \kappa N_A \Delta x = \kappa N_A \frac{\Delta E}{S]_A}; \quad Y_A(c) = \kappa N_A(c) \frac{\Delta E}{S]_A^c}$$

where κ is a geometrical factor which includes also the Rutherford cross section, N_A and $N_A(c)$ are the atomic density of A atoms in pure A and in the mixture respectively, ΔE is the energy width of the channel, $S]_A$ and $S]_A(c)$ are the backscattering energy loss parameters for A atoms in pure bulk sample and in the mixture respectively. In formulas

$$S]_A = N_A \left[k_A^2 \, \varepsilon_A \big|_{E_0} + (\cos\theta)^{-1} \, \varepsilon_A \big|_{k_A^2 E_0} \right]$$

$$S]_A^c = N_t(c) \left[k_A^2 \, \varepsilon(c) \big|_{E_0} + (\cos\theta)^{-1} \, \varepsilon(c) \big|_{k_A^2 E_0} \right],$$

k_A^2 is the billiard-ball kinematic factor, $N_t(c)$ is the atomic density of the mixture, θ is the laboratory backscattering angle, E_0 and $k_A^2 E_0$ are the incident and the backscattered particle energies respectively. From the ratio $Y_A(c) / Y_A$ one obtains

$$\frac{Y_A^c}{Y_A} = c \frac{\left[k_A^2 \, \varepsilon_A \big|_{E_0} + (\cos\theta)^{-1} \, \varepsilon_A \big|_{k_A^2 E_0} \right]}{\left[k_A^2 \varepsilon(c) \big|_{E_0} + (\cos\theta)^{-1} \varepsilon(c) \big|_{k_A^2 E_0} \right]}$$

The concentration c can be then obtained from the measured ratio of the yields; this procedure is shown in Fig.3a for Au in the Au-Al system.

b) Assuming again the additivity of the stopping cross section the mixture composition can be obtained by the ratio between the two yields $Y_A(c)$ and $Y_B(c)$ measured on the same sample at the same depth. In fact

$$Y_A(c) \propto N_A(c) \sigma_A \frac{\Delta E}{S]_A^c}; \quad Y_B(c) \propto N_B(c) \sigma_B \frac{\Delta E}{S]_B^c}$$

where σ_A and σ_B are the Rutherford cross sections for A and B atoms respectively. Then

$$\frac{N_A(c)}{N_B(c)} = \frac{c}{1-c} = \frac{Y_A^c}{Y_B^c} \frac{S]_A^c}{S]_B^c} \frac{\sigma_B}{\sigma_A}$$

The ratio $S]_A^c / S]_B^c$ is usually nearly independent of the composition; for the Au-Al system it ranges from 1.10 ($c_{Au} = 0$, $c_{Al} = 1$) to 1.13 ($c_{Au} = 1$, $c_{Al} = 0$). The results of this second procedure are shown in Fig.3b.

DISCUSSION

Q: (D. M. Mattox) Did you see any effect of Kirkendal diffusion and Kirkendal porosity on the diffusion measurements as diffusion progressed?

A: From backscattering measurements alone it is difficult to infer any porosity present in the films as diffusion progresses.

Q: (J. E. Westmoreland) Does the second phase go to completion (Fig. 6 has more of a slope than a step)? Also, can you determine the Kirkendal porosity by TEM?

A: The second phase goes to completion if sufficient aluminum is present and the slope of the plateau for the Au signal in the $AuA\ell_2$ compound as determined by the backscattering cross-section coincides with that of the virgin sample. The porosity produced by the Kirkendal effect can be seen by TEM. Preliminary measurements have been carried out and further work is in progress.

Q: (W. L. Brown) Do you think the second activation energy is indicative of a different species diffusing, or is it indicative of diffusion through a different lattice?

A: I do not know. Marker measurement could be helpful to clarify which species moves.

Q: (G. Dearnaley) Did you consider in the case of mixing in bimetallic thin films the consequences of a change in molar volume of the interdiffused layer, i.e., the analogue of the Pilling-Bedworth ratio in corrosion science? It seems to me that the resulting strain would have important influences on the migration of the mixing species.

A: I agree that strains could be relevant for the intermixing process. However Au and Aℓ have, within a few percent, the same interatomic distance and the same lattice structure.

THIN FILM INTERDIFFUSION OF CHROMIUM AND COPPER

J. E. E. **Bag**lin, V. Brusic, E. Alessandrini, J. Ziegler

IBM Thomas J. Watson Research Center

Yorktown Heights, New York 10598

INTRODUCTION

Motivated by the need to understand the behavior of the Cr-Cu system in applications requiring high temperature annealing, we have examined the interdiffusion of thin overlaid Cr and Cu films at temperatures up to 750°C.

Since no metallic compound phase of Cu-Cr is formed at these temperatures,[1] and the bulk solubilities of Cu in Cr and Cr in Cu are very small (<0.1 at. % and < 0.2 at. % respectively), we may expect this to be a case where bulk diffusion is negligibly small, and most interdiffusion observed must take place between individual grains.

PREPARATION OF FILMS

Films were made by evaporating a (1050 $\pm$ 50) $\overset{\circ}{A}$ layer of Cu on to a sapphire substrate at room temperature, followed promptly by a 1050 $\overset{\circ}{A}$ layer of Cr. Sapphire was chosen because of its freedom from interaction with Cu.

By use of oriented (normal to C-axis) and un-oriented substrate wafers, it was possible to obtain two sets of films -- one set epitaxially formed having single-crystal characteristics, the other being polycrystalline, (as determined by reflection electron diffraction tests.)

Film thicknesses were chosen to suit the 2.4 MeV [4]He-ion backscattering technique to be used for obtaining the metal

169

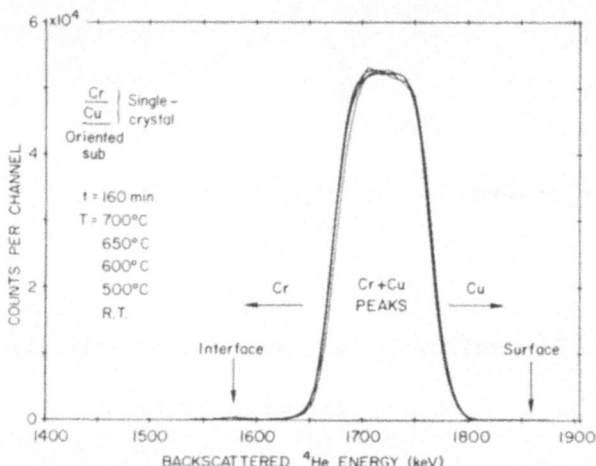

1. ^{4}He backscattering spectrum from Al_2O_3 /Cu(1050 Å)/
     Cr(1050 Å). Locations of surface and interface layers
     appearing after heat treatment are shown. System
     resolution is approx. 20 keV FWHM.

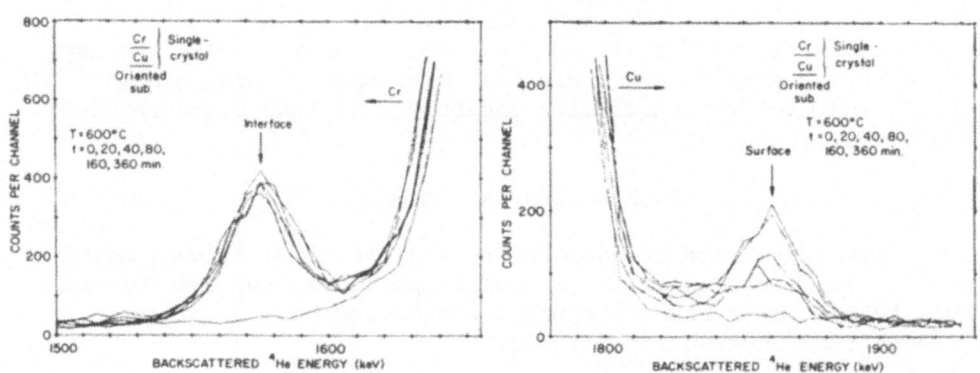

2.   Backscattering spectra from Cr/Cu (single-crystal) in the
     region of the Cr peak indicating an interface layer between
     Cu and substrate, and in the region of the surface Cu peak.

interdiffusion profiles after heat treatments. This choice ensured that the backscattering profiles from the as-deposited Cr and Cu layers coincided, thus clarifying the identification of migrated Cr or Cu. This is illustrated in Figure 1, where spectra from heat-treated samples show small extensions to the right (migrated Cu) and left (migrated Cr) of the main spectral peak.

HEAT TREATMENT

Both sets of substrates were heat treated in an ambient of He which was continuously passed through a LN_2 cold trap and a 900°C titanium getter, in order to remove oxygen from the system with high efficiency.

Annealing temperatures from 500°C to 750°C were chosen, and duration of actual heat treatments was stepped from 20 minutes to 420 minutes. Following heat treatment, samples were moved quickly to a cool part of the furnace and allowed to cool slowly for a period of several minutes. Each wafer was subjected to heat treatment once only.

BACKSCATTERING ANALYSIS

Following heat treatment of each sample, ^{4}He-backscattering at 170° (E = 2.4 MeV) was used to determine concentration profiles of Cr and Cu in the two film layers. Spectra were obtained using a 100 mm^2 surface-barrier Si detector (solid angle at target = 4.11 msr).

Absolute concentrations were deduced, based on the ^{4}He beam charge per run integrated from the target carrier (which was covered by a suppressor plate and ground plate forming a good Faraday cup).

SURFACE/INTERFACE LAYERS - SINGLE-CRYSTAL FILMS

Figure 1 shows spectra from the single-crystal samples which were treated at various temperatures for 160 minutes each. Typical portions of "single-crystal" spectra showing diffusion concentration profiles for samples annealed at 600°C for various times are expanded in Figure 2.

It is evident from these spectra that at 600°C, a 20-min. anneal causes Cr to migrate rapidly through the Cu film and form a stable layer of 4.5×10^{15} atom/cm^2 at the Cu-substrate

interface. This is presumably one or two monolayers, -- the
spectral peak width is entirely due to our system resolution.
Similarly, Cu migrates very rapidly to form a layer $\sim 1.5 \times 10^{15}$
atoms/cm^2 at the exposed sample surface.

The notable feature about both surface peaks is their
constancy after longer or higher-temperature annealing. At
600°C, saturation is reached after 80 min. by Cu, and it has
already been reached by the migrated Cr peak after only 20 min.

An additional feature of these spectra is the near-absence
of backscattering counts in the region representing the bulk of
each host film. No more than ~ 1 at. % of Cr is seen to lodge
in the body of the Cu film. For Cu, there does seem to be
~ 2 at. % deposition of Cu in the Cr layer. However, this turns
out to be an illusion, caused by pinholes in the Cr layer
exposing the virgin Cu film to about 2% of the incoming ^{4}He ions.
The holes have been observed by S.E.M. and by T.E.M. in surface
replicas made on heat-treated samples. The holes have faceted,
irregular shapes, and in S.E.M. photographs, they occupy a
constant 2% of the surface area independent of annealing
conditions. After subtracting the spurious counts in test spectra
arising from these gaps in the Cr film, only a surface peak
remains. We conclude that the amount of migrated Cu within the Cr
layer could not exceed 1 at. % in this case.

In the absence of data on shorter heat treatments, we may
set only a lower limit on the diffusivity of Cr in Cu needed in
order to produce the saturated interface layer so fast:
D(Cr in Cu, 500°C) $\geqslant 2.3 \times 10^{-13}$ cm^2/sec. Similarly,
D(Cu in Cr, 500°C) $\approx 2.3 \times 10^{-13}$ cm^2/sec. Presumably such
diffusion occurs along grain boundaries.

BULK ACCUMULATION - POLYCRYSTALLINE FILMS

Figures 3 and 4 show typical concentration profiles of Cr
and Cu in the polycrystalline samples following the heat
treatments indicated. Surface peaks still were identified at
low temperatures, saturating at identical levels to those of the
single-crystal samples. However, the striking feature of these
profiles is the massive interpenetration of the two films at
temperatures around 700°C.

In order to clarify the data for analysis, small surface-peak
profiles and no-anneal sample background have been subtracted
from the raw data, thus giving plots representing bulk concentration as a function of penetration distance from the source
layer interface. S.E.M. photographs of these films did not reveal

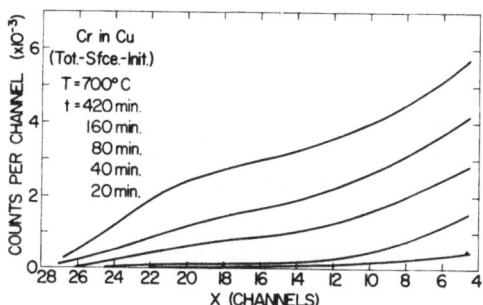

3. Concentration profiles of Cr diffused in Cu (poly-
    crystalline). Distance scale x originates at the edge of
    the main Cr peak. Interface peaks and no-anneal profile
    have been subtracted.

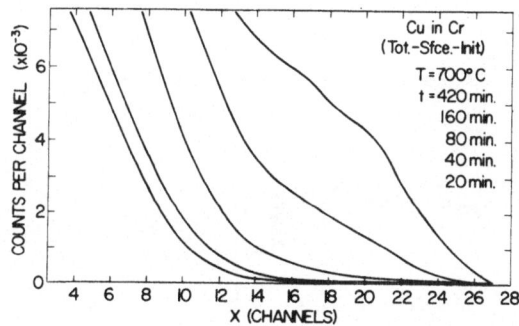

4. Concentration profiles of Cu diffused in Cr (polycrystalline).
    Distance scale x originates at edge of main Cu peak. Surface
    peaks and no-anneal profile have been subtracted. Separate
    [63]Cu and [65]Cu components are visible in the upper curve.

the faceted cavities in the Cr layer which appeared in the
single-crystal case.

A qualitative difference between these families of profiles
is immediately evident. It appears that the initial movement of Cr
through the Cu film provides a prompt "source" layer at the grain
boundary for further slower diffusion laterally into the Cu film.
This accounts for the fairly uniform concentration of Cr as a
function of depth in Cu. In contrast, the advancing Cu "edge"
seems to require a lateral diffusion of Cu whose rate at these
temperatures is comparable to that of the grain boundary diffusion
through the film, which supposedly supplies it.

DIFFUSIVITIES, ACTIVATION ENERGIES; THE DIFFUSION MECHANISM

For the polycrystalline case, a trial plot of log (counts
per channel, $C(x)$) as a function of (diffused depth, $x)^2$ indicates
immediately by its non-linearity for both Cu and Cr, that a simple
bulk diffusion model is inappropriate. In each case, however, plots
of log $C(x)$ vs. $x^{6/5}$ for a given anneal time proved to be linear
(Fig. 5) and we proceed to treat this as justification for
applying the form of Whipple's model[2] to our analysis.

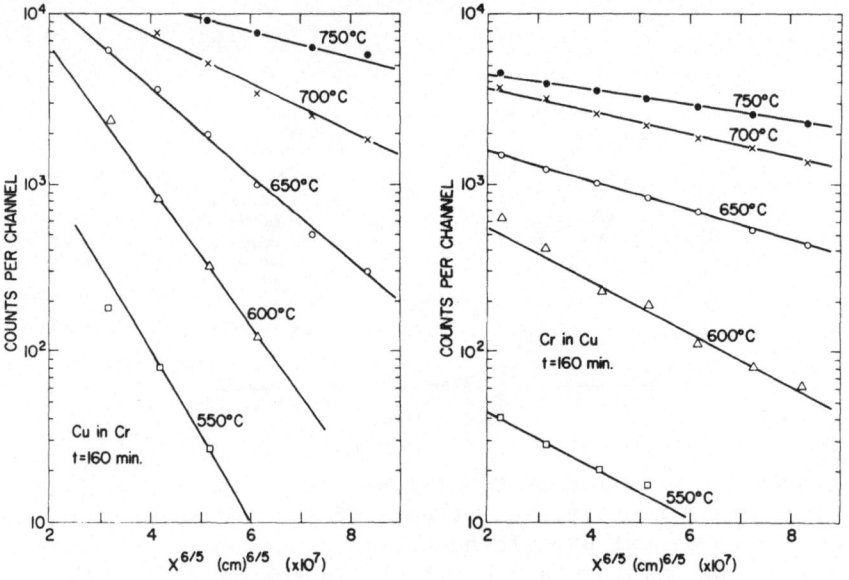

5.   Linear dependence of log (concentration of diffused Cr or Cu)
      upon $(x)^{6/5}$ for 160-minute anneal data.

The Whipple formalism treats the case of grain boundary diffusion through a film, coupled with a lattice diffusion whereby material within the boundary layer migrates laterally into the grains. Although lattice diffusion to form a Cr-Cu solution is negligible, we note that a similar formalism would be required to describe a simple lateral diffusion of any kind, such as the filling of grain boundaries which are not oriented normal to the film. Identifying Whipple's "D" simply as a "lateral" diffusivity D_ℓ, we proceed to evaluate parameters for such a process, using the Whipple result:[3]

$$\delta \cdot D_b = \left\{\frac{\partial (\ell n \ C)}{\partial (x^{6/5})}\right\}^{-5/3} \cdot \left\{\frac{4D_\ell}{t}\right\}^{1/2} \cdot (0.78)^{5/3} ,$$

(1)

where

D_b = diffusivity via the grain boundaries normal to the film surface

δ = mean width of those boundaries.

t = time of heat treatment

C = concentration of migrating species at depth x.

For δ we estimate a reasonable constant value of 5×10^{-8} cm, in order to assign a value to D_b. For D_ℓ, we use $D_\ell = D_\ell_o$. $\exp(- Q_\ell/RT)$, where Q_ℓ is the activation energy for bulk diffusion and $D_o = 0.20$ cm^2/sec. Q_ℓ for Cr in Cu has been observed by tracer techniques[4] to have the value (53.5 ± 5) Kcal/mole °C. In the absence of an independently measured value of Q_ℓ for Cu in Cr, we have used an estimate of 47 Kcal/mole °C, based on self-diffusion data. (In fact, our results and conclusions are not significantly affected by uncertainties in this value of Q_ℓ as large as $\pm$ 20%).

Substitutions of the above values in Eq. (1) led to the values of $\delta \cdot D_b$ shown in the Arrhenius plots for the two diffusion processes (Figure 6). The observed data points seem to fit the expected linear form of this plot excellently.

Activation energies for D_b derived from these data are: Q_b(Cu in Cr) = 46 Kcal/mole°C and Q_b(Cr in Cu) = 56 Kcal mole°C. By their surprising similarity to the activation energies for bulk migration, these results suggest that the "lateral" and "grain boundary" diffusion processes in this case have common origins -- for example, a lateral grain boundary diffusion coupled to a grain boundary diffusion normal to the film. Until firmer final evaluations are made, however, we must label this as "speculation".

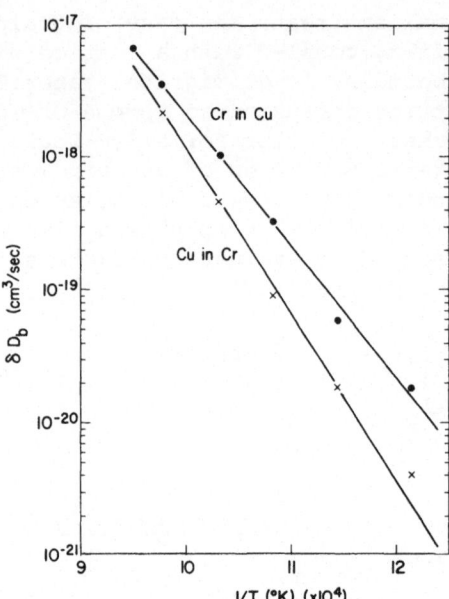

6.　Arrhenius plots for Cu in Cr and Cr in Cu.

The ratio D_ℓ/D_b may also be derived from this analysis.　For T = 650°C, for example, we obtain D_ℓ/D_b (Cr in Cu) = 0.1; D_ℓ/D_b (Cu in Cr) = 0.8.　This is consistent with our qualitative observations on the shapes of our diffusion profiles, where the transverse diffusion is much faster than lateral diffusion for Cr in Cu while the two diffusivities are comparable at 650°C for Cu in Cr.

No new mechanism is required to account for the initial surface (or interface) layer formation.　The grain boundary diffusivity values derived here for the migration within each layer satisfy our minimum D_b values needed to generate surface layers as rapidly as we observe.　For example, at 500°C, we derive

$$D_b \text{ (Cu in Cr)} = 4 \times 10^{-13} \text{ cm}^2/\text{sec}$$

$$D_b \text{ (Cr in Cu)} = 26 \times 10^{-13} \text{ cm}^2/\text{sec}$$

Surface peaks seen require (at 500°C)

$$D_b \text{ (Cu in Cr)} \approx 2.3 \times 10^{-13} \text{ cm}^2/\text{sec}$$

$$D_b \text{ (Cr in Cu)} \geq 2.3 \times 10^{-13} \text{ cm}^2/\text{sec}$$

We therefore conclude that the surface layers arise from the same grain-boundary diffusion that subsequently becomes the source for lateral diffusion in the polycrystalline samples.

GRAIN SIZE

In view of the very large inter-diffused concentrations reached in the polycrystalline samples, we must postulate substantial amounts of migrated material deposited between small grains. Replica micrographs of our samples show typical lateral grain dimension of $\sim$ 200 Å in the Cr layer before heat treatment. These grains appear to grow to $\sim$400-500 Å (with an occasional grain up to 2000 Å) following 700° heat treatment. Similarly deposited and treated Cu surfaces showed typical grain sizes before heat treatment of about 600 Å, and 500-1000 Å after heating. The relatively smaller grains of Cr (as deposited) might be expected since the substrate was not heated during evaporation of both films, and the melting point of Cr is substantially higher than that of Cu. At 750°C, inter-grain layers up to 120 Å thick would be needed to accommodate the observed diffused concentrations surrounding grains of the above sizes. This layer is remarkably thick, and further modeling of the diffusion mechanism in this case is clearly needed.

It is worth noting finally that in the Cr-Cu system we appear to have an interdiffusion process which is totally governed by the grain characteristics of each film. The lodgment of migrated material ranges from < 1% in the case of single-crystal films ("extremely large grains") to $\sim$ 25% in the case of $\sim$ 400 Å grains. The practical mechanism whereby such an inter-diffusion can be externally controlled may lie in control of grain size during the formation of the films.

ACKNOWLEDGMENTS

We wish to acknowledge with gratitude the valuable collaboration of K. N. Tu, D. Gupta, N. Chou, R. Hammer, W. Hammer and C. Aliotta.

REFERENCES

1.   M. Hansen, ed. "Constitution of Binary Alloys", McGraw-Hill, (New York), 1958; and R. Elliott, ed. (Supplement), 1965.

2.   R. T. Whipple, Phil. Mag. 45, 1225 (1954).

3.   D. Gupta, J. Appl. Phys. 44, 4455 (1973).

4.   G. Barreau, G. Brunel, G. Cizeron, Comptes Rendues Acad. Sci. (Paris) C272 618 (1971).

DISCUSSION

Q: (A. Hiraki) In your study of several temperature ranges, do
you assume that the sizes of the grains do not change as a function
of temperature?

A: The analysis does not explicitly involve grain dimensions, or
their form. Hence, our values of D_b do not depend on grain size
assumptions. However, if we are going to proceed to a more detailed model, I'm sure grain dimensions will need to be checked.

ION BACKSCATTERING STUDY OF WSi_2 LAYER GROWTH

IN SPUTTERED W CONTACTS ON SILICON[*]

J. A. Borders and J. N. Sweet

Sandia Laboratories

Albuquerque, New Mexico 87115

ABSTRACT

Helium ion backscattering has been used to study the reaction kinetics in samples composed of thin ($\sim$ 2200 Å) W films dc sputtered onto chemically cleaned P or B doped $\langle 111 \rangle$ single crystal Si substrates. At temperatures in the range 625 C to 750 C, Si is observed to migrate into the W films and form a well-defined WSi_2 reaction layer. Typically, at 650 C, about 750 Å of W is consumed in four hours. A detailed comparison of the backscattering spectra indicates that the WSi_2 reaction layer grows approximately as t^n where n $\approx$ 1 during the initial growth but then decreases to n $\approx \frac{1}{2}$ when the W films are over 30% reacted. An activation energy which characterizes the initial stage of WSi_2 layer growth is 63-71 kcal/mole for 35% W film reaction (2000 Å WSi_2 formed). This activation energy is somewhat higher than the 50 kcal/mole reported by other investigators for WSi_2 layer growth at temperatures over 850 C where layer growth was observed to be governed by $t^{\frac{1}{2}}$ kinetics. A preliminary estimate of the activation energy governing the beginning of the $t^{\frac{1}{2}}$ layer growth observed here is 60 ± 15 kcal/mole. Prior to complete reaction of W films aged in the temperature range 625 C to 730 C, an upper limit of 2 at.% Si is observed for the Si concentration in the unreacted portion of the W film.

---

[*]This work was supported by the U. S. Atomic Energy Commission.

INTRODUCTION

Tungsten has some unique advantages for use as a conductor and contact metallization for silicon devices. The Schottky barrier height between W and n-type Si is relatively low (0.65 eV)[1] and the thermal expansion coefficient for silicon is better matched by that of tungsten than by any other elemental metal. The silicon-tungsten phase diagram indicates that the lowest eutectic temperature in this binary system is 1400 C. These properties suggest the use of tungsten may be particularly advantageous in devices where high temperature operation or processing is necessary. If temperatures are too high, however, interdiffusion and reaction can result in formation of WSi_2 at the contact interface. This could affect the electrical and mechanical properties of the contact. For example, it was shown for tungsten contacts on silicon-germanium alloys[2] that tungsten silicide formation is accompanied by contact failure. In this paper we investigate the kinetics and energetics of WSi_2 formation in sputtered tungsten contacts on single crystal silicon using energetic ion backscattering as a tool to probe the depth distribution of atomic composition. From the time-temperature behavior of the depth distributions it is possible to extract information on the growth kinetics and the energetics which control the formation of a WSi_2 layer. In addition, we have been able to estimate an upper limit to the solubility of silicon in sputtered tungsten films.

EXPERIMENTAL TECHNIQUE

Unbiased dc sputtering was used to apply tungsten films to samples of ⟨111⟩ oriented wafers of boron-doped or phosphorus-doped silicon. Resistivities ranged from 0.003 Ω-cm to 30 Ω-cm. The slices were chemically cleaned and HF etched before sputtering, but a "native" oxide of 15-20 Å probably remained.[3] During the sputtering process, the Si substrate temperature was maintained near 350 C. The final tungsten film thicknesses were about 2200 Å. Between backscattering runs, the samples were annealed in an ion-pumped vacuum chamber with an ambient pressure of < 5 x 10^{-7} Torr.

The details of ion backscattering have been well documented and will not be repeated here. Suffice it to say that 1.83 MeV He^+ ions were used for all analyses and that scattering spectra were recorded at a laboratory scattering angle of 173° with a measured resolution of 11-13 keV FWHM. The beam spot was moved to a different position on the sample after each anneal to avoid any beam-enhanced effects.

EXPERIMENTAL RESULTS AND DATA ANALYSIS

The general features of the scattering spectra from an unannealed sample are shown in Fig. 1. The high energy edge of the scattering peak due to tungsten is resolution limited and the energy spread of the peak as measured between the half-height points of the high- and low-energy edges can be used with the known stopping cross section of helium ions in tungsten[4] to evaluate the tungsten film thickness of $\approx$ 2200 Å. The steep low energy edge of the tungsten peak and the high energy edge of the silicon spectrum indicate that there has been little or no interdiffusion during the sputtering process.

As the silicon and tungsten interdiffuse, the low energy edge of the tungsten peak and the silicon edge will move to lower and higher energies, respectively. Also, both of these edges will broaden. If an intermediate phase of uniform composition is formed in a well-defined layer, a "step" of relatively constant yield will be observed in the low energy edge of the tungsten peak. This has been discussed by Borders and Sweet,[2] and analyzed in more detail in the Appendix of a paper by Ziegler, et al.[5] The composition

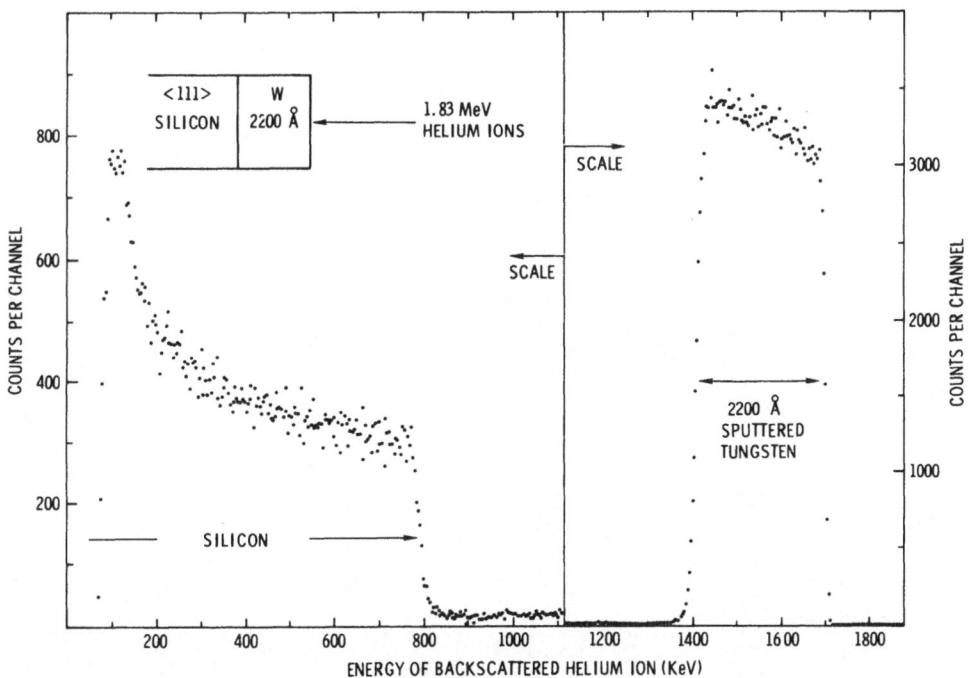

Fig. 1. Ion backscattering spectrum from an unannealed sample consisting of 2200 Å sputtered tungsten on single crystal ⟨111⟩ silicon.

information is all contained in the tungsten peak,[2,6] and the sili-
con/tungsten atomic ratio can be evaluated by measuring the scattering yields from tungsten in the portion of the peak corresponding
to pure tungsten and in the "step" portion where silicon contributes
to the stopping power. The energy width of the step in the tungsten-
scattering peak due to a layer of WSi_2 can be used to measure the
thickness of the WSi_2 layer. Prior to the formation of enough silicide for a well-defined step in the scattering spectrum, we can
still estimate the tungsten silicide thickness by the spreading of
the low energy edge of the tungsten peak.

In Fig. 2 are shown three spectra due to scattering from tungsten, after isothermal anneals at 675 C. The first anneal of 40
minutes caused about 850 Å of WSi_2 to form. At total annealing
times of 100 min and 160 min, WSi_2 layers of about 1600 Å and 2200 Å,
respectively, were observed. The steps in the spectra taken after
the second and third anneal are well defined, but the step in the
first spectrum is not. The edges of the step were sharp and the

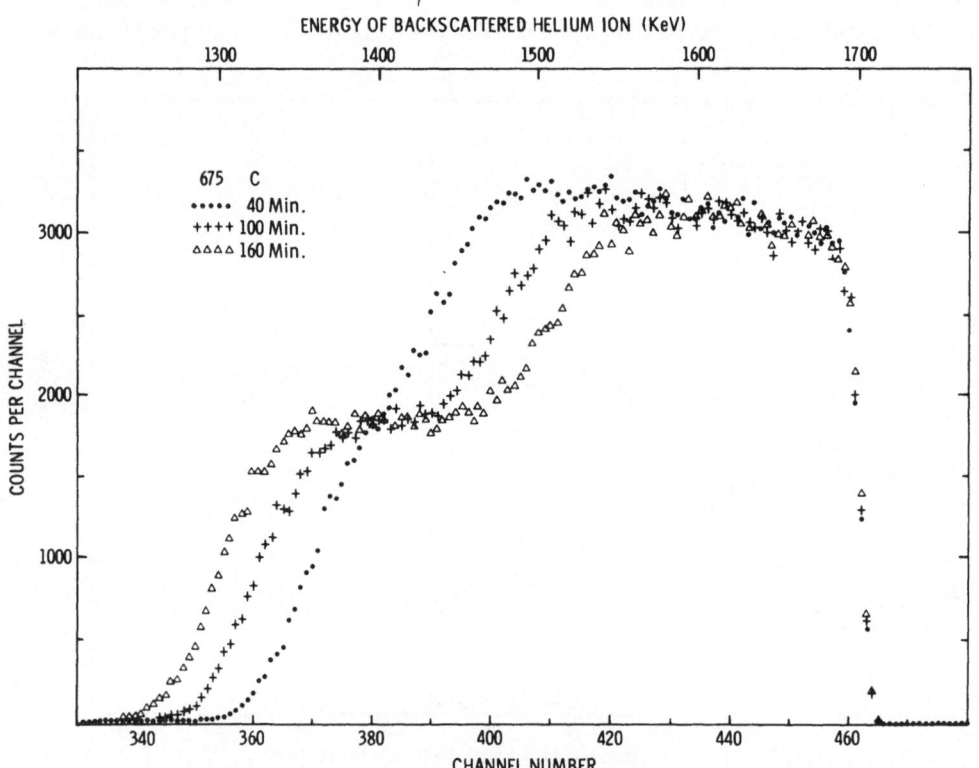

Fig. 2.   Portions of the scattering spectra due to scattering from
          tungsten atoms after 675 C anneals for 40, 100, and
          160 min.

scattering yield within the step was approximately constant. Some samples did not exhibit this behavior, but instead had a low energy edge of the tungsten scattering peak which varied monotonically from metallic W yield to zero yield with little sign of an inflection. We believe this is due to non-uniformity of WSi_2 formation over the beam analysis area.

Optical micrographs of two areas of the tungsten surface of a sample which had been annealed at 625 C for 1482 min are shown in Fig. 3. The backscattering spectrum for this sample indicated that about 20% of the W film had been converted to WSi_2. The photo on the left shows an area in which the reaction appears to have been relatively uniform with closely spaced protruding bumps 1-3 μ in size. The photo on the right of Fig. 3 shows another area of the same sample in which the reaction has been much less uniform. In the lower right-hand corner of the second photograph, the reaction is much more pronounced than it is in the upper left. Since the area of the ion-beam spot is about 80 x the area shown in each photograph, it can readily be seen that uneven reaction can contribute

20μ

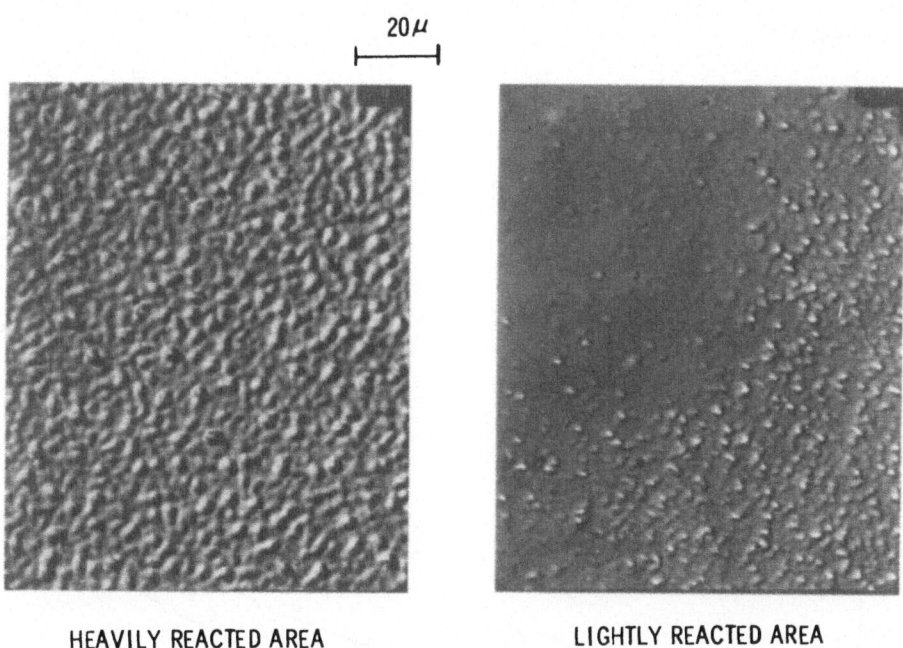

HEAVILY REACTED AREA   LIGHTLY REACTED AREA

Fig. 3.   Photomicrographs of two areas of a n-type Si:W sample annealed for 1482 min at 625 C. On the left is shown a heavily reacted area, and on the right is a lightly reacted area. The heavily reacted area shows very uniform reaction throughout the photograph area, whereas the lightly reacted area is very non-uniform.

to broadening of the backscattering spectrum edges and also to the
net uncertainty in WSi$_2$ thickness for a given annealing time and
temperature.

Because the beam spot was moved after each anneal, there is
quite a bit of scatter in the layer thickness versus time plots,
probably due to reaction non-uniformities across the sample, as discussed above. The general behavior is clear however. In Fig. 4
are shown the data on WSi$_2$ layer growth as a function of the square
root of time for five samples annealed at 625 C. If the layer
growth rate is controlled by planar interdiffusion of the reacting
species, the thickness of the reaction layer will be proportional
to the square root of time. As Fig. 4 clearly indicates, the layer
growth at 625 C is proportional to $t^{\frac{1}{2}}$ for times greater than about
625 min corresponding to the last three data points for all the samples in Fig. 4. At shorter times, the layer growth depends on a
higher power of time. Analysis of the short time data shows that
the initial growth is approximately linear in time. This general
behavior has been observed at temperatures ranging from 625 to 730 C.
Thus, the WSi$_2$ layer thickness initially exhibits a linear dependence
upon time, but after a definite thickness has formed, the reaction
rate gradually slows until a $t^{\frac{1}{2}}$ dependence is reached.

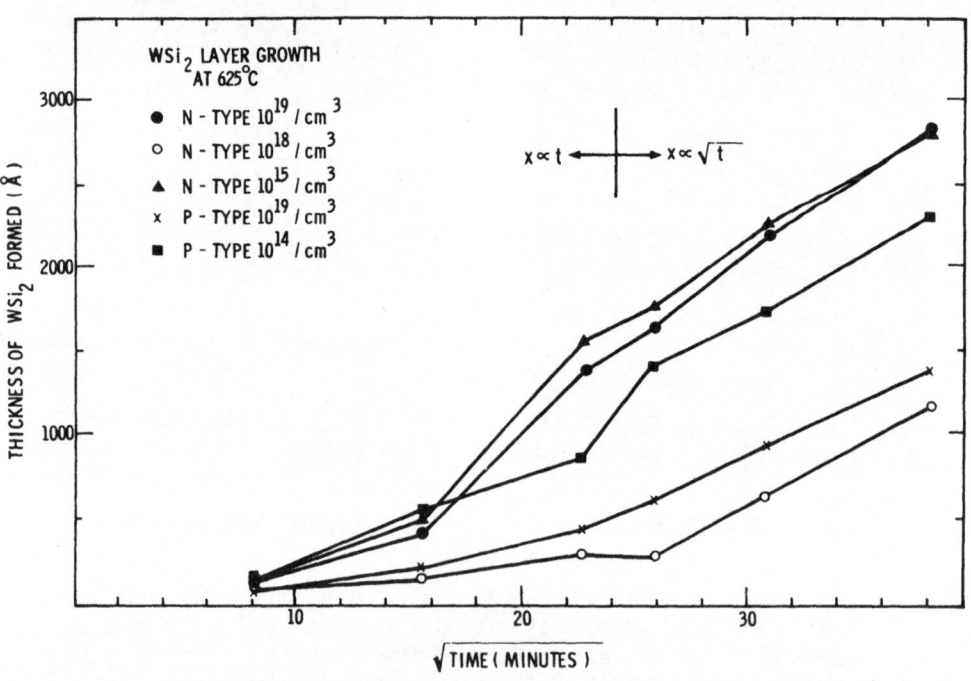

Fig. 4. Time dependence of the WSi$_2$ layer growth at 625 C.

Figure 4 indicates that three of the samples reacted at approximately the same rate, but that two samples, the $10^{19}/cm^3$ p-type and the $10^{18}/cm^3$ n-type, had a much smaller layer of WSi$_2$ formed than the other three samples for a given annealing time. Similar behavior was found at the other annealing temperatures. The samples in which this inhibited silicide growth occurred were always the more-heavily doped samples, although samples cut from the same wafer did not always behave similarly. As discussed later, we believe this effect is probably due to residual oxide at the W:Si interface acting as a diffusion barrier.

From the temperature dependence of the data we can determine the activation energy of the WSi$_2$ layer formation. To do this we have assumed that we can characterize the time to formation of a given thickness layer by an equation of the form,

$$t = t_o \, e^{Q_A/RT} \, , \qquad\qquad (1)$$

where Q_A = the "activation energy", R = the universal gas constant, and T is the absolute temperature. This is equivalent to assuming the layer growth rate is governed by a thermally activated process with one activation energy dominant throughout the temperature range of interest. We have, somewhat arbitrarily, chosen to fit the data to Eq. (1) at a time corresponding to the point at which 2000 Å of WSi$_2$ had been formed. This corresponds to 750 Å of W consumed, or about 35% reaction of the original W film. In fitting the data, we disregarded all the samples which seemed to react much more slowly than the majority of samples at a given temperature. The total spread in the time to form 2000 Å of WSi$_2$ for the remaining samples is shown plotted against $1000/T(K)$ on an Arrhenius plot in Fig. 5 for the test temperatures of 625, 675, 700, and 730 C. The three highest temperature points fall nicely on a straight line (solid) with a slope of 63 ± 1 kcal/mole and t_o = 2.42 x 10^{-13} min. A fit to all four points (dashed line) gives a slope of 71 ± 5 kcal/mole, and t_o = 4.42 x 10^{-15} min.

After a WSi$_2$ layer has formed, the backscattering spectra can be used to obtain additional information on the silicon-sputtered tungsten system. An estimate can be made of the amount of silicon dissolved in the unreacted metallic tungsten between the WSi$_2$ layer and the surface. The scattering yield in the energy region corresponding to silicon between the tungsten surface and the front of the WSi$_2$ layer (this is the energy region near 1000 keV such as shown in Fig. 1 for an unreacted sample) shows no change to within the accuracy of our measurements. From these results we estimate that no more than 2 at.% silicon is dissolved in the tungsten film at temperatures in the range 625 to 730 C. To our knowledge this upper limit on the solubility is the first information on the solubility of Si in W at these temperatures.

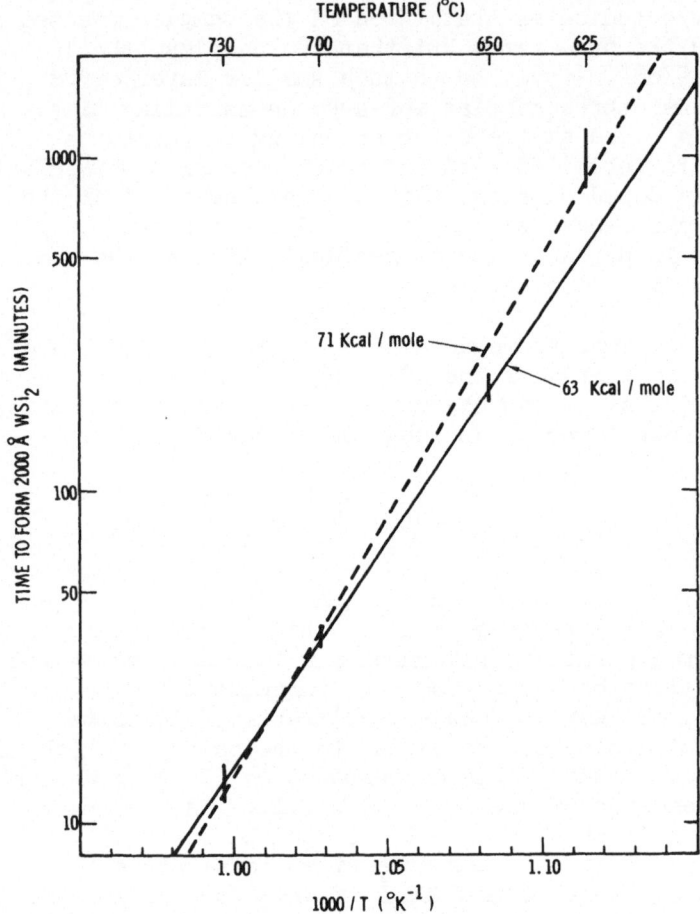

Fig. 5. Arrhenius plot of time to form a 2000 Å layer of WSi$_2$ vs. 1000/T, where T is the absolute temperature. The solid line is a least-mean-square error fit to the three highest temperature points. The dashed line is a fit to all four points.

DISCUSSION

The kinetics of silicide layer growth enables us to gain some insight into the mechanisms controlling the growth of the WSi$_2$ layer. A number of metal-silicon systems have already been studied by backscattering,[7] and for most of these systems the layer growth obeys $t^{\frac{1}{2}}$ kinetics characteristic of a diffusion dominated process. In two cases, molybdenum-silicon and chromium-silicon, growth of the silicide layer was found to be linear in time.[8] This was interpreted as being consistent with a reaction-rate-limited growth mechanism. The results reported here are the first to show both types of behavior in the same system.

The kinetic data obtained in this experiment suggest that different mechanisms control the rate of layer growth for very thin reaction thicknesses than for reaction thicknesses on the order of 200 Å and larger. The $t^{\frac{1}{2}}$ behavior observed for fairly thick reaction layers can be attributed to the reaction rate being controlled by planar diffusion of one of the constituents through the previously formed layer of WSi_2.

In similar experiments on the W:SiGe system,[2] it was determined that intermixing of the silicon and tungsten atoms occurred upon annealing, but that no intermixing of the tungsten and germanium was observed. In effect the 20 at.% germanium content of the SiGe alloy can be considered a marker. We do not expect the absence of germanium to affect diffusion in the WSi_2 since there is no Ge in the WSi_2. We therefore suggest that silicon is the mobile species in the W:Si system as well as the W:SiGe system. Thus, the $t^{\frac{1}{2}}$ behavior in W:Si couples is attributed to diffusion of silicon through the already formed layer of WSi_2, to the W:WSi₂ interface.

The linear portion of the growth curves observed for thin WSi_2 layers is more difficult to explain on the basis of kinetics alone. It may be due to a reaction-rate-limited growth or perhaps it characterizes localized penetration of the native oxide barrier and subsequent nucleation of WSi_2 regions. At any rate it indicates that the kinetics governing the formation of very thin layers of WSi_2 are not controlled by planar interdiffusion with constant atomic concentrations at the phase boundaries.

In order to compare activation energies for a process, one must be sure that the comparison is appropriate. There has been much confusion over this point, particularly when referring to diffusion-controlled processes. The growth law for a reaction zone of thickness x in a process governed by interdiffusion is,

$$x = k_1 t^{\frac{1}{2}} , \qquad (2)$$

where k_1 is a rate constant which will be of the form $k_1 = k_{1_0} e^{-Q_1/RT}$, if a single activation energy dominates the interdiffusion process.[9] The "activation energy" Q_1 is that which would be obtained by plotting parabolic growth constants on an Arrhenius plot. Another way of writing Eq. (2) is

$$x = (k_2 t)^{\frac{1}{2}} . \qquad (3)$$

This rate constant has the same form, $k_2 = k_{2_0} e^{-Q_2/RT}$, but simple algebra shows us that

$$Q_1 = Q_2/2 . \qquad (4)$$

If we plot the time to x = constant on an Arrhenius plot, the activation energy we will derive is Q_2. Note that k_2 has the units of
$[\ell]^2/[t]$, and is the appropriate activation energy to compare with
diffusion results. Activation energies of the type Q_1 have frequently[10-12] been reported in the literature on WSi_2 layer growth.
We have converted these activation energies to the type $Q_2 = 2Q_1$ in
order to compare them with our results and with other activation
energies reported in the literature.

Gage and Bartlett[10] and Zmii and Seryugina[11] observed $t^{\frac{1}{2}}$ layer
growth in bulk diffusion couples in the temperature range 855 to
1350 C. Both groups determined an activation energy of approximately
50 kcal/mole (2.2 eV/atom W) for the growth rate constant.
Hashimoto[12] studied WSi_2 layer growth in a ⟨111⟩ Si: chemically
vapor-deposited tungsten system in a temperature range 1040 to
1330 C and observed $t^{\frac{1}{2}}$ layer growth rate characterized by an activation energy of 44 kcal/mole (1.9 eV/atom W). Sinha and Smith[13] have
recently investigated WSi_2 layer growth in the ⟨100⟩ Si:PtSi(900 Å):
W (sputtered, 2000 Å) system in the temperature range 690 to 840 C.
They observe a very high activation energy, 102 kcal/mole (4.4 eV/
atom W), for the time at which 50% of the W film has reacted but the
layer growth did not go as $t^{\frac{1}{2}}$. This high activation energy is probably associated with the presence of the PtSi layer. Sinha and Smith[13]
quote unpublished work by Locker and Capio in which an activation
energy of 65 to 74 kcal/mole has been found for reaction in the ⟨100⟩
Si:W system but no temperature range is quoted for this experiment,
nor do they state which stage of layer growth this energy
characterizes.

Our activation energy range 63 to 71 kcal/mole for the time to
form 2000 Å of WSi_2 is somewhat larger than the range 44 to 50 kcal/
mole measured for activation energies characterizing $t^{\frac{1}{2}}$ growth at
higher temperatures. However, our method of determining an activation energy heavily weights the initial stages of layer growth when
$t^{\frac{1}{2}}$ growth has not been established. We have made a very preliminary
estimate of the activation energy characterizing the beginning of
the $t^{\frac{1}{2}}$ growth region by fitting parabolic growth rate constants to
the data points which fell in the $t^{\frac{1}{2}}$ growth regime. For example,
the last three data points for each sample shown in Fig. 4 were used
to determine the 625 C growth rate constant. Using data from 625,
650, 675, and 700 C, a preliminary activation energy of 60 ± 15 kcal/
mole was obtained. Since the kinetic data are from the beginning
of the $t^{\frac{1}{2}}$ growth regime, it is not unreasonable to expect that the
linear growth mechanism may still affect the measurements. Although
the errors on this activation energy are large, it appears to fall
between the activation energy for the linear growth region and that
found for $t^{\frac{1}{2}}$ growth where the effect of the linear growth mechanism
is negligible. The more-slowly reacting samples are somewhat puzzl-
ing. These samples are always highly-doped samples (10^{18} to 10^{19}/
cm^3), but other highly-doped samples react as rapidly as the low-

doped samples. We now believe that the slowly reacting samples have a larger "native" oxide and that this oxide barrier must be broken before uniform reaction can take place. An ellipsometric and Auger spectroscopic study of Si thermal oxidation by Chang, et al.[14] indicates that heavily doped Si oxidizes more rapidly than lightly doped Si.

Kinetic data of the sort we have generated are sometimes fitted to a phenomenological rate equation of the form,[13,15]

$$\Delta V_W(t) = 1 - \exp[-J(T)t^n] \quad , \qquad (5)$$

although it is somewhat difficult to interpret what the fitting value of n signifies.[15] In Eq. (5), ΔV_W = volume fraction of W transformed, at time t and $J(T)$ = temperature dependent "rate" constant. In the case where $J(T)t^n \ll 1$, Eq. (5) reduces to

$$\Delta V_W \approx J(T)t^n \quad , \qquad (6)$$

and hence n = power of the growth law in this case. Thus, if $n = \frac{1}{2}$, the process is probably controlled by the planar interdiffusion of reacting species.[9] Other rational values of n correspond to other types of diffusion and growth processes as discussed by Christian.[15] Experimental data are usually fitted to Eq. (5) by rewriting it as,

$$\ln\{\ln[1/(1 - \Delta V_W)]\} = n \ln(t) + \ln(J(T)) \quad . \qquad (7)$$

Sinha and Smith[13] found values of n in the range, n = 1.27 to 1.40, for their studies of WSi₂ growth in the W:PtSi: ⟨100⟩ Si system. Our data can be fit to Eq. (7) only over limited regions of ΔV_W. For the 625 C data we find that n = 1.0 ± 0.4 by least squares fitting using all the data points, but n ≈ 0.65 ± 0.05 using only the four data points for the largest values of t. This is in the time range where a $t^{\frac{1}{2}}$ growth rate was observed. In the 650 C test where all the data is for $\Delta V_W \geq 0.1$ we find n = 0.85 ± 0.15 with a definite nonlinear dependence of $\ln\{\ln[1/(1 - \Delta V_W)]\}$ on $\ln(t)$. We conclude that Eq. (5) (or Eq. (7)) is not a good way to parameterize our data except in a very limited reaction range.

For the early stages of layer growth values of n > 1 in Eq. (7) are appropriate while the later stages can be characterized by n < 1. This is consistent with our hypothesis that the initial stage of layer growth is dominated by a nucleation, reaction rate, or localized oxide breakdown process followed by a later stage of planar diffusion. Christian[15] has shown that values of n in the range $1 \leq n \leq 2.5$ are characteristic of initial nucleation or diffusion controlled growth of irregular shapes.

CONCLUSIONS

A preliminary study of WSi_2 formation in the sputtered W: single crystal silicon system indicates that WSi_2 formation cannot be explained simply on the basis of a reaction-rate limited or a diffusion limited growth process alone. However, the experimental data indicate that the latter stages of growth are probably diffusion controlled. An activation energy for the time to form 2000 Å WSi_2 has been found which is somewhat higher than that determined for $t^{\frac{1}{2}}$ growth at higher temperatures.

An upper limit of 2 at.% has been established for the solubility of Si in the unreacted portion of a partially reacted sputtered W film for temperatures of 625 to 730 C. Work is underway to establish the mobile species, determine any type dependences, and characterize the effect of oxide barriers at the W:Si interface on the WSi_2 reaction.

REFERENCES

1. C. R. Crowell, J. C. Sarace, and S. M. Sze, Trans. Met. Soc. AIME 233, 478 (1965).

2. J. A. Borders and J. N. Sweet, J. Appl. Phys 43, 3803 (1972).

3. F. Lukes, Surface Sci. 30, 91 (1972).

4. J. A. Borders (to be published) and J. A. Borders, Bull. Am. Phys. Soc. 17, 682 (1972).

5. J. F. Ziegler, J. W. Mayer, C. J. Kircher, and K. N. Tu, J. Appl. Phys. 44, 3851 (1973).

6. D. K. Brice, Proc. Intl. Conf. on Ion Beam Surface Layer Analysis, Thin Solid Films (to be published).

7. J. W. Mayer and K. N. Tu, J. Vac. Sci. Technol. (to be published).

8. R. W. Bower and J. W. Mayer, Appl. Phys. Lett. 20, 359 (1972).

9. G. V. Kidson, J. Nucl. Mater. 3, 21 (1961).

10. P. R. Gage and R. W. Bartlett, Trans. Met. Soc. AIME 233, 832 (1965).

11. V. I. Zmii and A. S. Seryugina, Zashch. Pokryt. Metal 2, 195 (1968); English translation, Prot. Coatings on Metals 2, 158 (1970).

12. N. Hashimoto, Trans. Met. Soc. AIME 239, 1109 (1967).

13. A. K. Sinha and T. E. Smith, J. Appl. Phys. 44, 3465 (1973).

14. C. C. Chang, P. Petroff, G. Quintana and J. Sosniak, Surface Sci. $\underline{38}$, 341 (1973).

15. J. W. Christian, The Theory of Transformation of Metals and Alloys (Pergamon Press, London, 1965), pp. 16-22 and 471-495.

16. Ref. 12, Table IX, p. 489.

DISCUSSION

Q: (W. L. Brown) If you extract the energetics at different re-action zone thicknesses do you get the same activation energy? It would be nice to determine if for t or $t^{\frac{1}{2}}$ dependencies the same activation energy is obtained.

A: This is a preliminary report and our data as yet are rather limited. We had more data at WSi₂ thicknesses of 2000 Å than any other reaction layer thickness. The activation energies deduced using 1500 Å and 2500 Å are not inconsistent with the results at 2000 Å but are subject to much larger errors.

REACTIONS OF THIN METAL FILMS WITH

Si OR SiO$_2$ SUBSTRATES

H. Kräutle, W.K. Chu, M-A. Nicolet, J.W. Mayer

California Institute of Technology

Pasadena, California   91109

and

K.N. Tu

IBM-Thomas J. Watson Research Center

Yorktown Heights, New York   10598

ABSTRACT

The changes in vacuum-evaporated films of Ti, V and Nb on Si and SiO$_2$ substrates after thermal anneals are investigated by backscattering and by x-ray spectrometry. Backscattering analysis provides the relative atomic composition as a function of depth with high sensitivity. Glancing angle x-ray spectrometry detects the chemical composition with high specificity. Combined, the two methods create a specific picture of the transformations induced in the films by the thermal treatment. Generally the reaction on a pure Si substrate produces a Si-rich silicide, and on a SiO$_2$ substrate a metal-rich silicide in the form of an intermediate layer largely free of oxygen. The oxygen originally bound to the Si in the SiO$_2$ is transferred to the remaining metal layer. Residual oxygen in the metal film and metal oxides on the metal film influence the silicide formation.

I.   INTRODUCTION

Transition metals, and Ti in particular, are used extensively in metallization schemes for integrated circuits and solar cells. Good adhesion and uniformity of the resulting contact are main reasons for their frequent application. The latter aspect has recently been investigated in connection with the Al-Ti metallization scheme.[1] The possibility to form superconducting silicides of transition metals by thin film reactions with the substrate has prompted Tu et al. to investigate the behavior of V on Si and SiO_2 substrates.[2] The strong affinity of Ti and V for oxygen is believed to be the reason why these metals make good blocking contacts to p-type semiconductor oxides and good ohmic contacts to n-type oxides.[3] Niobium (Columbium) maintains high mechanical strengths at elevated temperatures, but oxidizes readily unless protected by a coating. The disilicide constitutes such a protective layer, based on the ability of the coating to generate a silica-based glassy oxide as protection against the atmosphere.[4]

Preliminary data indicate that the heat treatment of these metal films on Si results in the formation of silicides, but that the same metal films deposited on SiO_2 forms both silicides and oxides, generally separated in distinct layers.[5] We present here a systematic study of samples annealed in both vacuum and in dry oxygen. Comparative investigations such as these should lead to insight in the dominant processes at work in these thin films, and to a better understanding of their applications.

II.   METHODS AND ANALYSES

A. Backscattering Spectrometry

In this study, the depth and mass perception of back-scattering analysis are used to determine the relative atomic composition of the films. The backscattering technique and the method of analysis are described in detail elsewhere.[6] In brief, a monoenergetic and well collimated beam of $^4He^+$ ions with energy ranging from 2.0 to 2.4 MeV impinges perpendicularly onto the sample. A small fraction of these ions is scattered back into a surface barrier Si detector which is mounted at an angle of 168° against the incident beam. The detector signals are amplified, shaped, and recorded in a multi-channel analyzer. The resulting energy spectrum (counts per energy interval versus energy) furnishes the information of atomic composition versus depth.

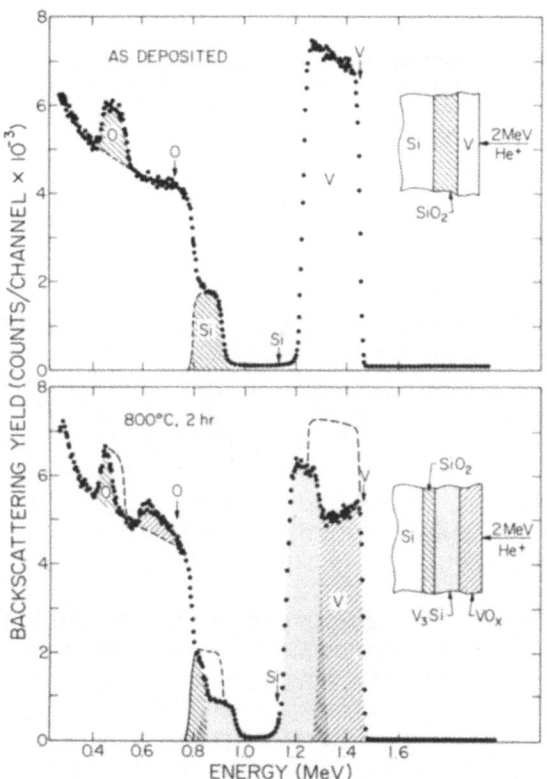

Figure 1.   Spectrum of 2 MeV He$^+$ backscattered from a V film on
an SiO$_2$ layer on Si as evaporated (top), and after
heat treatment in vacuum at 800°C for 2 hours (bottom).

As an example, Fig. 1 (top) shows an energy spectrum of 2 MeV ^{4}He$^+$ backscattered from 2100Å of vanadium deposited on 2700Å of thermally grown SiO$_2$ on a Si substrate. Helium ions backscattered from oxygen, silicon and vanadium atoms in the sample have energies which depend on the mass of the scattering atom and its location in depth. After the same sample   is annealed in vacuum at 800°C for two hours, the backscattering analysis is repeated and the energy spectrum given in Fig. 1(bottom) is obtained. For ease of interpretation, various areas under the energy spectra which correspond to different layers have been shaded (see also insert in the figures).   The two spectra clearly differ, as can be seen in Fig. 1 (bottom) where for comparison the spectrum of Fig.1(top) is replotted as a dashed line.   The formation of two distinct layers

(V_3Si and VO_x) and the remaining part of the SiO_2 is readily distinguished.

One can convert a backscattering energy spectrum into an atomic concentration profile with the energy loss factor $[S] = \Delta E/\Delta x$ which relates a change ΔE in backscattering energy to a change Δx in the depth of the sample.[6] The [S]-factor depends, among others, on the energy of the incident particle and the composition of the layer. The energy dependence is slow and for thin films, [S] can be assumed to be a constant. To a good approximation, the ratio of the [S] factors for the various elements in a film are also independent of composition. By measuring the ratio of the signal height generated in a spectrum by the various elements of a film, one therefore obtains the elemental concentration ratios of this unknown film. A correction factor must be applied to account for the different (but known) Rutherford scattering cross sections of elements. Alternatively, concentration ratios can also be obtained by comparing the signal heights of a element in two regions of a spectrum corresponding to layers of known and of unknown composition.[6]

All backscattering spectra reported here are presented after such a conversion to concentration profiles (see Fig. 3 to 9). In these profiles, the actual data points are provided at the interfaces only when clear evidence for non-abrupt interface was present. In all other cases, it was presumed that the details in the backscattering spectrum at the interface reflected the finite resolution of the system (detector resolution, energy straggling of the beam) rather than the steepness of concentration variations. The position of these interfaces is therefore indicated by sharp vertical lines only.

B. Seemann-Bohlin X-Ray Diffractometry

A limitation of backscattering spectrometry resides in the fact that the results give atomic concentrations only, without reference to the chemical constitution of the sample. X-ray diffraction analysis can readily identify chemical composition, and thus constitutes an excellent complementary technique to backscattering analysis. Because of the more limited sensitivity of x-ray diffraction, it is advantageous to lengthen the path of x-rays in the thin film by reducing the incident angle. For example, an incidence angle of 5° will increase the path length of the beam in the specimen to about 12 times its thickness. In the focusing Seemann-Bohlin arrangement the specimen is placed on the circumference of the diffraction circle. The angle of incidence is fixed and can be made as small as a few degrees. Based on this principle, R. Feder and B.S. Berry have designed and built a 20 inch diameter x-ray diffractometer specifically for thin film studies.[7] The diffractometer employs a pyrolytic

graphite monochromater crystal to obtain a high intensity mono-
chromatic Cuka radiation, which is incident upon the specimen
at an angle of 6.4°.

This diffractometer possesses sufficient sensitivity
to resolve the first five diffraction peaks of a copper film
of only 150Å thickness using a scan in steps of 0.2° (4θ) and a
counting time of 75 sec/step. The broadening exhibited by the
diffraction profile is consistent with a particle size of less
than 100Å. The diffractometer holds great promise in the investigation of the kinetics of reaction in thin films, because
the reaction products can be identified by their x-ray reflections very early in the diffusion cycle, and the rate of[8]
change can be measured by changes of their peak intensities.
The capability of the diffractometer to determine the lattice
parameter very precisely also allows the measurement of stress
in a thin film. The measured lattice parameter of a well annealed Ni powder sample was in excellent agreement with the value,
a = 3.5283Å (26°C) given in the ASTM file. Using the powder
sample as the reference, the stress state of a number of 1000Å
Ni films on glass substrates was determined and found to agree[7]
with the results obtained by the bending cantilever technique.

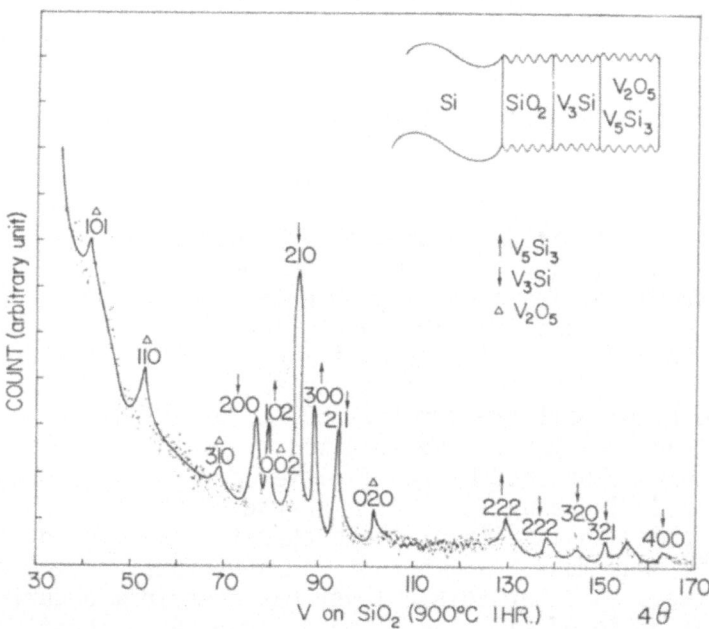

Figure 2.   X-ray diffraction pattern of the sample of Fig. 1 after
            heat treatment. Reflections of V_3Si, V_5Si_3 and V_2O_5
            are observed.

We have used the diffractometer to study the reaction
products of our transition metal films. The x-ray analyses were
performed partly on samples identical with those measured by
backscattering, partly on samples prepared and annealed under
closely similar conditions. For best results, samples of several
cm^2 area are required for x-ray measurements, which is typically 10 times the size needed for backscattering analysis.
Figure 2 shows a typical x-ray spectrum for a sample of 2000Å of
V on an oxidized Si wafer (5000Å of SiO_2) after anneal at 900°C
for 60 minutes in vacuum. The spectrum was obtained by scanning
the sample at steps of 0.15° (4θ) increment and with a counting
time of 30 sec/step. The peaks have been indexed as reflection
of VSi_3, V_5Si_3 and V_2O_5.

C. Sample Preparation

For the experiments on Si substrates, commercially
polished wafers of either <100> or <111> orientation and of
usual n- and p-type doping levels were employed. Orientation
and doping have no detectable influence on the results. Immediately before vacuum evaporation of the metal film, the wafers
were etched in a solution of 2 HNO_3:1 HF:2 acetic acid and
rinsed in deionized water. The SiO_2 substrates were obtained
by thermally oxidizing polished Si wafers at 1100°C in a atmosphere of wet oxygen. All metal films were deposited by vacuum
evaporation.

III. REACTIONS WITH Si AND SiO_2 SUBSTRATES

A. Titanium

The result of the reaction of Ti with a Si substrate
can be seen in Fig. 3. The zero in the depth scale is placed
at an arbitrary reference point within the substrate, and positions
between this point and the surface are indicated with thicknesses greater than zero. Figure 3 (top) shows 2700Å of Ti as evaporated on a Si substrate. The bottom part of this figure shows
this sample after heat treatment at 600°C for 20 min in a vacuum
of better than 10^{-5} Torr. The uppermost layer now contains
oxygen which was absorbed by the Ti layer during the storage in
air and also during annealing. The spread of the points exhibits the poor sensitivity of the backscattering method for
elements of lower mass than that of the substrate. Below the Ti
on the surface, an intermediate layer has been formed during
anneal. The ratio of Ti to Si in that layer can be easily extracted from this graph to be 1:2. The chemical structure has

been clearly identified by x-ray analysis as $TiSi_2$. By comparing these two figures, one observes that the Si concentration in the silicide layer is nearly as high as the concentration in the pure substrate. This, shows directly that the density of atoms in this compound is higher than that of Si and the total thickness of the system decreases due to the change in densities after compound formation.

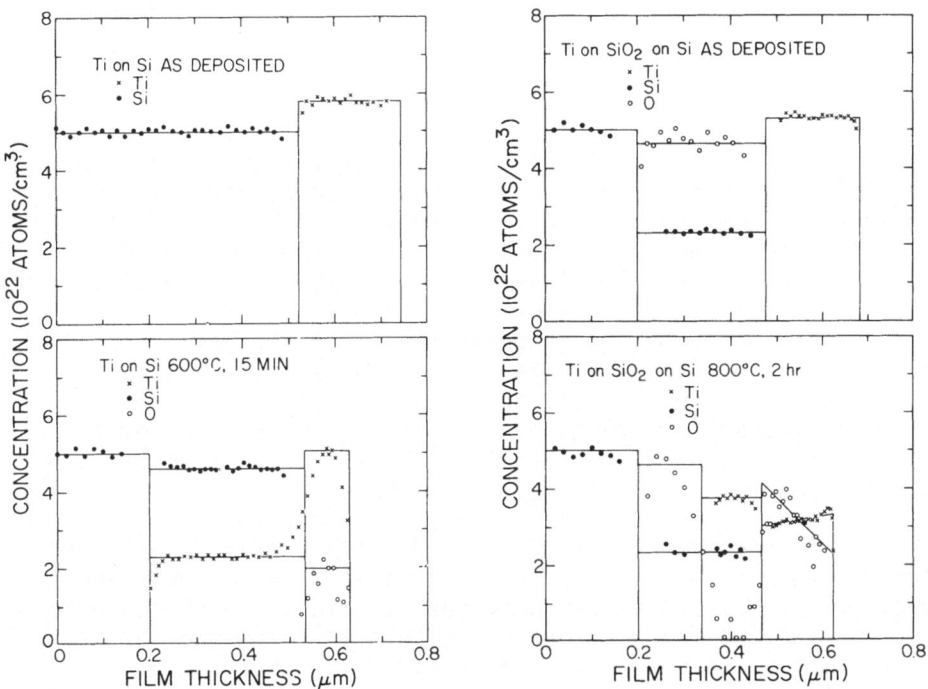

Figure 3. (left) Concentration profile of a Ti film on Si as evaporated (top), and after heat treatment in vacuum at 600°C for 15 minutes (bottom).

Figure 4. (right) Concentration profile of a Ti film as evaporated on an oxidized Si substrate (top), and after heat treatment in vacuum at 800°C for 2 hours (bottom).

The interaction of a vacuum-evaporated Ti film with an amorphous SiO_2 substrate is more complicated, since the metal reacts with a compound. A good example of the complexity of this system is shown in Fig. 4. The top figure gives the concentration profile of a sample after deposition of the metal film, and the figure below shows what happens to that sample after heat treatment at 800°C for 2 hours in vacuum. The surface layer, which was pure Ti before annealing, contains about 50% oxygen after annealing. The layer beneath consist mostly of Si and Ti in a ratio Si:Ti = 0.62. A small amount of oxygen is present also. Its concentration is difficult to evaluate by the backscattering method. Within an accuracy of about 10% the spectrum tells that the amount of oxygen in the surface layer is equal to that originally present in the SiO_2 which reacted with Ti to form a silicide. X-ray analysis could not determine the composition of this Ti-O layer.

B. Vanadium

Far clearer is the interaction of V films with Si and SiO_2. In Fig. 5 the reaction of V with pure Si can be seen. The intermediate layer which grows between the Si substrate and the V film shows two well delineated interfaces in the backscattering spectrum. There is no detectable oxidation of V from the air at room temperature or during annealing in our vacuum system. The backscattering spectrum indicates a ratio of V:Si = 1:2 in the intermediate layer. The x-ray analysis substantiates the result. Only one compound VSi_2 is formed over the whole temperature range from 500°C to 1000°C. In addition, the relative intensity ratios of the x-ray diffraction pattern indicate no preferred orientation in the silicide.

At temperatures of 700°C and above V films react with SiO_2, as can be seen in Fig. 6. The various layers observed after the reaction are very well resolved in the backscattering spectra (see Fig. 1). According to this spectrum the surface layer after reaction consists of V and O in the ratio of V:O about 1:1. According to the x-ray data, however, the layer contains V_2O_5 and V_5Si_3. No measurable amount of Si has been found in this layer by Rutherford backscattering analysis. The intermediate layer contains only V and Si. The composition is V:Si = 3:1 from backscattering data and is identified as pure V_3Si by x-ray analysis. The conservation of the total oxygen content in the system is verified within an accuracy better than 10% by comparing the total oxygen amount in layers before and after heat treatment, or by measuring the position of the rear edge of the oxygen signal of the SiO_2 layer. This edge must move if any substance evaporates on or off the surface, or diffuses in or out of the system. This indicates also that

no measurable amount of the substrate (Si) diffuses through
the SiO$_2$ layer into the substrate.

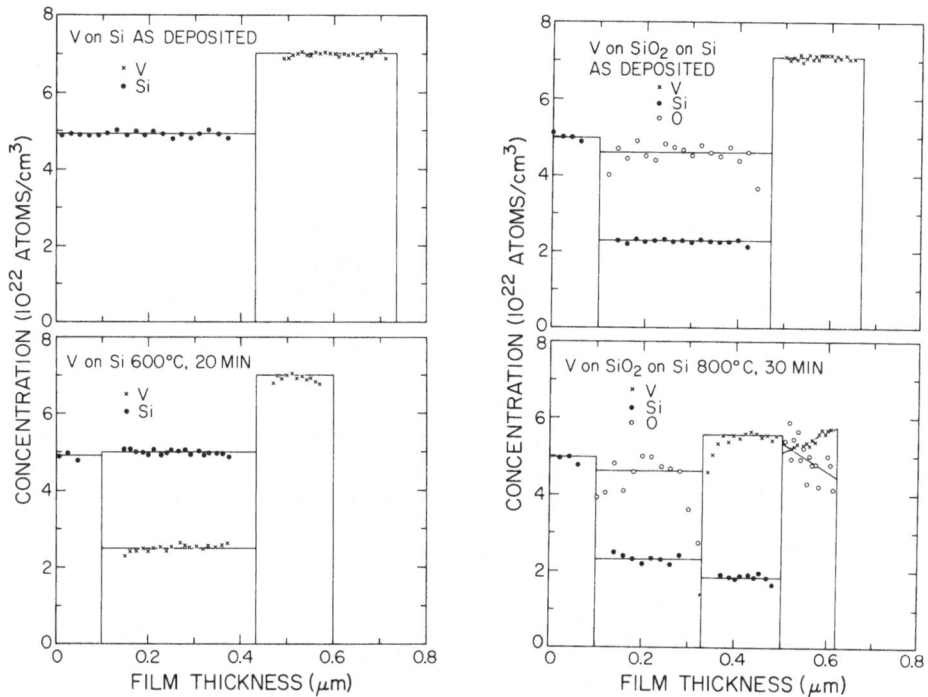

Figure 5. (left) Concentration profile of a V film on Si as
           evaporated (top), and after heat treatment in vacuum
           at 600°C for 20 minutes (bottom).

Figure 6. (right) Concentration profile of a V film as eva-
           porated on an oxidized Si substrate (top), and after
           heat treatment in vacuum at 800°C for 30 minutes
           (bottom).

C. Niobium (Columbium)

The reaction temperature for Nb films on Si and on SiO$_2$
prepared at Caltech are higher than 700° and 900°C respectively.
Unfortunately, several experimental flaws affect the results.
The evaporated layers always contain some oxygen. During heat
treatment the oxygen content increases further. From the best

evaporations, we determine for both substrates a ratio for
Nb:Si of about 5:3 for the intermediate layers. Since we did
not succeed in preparing or maintaining oxygen-free Nb layers,
the top layer of Nb has always a noticeable oxygen content.

IV. OXIDATION OF METAL FILMS ON SiO_2

The oxidation products of the metal layers in an oxygen
atmosphere have also been studied. The results can be seen in
Figs. 7-9.

Ti is known as a very reactive material able to in-
corporate a remarkable amount of oxygen even at room temperature.
We have observed that heat treatment in vacuum tends to homogeneously distribute oxygen throughout the Ti film, indicating
that oxygen is very mobile. Heat treatment in dry oxygen atmosphere at 400°C for 2 hours show a total oxygen concentration of
about 50% which, however, is not uniformly distributed in the
layer. But at 600°C, a rapid increase of the oxygen concentration
can be seen (Fig. 7, lower part). The layer can be divided into
two regions; the top layer shows an oxygen to Ti ratio of about
3:2 and the adjoining region below, which does not react visibly
with the SiO_2 layer at that temperature, has a ratio O:Ti = 1:1.

Vanadium seems to form a compound at relatively low
oxidation temperatures. In Fig. 8, the composition of the sample
can be seen before (top) and after (bottom) heat treatment at
400°C for 30 minutes in dry oxygen. The transition between the
oxide layer and the V metal layer is sharp. The ratio of the
oxidized top layer is V:O = 2:5, and is identified by x-ray
analysis as V_2O_5.

This compound, which is the same as that found by
x-ray diffraction in the product of the SiO_2-V reaction, has as
low melting point (660°C).[9] To evaluate the effects associated
with this melting point, the following experiment has been performed. A V_2O_5 layer, formed by complete oxidation of a V film
on SiO_2 at 600°C in dry oxygen, was further heat treated at
700°C in oxygen. The layer melts and forms drops on the surface. The same V_2O_5 layer annealed below 600°C in vacuum shows
no visible change. But during an anneal at 640°C in vacuum for
several hours, most of the V_2O_5 layer vaporizes. The behavior
of an only partly oxidized V film on SiO_2 as shown in Fig. 8 is
different. Up to 600°C in vacuum no change can be seen. But
at temperatures higher than 700°C the top layer changes composition. The V:O ratio after annealing for several hours is

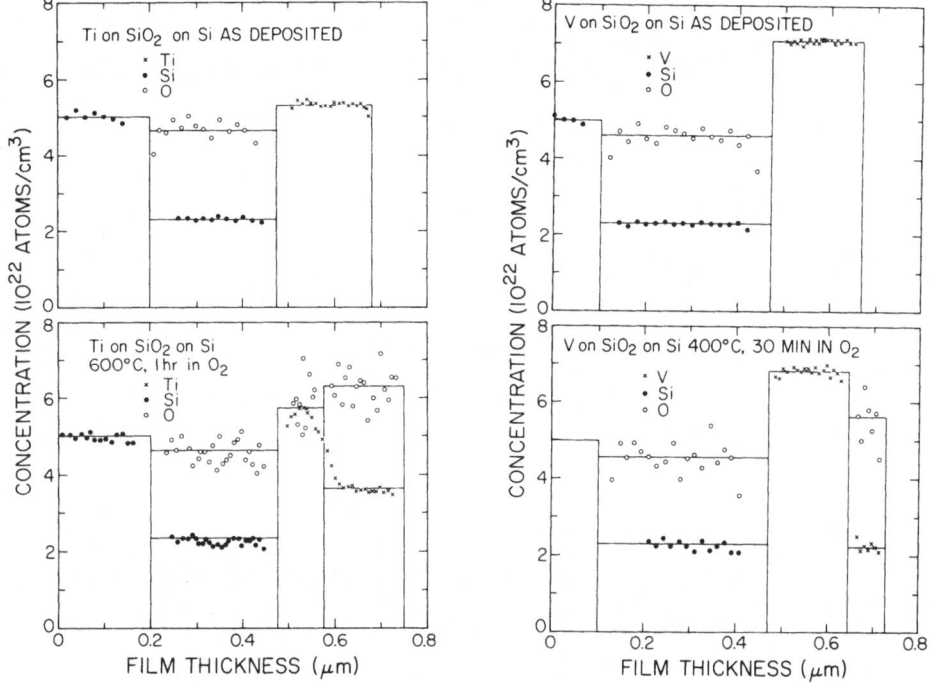

Figure 7.   (left)   Concentration profile of a Ti film as eva-
            porated on an oxidized Si substrate (top), and after
            heat treatment in dry oxygen at 600°C for one hour
            (bottom).

Figure 8.   (right)   Concentration profile of a V film as eva-
            porated on an oxidized Si substrate (top), and after
            heat treatment in dry oxygen at 400°C for 30 minutes
            (bottom).

about 1:1 and below this thick V-O layer is a thin V_3Si layer
from the reaction of SiO with the excess V. The V and oxygen
content is conserved within the accuracy of the measurement.
The same sample has been annealed again in a dry oxygen atmosphere at 400°C and the top layer oxidizes until a ratio of
V:O = 2:5 is reached.

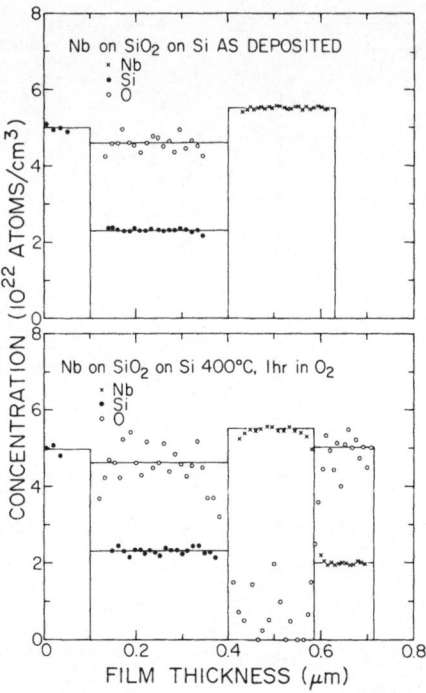

Figure 9.   Concentration profile of a Nb film as evaporated
            on an oxidized Si substrate (top), and after heat
            treatment in dry oxygen at 400°C for one hour
            (bottom).

        After annealing of a Nb film on SiO_2 in oxygen at 400°C,
a layer of the composition Nb:O = 2:5 can be found (Fig. 9).

V.   DISCUSSION AND CONCLUSION

        The binary phase diagrams of these metal-Si systems
show that a variety of silicides can exist. For the Si-Ti system
four compounds (Ti_3Si, Ti_5Si_3, TiSi, $TiSi_2$), for the Si-V system
three compounds (V_3Si, V_5Si_3, VSi_2) and for the Si-Nb system
four compounds (Nb_4Si, Nb_3Si, Nb_5Si_3, $NbSi_2$) are reported.[10,11]
No prediction can be made what compound will be formed first, if
any, by interaction of the elements or by interaction of different
compounds at temperatures below the melting point. The phase

diagrams of the metal-oxygen systems are even more complicated because of the fact that some of the many possible compounds have large regions of existence.

Table 1. Reaction products of Ti, V and Nb on Si and SiO_2 substrates heat treated in vacuum or in oxygen atmosphere.

Substrate		Si	SiO_2		
		interface layer (composition)	interface layer (composition)	surface layer (composition)	surface layer after oxidation
Ti	Back-scattering	Ti:Si=0.5	Ti:Si=1.6	Ti:0~1	Ti:0=0.6
	x-ray	$TiSi_2$	Ti_5Si_3	unidentified	unidentified
	reaction temp.	>500°C	>700°C		600°C
V	back-scattering	V:Si=0.5	V:Si=3	V:0~1	V:0=0.4
	x-ray	VSi_2	V_3Si	$V_2O_5+V_5Si_3$	V_2O_5
	reaction temp.	>500°C	>700°C		400°C
Nb	back-scattering	Nb:Si~1.7	Nb:Si~1.7	Nb:0~1	Nb:0=0.4
	x-ray	$NbSi_2$*	Nb_3Si*	unidentified	Nb_2O_5
	reaction temp.	>700°C	>900°C		400°C

A summary of the compounds observed in the reaction products between the metal and the Si or SiO_2 substrates, and with oxygen is given in the Table 1. We list the ratio of the elements contained in the films as derived from backscattering analysis. Beneath the backscattering data, the chemical compounds are listed which are identified by glancing angle x-ray measurements. An approximate temperature is also given above which a noticeable reaction rate is observed. Corresponding data of x-ray and of backscattering analyses were all obtained on the identical sample, except where marked by*. The results of the present experiments lead to the following conclusions:

1. Reaction between a SiO_2 substrate and the metal films requires temperatures which are about 200°C higher than those needed for the reaction of a Si substrate with the same metal film.

2. Silicides formed between metals and Si are generally silicon rich (e.g. VSi_2).

3. Silicides formed between metals and SiO_2 are generally metal-rich (e.g. V_3Si).

4. When metals and SiO_2 react in vacuum, metal silicides and metal oxides are both formed, but in separate layers.

5. Metal oxides produced by reactions with SiO_2 have less oxygen than the metal oxides formed in an oxidizing ambient.

6. The total amount of Si, oxygen, and metal present in a sample does not change when the reaction takes place in an inert surrounding.

7. No indication of reversibility of the reaction has been observed.

Questions such as how fast the reaction proceeds, what terminates it, and what happen when a given layer is completely consumed are difficult to answer. Preliminary experiments indicate that some of the complications are due to problems of the cleaness at the interface, and oxygen contamination throughout the layers.

ACKNOWLEDGEMENTS

This work was supported financially at Caltech by the Air Force Cambridge Research Center (D.E. Davies), and at the IBM Thomas J. Watson Research Center in part by ARPA. Contract No. F19628-73-C-006 administered by the Air Force Cambridge Research Laboratories. We also thank the Kellogg Radiation Laboratory at Caltech for the use of the 3 MeV accelerator, and in particular Dr. C.A. Barnes for his continuous assistance in the backscattering measurements.

REFERENCES

1. R.W. Bower, Appl. Phys. Lett. 23, 99 (1973).
2. K.N. Tu, J.F. Ziegler and C.J. Kircher, Appl. Phys. Letters (to be published).
3. B. Schwartz (Ed.), "Ohmic Contacts to Semiconductors," The Electrochemical Society, (New York, 1969).
4. L.J. Schwartz, Ph.D. Thesis, The City University of New York, 1970.
5. H. Kräutle, M-A. Nicolet and J.W. Mayer, Phys. Stat. Sol. (a) (submitted).
6. W.K. Chu, J.W. Mayer, M-A. Nicolet, T.M. Buck, G. Amsel and F. Eisen, Thin Solid Films 17, 1 (1973).
7. R. Feder and B.S. Berry, J. Appl. Cryst. 3, 372 (1970).
8. K.N. Tu and B.S. Berry, J. Appl. Phys. 43, 3283 (1972).
9. Gmelins Handbuch der anorganischen Chemie, Vol. 41, 48 and 49, (Verlag Chemie GMBH, Weinheim 1951).
10. M. Hansen, Constitution of Binary Alloys (McGraw-Hill, 1958).
11. F.A. Shunk, Constitution of Binary Alloys, second supplement (McGraw-Hill, 1969).

DISCUSSION

Q: (J. F. Ziegler) Where is the V_5Si_3 in the films? It shows up in the x-ray analysis, but not in the backscattering data.

A: We are not sure; the backscattering data indicate it is not a uniform layer thick enough to resolve.

ION BEAM INDUCED INTERMIXING IN THE Pd/Si SYSTEM *

W.F. van der Weg,** D. Sigurd,*** and J.W. Mayer

California Institute of Technology

Pasadena, California 91109

ABSTRACT

MeV He backscattering, scanning electron microscopy and x-ray diffraction analysis were used to investigate the effect of Ar ion bombardment on thin evaporated Pd layers on single crystal Si. Ar ions of energies between 40 and 400 keV were used and the thickness of the Pd films varied between 300 and 1100Å. Backscattering measurements showed that for cases where the range of the implanted ions was comparable to the Pd film thickness, pronounced intermixing occurred. X-ray diffraction analysis showed that the intermixing resulted in the formation of Pd_2Si. The intermixing was found to occur for ion doses $> 10^{16}$ cm^{-2}. No mixing was observed for ion doses $< 10^{15}$ cm^{-2} and for film thickness in excess of the ion range. We attribute the intermixing to a dynamic process of beam-enhanced diffusion. It does not appear to be a heating effect and it is also unlikely that the intermixing results from knock-on implantation.

---

*Work supported by Air Force Cambridge Research Laboratories (D.E. Davies) and Office of Naval Research (L. Cooper)
**Permanent Address: Philips Research Labs., Amsterdam, The Netherlands
***Permanent Address: Research Institute for Physics, Stockholm, Sweden

I. INTRODUCTION

Ion implantation into thin films can produce changes in structure and composition. The resistance of thin films can be increased orders of magnitude, their adherence to glass substrates increased significantly and electromigration properties can be influenced. The subject has a long history[1] and has recently been reviewed.[2]

Since thin film structures can have thicknesses less than the range of the implanted ions, atoms near the substrate-film interface can be transferred across the boundary during the implantation process. For example, it was found that for aluminum films on glass substrates after argon bombardment a significant enhancement of adhesion occurred.[3] It has also been reported that this recoil implantation process could lead to the introduction of significant amounts of oxygen into silicon by implanting through a thin oxide layer covering a silicon substrate.[4]

We have investigated ion beam induced intermixing in structures consisting of Pd films evaporated on single crystal Si. Some indication of intermixing in this system has been reported by Lee et al.[5] where it was found that a dose of 2×10^{15} phosphorus ions/cm^2 of 80 keV energy produced interdiffusion between Pd and Si. The purpose of the present work is to investigate in more detail the conditions under which this intermixing occurs. We have utilized MeV He backscattering, scanning electron microscopy and x-ray diffraction analysis to investigate this intermixing.

II. EXPERIMENTAL

Samples for this study were prepared by evaporation of Pd onto <111> and <100> oriented single crystal Si substrates. Prior to evaporation the Si wafers were treated in organic solvents, immersed in dilute HF and subsequently rinsed in water. The background pressure during evaporation was $\leq 3 \times 10^{-7}$ Torr.

Argon implantations were performed at the 100 KV accelerator at California Institute of Technology and at the 500 KV accelerator of Philips Research Laboratories, in Amsterdam. Implantation energies ranged from 40 to 400 keV. Doses from 10^{15} to 3×10^{16} ions/cm^2 were used at dose rates between 1.4×10^{11} and 1.8×10^{13} ions $cm^{-2} s^{-1}$. During implantation samples were tilted 5° to avoid channeling. The implantations were performed at room temperature.

Backscattering analyses were performed at the Kellogg accelerator using 1 and 2 MeV ^{4}He ion beams. Details of the equipment

are published elsewhere.[6] X-ray analysis was made by K.N. Tu of
IBM Research, using Seeman-Bohlin diffraction configuration. This
is basically glancing angle x-ray diffraction using a focused beam,
a fixed angle (6°) of incidence and a detector that moves on a
diffraction circle.[7]

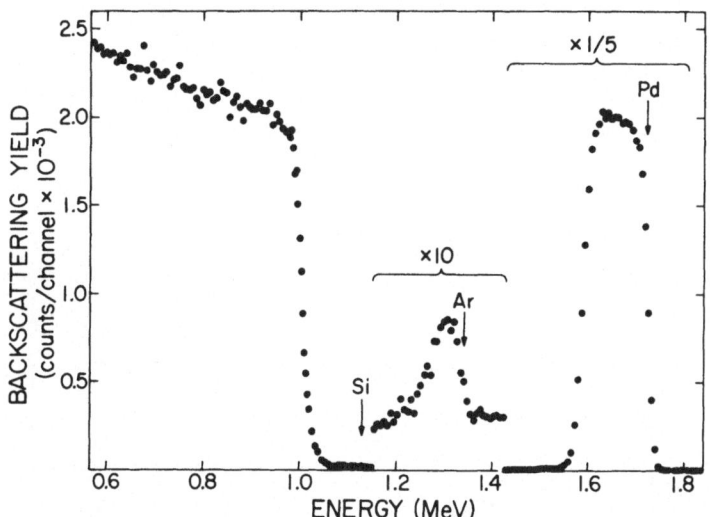

Figure 1.   Backscattering energy spectrum for a 1100Å Pd film on
            a <100> Si substrate after implantation with
            $1.3 \times 10^{16}/cm^2$ Ar ions of 40 keV. The energy of the
            analyzing He beam was 2 MeV.

III. RESULTS

     Figure 1 shows a backscattering energy spectrum for a 1100Å
Pd film on a <100> Si substrate after implantation with Ar ions
at 40 keV ion energy. The broad peak at approximately 1.7 MeV
represents scattering from Pd atoms. The narrow peak at approximately 1.3 MeV originates from the implanted Ar atoms and the
distribution below 1.1 MeV is the contribution from the underlying
Si substrate. The Pd peak has an energy width of 130 keV and the

peak in the Ar distribution is displaced 33 keV below its surface position, i.e. the peak in the Ar distribution occurs at
approximately 1/4 of the Pd film thickness. The full width at
half maximum of the Ar peak is 68 keV which is again substantially less than the Pd film thickness. From this we conclude
that nearly all of the Ar ions have stopped within the Pd film.
The integrated area of the Ar distribution corresponds to
1.3×10^{16} atoms/cm^2.

This spectrum is typical for cases where no intermixing
occurs. This can be inferred from the steep high energy edge of
the Si contribution and the steep low energy edge of the Pd distribution. If intermixing occurs one typically finds steps or
graded slopes in these edges.[8,9]

An example of a case where intermixing of Si and Pd has
taken place is shown in Fig. 2. In this case the Pd thickness
is 1/3 of the thickness of the sample shown in Fig. 1 and the

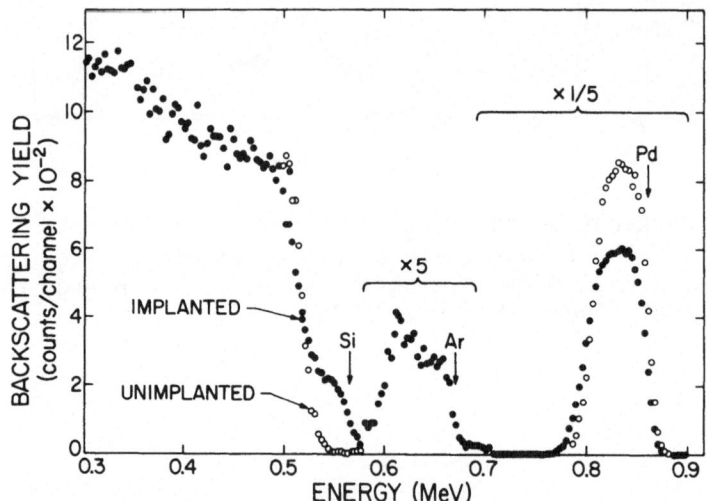

Figure 2.   Backscattering energy spectrum for a 350Å Pd film on a
            <100> Si substrate after implantation with 3×10^{16}/cm^2
            Ar ions of 40 keV. The 1 MeV analyzing the beam was
            incident 40° off the surface normal.

implantation dose was 3×10^{16} Ar ions/cm^2 of 40 keV. The data
in this spectrum were taken with 1 MeV He ions and the sample
tilted 40° (giving an apparent thickness-broadening of approximately 30%). There is a pronounced step in the high-energy edge
of the silicon component of the spectrum. The appearance of this
step shows that some intermixing of Si and Pd atoms has taken
place. In this example the Si has penetrated to the top surface
of the sample, i.e. through the Pd layer. The height of the Si
step, compared to that of the Pd peak corresponds to a concentration in the mixed layer of 1.4 Pd to 1 Si. This ratio indicates a more Si-rich composite layer than is usually found when
Pd-Si structures are annealed at 250°C, where Pd_2Si is formed.[10]
To determine whether the apparent intermixing was due to cracks
or pinholes in the film the sample was examined with a scanning
electron microscope. At magnifications up to 20,000 no unusual
features could be detected in the surface topography.

In this spectrum the Ar-component was expanded by a factor
of 5. The Ar distribution is flat-topped with an energy width at
half maximum of 69 keV. This energy width is comparable to the
width of the Pd peak. Also, the energy at half height of the leading edge of the Ar distribution corresponds to scattering from
Ar atoms situated at the surface. Hence, the Ar distribution
extends throughout the whole mixed Pd/Si layer. This Ar distribution is different from the one observed in Fig. 1, where a
relatively thick layer of Pd was implanted. In this case, as
noted earlier, the Ar distribution is peaked at a depth of 300Å
below the surface.

We found this intermixing to occur in samples with thin Pd
layers (<400Å) for various implantation conditions. Implantation
energies of 40 and 80 keV produced the same results. Mixing
occurred for implantation doses of $\geq 10^{16}$ ions/cm^2 while no
effect was observed for doses $\leq 10^{15}$ ions/cm^2. Intermixing was
observed for all implantation dose rates used for these samples,
i.e. 3×10^{12} to 1.8×10^{13} ions cm^{-2} s^{-1}. The effect was also
observed on both samples prepared on <111> and <100> oriented
Si substrates.

Samples with thicker Pd layers (950Å) showed no intermixing for implantation energies up to 100 keV. However, implantation at 200 and 400 keV Ar energy showed pronounced intermixing as is evidenced in Fig. 3. This figure shows backscattering energy spectra for 950Å thick Pd layers implanted with Ar at
100, 200 and 400 keV to a dose of 10^{16} ions/cm^2. It is seen that
in the 200 keV case there are steps both in the Si contribution

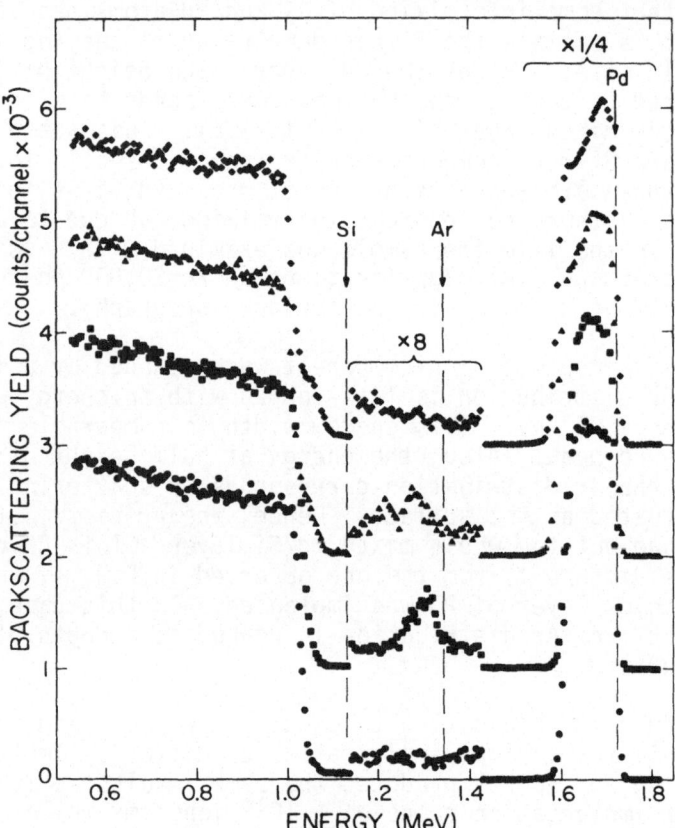

Figure 3.   Backscattering energy spectra for a 950Å Pd layer on a
                   <100> Si substrate. The unimplanted case and spectra
                   for implantations with $10^{16}/cm^2$ Ar ions of 100, 200
                   and 400 keV are shown. The lower spectrum represents
                   the unimplanted case. The 100, 200 and 400 keV im-
                   planted cases have been displaced by respectively
                   1×10^3, 2×10^3 and 3×10^3 counts/channel. The
                   energy of the analyzing He beam was 2 MeV.

and in the Pd peak of the spectrum. These steps are even more pronounced in the 400 keV spectrum. In contrast with the results
obtained for the thin samples the concentration ratio of Pd to
Si is here 2 to 1. X-ray diffraction analyses of the sample
implanted at 400 keV showed the presence of both Pd and Pd_2Si
phases. The latter was estimated to be about 400Å thick and
polycrystalline with weak epitaxy on the wafer. The observed
diffraction peaks did not coincide with indexed reflections for
PdSi or Pd_3Si.

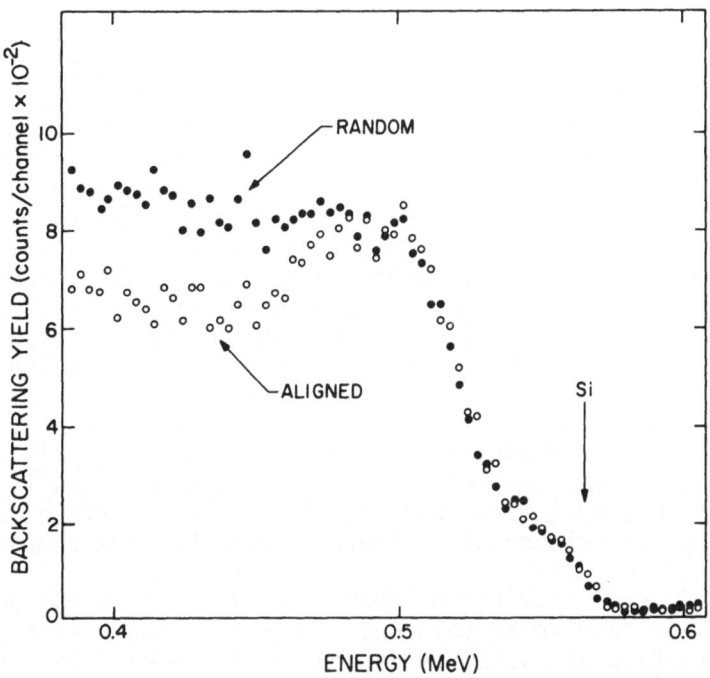

Figure 4.   Aligned and random backscattering energy spectra for a
            350Å Pd layer on a <100> Si substrate, implanted with
            $10^{16}/cm^2$ Ar ions of 80 keV. The alignment of the
            crystal with respect to the 1 MeV analyzing beam was
            performed on a portion of the substrate not covered
            with Pd. In the figure only the Si contribution is
            displayed.

In order to obtain information about the depth distribution of radiation damage in the implanted structure, channeling measurements were performed. Fig. 4 shows aligned and random backscattering energy spectra for a 350Å Pd layer on a <100> Si substrate, where intermixing had occurred after Ar implantation at 80 keV. The channeling spectrum shows that the distribution of damage extends 900Å into the underlying substrate. This damage is caused by the tail in the Ar distribution since an amorphous region can be created[11] with a dose of approximately 5×10^{14} ions/cm^2.

IV. DISCUSSION AND CONCLUSIONS

These results show that intermixing occurs when the range of the Ar ions is comparable to the thickness of the Pd layer and the ion dose is of the order of 10^{16} cm^{-2}. After intermixing, a number of Pd atoms in excess of 2×10^{17} cm^{-2} appears intermixed with the Si atoms. This number seems substantially greater than what can be accounted for by recoil implantation.

We do not believe that the intermixing is caused by overall sample heating during implantation for two reasons. In the first place, a high dose implantation in a thick Pd layer, in which the Ar ions are completely stopped in the layer, did not produce any mixing. Secondly, areas immediately adjacent to the implanted region did not show intermixing. In addition, the same effect was found for samples implanted with the different dose rates used. In fact, in order to keep the energy supplied by the argon beam low, (< 10 mW/cm^2) the high energy implantation required 20 hours. The observed effects cannot be ascribed to the formation of amorphous silicon during the implantation. In previous work[10] it was found that there is no intermixing in sequentially evaporated Si and Pd layers. The evaporated layer of silicon was amorphous and oxide growth at the interface was minimized by evaporating both layers during the same pumpdown.

There is evidence that the observed radiation enhanced mixing is a formation of a Pd/Si compound. This is shown directly by the x-ray diffraction analysis, performed on the sample implanted at 400 keV. Also, the backscattering analysis indicates that there is a sharp interface between the mixed Pd/Si and the underlying Si substrate. This is similar to earlier results of Pd$_2$Si formation during heat treatments.[10] Therefore a possible explanation of the observed phenomenon is that the implantation enhances both the dissolution of Si into the Pd film and also the subsequent diffusion. The occurrence of an enhanced diffusion in the bombarded layer can be inferred from the spreading of the implanted Ar profile.

ACKNOWLEDGEMENTS

We are indebted to our colleagues at Philips Research Lab, who performed the high energy implantations. We thank K.N. Tu at IBM Labs. for performing the x-ray analysis.

REFERENCES

1. J.J. Trillat, "Ionic Bombardment, Theory and Application," Gordon and Breach, New York (1962).
2. P.T. Stroud, Thin Solid Films 11, 1 (1972).
3. L.E. Collins, J.G. Perkins and P.T. Stroud, Thin Solid Films 4, 41 (1969).
4. R.S. Nelson, Rad. Eff. 2, 47 (1969).
5. D.H. Lee, R.R. Hart, D.A. Kiewit and O.J. Marsh, Phys. Stat. Sol.(a) 15, 645 (1973).
6. E. Rimini, E. Lugujjo and J.W. Mayer, Phys. Rev. B6, 718 (1972).
7. K.N. Tu and B.S. Berry, J. Appl. Phys. 43, 3283 (1972).
8. W.K. Chu, J.W. Mayer, M-A. Nicolet, T.M. Buck, G. Amsel and F. Eisen, Thin Solid Films 17, 1 (1973).
9. J.A. Borders, Conf. on Ion Beam Analysis of Surface Layers, June 1973, (to be published in Thin Solid Films).
10. R.W. Bower, D. Sigurd and R.E. Scott, Solid State Elec. (in press).
11. J.W. Mayer, L. Eriksson and J.A. Davies, Ion Implantation in Semiconductors, Academic Press (1970).

DISCUSSION

Q: (F. W. Saris) You discussed the possibility of recoil-implantation of Pd into Si. Could you also comment why the sputtering of Si into the Pd layer can be ignored?

A: The sputtering could contribute to the transfer of atoms across the Pd/Si interface. It is unlikely, however, that the intermixing which results in a compound layer of ~ 400 Å thickness can be attributed to this effect.

Q: (R. Bower) [1] Did you look for Pd_2Si with x-ray diffraction in the thin film case? [2] Did you look for preferential orientation of Pd_2Si in thicker films?

A: [1] No. [2] Yes, some degree of orientation was noted in the x-ray diffraction analysis.

Q: (J. M. Poate) We have looked at the Pt-Si system and can apparently produce mixing over ~ 200 Å at the interface with Ar bombardment. What we have found however is that the mixed layer in-
hibits subsequent silicide formation on annealing. Did you try
annealing after bombardment to see how the kinetics were affected?

A: We did not study the reaction kinetics on the implanted
samples.

Q: (E. V. Kornelsen) Why does the Ar depth distribution become
wider only when the intermixing occurs?

A: Since in the intermixing process large numbers of Si and/or Pd
atoms are transported it is likely that during the process the
depth distribution of the implanted Ar atoms also is affected. At
present, we cannot identify the mechanism by which both the intermixing of Pd and Si and the redistribution of Ar takes place.

CHAPTER IV

ALLOYING AND MIGRATION IN HIGH FLUENCE IMPLANTS

PRECIPITATION DURING ION BOMBARDMENT OF METALS

R.S. Nelson

Metallurgy Division, AERE, Harwell, Berks.

ABSTRACT

During the ion implantation of a metal, two fundamental changes occur. First, the gradual addition of impurity atoms results in a gradual change in chemical composition; and second, atomic collision results in radiation damage. The final physical state of the metal then depends on the interplay between these two processes.

This paper reviews the physical mechanisms which influence the stability of precipitates under irradiation. A simple theoretical treatment will be outlined to describe the equilibrium conditions which are expected to exist during irradiation. The model will be applied to the case of alloy formation during ion implantation and the predictions illustrated by a number of experimental observations using the electron microscope.

1. INTRODUCTION

During the last decade the application of ion beams to materials research has attracted substantial study. In particular in connection with the doping of semiconductors by ion implantation, in the study of radiation damage in reactor materials, and more recently in the controlled modification of surface properties of metals. During the ion implantation of a metal two fundamental changes occur. Firstly, the gradual addition of impurity atoms results in a gradual change in chemical composition; and secondly, as a consequence of the energetic recoils which occur as the ions slow down in the solid, atomic displacement and

rearrangement occur.   It is the subtle interplay between these
two processes which determines the final physical state of the
implanted solid.

In this review we will discuss the basic physical processes
which occur during the bombardment of an alloy and during the
implantation of metals to ever increasing doses.   For clarity of
presentation we will approach the final state in a series of
logical steps.

2. SOLUBILITY CONSIDERATION

At very low doses, the implanted atoms will simply remain as
isolated impurities within the host lattice, and in principle
these may occupy vacancy or interstitial sites, depending on which
corresponds to the lowest free-energy state, this in turn depends
on parameters such as the size and electronic configuration of the
atoms.   In an irradiation environment, however, the situation can
be quite different.   For instance, the majority of implanted
atoms come to rest in an environment where initially both lattice
vacancies and interstitials are plentiful.   Calculation suggests
that within a random collision cascade the incident ion has a high
probability of finally coming to rest as a consequence of a
replacement collision and therefore remains in a substitutional
site.   If the surrounding lattice interstitials and vacancies
produced by the irradiation are mobile, this atom may subsequently
move by vacancy exchange, become trapped at a lattice interstitial,
be thermally activated into a free interstitial configuration, or
in some case can be directly replaced by a lattice interstitial.
Generally speaking, free interstitials have low activation
energies for migration and move off rapidly through the lattice
until they become trapped or cluster.   However, this is not
always the case; for instance, small impurity atoms such as
carbon and nitrogen quite happily occupy stable interstitial
positions within the more open b.c.c. lattices such as iron and
molybdenum at room temperature.

At elevated temperatures thermal activation can result in a
variety of modifications.   For instance, the break up of pairs
and the release from traps, as well as the diffusion of isolated
substitutional impurities to extended defects such as dislocation
lines and grain boundaries and the free surface.

As the ion dose is increased, more and more atoms are introduced into the lattice and it is of considerable interest to
determine whether or not the alloy so formed remains in a solid
solution indefinitely.   In comparison with conventionally
prepared alloys, if the concentration of impurity exceeds the
maximum solubility limit in the target, and the temperature is

such that diffusion occurs, the precipitation of a second phase
may result.   Unlike conventionally prepared systems where
precipitation effects can only be observed in a limiting number of
alloy systems, due to the supersaturation which occurs on the
reduction in temperature of a solid solution; there is, in
principle, no such limit in the case of ion implantation, as ions
of anelement can be injected into any solid irrespective of
solubility considerations.   However, there is a limit to the
maximum concentration of impurity atoms that can be implanted into
a metal, due to the fact that the free surface is continually
receding as a consequence of sputtering.   Ignoring diffusion and
precipitation effects this maximum concentration is given
approximately as the inverse of the sputtering ratio (the number
of atoms ejected per incident ion).   Thus for instance for an
ion – target combination having a sputtering ratio of 5 this
yields a maximum implanted concentration of 20%.

The precipitation in conventionally prepared alloys has been
well studied and the conditions pertaining to thermodynamical
equilibrium are reasonably well defined, see for instance Kelly

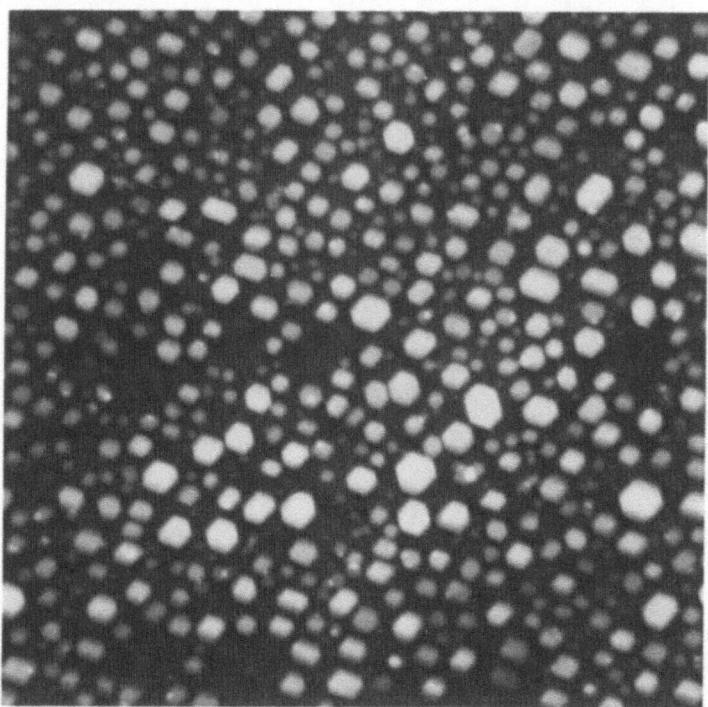

Fig. 1.   Argon gas bubbles in Cu bombarded with a saturation dose
of 40 keV A^+ and annealed to 700°C.

and Nicholson (1963).   For instance, the inert gases are known to
have vanishingly low solubilities in solids and, provided their
mobility is adequate, precipitation will occur for even the smallest
concentrations.   In this case, due to the mobility of the inert
gases to form compounds, precipitation takes the form of an
agglomerate of gas atoms which can best be described as a bubble
(see fig. 1).   However less inert impurities precipitate out to
form a compound on attaining their solubility limit, for example
Cu-Al forms $CuAl_2$ and Al in Ni forms small coherent Ni_3Al
precipitates which have the well known Cu_3Au structure (see fig. 2).

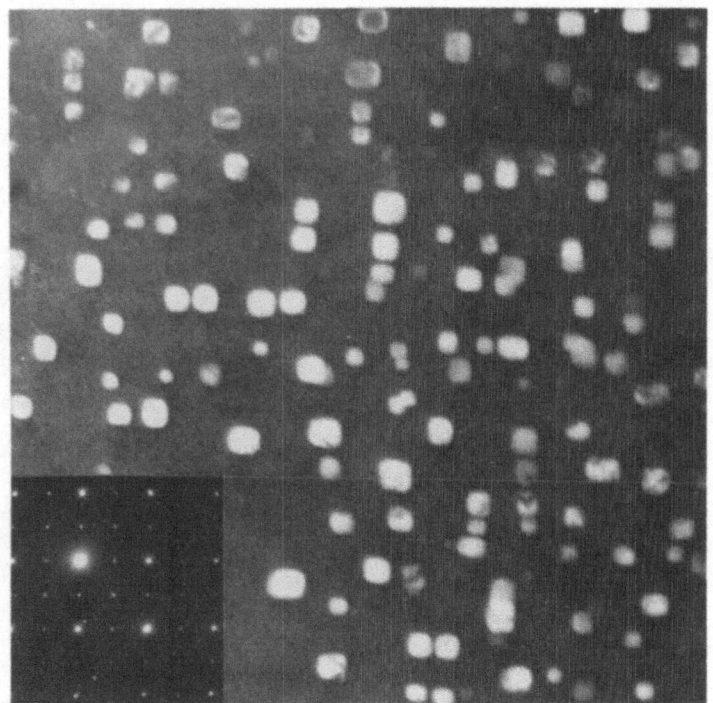

Fig. 2. $Ni_3Al(\gamma')$ precipitation in a Ni-Al alloy.

However, in some special cases - such as Ag in Au - the one
species is quite stable within the other and remains in
substitutional solution at all temperatures and concentrations.
We must now consider whether the general arguments and facts
which are pertinent to thermodynamical equilibrium can be applied
directly to the case of ion implanted solids, or whether the non-
equilibrium conditions created by the accompanying radiation
damage can introduce complicating effects.

3. IRRADIATION ENHANCED DIFFUSION

We have already stated that the ability of implanted impurities to precipitate depends on their diffusibility in the target.   However, it is well known that due to the irradiation induced vacancy concentrations in excess of those dictated by thermal equilibrium, diffusion processes can be enhanced during irradiation.   In the particular case of vacancy controlled diffusion of solute atoms the enhanced diffusion coefficient will depend intimately on the dynamic vacancy concentration which pertains during the irradiation.   Theoretical treatments of the problem, e.g. Damask and Dienes (1958) and Sharp (1972), identify two processes which influence the vacancy concentration during irradiation;   namely the loss of vacancies to fixed sinks such as dislocation lines and the loss of vacancies by recombination with interstitials.   In practice it is possible to experimentally separate these two regimes as conditions can be adjusted so that either one or the other dominates.

In the particular case of vacancy controlled diffusion of solute atoms, where the dominant loss of vacancies is to fixed sinks, the radiation enhanced diffusion coefficient is simply given by:-

$$D' \; = \; K/\alpha \; cm^2 \; sec^{-1} \tag{1}$$

where K is the defect production rate in displacements/atom/sec and α is the dislocation density or fixed sink density.   On the other hand where the vacancy jump rate is slow and the major loss of vacancies is by vacancy-interstitial recombination, the enhanced diffusion coefficient is given by:-

$$D' \; = \lambda^2 \left(\frac{K\nu_s}{Z} \right)^{\frac{1}{2}} \tag{2}$$

where λ is the jump distance (e.g. 1 lattice spacing), ν_s is the solute-vacancy jump frequency and Z is the recombination probability ~ 1.

However, perhaps the best estimate of enhanced diffusion can be obtained from the computer calculations of Bullough and Perrin (1971) concerned with void growth in irradiated metals.   In this treatment the drift interaction of vacancies and interstitials with dislocation lines together with their recombination is specifically included in a calculation to find the dynamic vacancy concentration existing during irradiation.   It is then a simple matter to convert such vacancy concentrations to enhanced diffusion coefficients.   A typical example is shown in fig. 3. for

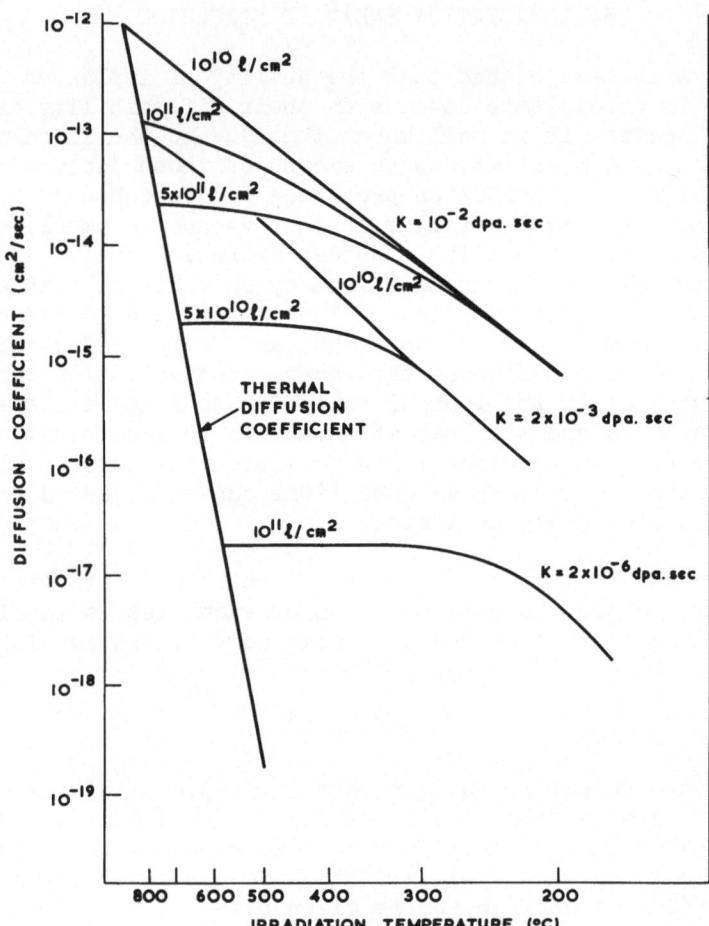

Fig. 3. Radiation enhanced diffusion coefficient for different dislocation densities at damage rates of 10^{-2}, 2×10^{-3} and 2×10^{-6} displacements/atom/sec.

nickel corresponding to a damage rate typical of that used during ion implantation; the thermal self-diffusion coefficient for nickel is shown for comparison. It is seen that at the highest dislocation density of 5×10^{11} lines cm^{-2} the enhanced diffusion coefficient approaches a constant value at about 400°C of 2×10^{-14} cm sec^{-1}, essentially identical to that which would have been given by equation (1). On the other hand, at a lower dislocation density of 5×10^{10} lines cm^{-2} the enhanced diffusion coefficient depends on temperatures as defined by equation (2).

In order to ascertain the enhanced diffusion coefficient pertaining during a particular ion bombardment, it is first of all essential to consider the magnitude of the dislocation densities which might exist; and furthermore, if the implantation is carried out with low energy ions (< 100 keV), then the influence of the

free surface as a major sink must be accounted for.   A typical
bulk dislocation density in a well annealed metal might be $\sim 10^8$
lines cm^{-2}.   However, during irradiation the clustering of
interstitials and vacancies created by the irradiation causes a
dislocation structure to grow.   In this case the dislocation
density depends on the ion dose and the irradiation temperature,
see for instance fig. 4.   However, eventually the dislocation

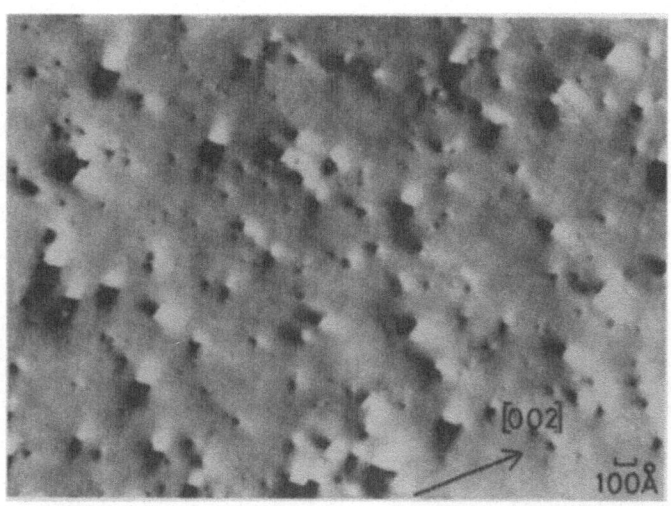

Fig. 4.   Black spot defects in Cu irradiated with 30 keV Cu$^+$ ions,
each spot corresponds approximately to one incident ion.   Analysis
of the diffraction contrast suggests that such defects are vacancy
type dislocation loops.

density saturates at some value; for instance in soft metals like
Cu, Au or Ni irradiated at about 150°C, 200°C, or 500°C
respectively this might be $\sim 5 \times 10^{10}$ lines cm^{-2}, whereas in the
case of hard metals such as steel the dislocation density might
build up to 5×10^{11} lines cm^{-2} after a dose equivalent to 5
displacements/atom.   However, as the defect production rate and
irradiation produced dislocation structure are both not uniform
within the irradiated area, together with the fact that the surface
sink will influence the situation, the irradiation enhanced diffusion
coefficient will suffer some complex variation throughout the bombarded region.

4.   RADIATION ENHANCED PRECIPITATION DISSOLUTION

It has been known for sometime that irradiation can both order
and disorder alloys, for instance Cu$_3$Au is readily disordered
during irradiation at room temperature (see Damask, 1965).   It is
therefore reasonable to assume that ordered precipitates will

suffer disordering during irradiation with subsequent dissolution
into the surrounding matrix.   Fig. 5 illustrates the effect of
precipitate dissolution in a NiAl alloy, both bright and dark
field micrographs are shown and the regions still containing
precipitates were shielded during irradiation.

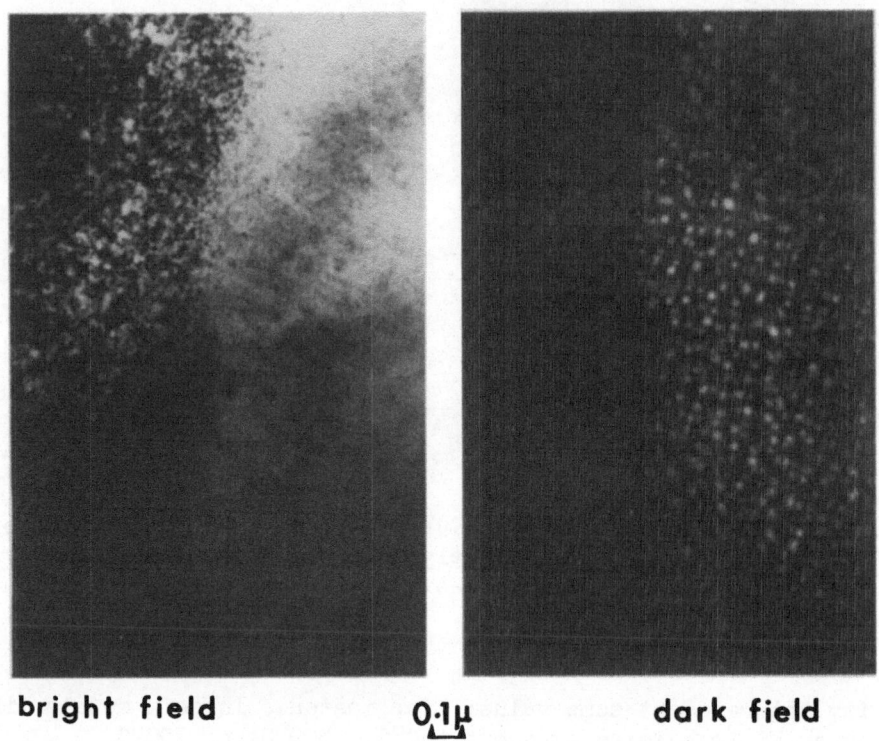

bright field   0.1μ   **dark field**

Fig. 5.   Bright and dark field pictures showing the dissolution
of γ' precipitate in Ni-Al alloy.   The region still showing
precipitates was shielded during irradiation.

4.1   Recoil Dissolution

The dynamic collision events which occur as a result of
atomic displacement within collision cascades cause atoms within
the precipitate to recoil into the surrounding matrix.   From our
knowledge of displacement processes within solids it is reasonable

to assume that every atom crossing the precipitate interface into the surrounding metal with an energy of about 25 eV is dissolved. The flux of such recoils is readily estimated from our knowledge of the energy spectrum within collision cascades and the number of atoms sputtered from solid surfaces during irradiation. For convenience they can be related to the damage rate; calculations suggest that for a damage rate of K displacements/atom/sec this flux of atoms is $\phi \sim 10^{14}$ K/cm^2/sec (Nelson 1969). For convenience we will assume that the precipitates are spheres of radius r, when the dissolution rate is simply

$$\frac{dV}{dt} = -\frac{4\pi r^2 \phi}{N} \tag{3}$$

where N is the number of atoms per unit volume.

4.2   Disorder Dissolution

An alternative mechanism for dissolution particularly pertinent to ordered precipitates is that the disordering effect of the displacement cascades essentially destroys the ordered precipitate so that localised regions of high solute concentrations are created. In the absence of diffusion such a state will persist, as for example during the irradiation of Ni-Al alloys at room temperature. However, when diffusion occurs, the small disordered regions created within the precipitate will re-order whilst those near to the interface can diffuse to the surrounding matrix.

An approximate expression for the dissolution rate of precipitates resulting from disorder dissolution has been derived elsewhere (Nelson et al, 1972).

$$\text{i.e.} \quad \frac{dV}{dt} = -4\pi r^2 \gamma K \tag{4}$$

where V is the precipitate volume and γ is a dissolution parameter $\sim 10^{-6}$ cm.

5.   EQUILIBRIUM CONDITIONS

Under thermal activation the stability of precipitates is controlled by the balance between the diffusion of solute atoms to the precipitate and the thermally activated release of atoms from the precipitate surfaces. When the rate of the dissolution process exceeds the rate of diffusion to precipitates, the precipitates disappear to maintain the solute atoms in solution. However, at lower temperatures where a reduced solute concentration

is in equilibrium with precipitates, the system attempts to lower
its total free energy by the growth of large precipitates at the
expense of smaller ones.   In an irradiation environment, thermal
equilibrium is disturbed in as much as the irradiation can
influence both the precipitate dissolution rate and the solute
diffusion coefficient.   New equilibrium conditions are therefore
created.

Equilibrium conditions will prevail when the rate of
dissolution equals the growth rate due to the diffusion of solute
atoms to the precipitates.   For simplicity let us assume that
every precipitate has a radius r and that there are n precipitates
distributed uniformly throughout the irradiation volume.   If the
total concentration of solute atoms at any one instant is C, we
can define a boundary condition such that the total solute
contained in precipitates and in solution equals C,

$$\text{i.e.} \quad C = \frac{4}{3}\pi r^3 pn + c \qquad (5)$$

where p is the atomic fraction of solate atoms constituting the

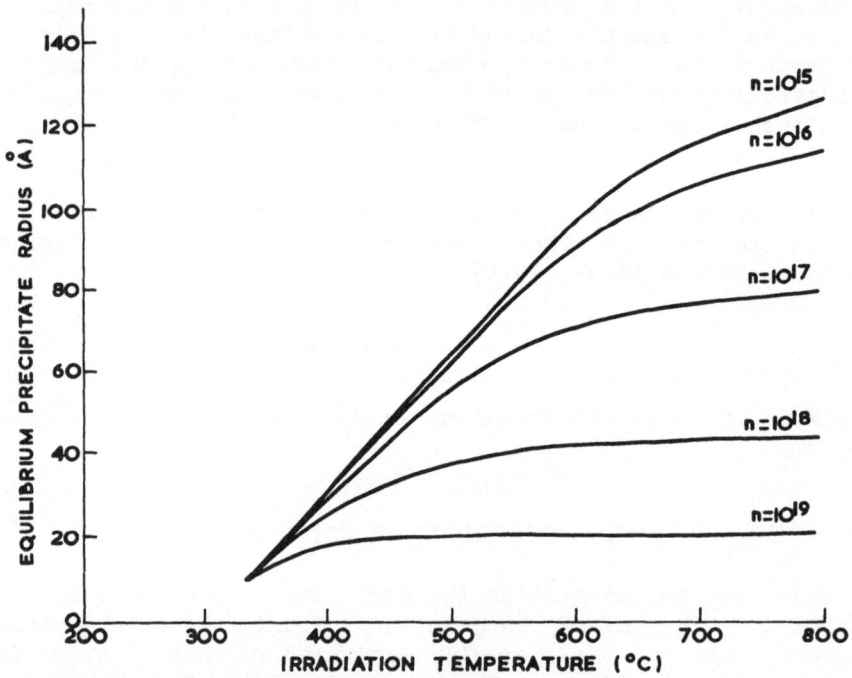

Fig. 6.   The equilibrium radius of precipitates in Ni as a function of irradiation temperature, with C = 0.135 and γ = 10^{-6} cm.

precipitate phase and c is the concentration held dynamically in solution during the irradiation.   From the work of Ham (1958) the growth rate of precipitates from an atomic concentration c in solution is given by

$$\frac{dV}{dt} = 3 \, (D + D') \, cr/p \qquad (6)$$

where (D + D') is the sum of the thermal and irradiation enhanced diffusion coefficients.   If we assume that the dissolved atoms simply contribute uniformly to the concentration in solution, equating say equations (4) and (6) then yields the following expression for the net rate of change of precipitate radius during disorder dissolution;

$$\frac{dr}{dt} = -\gamma K + \frac{3(D+D')}{4\pi pr} \, C - (D+D') \, r^2 n \qquad (7)$$

This cubic equation cannot be solved simply but from a computational solution we can derive typical curves as shown in fig. 6, which show the equilibrium radius of Ni_3Al precipitates in Ni as a function of irradiation temperature for different precipitate densities, with C = 0.135.   Unfortunately a major uncertainty is in estimating the precipitate nucleation density during irradiation, and at the present time no reliable calculations exist.

Clearly as ion implantation proceeds the total concentration steadily increases with the bombardment dose.   Let us assume that for low energies the implanted atoms come to rest in a uniform layer equivalent to the range of the incident particles, then to a first approximation C = $\emptyset$t/RN, where $\emptyset$ is the flux of bombarding ions and R is the range of the incident particles.   Fig. 7 then shows the equilibrium radius for NiAl precipitate in Ni as a function of Al ion dose for different precipitate densities at 500°C.   At doses much greater than 20 x 10^{16} ions/cm^2 the total solute concentration will saturate as a consequence of sputtering, under these circumstances the precipitate distribution will itself approach equilibrium.

In the case of incoherent precipitates formed with atoms which are generally relatively insoluble, dissolution can only occur as a result of recoil dissolution.   Then equating equations (3) and (6) leads to a similar expression to equation (7).   Under these circumstances however, the dissolution rate is about two orders of magnitude smaller than for disorder dissolution with the result that the precipitates will grow leaving a relatively small concentration in solution.

6. SOME OBSERVATIONS OF PRECIPITATION EFFECTS

It is convenient to separate the observation of precipitation effects into two sections, namely those which occur as a consequence of simply irradiating an alloy and those which occur during the implantation of a pure metal with impurity ions.

6.1 Precipitation in ion bombarded alloys

The majority of data concerned with radiation enhanced precipitation in alloys during ion bombardment has been obtained on materials such as steel which have been studied for their relevance to reactor systems.   For this reason, precipitation phenomena have mainly been concerned with carbide formation.   For instance, ASAI 316 steel is often prepared for use in a solution treated form, i.e. heated at 1050°C for 1 hr, such that all the alloying elements are maintained in solid solution.   If such an alloy is simply heated to say 600°C precipitation is not expected;

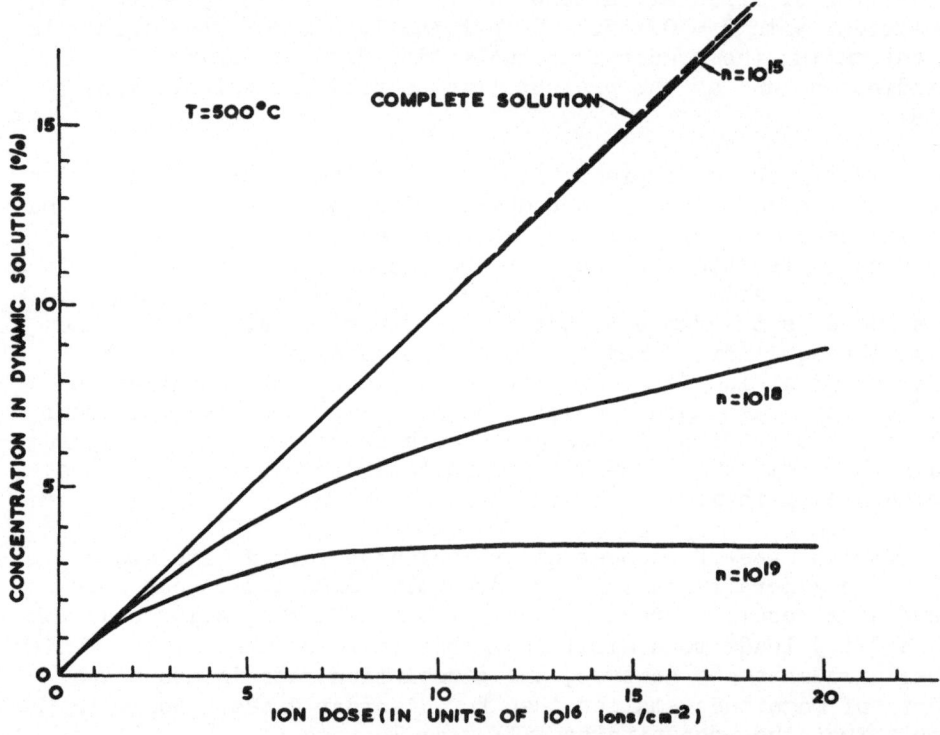

Fig. 7.   The fraction of solute contained in dynamic solution as a function of ion dose at 500°C with ψ = 10^{-6}cm.

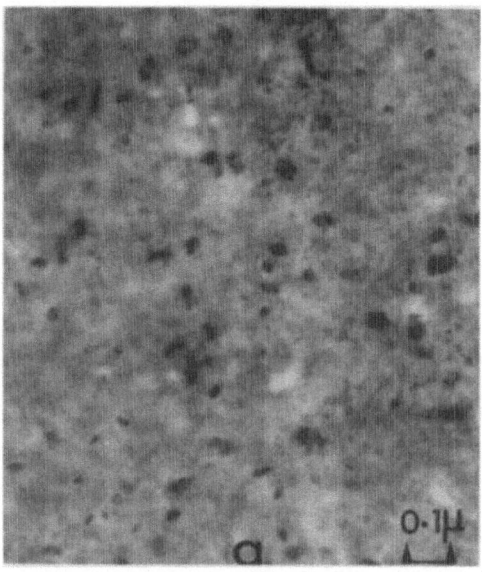

Fig. 8.   The formation of $Cr_{23}C_6$ carbide in PE16 during 45 MeV
Ni-ion irradiation at 570°C.

however, if the alloy is irradiated at 600°C for a similar length
of time, precipitation results. It should be noted that such irradiation induced precipitates appear fairly uniformly throughout
the bombarded sample; this is in contrast to thermally induced precipitation where $Cr_{23}C_6$ is often formed preferentially at grain
boundaries. Similar precipitation effects have also been observed
in the nimonic alloy PE16 which contains up to 0.2 at %, Fig. 8.
It should be noted that in both cases the carbides are readily dissolved during irradiation at room temperature, presumably due to
disorder dissolution.

Apart from carbide formation, another precipitate phase which
has been studied during ion bombardment is the so called γ'
precipitate in Ni alloys.   In the binary alloy Ni-Al, the
precipitate takes the form of an ordered Ni_3Al phase having a high
degree of coherency with the matrix.   Whereas in the nimonic alloy
PE16 the precipitate is $Ni_3(Ti,Al)$.

We have already shown that the γ' precipitate in NiAl is
readily dissolved at room temperature (fig. 2), after a damage
dose in excess of about 0.1 dpa.   However above some critical
temperature - which depends on the damage rate - such precipitates
persist, but often in a modified form.   In the case of Ni-Al

alloys irradiated at a damage rate of $\sim 10^2$ displacements/atom/sec, the critical temperature was 300-320°C. In other words, at this temperature the radiation-enhanced diffusion coefficient was just sufficient to allow the diffusion of solute atoms to the precipitate to balance the loss by dissolution. Fig. 9 shows a series of micrographs of Ni-Al initially containing 250A γ'

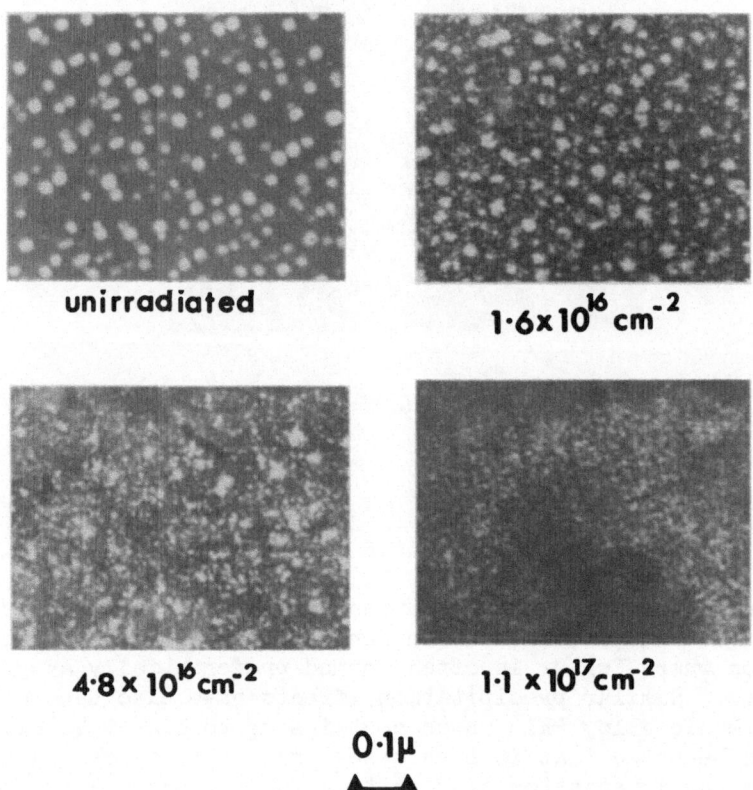

unirradiated 1.6×10^{16} cm^{-2}

4.8×10^{16} cm^{-2} 1.1×10^{17} cm^{-2}

0.1μ

Fig. 9. The change in precipitate size and density in Ni-Al as a function of dose K = 10^{-2} displacements/atom/sec, T = 550°C.

particles, taken as a function of increasing temperature somewhat above the critical temperature. It is readily apparent that the new precipitates are nucleated on a very fine scale, and at the same time the larger pre-existing precipitates are dissolved, so that ultimately all precipitates have obtained approximately the same size. It is interesting to note that even in the case of alloys which have been solution treated and quenched, the final state of precipitation is very similar to that shown in Fig. 9. In other words, it appears that whatever the initial starting configuration the final state is one which is set solely by the irradiation conditions.

Similar results have been obtained for the commercial alloy PE16, however in this case – as can be seen from Fig. 10 the γ'

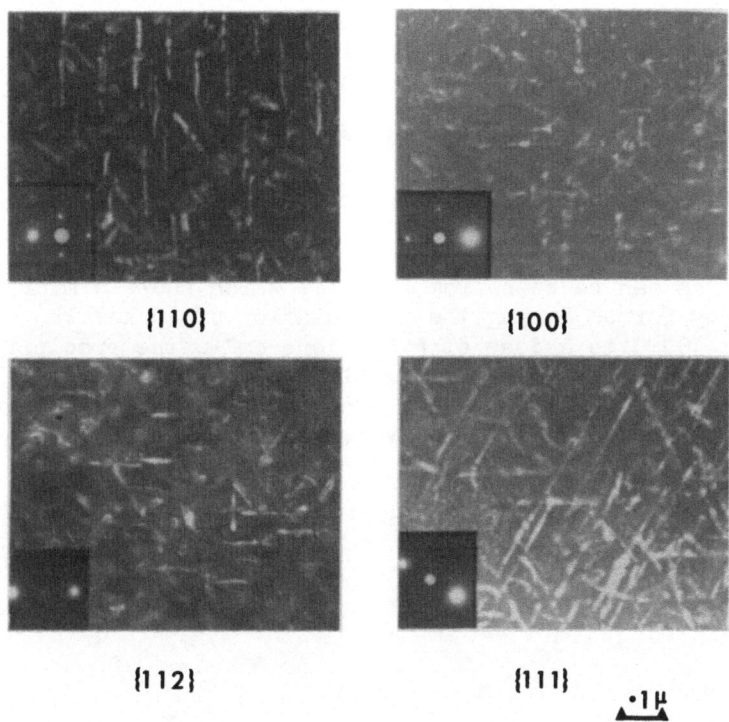

{110}   {100}

{112}   {111}

Fig. 10.   Lenticular γ' precipitates produced during irradiation of PE16 with 45 MeV Ni ions at 600°C; note the crystallographic orientations.

interacts with the radiation induced dislocation structures to form, as well as the background precipitates, crystallography oriented lenticular γ'.

6.2 Precipitation in implanted pure metals

Precipitation which occurs during ion implantation of pure metals has been studied for a few specific systems.   For instance,

a particular example is that relating to the implantation inert
gases, which, as we have already stated, precipitate as gas
bubbles.   The behaviour of such bubbles in solids is fairly well
understood – as they are of immense importance to the physical
and mechanical integrity of nuclear fuels and cladding materials.
For instance the size of a bubble is defined by free-energy
considerations and essentially occurs when the decrease of free
energy of the gas with volume is equal to the increase of surface
free energy of the bubble with volume.   Furthermore, the total
energy of the system is again minimised by the bubble developing
facets corresponding to faces exhibiting the lowest surface free-
energy.   Bubbles of inert gases can migrate through the solid by
processes such as surface diffusion, consequently can either burst
at the surface or agglomerate and grow.

        A further example of the implantation of a highly insoluble
element into a metal is the case of Pb in Al.   There being no
intermediate phase the Pb precipitates out as pure Pb in the Al
matrix and as can be seen from Fig. 11, which shows a micrograph
of the alloy formed during the implantation of 70 keV Pb$^+$ ions
into Al at 400°C to a dose of $\sim 10^{16}$ions cm^{-2}, the precipitates
crystallographically oriented parallel to the host lattice.   It
should be noted that the precipitates shown in the figure are in
fact all within the matrix, however, during the bombardment a

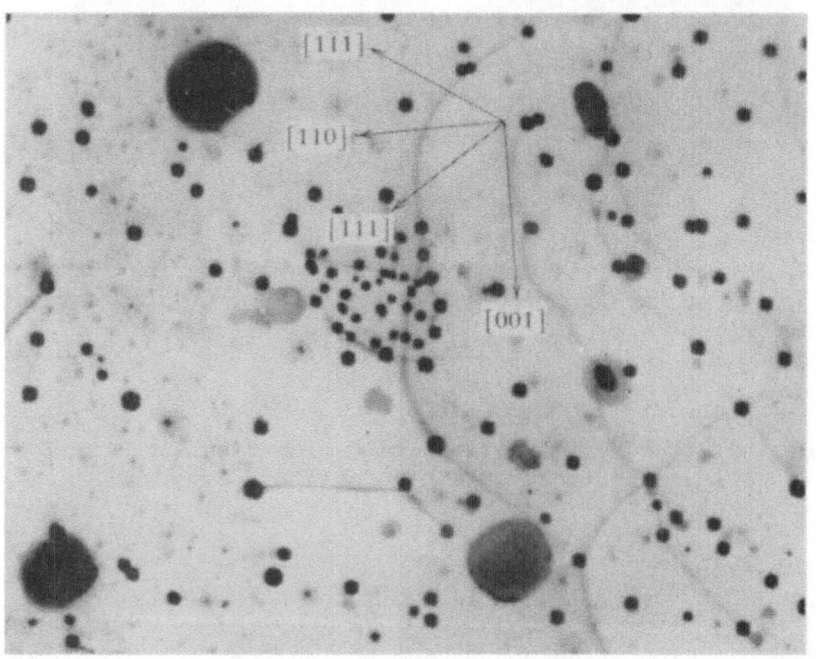

Fig. 11.   Crystallographic precipitates of Pb in Al at 400°C.

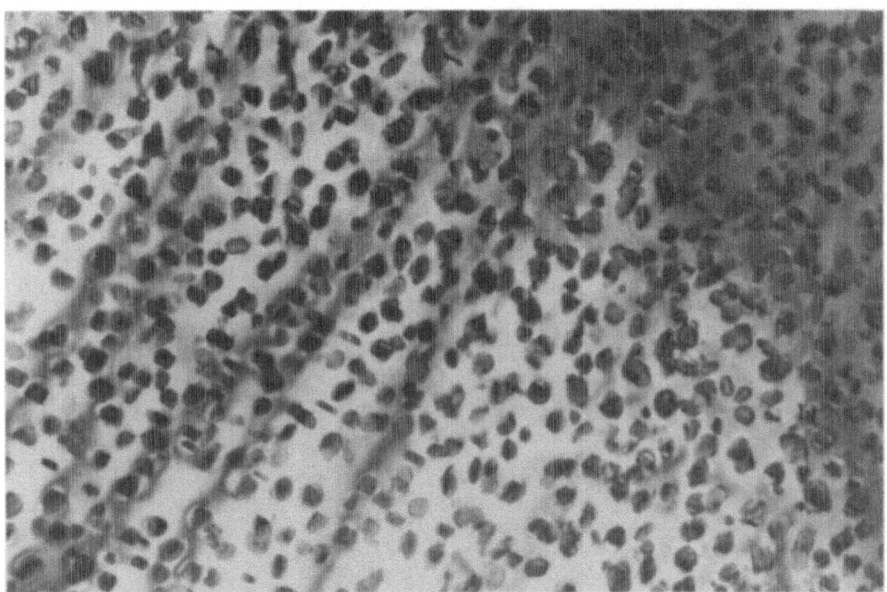

Fig. 12. Micrograph of 80 keV Sb$^+$ implantation of Al at 330°C to a dose of 5 x 10^{16} ions cm^{-2} showing AlSb precipitates.

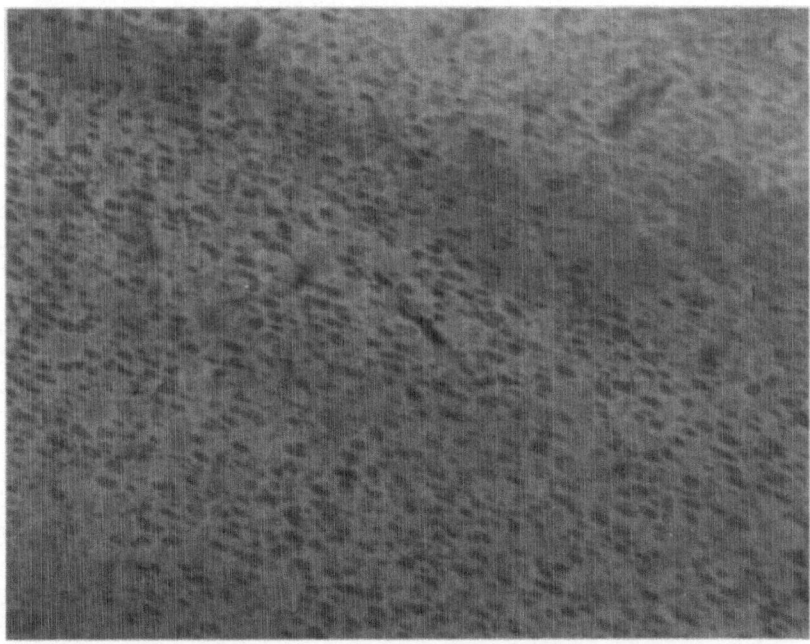

Fig. 13. Al implanted with 70 keV Cu$^+$ ions at 40°C to a dose of 4 x 10^{16} ions cm^{-2} showing CuAl$_2$ precipitates.

significant fraction of the Pb diffuses to the surface where it
agglomerates to form a Pb surface phase which can readily be
removed subsequent to irradiation.

Two further examples of precipitation effects which have been
studied are the result of implanting Al with Sb^+ and Cu^+ ions.
Figures 12 and 13 show typical micrographs.   In the case of Cu-Al
the larger $CuAl_2$ precipitates have been identified as being at
the surface, the smaller precipitates being within the bulk.

REFERENCES

1.   Bullough, R. and Perrin, R.C. (1971), BNES Conf. on Voids
     (Reading UK)
2.   Damask, A.C. and Dienes, G.J. (1958), J. App. Phys. 29, 1713
3.   Damask, A.C. (1965), AIME Symposium on Radiation Effects,
     Asheville, North Carolina p. 77
4.   Ham, F.S. (1958), J. Phys. Chem. Solids 6, 335
5.   Kelly, A. and Nicholson, R.B. (1963), Prog. Mat. Sci. 10 (3)
6.   Nelson, R.S. (1969), Proc. Roy. Soc. A 311, 53
7.   Nelson, R.S., Hudson, J.A. and Mazey, D.J., J. Nucl. Mat. 44,
     318 (1972)
8.   Sharp, J. V. (1972), Radiation Effects, to be published.

DISCUSSION

Q: (K. L. Merkle) One would expect that the dissolution of precipitates would depend on the cascade energy and size. Has this
been observed?

A: There are some experiments which have been performed in the
HVEM where the recoil energies are very low, just above displacement energies. In this case γ' precipitates in Ni-Aℓ alloys observed in the microscope are less susceptable to dissolution
during irradiation.

Comment: (John A. Spitznagel) We have been following the effects
of heavy ion bombardment on precipitation in a Ni-base alloy containing γ' particles which exhibit a large positive lattice mis-
match with the surrounding matrix. For this alloy the same general
phenomena you have outlined occur except that "dissolution" of
these 200 Å - 600 Å diameter particles does not seem to occur.
Instead, loss of coherency proceeds by the formation of interfacial

dislocations. Continued disordering is accomplished by the formation of dislocations within the particles until the particle has
disappeared and only a dense tangle of dislocations remains. It
would seem that a number of complex "dissolution" mechanisms may
be operative for coherent precipitate particles.

A: The dissolution of ordered precipitates with small coherency
strains such as γ' in Ni-Aℓ occurs about 10^3 times faster than for
incoherent precipitates or precipitates with large coherency
strains. In such cases it is necessary to physically knock out
atoms from the precipitants.

Q: (Wilkens) You have shown that the order of the Ni$_3$Aℓ phase is
completely destroyed at doses $\sim$ 0.1 dpa. Can this value be quan-
tatively related to the range of displacement collision sequences?

A: One expects the number of replacement collisions to be about
10 times the number of displacement events, so the value of 0.1 dpa
is approximately consistent with the current ideas of cascade
theory.

RADIATION DAMAGE AND ION BEHAVIOUR IN ION IMPLANTED VANADIUM AND NICKEL SINGLE CRYSTALS

M. Gettings, K.G. Langguth and G. Linker

Institut für Angewandte Kernphysik

Kernforschungszentrum Karlsruhe

ABSTRACT

Distributions and annealing behaviour of heavy ion induced radiation damage in single crystal nickel and vanadium are compared. Sharp annealing stages are reported for Ni while for V the production of a polycrystalline layer, ascribed to the action of precipitates, prevented the annealing of damage after high dose implantations. The use of $^4He^+$ ion channelling revealed disorder at depths much greater than the ions projected range, an observation that was supported by electron microscopy measurements. Implanted ion diffusion in vanadium was found to be dependent on the ion species used and the annealing behaviour of precipitates. Preliminary quantitative measurements indicate that diffusion coefficients are low.

INTRODUCTION

The investigation of radiation damage in transition metals by heavy ion bombardment is rapidly becoming a useful means of simulating long term damage in reactor environments. Further, there appears to be a growing interest in alloying and isotope enrichment using high dose implantation. In this present study, radiation damage and ion diffusion have been studied in the reactor base metals nickel and vanadium.

For fcc structured nickel, a number of electron microscopy investigations[1-3] dealing with radiation damage have been reported. At low ion doses in this metal, only vacancy clusters have been detected while for higher doses interstitial clusters are dominant. In general, the majority of damage has been detected within the ion projected range, although in other fcc metals some interstitial clusters at depth greater than the ion ranges, have been observed and explained by focussed collision sequences[4].

For bcc vanadium, however, only few results are presently available[5]. In previous publications[6,7] dealing with single crystal vanadium, we have reported enhanced damage depths with increasing ion energy and dose, a small dependence of the magnitude of damage on these parameters, the generation of a polycrystalline layer in the ion range for high dose implants and values of substitutional components dependent on the implanted ions ionic radii.

Results reported in this present work have been obtained using the backscattering technique from single crystal nickel implanted with bismuth, and single crystal vanadium implanted with gallium, bismuth and selenium. Some preliminary electron microscopy measurements have been taken for gallium ion induced damage in vanadium. The backscattering technique has also been employed to monitor the distribution of the implanted ions and its subsequent variations during heat treatment.

EXPERIMENTAL

Our backscattering system and the surface preparation of vanadium single crystals have been described in detail   elsewhere [7]. Like vanadium, nickel single crystals produced by electron beam zone refining have been purchased from Metals Research Corporation. Nickel samples were cut perpendicular to the <110> direction with a continuous wire saw and subsequently lapped with 15 μ , 7 μ , 1 μ and 0.25 μ diamond pastes. Samples were then etched for 5 second periods in a solution consisting of 30 cc HNO_3, 10 cc H_2SO_4, 10 cc H_3PO_3 and 50 cc glacial CH_3COOH held between 85 - 95ºC. Vanadium samples for our electron microscopy studies were prepared by mechanically thinning from the not implanted side to 0.1 mm by careful polishing on SiC paper followed by a vibratory polish with γ-Al_2O_3 of 0.05 μ diameter. Final thinning was done electrolytically with a mixture of 80 % acetic acid and 20 % perchloric acid. With this technique depth determination was estimated to be about ±200 Å.

Implantations were performed at room temperature with a scanned ion beam from a heavy ion accelerator. Metal samples were bombarded over an energy range from 20 - 360 keV and a dose range of 10^{15} - 10^{17} ions/cm^2. Irradiated specimens were annealed in a stainless steel tube under a vacuum of ≤ 6 × 10^{-7} torr, isochronally, to temperatures of 1000ºC and isothermally for times lasting to 18.5 hours.

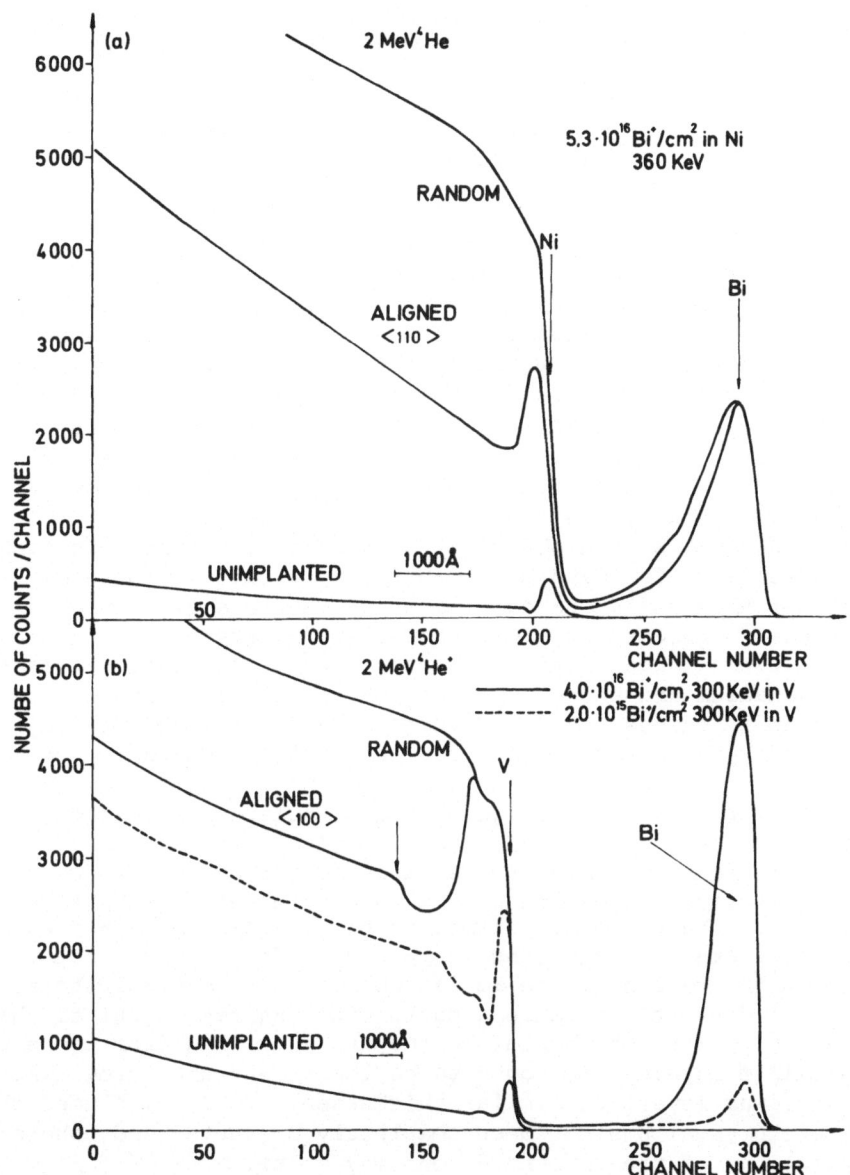

Fig. 1 Backscattering spectra from aligned and randomly oriented samples showing the damage produced in nickel and vanadium single crystals by bismuth ion bombardment.

RESULTS

The results reported in this paper will be presented in two
distinct sections:-

       a) A comparison of damage and annealing in
          nickel and vanadium
       b) Diffusion of implanted components in vanadium

a): In an attempt to compare radiation damage produced
during ion implantation in vanadium and nickel, single crystals
were bombarded with similar doses of Bi ions having the same energy. Typical aligned and random backscattering spectra from both
nickel and vanadium are presented in figure 1 to show differences
in the generated damage distributions. In nickel close to the surface, in a depth region between 350 and 450 Å for nonchannelled implants, a surface peak is found for all doses and energies under
consideration. The area of this peak increases with ion dose and
decreases with energy. In the case of channelled implantations only
the deep damage has been observed. Near surface disorder in vanadium however, was found to be strongly dependent on ion dose. For
doses lower than 10^{16} ions/cm^2 , a narrow surface peak (<200 Å)
was detected, whereas for higher doses a polycrystalline layer was
formed over the range of the implanted ions. These surface peaks
in both vanadium and nickel are thought to be disordered layers
rather than   oxides, as no enhanced oxygen peaks were found using
the helium-oxygen resonance at 3.06 MeV ^{4}He$^+$ ion energy.

The layer produced by high dose implantations in vanadium has
been investigated by X-ray diffraction and was found to be of a
polycrystalline nature having the vanadium bcc structure. We
ascribe this layer formation to the generation of precipitates
produced when the solubility limit of Bi in vanadium is exceeded.
Such precipitates are expected to partially destroy the host
lattice in the region of the implanted ions. These precipitates
and lattice distortions cause a backscattering rate equal to that
of a randomly oriented sample. As the dechannelling from these polycrystalline layers was found to be lower than that from evaporated vanadium layers of similar thicknesses, it is concluded that
the single crystal has not been completely disrupted and coherent
precipitations of the implanted ions may occur.

For nickel, the absence of a polycrystalline layer is attributed to the lower concentration of bismuth ions in this material,
as compared with vanadium for similar total doses. These lower
concentrations and ion distribution broadening can in turn be explained by the high sputtering yields of Bi ions in nickel compared to Bi in vanadium[8]. Besides the near surface damage,
damage at depths considerably greater than the ion projected range
was observed in both nickel and vanadium. Differences in the struc-

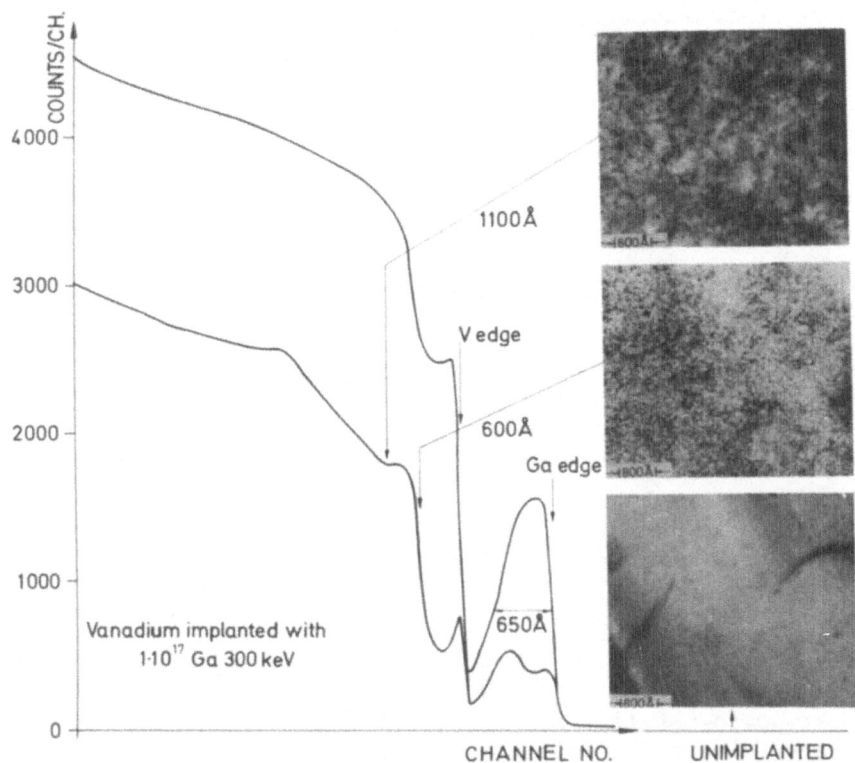

Fig. 2 Electron microscopy photographs from unimplanted and implanted (10^{17} Ga/cm^2, 300 keV) vanadium samples together with a backscattering spectrum indicating the depths at which photographs were taken.

ture of this deep damage, as revealed in the backscattering spectra, were observed for the two metals investigated. For vanadium, in the absence of a polycrystalline layer, a relatively damage-free region was observed behind the surface disorder peak and a characteristic knee was evident in the disorder distribution deeper in the crystal - as indicated in figure 1(b) by the arrow. In nickel, the deep damage distribution, as observed by channelling, exhibited no structure and was manifested by a growing dechannelling rate behind the surface damage peak.

As the backscattering method does not give any information about the nature of the generated damage, preliminary electron microscopy studies have been performed to characterize the deep damage in vanadium. The results for a sample implanted with 10^{17} ions/cm^2 of Ga are presented in figure 2 together with a backscattering spectrum indicating the two depths at which microscope photographs were taken. Also included is a picture taken with the same magnification from an unimplanted vanadium single crystal.

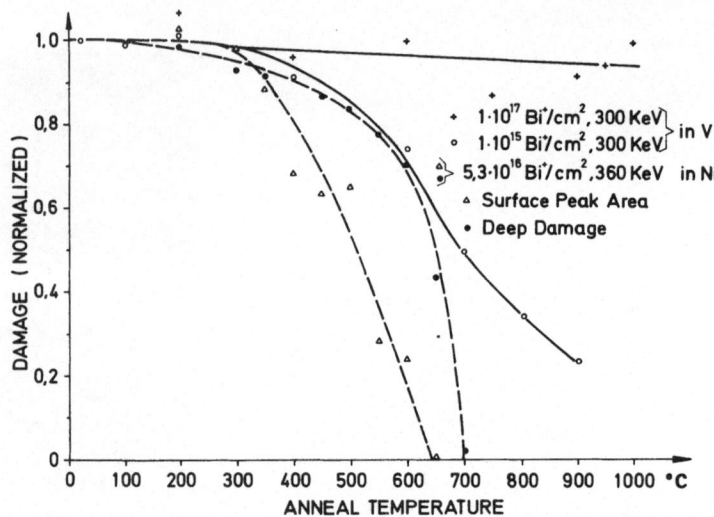

Fig. 3 Annealing curves for different types of damage generated by Bi implantations in vanadium and nickel.

At depths of 600 and 1100 Å within the crystal a fine grain structure was detected, which was observed to have more contrast at the shallower depth. At 1100 Å, voids with approximately 30 - 80 Å diameter were evident. In addition, in the diffraction patterns taken, we observed weak reflections which were attributable to small particles in a low concentration having an incoherent but ordered distribution in the host lattice. While no electron microscopy experiments have been performed for nickel in this study, others [4] have reported damage considerably beyond ion ranges. The enhanced penetration was explained by focussed collision sequences and the damage identified as interstitial clusters.

Annealing behaviour of the radiation damage in vanadium was found to be dependent on ion dose and the formation of a polycrystalline layer, whereas for nickel no such a dependence has as yet been found. These differences in annealing behaviour are illustrated in figure 3. In this figure, relative damage quantities measured by different methods are normalized to unity from samples that have not been annealed. In nickel, an annealing stage between 350° and 550°C was observed for the near surface disorder, and a stage between 500° and 700°C was found for the deep damage. Annealing of low dose vanadium implants produced , for both surface disorder and deep damage, a broad annealing stage between 500°C and 900°C. Little variation in the amount of radiation damage has been observed for the high dose implants, besides those crystals implanted with gallium, where the damage annealing is accompanied by   a complete recrystallisation of the polycrystalline layer.

b): Ion behaviour in vanadium has been studied after both
low and high dose implantations for several different ions. For
all systems studied, implantation profiles did not exhibit a sym-
metrical gaussian distribution, as predicted from LSS-theory, and
a deeply penetrating tail has been observed. As sputtering coefficients for vanadium are low this tail may be due to radiation enhanced diffusion. Such a process has been postulated by Smith[9]
for ion implantation in metals.

With low dose implantations, where a low damage level is observed in the range of the implanted ions, mainly indiffusion of
the implanted species occured. The diffusion starts at about $900^{\circ}C$,
although starting temperatures and diffusion rates were found to be
dependent on ion species.

The ion behaviour, after high dose implantations, has been
studied in some detail employing both isochronal and isothermal
temperature treatments. Diffusion characteristics have been found
to be dependent on the atomic radii of the implanted ions. It has
been previously reported[7] that substitutional solubility levels
show a strong dependence on this parameter and therefore ion species were selected to encompass diffusion mechanisms over a wide
atomic radii range.

For Ga, whose atomic radius "fits best" into the vanadium
lattice, diffusion from the polycrystalline layer is closely rela-

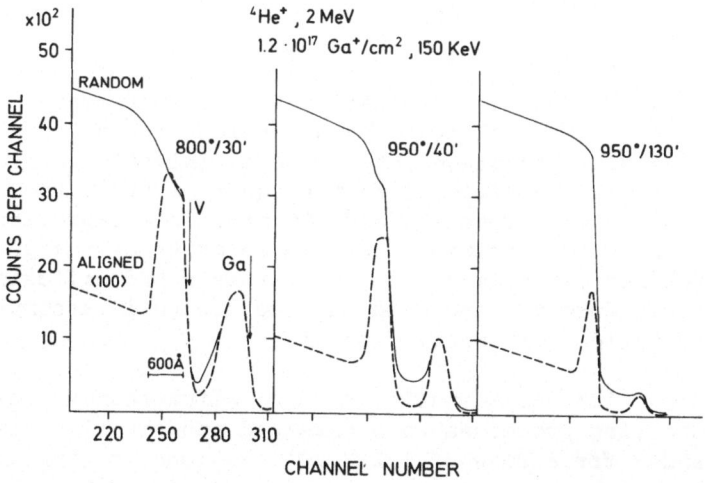

Fig. 4 Polycrystalline layer for 1.2×10^{17} Ga^+/cm^2 150 keV implantation and its recrystallisation, accompanied by substitutional
Ga indiffusion in an isothermal annealing process.

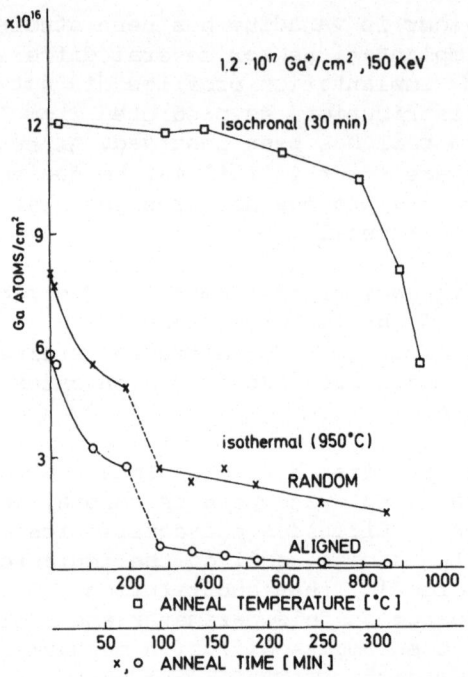

Fig. 5 Number of implanted Ga atoms/cm² remaining in vanadium as a function of temperature and time in isochronal and isothermal annealing processes showing different stages in Ga diffusion.

ted to this layer recrystallisation. Substitutional indiffusion of the Ga atoms, starting between 800 and 900°C, has been observed and backscattering spectra from this process are shown in figure 4. Here a prediffusion stage is compared with two steps in the isothermal process at 950°C. More quantitative measurements, as recorded in figure 5, reveal that the diffusion process is not uniform and three distinct stages are evident. Initially diffusion from the layer was proportional to the square root of the diffusion time. In a step between 75 and 100 minutes a rapid indiffusion occured; it was in this period that the majority of the polycrystalline layer recrystallized. Following this rapid diffusion, little Ga movement was detected and complete reordering is accomplished after the total loss of impurity atoms.

With Se, which has a smaller atomic radius than vanadium, two distinct annealing processes were observed and are illustrated in figures 6 and 7 for a dose of 1.2×10^{17} Se ions/cm² implanted at an energy of 200 keV. Isochronal annealing produced little loss of the implanted constituent as can be seen in figure 6. Movement of Se was detected however in this annealing process, as is illustrated in figure 7(a) by a shoulder generated in the Se distri-

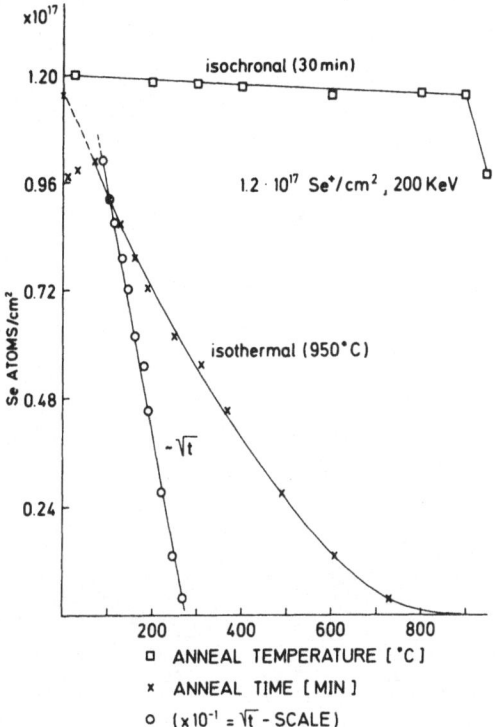

Fig. 6 Number of implanted Se atoms/cm^2 remaining in vanadium as a function of temperature and heating time, indicating the Se atom loss in an isothermal process at 950°C.

bution. This shoulder is thought to be caused by a fast diffusing component and its removal at about 950°C accompanies a break-up of a part of the polycrystalline layer. Isothermal annealing at 950°C revealed a selenium loss which, as can be seen from figure 6, was proportional to the square root of the annealing time. The loss process as depicted in figure 7(b) seems to have the characteristics of a diffusion from an infinitesimally thin layer into a body of finite dimensions with capturing boundaries. We believe that both diffusion processes are strongly dependent on the release of atoms from precipitations. This release from precipitations is thought to be the cause of the partial break-up of the polycrystalline layer.

Bismuth ions, with an atomic radius greater than that of vanadium, again show some specific diffusion characteristics. Isochronal annealing to 600°C produced little diffusion. Between 600°C and 1000°C however, loss of bismuth was accompanied by a deeply

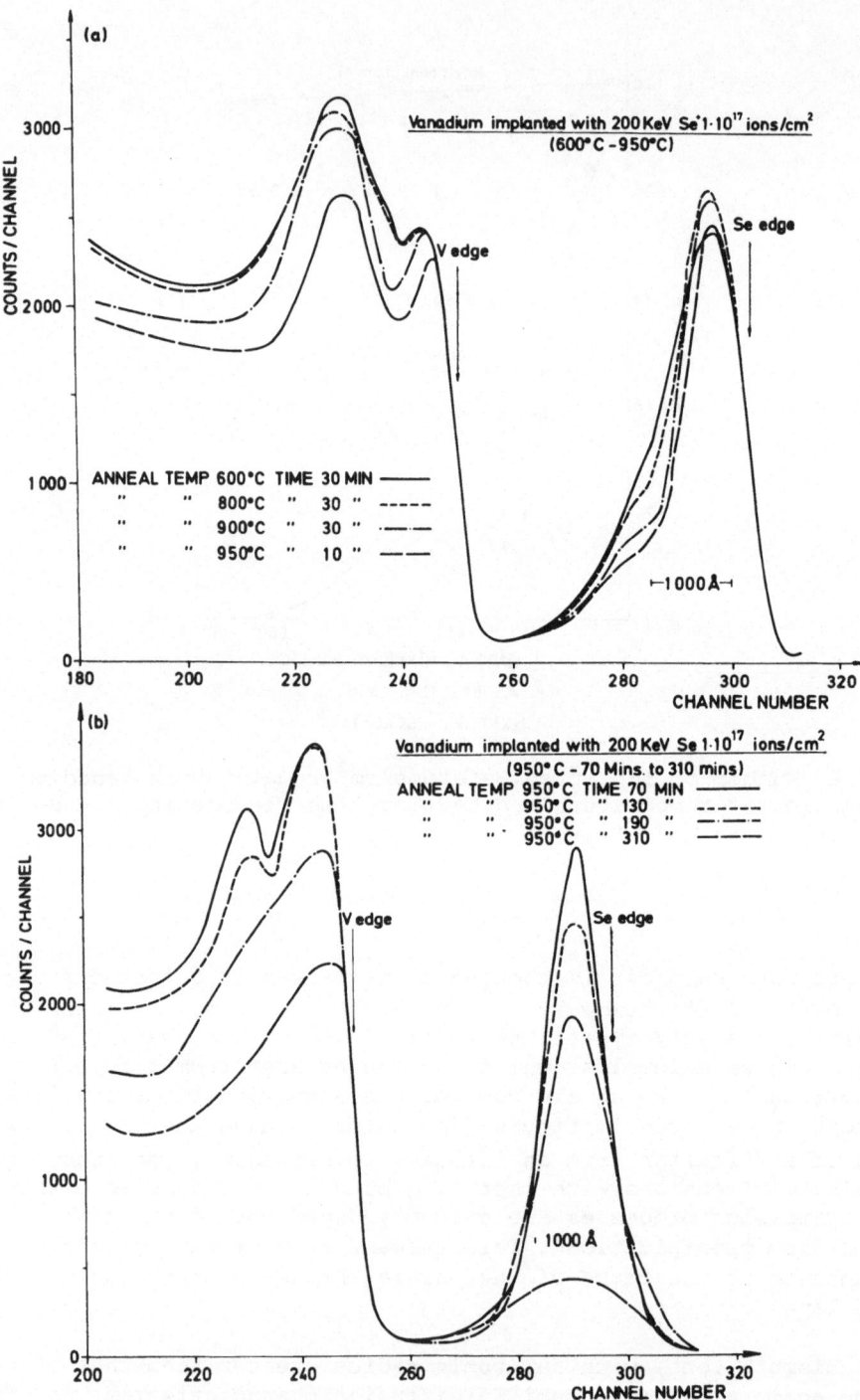

Fig.7   Se-peaks from backscattering spectra showing the two different stages in the Se diffusion process.

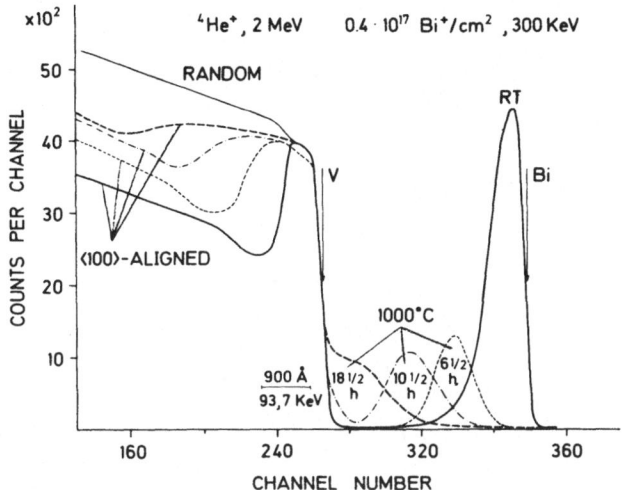

Fig. 8 Backscattering spectra from an as implanted (0.4 × 10 [17] Bi/cm[2], 300 keV) vanadium sample and from different stages in an isothermal heat treatment process at 1000°C. The main characteristics shown are an increase of the Bi distribution width and a shift of this distribution towards the vanadium edge, together with a broadening of the polycrystalline layer.

penetrating tail in this ions distribution. In an isothermal study at 1000°C, no appreciable loss of implanted ions occured although an increase in the half width of the bismuth distribution was evident. This process is illustrated in figure 8. The apparent movement of the whole Bi distribution towards the vanadium edge in the spectra can be explained by a vanadium atom migration to form an effective thickening of the polycrystalline layer at the surface. This explanation is supported by an observed broadening of this layer in the channelled backscattering spectra. The assumed vanadium migration to the surface was not accompanied by an appreciable specimen oxidation, as the use of the helium-oxygen resonance with this sample revealed the same oxygen content as was found in freshly etched specimens.

For the Bi ions implanted in nickel a rapid loss of this constituent was observed during the annealing stage of the deep damage. In a first attempt to obtain quantitative results for diffusion processes observed in our study, experimental ion distributions were compared with theoretical distributions calculated using various

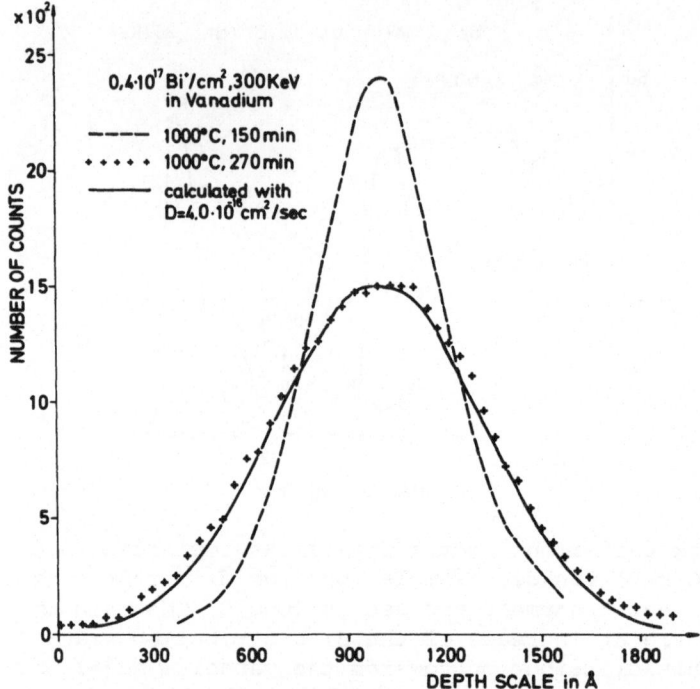

Fig. 9 Experimental and calculated Bi distributions together with the initial distribution used in the calculations.

values of the diffusion coefficient. The "best fit" for the broadening of the Bi peak at 1000°C was obtained using a diffusion coefficient of $D = 4.0 \times 10^{-16}$ cm²/sec. Experimental and theoretical profiles, together with the initial distribution used in these calculations, are shown in figure 9. The solution of the diffusion equation used for this example was for diffusion from an arbitrary distribution into an infinite body.

CONCLUSIONS

The results of this study have shown that Bi ion implantation in both nickel and vanadium generates damage components within the ion range and at much deeper depths. We believe that the deep disorder could be due to the migration of interstitials (possibly assisted by focussed collisions) which form clusters at depths 2 to 3 times greater than the ion range. At high implantation doses, a polycrystalline layer was detected in vanadium over the range of the implanted ions and ascribed to the action of precipitates.

No such layer was detected in nickel, as the high sputtering yields were thought to prevent solubilities being exceeded.

Annealing of radiation damage in vanadium was found to be dependent on ion dose and while a very broad annealing stage was observed for low ($< 1 \times 10^{16}$ ions/cm^2) doses, no significant annealing was evident for doses where the polycrystalline la yer was generated. For nickel however, little dependence of annealing on dose was found and the shallow and deep disorder components annealed between 350 - 650°C and 500 - 700°C respectively.

Characteristics of diffusion of implanted ion species were found to depend on the ion species under consideration. Gallium ions for example, because of their "good fit" in the vanadium lattice, diffused by a substitutional process while the observed Bi and Se diffusion was by an interstitial process. Diffusion mechanisms were thought to be strongly dependent on the behaviour of precipitates within the polycrystalline layer, such that break-up of this layer produced both a rapid in- and outdiffusion of implanted ion species. Preliminary quantitative results, which were observed by fitting theoretical and experimental diffusion profiles, indicate that diffusion coefficients are comparatively low.

While diffusion in vanadium was a comparatively slow process, bismuth loss in nickel was found to occur over a narrow temperature interval. Further work is at present being conducted to obtain more quantitative results on diffusion of ions implanted in both metals.

ACKNOWLEDGEMENTS

The authors would like to thank Dr. O. Meyer for his many valuable discussions, Mr. R. Smithey for his sample preparations and Mr. M. Kraatz for careful implantations.

REFERENCES

1. NORRIS, D.I.R., Phil. Mag. 19, 527, 653 (1969).

2. CHEN, C.W., MASTENBROEK, A., ELEN, J.D., Rad. Eff. 16, 127 (1972).

3. CHEN, C.W., Phys. Stat. Sol. (a) 16, 197 (1973).

4. DIEHL, J., DIEPERS, H., HERTEL, B., Can. J. Phys. 46, 647 (1968).

5. RAU, R.C., LADD, R.L.,J.Nucl.Mat. 30, 297 (1969).

6. LINKER, G., GETTINGS, M., MEYER, O., III Int. Conf. on Ion
    Implantation in Semiconductors and other materials, Yorktown
    Heights Dec. 1973.

7. GETTINGS, M., MEYER, O., LINKER, G., to be published in
    Radiation Effects.

8. GETTINGS, M., LANGGUTH, K.G., MEYER, O., Verh. DPG 3, 367
    (1973).

9. SMITH, H.J., Rad. Eff. 18, 65 (1973).

DISCUSSION

Q: (J. A. Davies) Did you check your polycrystalline foils for
preferred orientation? This might possibly be a contributing factor to the broader penetration profiles that you observe.

A: We did not check the foils for preferred orientation. But we
have also performed implantations into evaporated layers with similar results. Even here for randomly oriented crystallites, channeling cannot be completely excluded. But in agreement with the experiments performed by H. J. Smith, recently published in Radiation
Effects, we think that this effect is of minor importance.

Q: (H. Bernas) Referring to your observation of broadening in the
Bi distributions after room temperature implantation. The characteristic channeling angles are large at these energies; we found it
necessary to set up a goniometer in the implanter to reduce (yet
not totally eliminate) channeling for 400 keV heavy ions in Ni or
Fe. Did you check this point?

A: Our samples were fixed during implantation and in principle we
cannot exclude channeling. But as we obtained similar distribution
for implantations into foils, we believe that channeling is of
minor importance and that the distribution broadening is mainly due
to radiation enhanced diffusion.

Q: (L. C. Feldman) What is the lattice location of the implanted
ions for the channeling and non-channeling case?

A: In the non-channeled case for Ni and V comparable numbers of
about 50% of Bi atoms on lattice positions have been detected. The

enhanced number for channeled implants in Ni was about 70%. But this is a preliminary result and further channeled implants are necessary for confirmation. Up to now we have not performed channeled implants in V.

Q: (E. N. Kaufmann) With regard to your observation that Bi channeled in Ni shows a higher substitutional fraction than a random implant; an anomalous behavior for channeled implants in Ni has been reported before for Hf ions by Odden, Bertier, et al. It might be worthwhile to compare your result with theirs. I also would like to know below what dose of implant ions do you no longer observe the polycrystalline layer?

A: The formation of the polycrystalline layer was dependent on the implanted ion species and on dose. For example, for gallium, which has a good fit into the V lattice this layer was not observed in as-implanted samples for doses up to 2×10^{17} ions/cm^2 and appeared only on heat treatment. But for all ions under investigation, i.e., Se_2, Ga, In, Kr, and Cs, it can be stated that below 5×10^{15} ions/cm^2 the formation of a polycrystalline layer has not been observed.

Q: (G. Dearnaley) What was the dose rate and the beam heating on the 360 keV Bi^+ implantations?

A: About 10 $\mu A/cm^2$ and there would be a small amount of beam heating. The specimens were 2 mm in thickness.

Sb-IMPLANTED Aℓ STUDIED

BY ION BACKSCATTERING AND ELECTRON MICROSCOPY*

G. J. Thomas[†] and S. T. Picraux

Sandia Laboratories, Albuquerque, New Mexico 87115

ABSTRACT

Ion channeling and backscattering techniques, in conjunction with transmission electron microscopy, have been used to study the Sb-implanted Aℓ system at high fluences (∿ 1 at.% Sb). Implantations of 200 keV Sb into single crystal Aℓ samples were carried out at room temperature and at 300°C to fluences of $5 \times 10^{15}/cm^2$. Extensive Aℓ lattice disorder is present in the room temperature implant, whereas little disorder is present after 300°C implantation. Upon annealing of room temperature implanted samples, Aℓ defect clusters produced during implantation were removed before reaching 150°C; Sb clustering within the implanted region occurred at ∿ 250°C and cluster breakup was observed above 400°C. In contrast, implantation at 300°C produced triangular precipitates which were aligned with the Aℓ lattice and exhibited diffraction consistent with the AℓSb structure. Thus, the importance of temperature during implantation for achieving compound phase formation by ion implantation is demonstrated.

*This work was supported by the United States Atomic Energy Commission.

[†]Present address: Sandia Laboratories, Livermore, California 94550.

INTRODUCTION

Ion implantation offers a means of introducing high concentrations of an impurity into the near-surface region of a solid.
Certain limitations of a metallurgical nature such as solubility
limits ($\sim$ 0.03 at.% for Sb in Aℓ below 600°C[1]) can be overcome by
implantation since the concentration of implanted ions is not controlled by a thermal equilibrium process. Thus, there exists the
possibility of forming metastable or equilibrium phases by implantation which may be difficult to achieve by other means. There has
been some early work demonstrating compound formation in semiconductors[2] and in metals,[3,4] including the Sb-Aℓ system.[3] However,
there has been little detailed investigation of the limits and
requirements of compound formation.

The Sb-Aℓ system was chosen for this preliminary study because
it has a relatively simple phase diagram with a single intermetallic,
AℓSb, of high stability and similar crystal structure to that for
Aℓ. Also, the low equilibrium solubility of Sb in Aℓ makes the 1
at.% Sb region studied here inaccessible by traditional techniques
of precipitation from solid solution. Finally, the sputtering coefficients for Aℓ and Aℓ_2O$_3$ are relatively low, thereby allowing
high concentrations of impurities to be introduced by implantation.

High fluence Sb implantations into single crystal Aℓ were
studied by means of transmission electron microscopy and by ion
backscattering and channeling. The results demonstrate the highly
complementary nature of these techniques for studying near-surface
modification of metals by high-fluence implantation. Precipitation
clustering, compound formation and dissolution effects are observed.
In addition, the importance of implantation temperature as opposed
to post-implantation annealing on new phase formation is demonstrated.

EXPERIMENTAL TECHNIQUE

Implantations were performed on (110) single crystal Aℓ at
room temperature and at 300°C with 200 keV Sb to fluences up to
5 x 10[15] Sb/cm^2, corresponding to $\sim$ 1 at.% Sb within the implanted
region. Implantations were along a nonchanneling direction 7° from
the $\langle 110 \rangle$ axis and typical implantation fluxes were $\approx$ 4 x 10[11] ions/
cm^2-sec.

The annealing characteristics of room temperature implanted
samples were studied _in situ_ by 2 MeV He ion channeling and backscattering measurements. Additional samples implanted in the bulk
were subsequently thinned for 100 kV transmission microscopy and
also annealed _in situ_. In the next section, the annealing results
on room temperature implanted samples will be discussed first,
followed by the results for 300°C implantation.

RESULTS

Room Temperature Implants

Figure 1 shows the Sb part of the spectrum for 2 MeV He back-scattering measurements on a sample implanted at 23°C to 5×10^{15} Sb/cm^2. The scattering yield versus energy indicates the Sb depth profile; the corresponding depth scale is shown above the profile. The measured projected range, R_p, and spread, ΔR_p, of 750 Å and 275 Å, respectively, agree well with calculated values of $\sim$ 800 Å and $\sim$ 230 Å. In addition, an Sb concentration within the peak of $\sim$ 0.9 at.% was found which is close to the predicted value. A small enhanced tail is observed, however, on the deeper side of the profile.

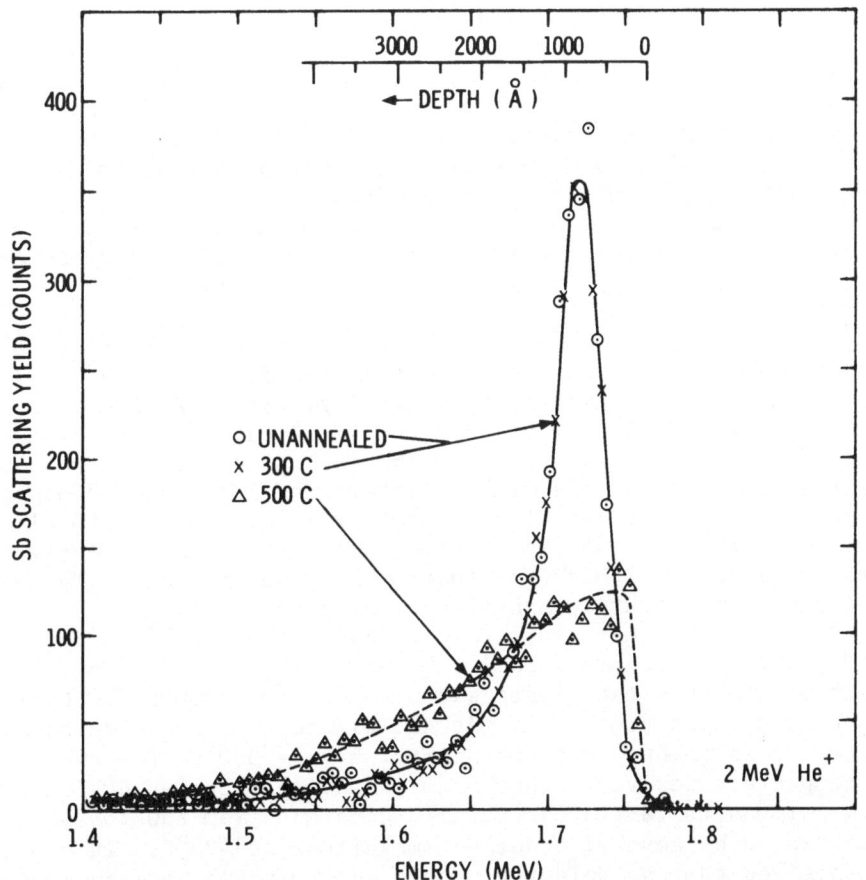

Fig. 1.   Sb portion of the backscattering spectrum for 2 MeV He incident on Al implanted with 200 keV Sb to a fluence of $5 \times 10^{15}/cm^2$ at 23°C and after 15 min anneals at 300 and 500°C.

As shown in Fig. 1 essentially no change in the total Sb depth distribution is observed upon annealing 15 min at 300°C. In fact, the profile remained the same for anneals through 450°C. At 500°C, a significant broadening and reduction in the Sb distribution occurred, indicating the onset of long range migration. The sharp edge on the high energy side of the Sb distribution after the 500°C anneal for 15 min indicates that some Sb has reached the surface of the sample. The annealed profile did not change with an additional 15 min anneal (at 500°C), suggesting clustering effects. Additional migration and loss of Sb from the profile was observed at 550°C.

Ion channeling measurements were also performed on room temperature implanted samples as a function of anneal temperature. They indicate ≤ 10% of the Sb atoms are located along ⟨110⟩ rows, and suggest that the Sb is primarily nonsubstitutional. Disorder-induced dechanneling in the ⟨110⟩ Aℓ spectrum occurs at depths corresponding to the Sb range distribution. The relative "channeling-disorder" was determined by the increase in the dechanneled level relative to that before implantation at a depth beyond the Sb distribution (≈ 2300 Å). Annealing to 150°C produces some reduction in the dechanneling, while further reduction occurs above 300°C. Thus, the ion channeling and backscattering measurements indicate some annealing of Aℓ disorder occurring at relatively low temperature, but no long range migration of the Sb until 500°C.

In Figs. 2(a) and 2(b) transmission electron micrographs are shown of a sample implanted at room temperature to 5×10^{15} Sb/cm^2. The as-implanted condition is shown in Fig. 2(a). A high density of dislocations, dislocation loops and small clusters can be seen. Figure 2(b) shows the same area of the foil after heating to 150°C. It is seen that all of the small clusters and loops have disappeared; however, the large dislocation segments remain. No additional changes were observed as the specimen temperature was linearly increased at a rate ~ 10–15°C/min until 250°C was reached. The results of the higher temperature annealing can be seen in Fig. 3. A high density of small defect clusters was observed by 250°C as shown in Fig. 3. The defects are ~ 20 Å in diameter and have a density > 10^{17}/cm^3. At 300°C the clusters coarsened and through 390°C, this coarsening process continued. At 425°C, the clusters were observed to decrease in number as shown in Fig. 3. By 500°C, the clusters were completely removed. The volume fraction occupied by the clusters was found to remain essentially constant from 250 to 390°C, but decreased by several orders of magnitude at 425°C. It should be emphasized that no defect clusters were visible between 150°C and approximately 225–250°C at the highest magnification attainable on the microscope. The onset of cluster visibility was difficult to ascertain.

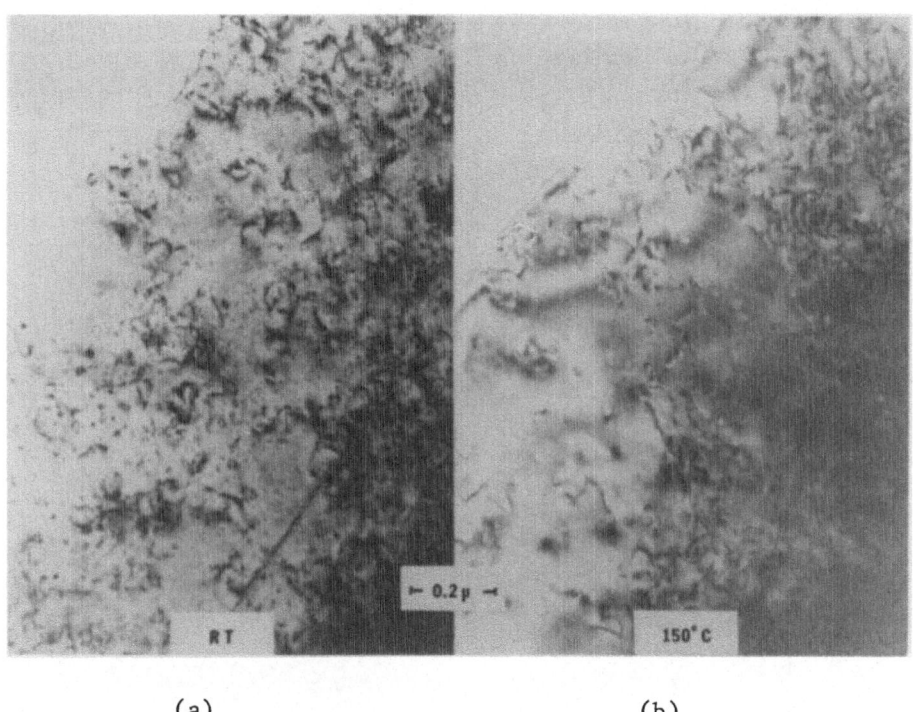

(a)   (b)

Fig. 2. Transmission electron micrograph of a 200 keV Sb implant
of 5 x 10^{15}/cm^2 in Al at 23°C. (a) as implanted and (b)
after 150°C anneal.

300°C Implants

Samples were implanted at 300°C in an attempt to enhance the
precipitate size and ordering, and to decrease lattice damage. The
previous results indicate that the implanted Sb was mobile in Al by
250°C, and that significant annealing of implantation-produced defect clusters occurred at lower temperatures. Ion backscattering
and channeling measurements for the 300°C implant showed the peak
of the Sb depth profile to be shifted closer to the surface and that
more Sb atoms were contained in the tail of the distribution compared to the room temperature implants. No significant change ($\sim$ 1%)
in the Al $\langle 110 \rangle$ dechanneling was observed after 300°C implantation.

The microstructure observed by TEM after 300°C implantation is
clearly different from the room temperature case. First, there is
no dense dislocation structure observed, and second, one finds triangular shaped precipitates. These precipitates are shown in Fig.
4, which is a dark field micrograph of a sample implanted at 300°C
to a dose of 5 x 10^{15} Sb/cm^2. The triangular faces appear to be on

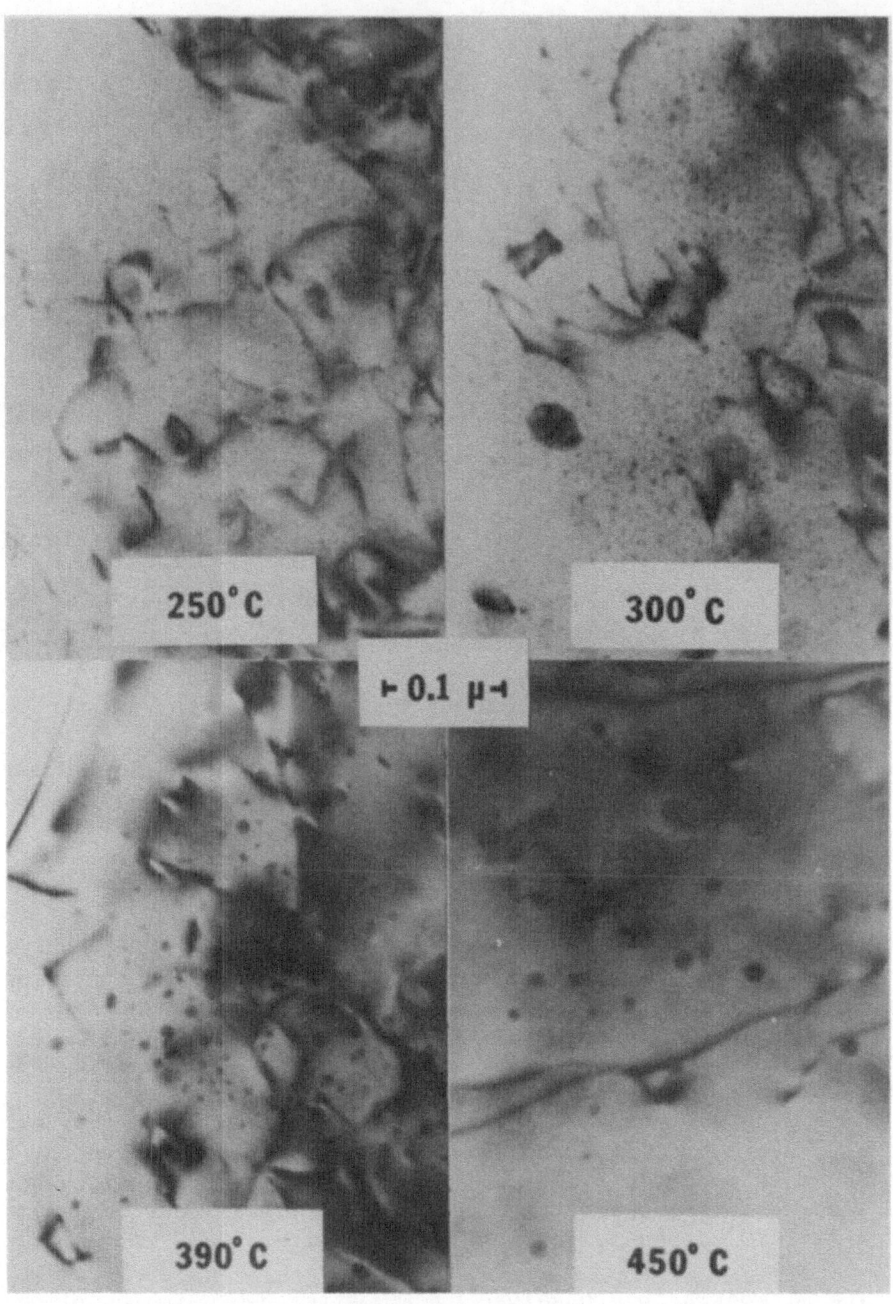

Fig. 3. TEM micrographs of same sample shown in Fig. 2 after higher anneal temperatures.

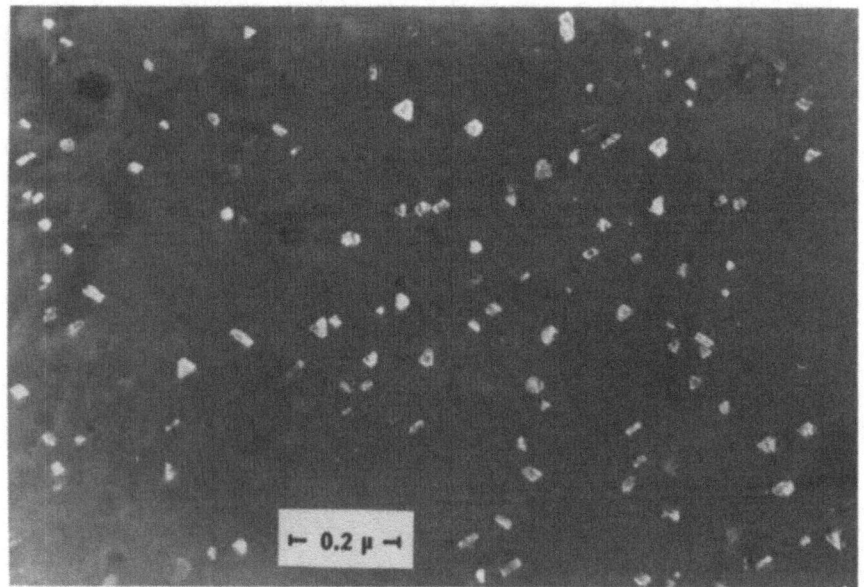

Fig. 4. Dark field TEM micrograph of Aℓ implanted to 5 x $10^{15}/cm^2$
         23°C.

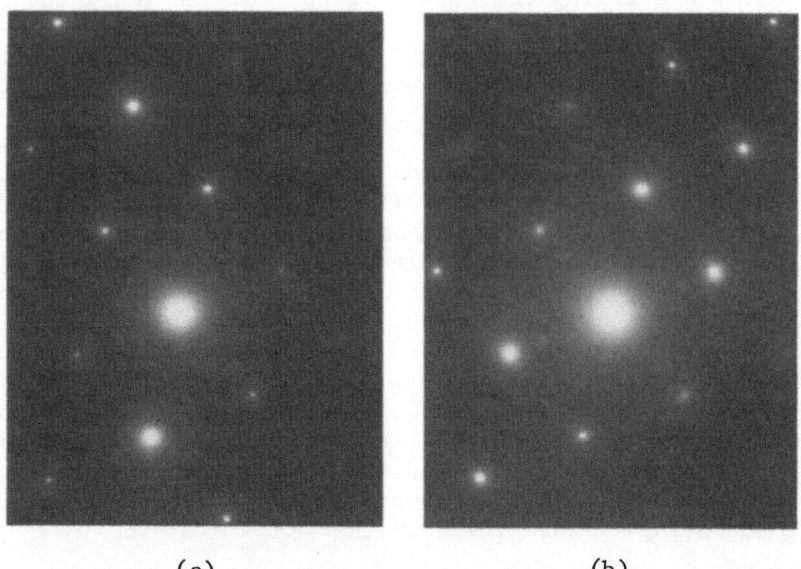

(a)   (b)

Fig. 5. Electron diffraction patterns. (a) sample implanted at
         room temperature. (b) sample implanted at 300°C. The
         diffraction spot used to image Fig. 4 is indicated.

(111) planes, with the edges along ⟨110⟩ directions. The precipitates are ∿ 300 Å along the edges and ∿ 100 Å thick with relatively little strain contrast in the Aℓ lattice. Thackery and
Nelson[3] observed a similar structure for 330 °C Sb implants in Aℓ.
The operating reflection used in Fig. 4 was a non-Aℓ lattice diffraction spot indicated in Fig. 5. Figure 5(a) is a diffraction
pattern of an Aℓ sample implanted at room temperature showing a
typical (110) zone pattern. In comparison, Fig. 5(b) shows the
same zone after 300°C implantation. The additional reflections
due to the precipitates are clearly evident. The spacings of the
diffraction spots were found to be consistent only with an AℓSb
lattice structure. In addition, the patterns were aligned with
the Aℓ lattice.

DISCUSSION

It is clear by comparing the backscattering and TEM results
that the defect clusters produced during implantation at room temperature are Aℓ defects and that Sb remains sub-microscopically
dispersed until about 250°C. At this temperature, a new distribution
of small clusters was formed with an average spacing ∿ 200 Å. This
is consistent with high temperature dilute diffusivities of Sb in
Aℓ which when extrapolated to 250°C indicate a diffusion length ≈
400 Å for times ≈ 5 min. The Sb backscattering yield remains essentially the same upon annealing below 500°C because agglomeration
occurs, thus preventing the escape of Sb from the implanted region.
The clusters observed by TEM are therefore believed to be Sb-rich
regions. In addition, the lack of continued Sb migration with
isothermal annealing at 500°C, as might be expected for dilutely
distributed Sb, suggests Sb clusters are present.

The disorder and Sb migration results of ion channeling, backscattering and TEM observations with annealing are summarized in
Fig. 6. It is seen that the significant annealing in Aℓ disorder
at ≈ 100°C correlates well with the TEM observed loss of implantation-produced defect clusters and loops. The Sb distribution remains
unchanged in this temperature region. Near 250°C Sb clustering
occurs while the overall implanted depth distribution is maintained.
A further reduction in the Aℓ lattice disorder above 300°C is seen
to accompany coarsening of the Sb-rich clusters as observed by TEM.
The observed breakup of Sb-rich clusters in the TEM at somewhat
lower temperatures (≈ 425°C) than the backscattering data (≈ 500°C)
is believed to be due to the close proximity of the surface in the
in situ annealing of thin foils in the TEM.

In contrast to the room temperature implantation, 300°C implantation produced large crystalline AℓSb precipitates aligned with the
Aℓ lattice and accompanied by little lattice damage. Since the
solubility of Sb in Aℓ is far below these implant concentrations,

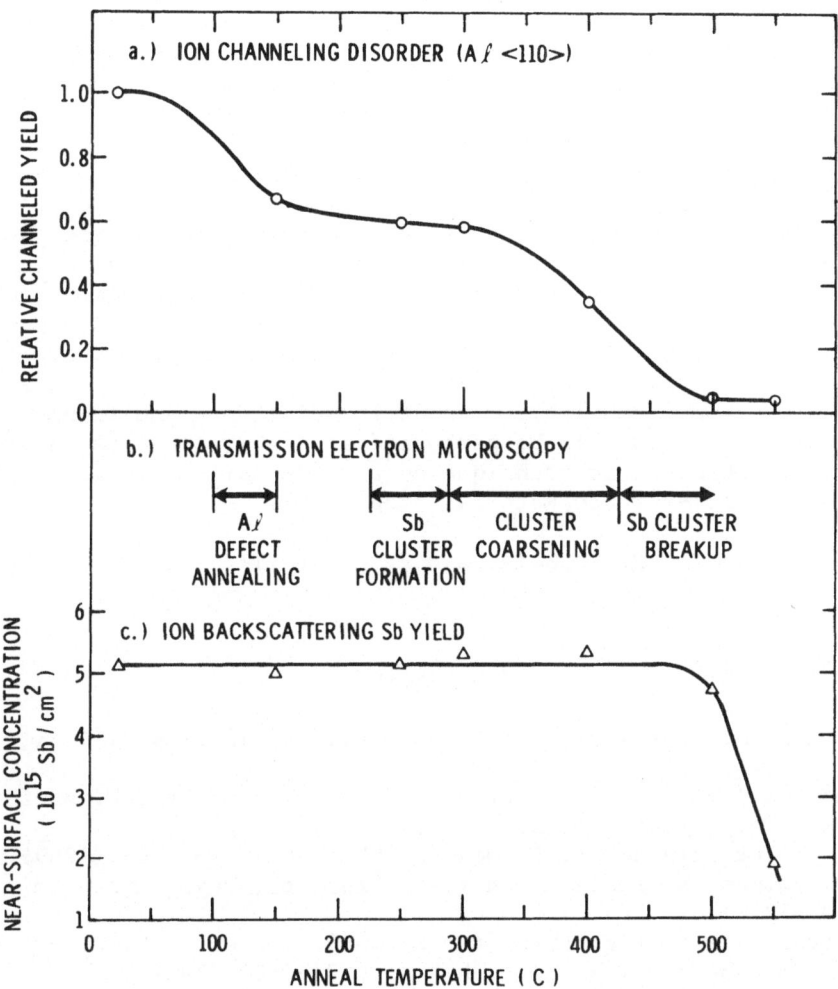

Fig. 6.   Summary of anneal data for room temperature 5 x 10^{15}
          Sb/cm^2 implanted in Aℓ.

one would expect the Sb to be removed from solution in the Aℓ lattice with second phase formation as soon as the Sb became sufficiently mobile. It is quite surprising, however, that when the Sb is mobile _during_ implantation, AℓSb precipitates form, whereas annealing of room temperature implants to corresponding temperatures results in Sb clustering. Although the reason for this behavior is not presently understood, the influence of the defect generation during implantation may be a key factor in the AℓSb compound formation. Consistent with these ideas are the early results of Thackery and Nelson[3] which indicated AℓSb precipitate formation by Sb implantation at only 40°C using much higher Sb implantation fluxes (6 x 10^{13} ions/cm^2-sec compared to the 4 x 10^{11} ions/cm^2-sec used here). In this case, the higher vacancy concentration with higher ion flux could have resulted in sufficient enhanced diffusion of the Sb for AℓSb formation to occur during implantation. A valuable additional experiment to help clarify these concepts would be to heat a low flux room temperature Sb-implanted Aℓ sample to 300°C while _under_ Aℓ ion irradiation. Also, Aℓ ion irradiation after Sb clustering at 300°C had occurred should help to clarify if the AℓSb phase is the one in dynamic equilibrium under irradiation. Additional implantation studies will be performed at higher fluences in order to further explore this technique as a means for controlled modification of near-surface material properties.

Assistance by R. G. Swier and F. A. Gruelich is gratefully acknowledged.

REFERENCES

1. M. Hansen, _Constitution of Binary Alloys_, (McGraw-Hill, New York, 1958) p. 130.
2. See for example, J. A. Borders, S. T. Picraux, and W. Beezhold, Appl. Phys. Lett. _18_, 509 (1971).
3. P. A. Thackery and R. S. Nelson, Phil. Mag. _19_, 169 (1969).
4. E. Arminen, A. Fontell and V. K. Lindroos, Phys. Stat. Sol. _4_, 663 (1971).
5. Approximate values taken from Sb in Si tables by D. K. Brice, Sandia Laboratories Research Report 71-0599 (1971).
6. S. Badrinarayanan and H. B. Mathur, Intl. J. Appl. Rad. and Isotopes _19_, 353 (1968).

DISCUSSION

Q: (B. Hertel) Do you know of which type the very small defects
are?

A: We have not analyzed the small defects and dislocation loops
produced in the room temperature implantations, since we were most
interested in following the Sb distribution. It is clear from our
results, however, that the room temperature clusters consist of Al
defects.

Q: (J. A. Sprague) Have you looked at the 500°C implant by microscopy to see if the AlSb precipitates are visible after this
treatment?

A: No, we've not done this yet.

Q: (J. E. Westmoreland) In addition to increased dechanneling
near R_p of Sb, you had an enhanced surface peak. Was the nature
of this investigated?

A: No. It is presumably due to surface damage and surface oxide
growth.

THE CHANGES IN ELECTRICAL PROPERTIES OF TANTALUM THIN FILMS FOLLOWING ION BOMBARDMENT

I. H. Wilson, K. H. Goh and K. G. Stephens

Department of Electronic and Electrical Engineering

University of Surrey, Guildford, Surrey, England

INTRODUCTION

Tantalum based thin films are currently used as resistors in hybrid integrated circuits because of their excellent stability and reliability. For high accuracy resistors it is required that the thermal coefficient of resistivity (TCR) is very low, whilst for a highly stable RC circuit the TCR must balance the temperature coefficient of capacitance.

In both cases a high sheet resistance is required to minimize area.

The films are normally deposited by sputtering using a reactive atmosphere in the case of the compounds and typical properties are given in Table 1.

TABLE 1
Properties of Typical Sputtered Films

Film	ρ ($\mu\Omega$cm)	TCR ppm/$^{\circ}$C	REF
α-Ta	50	+650	
β-Ta	170	-160	1
Ta+N	200	-100	
Ta+O	10000	-1500	2
Ta+N+O	700	-100	3
Ta+Al	300	<±100	4

The resistivity and TCR cannot be varied independently of one another and can only be varied over a very small range by altering the deposition parameters.

Tantalum nitride and tantalum/aluminium alloy films are the ones that are used most commonly.

The results of a preliminary investigation of the use of ion bombardment to alter the electrical properties of tantalum thin films are presented in this paper. The aim has been to investigate the amount of control of parameters such as resistivity and TCR that can be achieved by this technique. The long-term stability of the films has not been investigated.

EXPERIMENTAL

Details of the films reported in this paper, their starting properties and the ion bombardment are given in Table 2.

TABLE 2

Details of Thin Films Bombarded

Sample	Thickness Å	Ion	Energy keV	Resistance $\Omega/\square$	Resistivity $\mu\Omega$cm
Sputter Deposited Ta	700	N_2^+	40	34	230
Sputter Deposited Ta/Al Alloy	400	Ar^+ N_2^+	40 40	81 74	320 300
Evaporated Ta + Ta_2O_5	550 400 560 1400 1800	Ar^+ O^+ N_2^+	40 20 40 200 220	1700 650 1100 440 170	9100 2600 9500 5700 3000

The energy was chosen such that the projected range would be between $\frac{1}{4}$ and $\frac{1}{2}$ the initial film thickness. In all cases the substrates were Corning 7059 glass and all the normal cleaning procedures were followed in a clean room environment.

The sputter deposited films were prepared by Ultra Electronics Limited. The D.C. diode technique was used with an electrode spacing of $1\frac{1}{4}$ in., an argon gas pressure of 0.05 torr and a

sputtering voltage of 3.5 kV. The temperature of the cathode was
carefully controlled and a half-hour pre-sputter preceded deposition
of a film. Uniformity was checked by comparing films successively
deposited on 9 substrates. In the case of the Ta/Al films an
outer ring of aluminium was attached to the cathode, and the films
used were typical of those used in normal production by Ultra
prior to final oxidation of the surface. The area of the
sputtering cathodes was designed to produce 79 At% Ta/21% At% Al
but wet analysis gave a composition of 88%/12%. Substrates were
water cooled during sputtering.

Samples of the pure tantalum film from the same batch as those
used in this experiment were analysed by X-ray diffraction and the
results indicated that the films are composed of a mixture of b.c.c.
and β-tantalum. Electron diffraction of the Ta/Al alloy films
revealed an amorphous structure of ill defined nature. A depth
profile of a pure tantalum film was obtained using an electron
spectrometer for chemical analysis (ESCA) coupled with argon ion
etching[8] . This revealed an oxidised layer of about 40 Å in
thickness and a ratio of tantalum metal to oxidised tantalum of
3:1 in the bulk of the film.

The evaporated films were prepared by Edwards High Vacuum
Research Laboratories (Crawley) using electron beam evaporation in
a liquid nitrogen trapped oil diffusion pumped system. Background
pressure during evaporation was $< 1 \times 10^{-5}$ torr. Qualitative
X-ray and Rutherford backscattering analysis of these films
revealed the presence of a large amount of oxide. Using published
data[2] it is estimated from the resistivity of the films that the
oxygen content of the films ranges between 35 and 40 At%. If this
is in the form of Ta_2O_5 there would be 45 to 50% free tantalum.

The resistors were in the form of a strip typically 8 mm × 25 mm
with evaporated titanium/gold contact pads on each end. The
bombarded area was 12 mm across the film strip × 8 mm along the axis
so that the change in resistance of one square was measured.

The ion bombardment was carried out in a 600 keV heavy ion
accelerator with magnetic mass analysis[5] . Uniformity of dose
in the bombarded area was ensured by electrostatically scanning a
de-focused beam over a much larger area. The pressure in the ion
pumped target chamber was usually $< 1 \times 10^{-6}$ torr during bombardment.

A Marconi bridge type TF 2700 (frequency 1 kHz) was used to
measure the resistance in some cases but usually a Wheatstone
bridge accurate to 0.03% was used.

The resistance was measured at a number of fixed time intervals

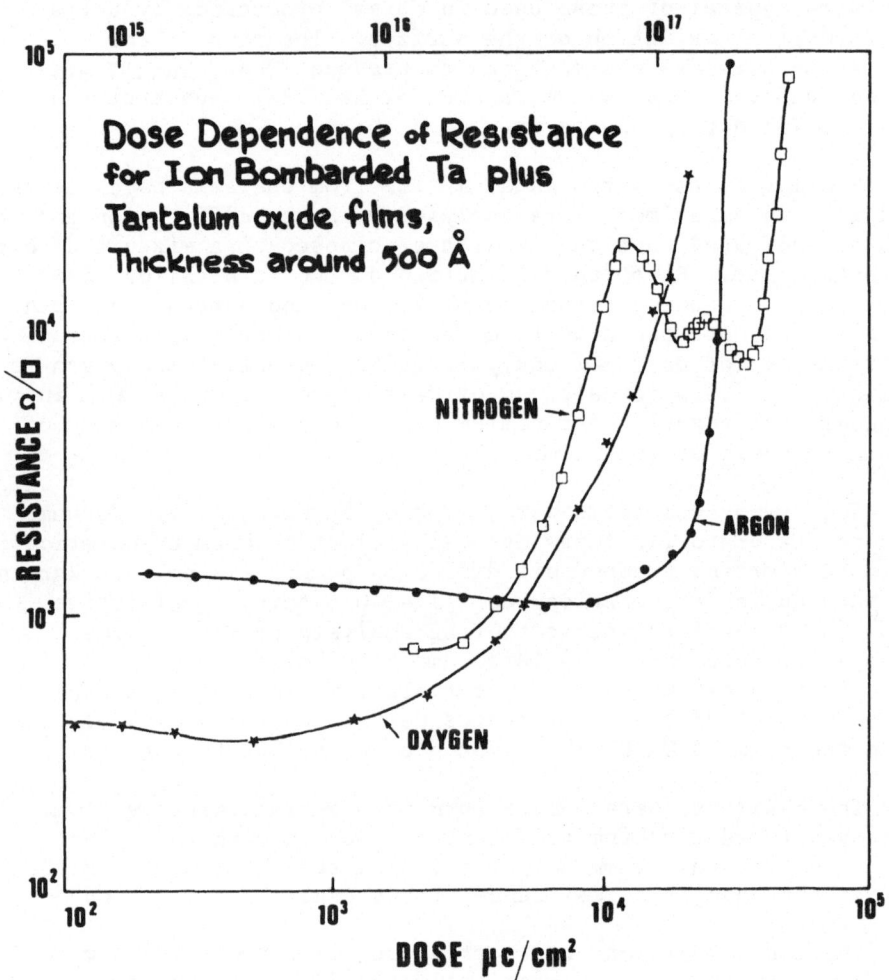

FIGURE 1

The Variation of Resistance with Ion Dose for Evaporated Films

after isolation from the beam to obtain a relative measure of the
TCR, from the change in resistance as the specimen cools from the
equilibrium temperature obtained during bombardment (100°C to 200°C
depending on ion energy).

A vacuum furnace was used for accurate TCR measurements over
the range 20° to 100°C at a pressure of $< 1 \times 10^5$ torr; this was
also used to check the changes in room temperature resistance that
would result from the beam heating.

No changes were detected below 300°C and the change in
resistance at 400°C showed a linear dependence on $(time)^{\frac{1}{2}}$
suggesting parabolic oxide growth kinetics[6] .

No changes in resistivity were measured after exposure of the
films to the atmosphere.

RESULTS

The dose dependence of resistance for argon, oxygen and
nitrogen bombardment is illustrated in Figure 1. In this case
the targets are evaporated films of around 500 Å in thickness,
but similar behaviour was shown by all the three types of film
studied.

The argon bombarded specimens show a slight drop in resistance
at low doses followed by a rise in resistance which becomes very
sharp increasing by up to five orders in magnitude for a 1%
increase in dose until the film becomes open circuited. In the case
of the oxygen bombarded specimens the rise in resistance starts
much earlier but is much more gradual than that seen for argon
bombarded specimens.

The resistance of the nitrogen bombarded specimens rises with
dose in a similar fashion to the oxygen bombarded specimens until
a dose is reached where the resistance starts to fall sharply. In
all cases the nitrogen bombarded specimens showed one or two peaks
in variation of resistance with dose.

The sensitivity of the initial change in resistance to vacuum
conditions is illustrated in Figure 2 for nitrogen bombardment of
Ta/Al alloy films, where it can be seen that the initial drop in
resistance does not occur if the specimen is left for some time in
a good vacuum. The dose dependence of resistivity for nitrogen
bombardment of sputter deposited tantalum is shown in Figure 3. The
resistivity is calculated assuming a constant rate of sputtering.
The dose dependence of TCR is also shown.

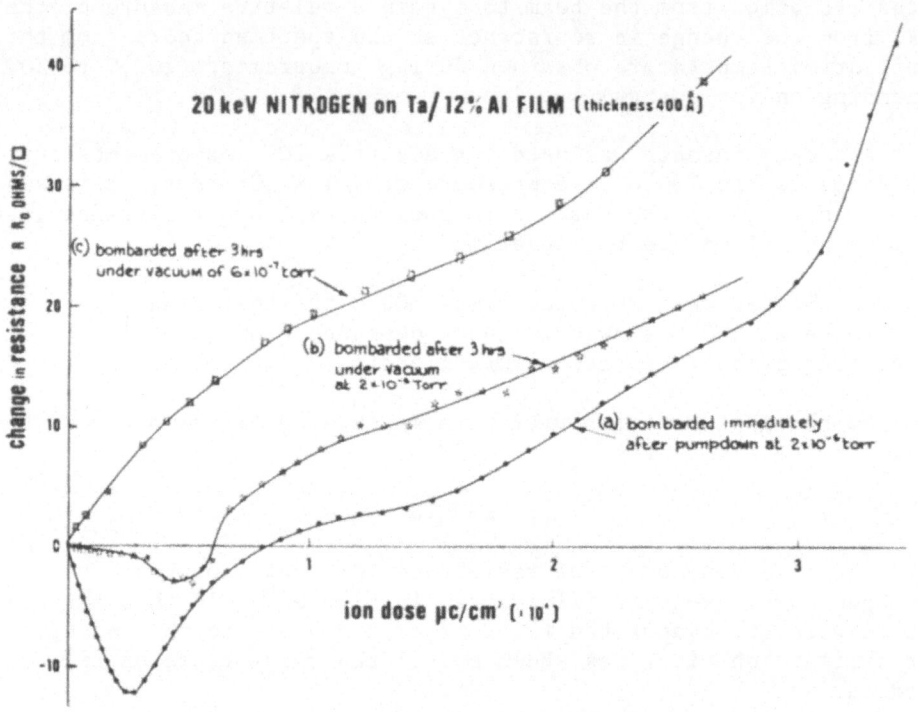

<u>FIGURE 2</u>

The Effect of Vacuum Conditions on the Change in Resistance
with Dose at Very Low Doses

The peaks in resistance are much less pronounced than in the
case of the evaporated films. The TCR starts positive but soon
goes negative and the dose dependence is very similar to that of
the resistivity showing a small step followed by a more pronounced
peak. The arrow marked $R_{N_2^+}$ indicates the dose at which it is
estimated that the thickness of the film equals the projected
range of the nitrogen ions. The dose dependence of resistivity
for argon and nitrogen bombarded Ta/Al alloy films is shown in
Figure 4. The nitrogen bombarded film shows a sharp peak in
resistivity and TCR at a dose where it is estimated that the film
thickness equals the projected range. There is also a step at a
similar dose for the argon bombarded specimen.

The dose dependence of resistance for nitrogen-bombarded
evaporated films of 1400 and 1800 Å thickness is shown in
Figure 5 together with that for the 560 Å film shown in Figure 1.
The two thinner specimens exhibit two sharp peaks whereas the
thickest specimen shows one broad peak in resistance with a very
large drop on the high dose side. The result for the thickest

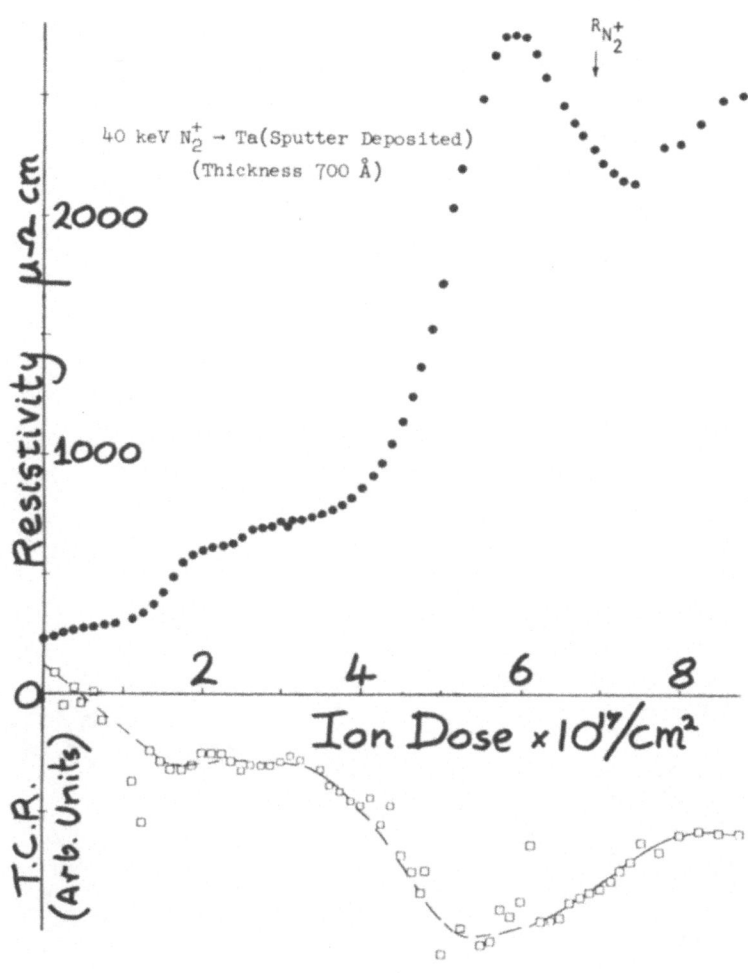

FIGURE 3

Dose Dependence of Resistivity and TCR for Nitrogen
Bombardment of Sputter Deposited Tantalum

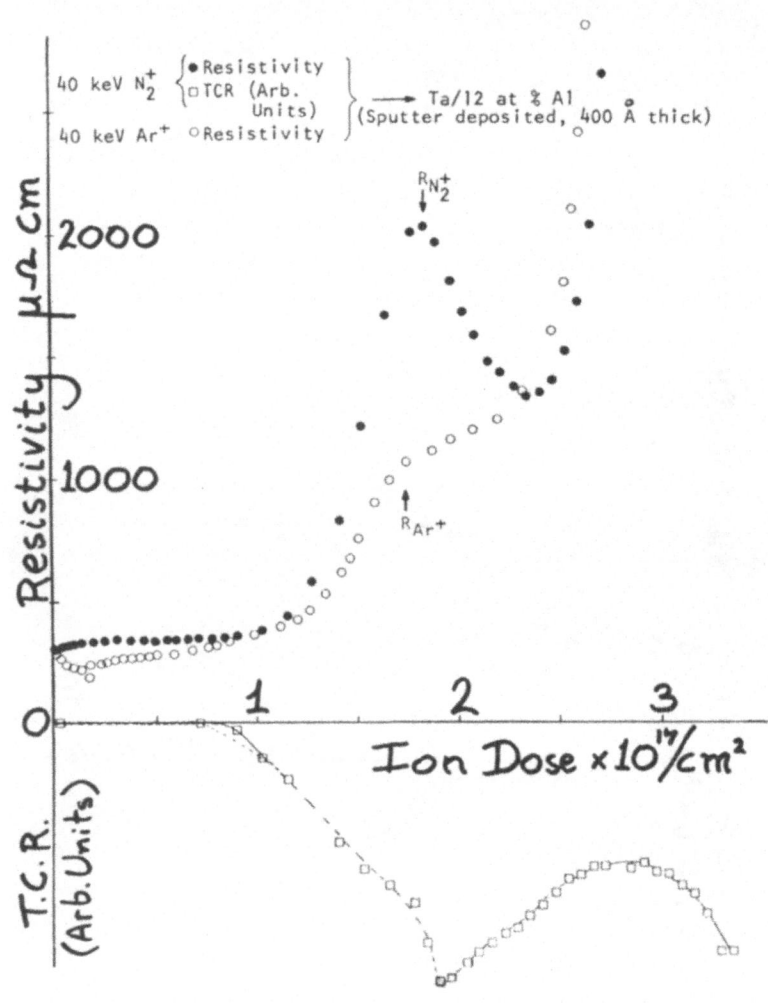

<u>FIGURE 4</u>

Dose Dependence of Resistivity and TCR for Nitrogen Bombardment & the
Dose Dependence of Resistivity for Argon Bombardment of Ta/Al Alloy

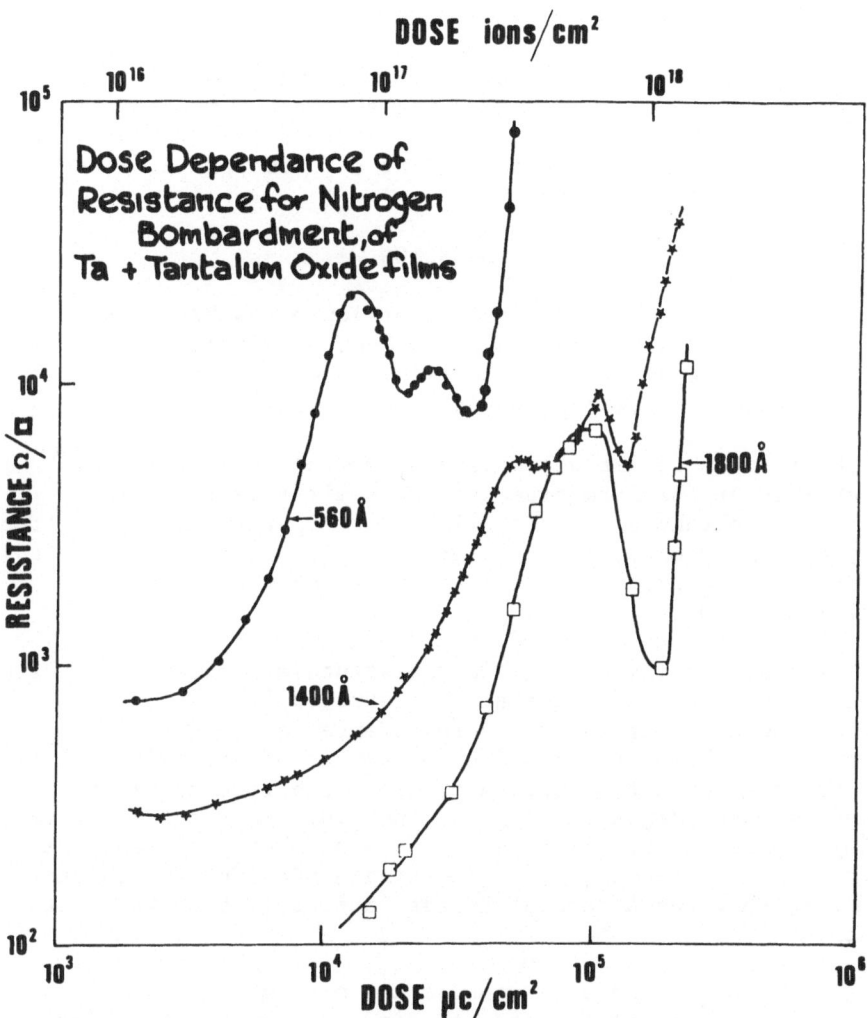

FIGURE 5

Dose Dependence of Resistance for Nitrogen Bombardment
of Tantalum + Tantalum Oxide Evaporated Films

specimen is re-plotted in Figure 6 as a dose dependence of resistivity for comparison with the TCR. The plotted points are the absolute values determined in the vacuum furnace and the justification for drawing the dotted curve to these points was supported by relative measurements of TCR taken by the beam heating method. The dose dependence of the TCR follows that of the resistivity, showing a very sharp dip and going to a small positive coefficient at the same dose as the minimum in resistivity. It is interesting from the applications angle that near the minimum the film has a TCR of $< \pm 100$ ppm per $^{\circ}$C and a resistivity $\gtrsim 5000$ $\mu\Omega$ cm.

DISCUSSION

The initial fall in resistance at low doses has been observed by other workers[7] and has been attributed to desorption of gases from the surface of the film. Our observations of the effect of vacuum conditions would support this view.

The rise in resistance with dose of the argon bombarded specimens is of the form expected to result from a uniform reduction in thickness in the film due to sputtering. The effect of oxygen bombardment, namely a gradual climb in resistance from low doses can be interpreted as a gradual conversion of the tantalum to Ta_2O_5.

The doses at which peaks in resistance are observed for the nitrogen bombarded specimens are illustrated in Figure 7. The ratio of the dose for a peak in resistance to the dose where the resistance is rising rapidly due to removal of the film (D_T) is shown together with the dose ratio where the film thickness equals the mean projected range of the ion. In the case of films showing two peaks the first peak occurs at about $\frac{1}{4}$ D_T and the second occurs at about $\frac{1}{2}$ D_T. The single peak for the thickest evaporated specimen occurs at a dose intermediate between these two doses.

An attempt was made to measure the nitrogen profile by E.S.C.A. coupled with argon sputter etching for a 700 Å pure tantalum specimen bombarded to a dose in the region of the first peak[8] This indicated with a $\pm$ 10% accuracy that the nitrogen concentration was uniform throughout the film.

Assuming that the implanted nitrogen is distributed evenly throughout the film then the peaks in resistance for the 560 Å evaporated film occur at doses of 0.53 and 1.1 × the number of free tantalum atoms remaining in the film (assuming 45% free tantalum). This would suggest that the first peak in resistance is due to a change from a mixed phase region (Ta + Ta_2N) to single-

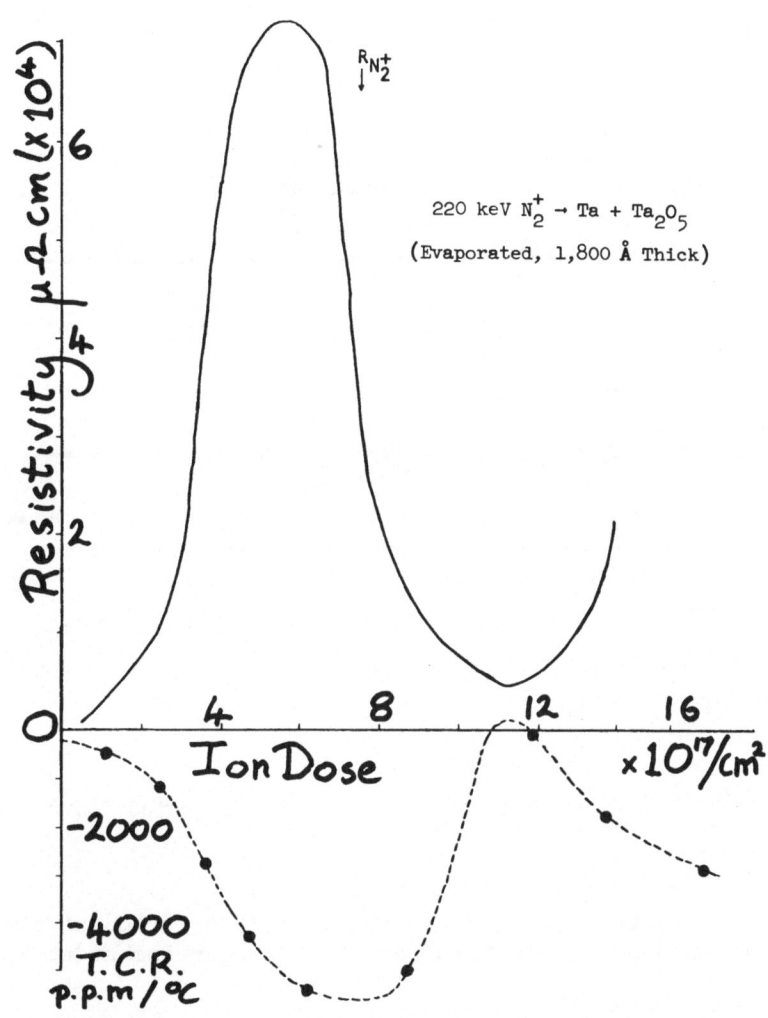

220 keV N_2^+ → Ta + Ta_2O_5

(Evaporated, 1,800 Å Thick)

FIGURE 6

Dose Dependence of Resistivity and TCR for a Nitrogen Bombarded
1800 Å thick, tantalum + tantalum oxide evaporated film

phase Ta_2N and the second peak is due to the change from Ta_2N + TaN to single-phase TaN. The single peak in the case of the evaporated specimen is possibly due to non-uniformity in doping resulting in both nitrides being formed in different parts of the film at the same time. The scatter in the positions of the peaks as shown in Figure 7 possibly reflects the different sputtering rates of the films.

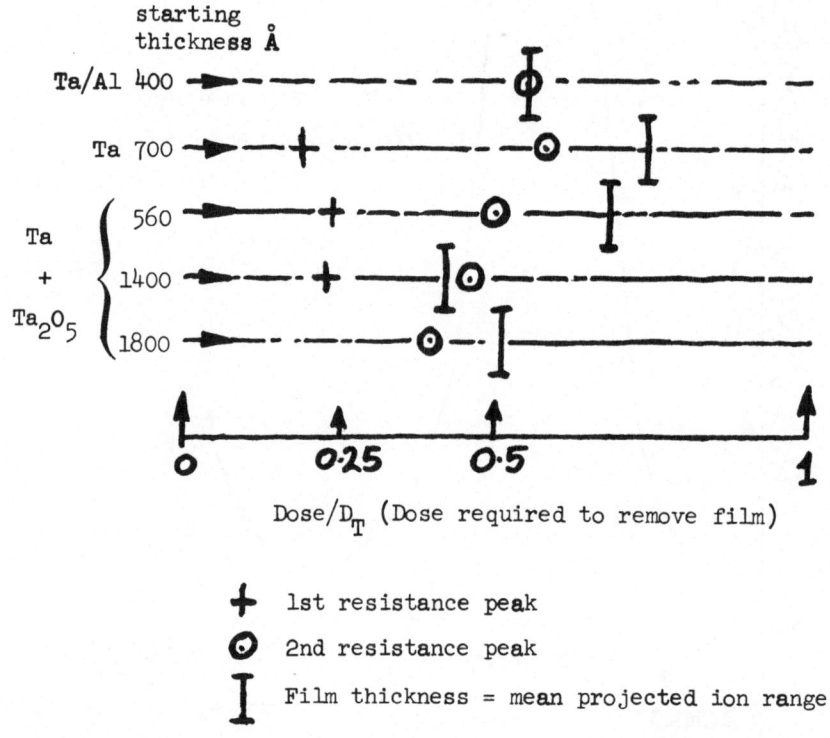

FIGURE 7

Diagram of the Resistance Peak Positions in Relation to the Dose Required to Remove all the Film for the nitrogen Bombarded Specimens

Several questions cannot be explained in the present work. No similar fluctuation in resistivity on changing structure from Ta_2N (h.c.p.) to TaN (f.c.c.) has been reported by workers using deposition by reactive sputtering; in fact there is usually a plateau in both the resistivity and TCR at about 200 $\mu\Omega$ cm and – 100 ppm/$^{\circ}$C respectively[1,9]. This could be related to the fact that implanted ions are much more energetic than sputtered atoms, leading to different structures in the two cases.

The resistance peaks for the heavily oxidised films are much

more pronounced than those for pure tantalum. This could be due to
the effect of the regions of oxide on the sputtering rate, the
entrapment and diffusion of nitrogen or the structure of the resulting nitrides.

The single peak for the Ta/Al alloy films is just where the
range equals the film thickness, and an argon bombarded specimen
shows a step at a similar position. This could be due to radiation
enhanced annealing of damage only detectable when the damage peak
moves into the substrate, this effect being greatly enhanced when
nitrides are being formed. No such dependence on ion range was
found with the evaporated films as increasing the beam energy did
not alter the peak position but just reduced the size of the peak.

CONCLUSIONS

The resistivity versus dose curve of ion bombarded tantalum,
tantalum + tantalum oxide and tantalum/aluminium alloy films show
three regions. At low doses the resistance change is sensitive to
absorbed gases on the surface, whilst at high doses the effect of
reduction in thickness by sputtering swamps all other effects.

At doses between these two extremes the chemical effects of
reactive ions can be observed. For oxygen bombardment there is a
gradual and continuous rise in resistance, probably as the free
tantalum is converted to Ta_2O_5.

For nitrogen bombardment, peaks in resistance and TCR are seen
which are probably associated with the formation of Ta_2N and TaN.
The peaks are much more pronounced for the evaporated films containing
large amounts of oxide and in this case it is possible to achieve
resistivity of > 5000 $\mu\Omega$ cm coupled with a TCR of less than
± 100 ppm/°C which is a considerable improvement on what has been
achieved previously for tantalum oxide and/or tantalum nitride
resistors.

Some evidence of sensitivity to radiation damage is seen for
the Ta/Al alloy films. Clearly a great deal more work is needed,
especially in analysis of the composition and structure of the
implanted films, to explain the mechanisms involved in these effects.

REFERENCES

1. A. Schaver & M. Roschy, Thin Solid Films, 12(1972)313.
2. D. Gerstenberg & C. J. Calbick, J. Appl. Phys., 35(1964)402.
3. W. R. Hardy et al., Thin Solid Films,8(1971)81.
4. R. G. Duckworth, Thin Solid Films, 10(1972)337
5. P. J. Cracknell et al., Nucl. Instr. Methods, 92(1971)465.
6. C. A. Steidel & D. Gerstenberg, J. Appl. Phys., 40(1969)3828.

7.   B. Navinsek & G. Carter, Can J. Phys., 46(1968)719.
8.   P. L. F. Hemment, University of Surrey, private communication.
9.   M. Nakamura et al., Jap. J. Appl. Phys. 12(1973)30.

IMPLANTATION AND DIFFUSION OF Au IN Be:

BEHAVIOR DURING ANNEALING OF A LOW-SOLUBILITY IMPLANT[*]

S. M. Myers and R. A. Langley

Sandia Laboratories

Albuquerque, New Mexico 87115

ABSTRACT

The behavior of implanted gold in single-crystal α-beryllium has been studied during annealing, under conditions where the local concentration of the implant was much greater than the solid solubility. Room temperature implants of $\simeq 1.0 \times 10^{17}$ Au atom/cm^2 at 100 keV resulted in a concentration peak of $\simeq 14$ at.% at a depth of $\simeq 700$ Å. Energy spectra of backscattered 2 MeV He$^+$ ions were used to determine the Au concentration versus depth during isothermal anneal sequences at 665 and 780 C. It was found that the implantation peak in the Au concentration profile did not broaden with annealing, but rather decreased continuously in amplitude. Gold diffused from the implanted layer into the bulk of the sample, where its concentration was $\leqslant 0.1$ at.%. A quantitative description of this behavior was achieved by using the diffusion equation and by assuming the concentration outside the implanted layer to be limited by the solid solubility of Au in α-Be. Diffusion coefficients D and solid solubilities C_o were extracted from the analysis with uncertainties of about ±30% in D and ±10% in C_o: at 665 C, $D_A = 2.8 \times 10^{-12}$ cm^2/sec, $D_C = 1.5 \times 10^{-12}$, and $C_o = 0.043$ at.%; at 780 C, $D_A = 6.5 \times 10^{-11}$, $D_C = 4.3 \times 10^{-11}$, and $C_o = 0.10$ at.%. The absolute magnitude, temperature dependence, and anisotropy of D ($D_C/D_A \sim 0.6$) were found to be similar to those reported for Cu in Be. This consistency contrasts with the findings of Naik, Dupouy, and Adda for Ag in Be, where the anisotropy was reversed ($D_C/D_A \sim 1.8$).

---

[*]This work was supported by the U. S. Atomic Energy Commission.

I. INTRODUCTION

We have implanted gold into single-crystal beryllium at room
temperature, and have measured the time-dependent depth profile of
the Au during subsequent annealing above 600 C. The energy spectrum
of backscattered 2 MeV He[+] was used to determine concentration versus depth. One objective of this work was to examine the thermal
kinetics of a metal-in-metal implant where the solubility limit is
greatly exceeded. Such a condition is reached frequently, for
example, in high-fluence implants whose objective is to modify surface properties. The results for Au were compared to those for Cu
implants in Be, where the relatively high Cu solubility was not exceeded.[1] A second objective was to measure the diffusion coefficient D and solid solubility C_0 for Au in α-Be, since neither of
these metallurgical parameters have been reported.

The normally occurring α-form of Be has the hcp structure with
the smallest value of c/a of all the hexagonal metals: 1.567. The
small interatomic spacing (2.287 Å to a neighbor in the same basal
plane and 2.225 Å to one in the adjacent plane) and its position in
the periodic table lead one to expect very low solid solubilities
for most metals, in general agreement with experiment.[2] An exception is Cu, whose solubility can exceed 5 at.% at high temperatures.[3]
Previous studies of the solubility of Au have involved melting measured amounts of Au and Be together, and characterizing the results
by x-ray diffraction.[4,5] Introduction of small amounts of Au under
these conditions resulted in the appearance of an allotropic form
(β) of Be, still with the hcp structure but with a unit cell volume
about 29 times that of the α-form. The solubility of Au was then
reported to be about 2-3 at.% in Ref. 4 and 0.25 at.% in Ref. 5,
with most of the Au residing in the β-form. Greater amounts of Au
yielded the intermetallic compound $AuBe_5$. Thus, the solid solubility of Au in α-Be was not determined.

The diffusion coefficient D has previously been measured for
Be,[6] Cu,[1,7] and Ag[8] in Be. The activation energies for diffusion
of 1B metals in Be are somewhat greater than for Be self diffusion,
as would be expected for monovalent impurities in a divalent host,
but all are within a few tenths of an eV of 2.0 eV. The pre-
exponential factors are all similar, differing by less than a factor of 5. These results are not surprising, and suggest that Be
and the first two 1B metals all diffuse in Be by the vacancy mechanism. Anomalies appear, however, when one considers the anisotropy
of diffusion. (Anisotropy is permitted in the hcp structure, since
the lattice symmetry is non-cubic.) First, the anisotropy is nearly
the same for Be in Be as for Cu in Be; for example, at 700 C,
$D_C/D_A = 0.5$ in both cases. From the electrostatic model of
LeClaire,[9] one predicts that the ratio D_C/D_A for self diffusion
should be about three times that for Cu diffusion. More importantly,

for Ag in Be, D_C/D_A has been reported to be $\simeq 1.8$ at this temperature.[8] Not only does this result differ strongly from the prediction of the LeClaire model, but it is also opposite to the anisotropy
for Cu.[1,7] Therefore, it seemed desirable to measure D_C and D_A for
Au in Be.

II. EXPERIMENTAL PROCEDURES

Beryllium single crystals approximately 10 mm x 10 mm x 3 mm
were obtained from the Franklin Institute. The large faces had been
planed and then electropolished to remove the cutting damage. X-ray
measurements on one of the samples indicated a mosaic spread of
$\simeq 0.6°$ FWHM over the 1 cm^2 face, which should have a negligible effect on the experimental results. Implantations were performed on
four samples at room temperature using 100 keV Au^+. During implantation the crystal faces, which had been cut perpendicular to either
the C or A crystalline axis, were tilted 8° from the normal to the
beam to avoid channeling effects. The beam current density was
~ 1 $\mu A/cm^2$, the fluence $\simeq 1.0$ x 10^{17} Au/cm^2. The anneals were done
in a tubular furnace under flowing dry nitrogen. The sample was
held on a movable quartz support, which had a Chromel-Alumel thermocouple junction in contact with the specimen. Rapid thermal equilibration (usually within 20 C of final temperature in less than 3
min) was achieved by inserting the holder with sample after the furnace had been brought to temperature. The anneal was terminated by
removing the sample from the furnace. The temperature of the furnace was regulated to better than ±5 C.

Ion backscattering provided a straightforward and nondestructive means of determining composition versus depth.[10] These measurements were made using a 2 MeV He^+ ion beam from a Van de Graaff
accelerator, and alpha particles backscattered at an angle of 170°
were energy-analyzed with a surface-barrier Si detector. The energy
resolution of the detection system was $\simeq 13$ keV FWHM. From the energy spectrum of backscattered particles, the Au concentration could
be established with a depth resolution of $\simeq 0.03$ μm for depths up
to ~ 2 μm. The analysis was made quantitative by using published
stopping rates for alpha particles in Be,[11] Au,[12] and O,[13] and by
taking these rates to be additive in a compound (Bragg Rule).[14]

III. RESULTS AND INTERPRETATION

Figure 1 shows two representative energy spectra of incident
2 MeV He^+ backscattered from Au-implanted Be. One spectrum was
taken following the implantation, the other on the same sample after
a 30-min anneal at 780 C. The pre-anneal spectrum has peaks corresponding to the implanted Au (1.1 x 10^{17} Au atom/cm^2), surface oxide
on the Be host (1.4 x 10^{17} O atom/cm^2), and some surface C (3 x 10^{16}

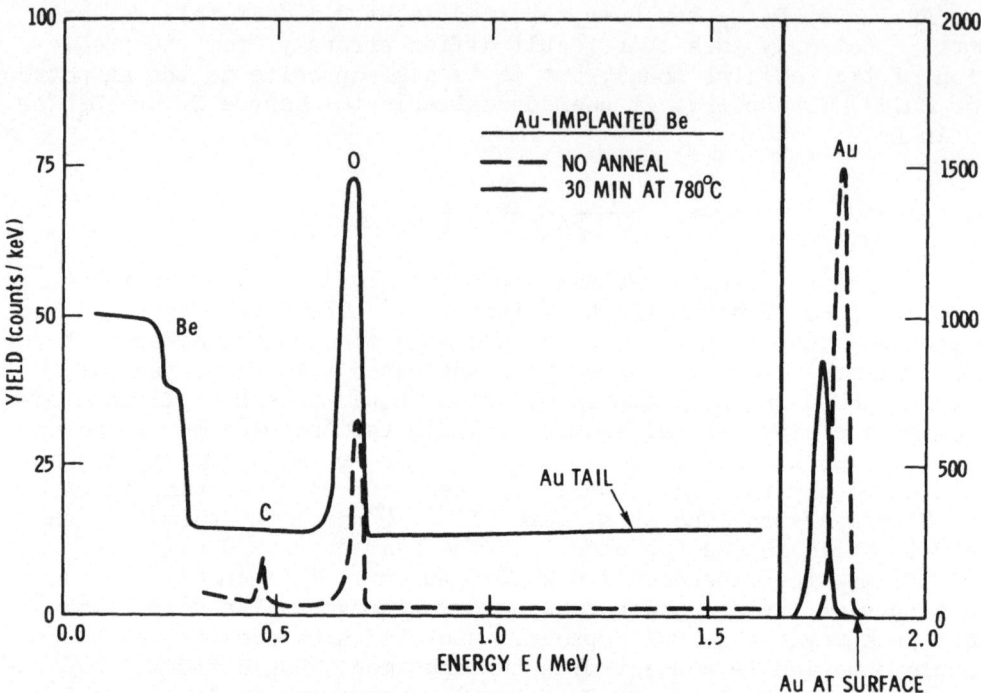

Fig. 1. Spectra of backscattered 2 MeV He⁺ from the C-axis face of
Au-implanted Be. The vertical scale is expanded by x20 to
the left of the divider. The implantation fluence is
1.1×10^{17} Au atom/cm², the implantation energy 100 keV.
Details of the spectra are discussed in the text.

C atom/cm²). The C is probably introduced by breakdown of residual
hydrocarbon gases during implantation. For a given atomic species,
decreasing energy corresponds to increasing depth. The energy corresponding to Au at the surface, which was determined by backscattering at the same incident energy from a Au foil, is indicated on
the figure. The relation between backscattered energy and depth
for Au is approximately linear, with a scale factor of 22.5 Å/keV
near the surface.

The anneal produced several changes in the backscattering spectrum of Fig. 1. Most important is a reduction in the amplitude of
the Au peak accompanied by the appearance of a tail at lower energies indicating that Au is diffusing from the implanted region into
the bulk of the Be host. (Loss of Au at the sample surface should
be negligible throughout, since pure Au has a vapor pressure of
$\sim 3 \times 10^{-9}$ Torr at the maximum temperature used in this experiment.[15])
In most cases the depth penetration of the tail is large compared to
the probing depth of ~ 2 μm, so that only the portion near the implanted region can be detected by He backscattering. This is clearly

seen in Fig. 1. The only exception was for the first anneal at the lower temperature, 665 C. Consequently, there exist two varying experimental quantities: the area under the Au peak, and the amplitude of the Au tail immediately beneath the implanted layer. Such behavior contrasts strongly with results reported for diffusion of Cu implants in Be.[1] In that experiment the relatively high solid solubility ($\sim$ 5 at.%) was not exceeded, and annealing simply caused the concentration profile to broaden while retaining its Guassian shape.

A second change in the backscattering spectrum caused by annealing is the increase in oxygen to 6×10^{17} O atom/cm^2, which is the result of Be oxide formation. The Au is excluded from the Be oxide layer, as seen from the shift in the Au peak in Fig. 1. We note that such an exclusion also was found for implanted Cu in Be after oxidation. Since the thickness of the oxide layer in the present case was always small in comparison to the diffusion distances, and since Au is not taken into the oxide, the experimental results should not be significantly affected by its presence.

In all of the cases studied we found that after a 30-min anneal, which was the shortest time used, the region of the Au tail immediately beyond the Au peak had reached a limiting amplitude. This level was retained until $\geqslant$ 70% of the Au had diffused from the implanted region. Continued annealing beyond this point at 780 C resulted in a decrease in the Au amplitude. At 665 C the sample was not carried into this second regime because of the excessive anneal times which would have been necessary. The limiting amplitude of the tail was found to be temperature-dependent, and corresponded to a concentration of 0.043 ± 0.005 at.% at 665 C and 0.10 ± 0.01 at.% at 780 C. At a given temperature, this amplitude was the same for faces perpendicular to both the A and C crystalline axes.

It seems clear from the above behavior that two distinct regions are present in the Au-implanted Be samples. In the implanted region, where the solid solubility is greatly exceeded, we infer that the Au is not substitutional in α-Be. The most probable configuration is a precipitate of some Au-Be intermetallic compound, $AuBe_5$, for example. We believe that the remainder of the Be host retains the α-form, and that when Au diffuses from the implanted region it is substitutional and moves by the vacancy mechanism. The bases for the latter assertion are, first, that the Au concentration remains quite low; and, second, as will be discussed, that the observed diffusion parameters are very similar to those for Cu in Be. In order to describe the thermal kinetics of this situation, we apply the diffusion equation

$$\frac{\partial}{\partial t} C(x,t) = D \frac{\partial^2}{\partial x^2} C(x,t) \quad , \tag{1}$$

with the initial condition

$$C(x,0) = 0 \quad , \tag{2}$$

to the region of the sample beyond the deeper edge of the implant. Here $C(x,t)$ is the local Au concentration, x is measured from the above edge, and t is the total anneal time. It then remains to apply a boundary condition at this edge which correctly describes the flow of Au from the implanted layer.

We shall consider two possible limitations on the flow $F(t)$ of Au from the implanted region. The first is simply the breakup rate of the precipitate or whatever other form the Au-Be mixture may have. Without more detailed information it is not possible to specify $F(t)$ in this case. However, for anneal times such that only a small fraction of the implanted Au has diffused, it should be a good approximation to take $F(t) = F_O = $ constant. The boundary condition is then

$$-D \frac{\partial}{\partial x} C(x,t) \Big|_{x=0} = F_O \quad , \tag{3}$$

which has the solution[16]

$$C(x,t) = \frac{F_O}{D} \left\{ \sqrt{\frac{4Dt}{\pi}} \exp\left(-\frac{x^2}{4Dt} \right) \right.$$

$$\left. - x \left[1 - \mathrm{erf}\left(\frac{x}{\sqrt{4Dt}} \right) \right] \right\} \quad . \tag{4}$$

The experimentally measured parameters are the loss of Au from the implanted layer and the concentration at the lower edge of this layer. These are given by

$$\int_O^t F(\tau)d\tau = t\, F_O \tag{5}$$

and

$$C(0,t) = 2F_O \sqrt{\frac{t}{\pi D}} \quad , \tag{6}$$

respectively.

The second possible limit on Au flow is its solid solubility C_O in α-Be. In this case the boundary condition is simply

$$C(0,t) = C_o \quad , \tag{7}$$

which yields the solution[16]

$$C(x,t) = C_o \left[1 - \text{erf}\left(\frac{x}{\sqrt{4Dt}} \right) \right] \quad . \tag{8}$$

One then has for the loss of Au from the implanted layer

$$\int_o^t F(\tau)d\tau = -D \int_o^t \frac{\partial}{\partial x} C(0,\tau)d\tau$$

$$= 2C_o \sqrt{\frac{Dt}{\pi}} \quad . \tag{9}$$

Of the two proposed models, it is clear the first does not apply since Eq. (6) predicts a continuous rise in $C(0,t)$ with anneal time. In contrast, we found experimentally that $C(0,t)$ had always reached its limiting value after one 30-min anneal. This was true even for a C-axis face at 665 C, where the loss of Au from the implanted layer was only about 4%. Eq. (7), on the other hand, is quite consistent with the immediate rise of $C(0,t)$ to a limiting value when annealing is started. It remains to be determined whether Eq. (9) is obeyed. This equation predicts that the square of the number of Au atoms per unit area lost from the implanted region is proportional to anneal time. Appropriate plots of the experimental data at 665 C are shown in Fig. 2, and their linearity indicates good agreement with Eq. (9). Similar behavior was found at 780 C. We conclude that, for the implant fluences of this experiment, the diffusion of Au from the implanted region up to $\simeq 70\%$ of the initial implant is limited by its solubility in α-Be. At 780 C, continued annealing results in a decrease in $C(0,t)$, indicating the predominance of another limitation. We note that as annealing proceeds the Au containing regions in the implanted layer must necessarily grow smaller and/or more sparse. Hence, the dissolution of these regions will at some point limit the Au flow, and quite possibly that has occurred here.

One feature of the experimental data in Fig. 2 does not agree completely with Eq. (9): The data do not extrapolate to the origin. The departure is small in comparison to the total variation, and in fact is comparable with the estimated experimental uncertainty. Nevertheless, a number of data points too closely spaced to be shown in Fig. 2 consistently showed an initial rise with slope greater than the final slope. At 665 C the final slope was reached after about 100 min. At 780 C the first anneal resulted in too large a Au loss to permit any conclusion about the behavior very near the

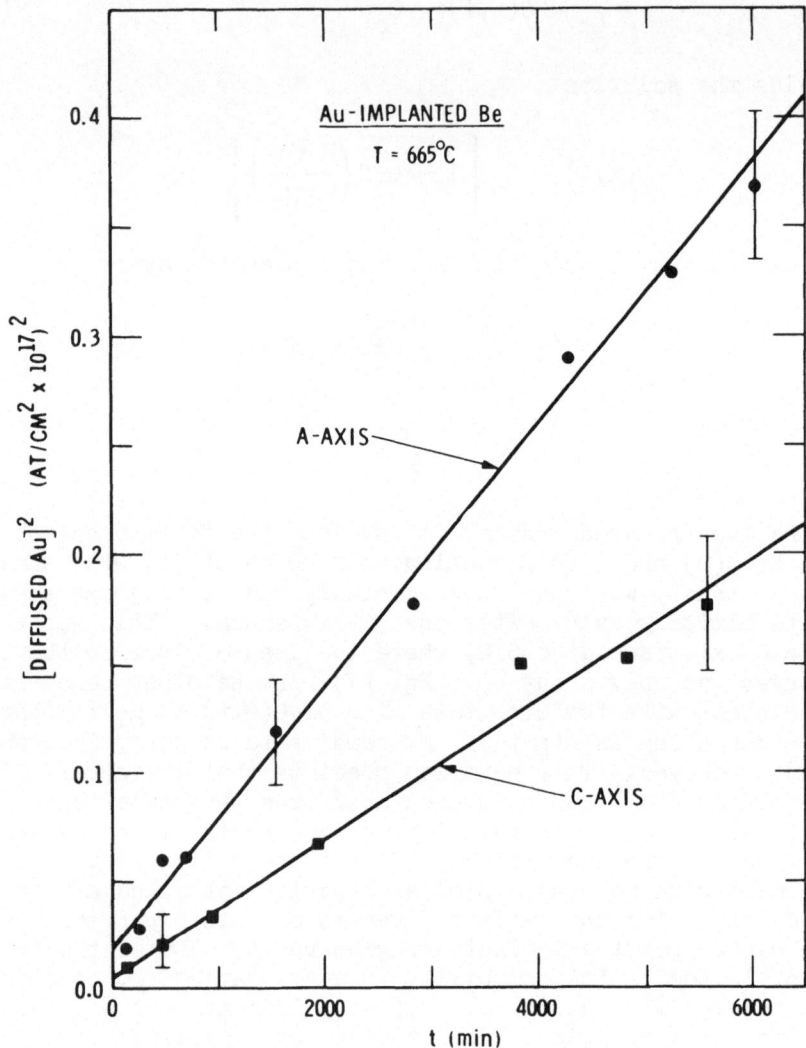

Fig. 2.   Time dependence of Au diffusion from the im-
          planted layer at 665 C.

origin. One possible explanation is an initial enhancement of the
diffusion by damage introduced during implantation. Enhanced diffusion of Cu implanted at room temperature into Be has already been
observed during initial annealing stages.[1]

   Assuming that the second proposed model correctly describes
the experimental results, we immediately obtain the solid solubility
C_o from the limiting amplitude of the Au tail. The slope of the
plot in Fig. 2 then gives the diffusion coefficient D via Eq. (9).

TABLE I. Diffusion coefficient D and solid solubility C_0 for dilute Au in Be. The experimental uncertainty is ±30% in D and ±10% in C_0.

| T(C) | D(cm^2/sec) | | C_0(at.%) |
	A axis	C axis	
665 C	2.8×10^{-12}	1.5×10^{-12}	0.043
780 C	6.5×10^{-11}	4.3×10^{-11}	0.10

The results are listed in Table I. The relatively small solubilities obtained here for Au in α-Be (≤ 0.10 at.%) are not directly comparable with the results of Refs. 4 and 5: 2-3 at.% and 0.25 at.%, respectively. In those experiments, the Au and Be were melted together, and most of the Au was found to occupy the allotropic β form of Be. The presence of the Au apparently stabilizes the more open β structure. However, one would expect the solubility of Au in α-Be to be less, which we find to be the case.

In Fig. 3 the diffusion coefficients for Au in Be are compared to those reported previously for Cu in Be.[1,7] The former are very similar to the latter in absolute magnitude, temperature dependence, and anisotropy ($D_C/D_A \sim 0.5$ in both cases). This gives credence to the assertion that we have indeed obtained the diffusion rate for the α-form of Be, and have not driven it to the allotropic state. The reported anisotropy for Ag in Be,[8] where $D_C > D_A$, now seems especially anomalous since $D_C < D_A$ for Be, Cu, and Au in Be.

IV. CONCLUSIONS

Annealing studies have been made on Au-implanted Be single crystals, for fluence levels such that the solid solubility of implant in host is greatly exceeded. A mathematical description of the Au diffusion has been developed which agrees with the experimental results. From the analysis we have obtained, for the first time, solid solubilities and diffusion rates for Au in α-Be. We have shown that the Au occupies two distinct regions in the Be host. In the implanted layer it is not substitutional in α-Be but instead probably exists in precipitates of some intermetallic phase. However, we have inferred that Au which diffuses from the implanted layer is substitutional in α-Be, and moves by the vacancy mechanism. At the fluence level used, diffusion from the implanted layer was initially limited by the solubility of Au in α-Be. However, at some point in the annealing sequence another limitation on the Au diffusion became predominant, presumably the dissolution rate of a Au-Be precipitate.

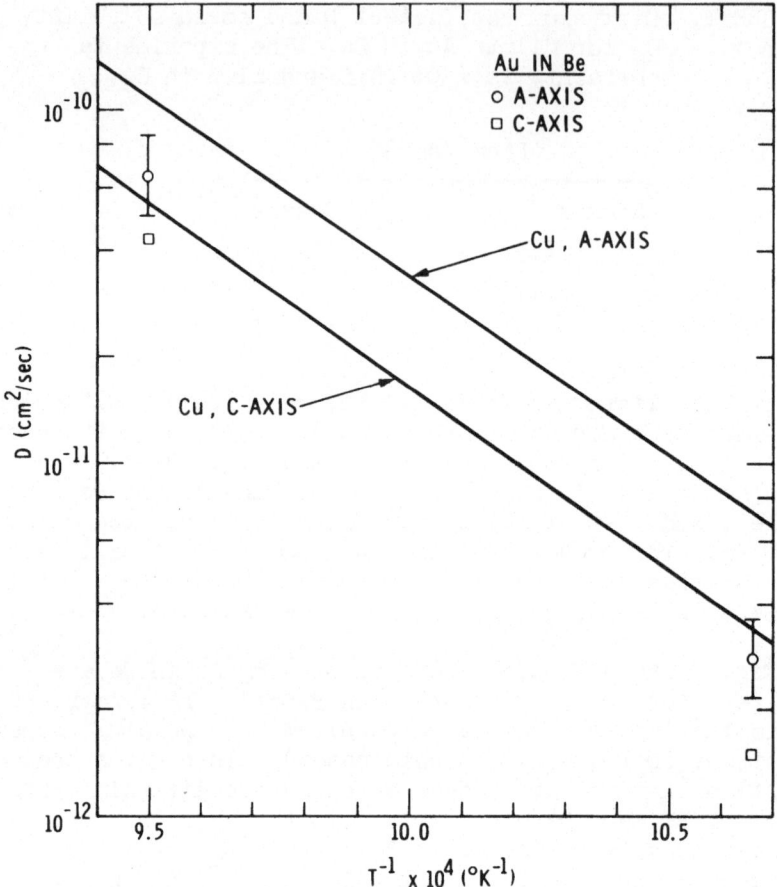

Fig. 3. Temperature dependence of D for Au in Be,
         compared to that for Cu in Be.

        We anticipate that the kind of behavior observed in this ex-
periment will occur frequently for metal-in-metal implantations
where a solid solubility is exceeded. Hopefully, the approach used
here will provide a useful complement to other techniques in characterizing such systems. From a different point of view, the present results illustrate how ion implantation and ion backscattering
may be combined to yield basic metallurgical information not readily
obtained by other means.

                              ACKNOWLEDGMENTS

        The authors acknowledge the valuable technical assistance of
J. M. McDonald in the annealing and backscattering measurements, and
the performance of the implantations by C. T. Fuller and N. D. Wing.

REFERENCES

1. S. M. Myers, S. T. Picraux, and T. S. Prevender (submitted to Physical Review).

2. See, for example, A. R. Kaufmann and P. Corzine, The Metal Beryllium, edited by White and Burke (Am. Soc. Met., Cleveland, 1955), Chap. 10.

3. See, for example, M. Hansen, Constitution of Binary Alloys (McGraw-Hill, N. Y., 1958), 2nd ed., p. 282.

4. S. S. Sidhu and C. O. Henry, J. Appl. Phys. 21, 1036 (1950).

5. G. P. Chatterjee, J. Mines, Metals, and Fuels (India), June 1962, p. 20.

6. J. M. Dupouy, J. Mathie, and Y. Adda, Mémoires Scientifiques Rev. Métallurg. 63, 481 (1966).

7. J. M. Dupouy, J. Mathie, and Y. Adda, Proc. Conf. Intern. sur la Métallurgie du Beryllium, Grenoble, 1965 (Presses Universitaires de France, Paris), p. 159.

8. M. C. Naik, J. M. Dupouy, and Y. Adda, Mémoires Scientifiques Rev. Métallurg. 63, 488 (1966).

9. A. D. LeClaire, Phil. Mag. 7, 141 (1962). A detailed discussion of the application of this theory to diffusion in the hcp lattice has been given by P. B. Ghate, Phys. Rev. 133, A1167 (1964).

10. See, for example, S. T. Picraux and F. L. Vook, Appl. Phys. Lett. 18, 191 (1971); and J. F. Ziegler and J.E.E. Baglin, J. Appl. Phys. 42, 2031 (1971).

11. W. K. Chu and D. Powers, Phys. Rev. 187, 478 (1969).

12. J. A. Borders, Rad. Effects 16, 253 (1972).

13. P. D. Bourland, W. K. Chu, and D. Powers, Phys. Rev. B3, 3625 (1971).

14. See, for example, P. D. Bourland and D. Powers, Phys. Rev. B3, 3635 (1971); and D. Powers, A. S. Lodhe, W. K. Lin, and H. L. Cox, Proc. Intl. Conf. on Ion Beam Surface Layer Analysis, Yorktown Heights, N. Y., 1973 (in press).

15. See, for example, R. E. Honig, RCA Review 23, 567 (1962).

16. See, for example, R. V. Churchill, Operational Mathematics (McGraw-Hill, N. Y., 1958), Chap. 4.

DISCUSSION

Q: (J. W. Miller) [1] What is the depth to which you are able to measure backscattering from Au separately from the backscatter from Be? [2] Is the apparent flatness of gold concentration vs. depth over this comparatively great depth consistent with the diffusion constants you have deduced for the gold?

A: [1] About 2.2 μm. [2] Yes, it is. In Fig. 1 the penetration of the Au tail is about 30 μm. Therefore, only the upper extreme of this tail is probed by the He ions.

Q: (J. M. Poate) Did you try channeling to see if your Au tail was substitutional?

A: We tried channeling, but the axial dips were very weak. This is probably due to the mosaic spread of the crystals, typically ~ 0.6°.

Q: (W. K. Chu) [1] There is a "knee" at the Be-step of your backscattering spectrum. Does this knee change its shape during heat treatment of your sample? [2] You suspect the formation of an intermetallic compound between Au and Be. Can this be easily checked by x-ray diffraction or other methods?

A: [1] Yes, it does. This "knee" is caused by the higher stopping rate per Be in the surface region, due to the presence of Au and O. A quantative interpretation of its amplitude is not simple because a) the Be spectrum rides on the tail of the Au spectrum, b) both the Au and the O affect the amplitude of the step, and c) the nuclear scattering cross-section for Be has some structure in the vicinity of 2 MeV. [2] The fluence level ($\simeq 10^{17}$ Au atom/cm^2) is such that an x-ray determination would be difficult. However, we plan to make transmission electron microscopy measurements to characterize the precipitates.

ANOMALOUS ROOM TEMPERATURE DIFFUSION OF ION-INJECTED Ni IN Zn TARGETS

H.J. SMITH and G.N. VAN WYK[*]

National Physical Research Laboratory

CSIR, Pretoria, South Africa

1. INTRODUCTION

It has usually been found that when energetic ions are impinging on the surface of a metal target at an energy at which the selfsputtering coefficient $S_e = 1$ (the sputtering coefficient S being the number of atoms sputtered per impinging ion, and self-sputtering referring to sputtering of a metal by ions of the same element), the total amount of projectile material trapped in the target soon reaches a saturation value of some μgram per cm^2.

This is explained by the advancing of the target surface due to sputtering, leading to the sputtering of previously trapped projectile atoms. Assuming that no movement of a trapped atom relative to the matrix takes place, i.e. no diffusion, the sputtering model of saturation [1-3] describes this behaviour quantitatively. The saturation value q_s is obtained as

$$q_s = \frac{m_1}{m_2} \frac{\eta}{S} d \text{ μgram per } cm^2 \qquad (1)$$

where m_1 and m_2 are the atomic masses of projectile and target atoms respectively, η is the sticking factor (of the order of unity) and d is a measure of the range of the ions.

Almen and Bruce[1] investigated trapping and sputtering for noble

---

[*] Permanent address: Department of Physics, University of Pretoria, Pretoria.

gas ions bombarding metals and found saturation values of the order of some μgram per cm^2.

Smith[4,5] has measured trapping and sputtering for 13 combinations of metal ions on metal targets and demonstrated that in all the cases diffusion of trapped atoms in the target takes place at room temperature during bombardment. This gives rise to trapped amounts substansively higher than predicted by the sputtering model of saturation.

As was reported before [6,7] this effect is very prominent on targets of Zn and indications are that diffusion is occuring even at 77K.

In the present paper a further investigation of this effect for Ni ions impinging on polycrystalline Zn targets at room temperature is reported.

2. EXPERIMENTAL METHODS

The apparatus and techniques employed are the same as used before and are described in detail elsewhere[8].

Polycrystalline Zn targets in the form of discs are homogeneously irradiated with Ni ions over an area of 4 x 10 mm^2 in an electromagnetic isotope separator. The design of the irradiation system is such that the actual ion current striking the target is measured and integrated correctly to within $\pm$ 5%, yielding the ion dose.

After a target has been irradiated, the amount of Ni trapped in it is determined by means of x-ray fluorescence analysis or atomic absorption spectrophotometry.

Targets are weighed before and after irradiation to determine the amount of material sputtered from them.

Unless otherwise stated, the Zn used was commercially obtained 99.999% pure Zn foil of thickness 37 mgram per cm^2. The average grain size in this Zn is between 3 and 4 micron. X-ray diffraction revealed no alignment of the microcrystals.

The normal beam intensity was 60 to 70 μA per cm^2 on the area actually being struck by ions, taking into account the size of the image of the ion beam on the target. But as the beam is moved across the target to obtain homogeneous irradiation, the actual average ion current was about 28 μA per cm^2 (see reference[8]).

The temperature of the targets is maintained at 294 $\pm$ 5K and the pressure in the irradiation chamber is 1 x 10^{-6} torr.

3. RESULTS

3.1. Trapping and Sputtering under Continuous Bombardment

In Figure 1 the loss of weight of a series of targets which have received successively increased doses of 39 KeV Ni ions is shown.   For every measurement the amount of Ni found in the target was added to the weight loss, thus obtaining the actual loss of weight in Zn.

This yields as expected a straight line, the gradient of which multiplied by m_1/m_2 yields the sputtering coefficient S.   This is found to be 18 atoms per ion.

In Figure 2 the trapped amounts of Ni in the same series of targets are shown as a function of dose (this is referred to as a trapping curve).   It was previously found that up to a dose of 100 μgram per cm^2 the trapped amount increases linearly with dose. Although there is a slight deviation, this is now shown to be the case up to 400 μgram per cm^2, with a gradient of 0.8 mass trapped per mass injected.

The most probable projected range of 39 keV Ni ions in Zn is of the order of 10 μgram per cm^2 (according to the Lindhard, Scharff and Schiøtt[9] calculations).   Thus most Ni ions should be stopped within about 30 ugram per cm^2.   If no movement of Ni occurs in the target matrix the sputtering model of saturation would apply and saturation should set in after this amount has been sputtered. According to equation 1 the saturation value should be of the order of 1 μgram per cm^2.   ($\eta \sim 1$, $m_1 \sim m_2$, S = 18, d $\sim$ 10 μgram per cm^2) (Zn has a particularly high sputtering coefficient and therefore should yield a low saturation value).

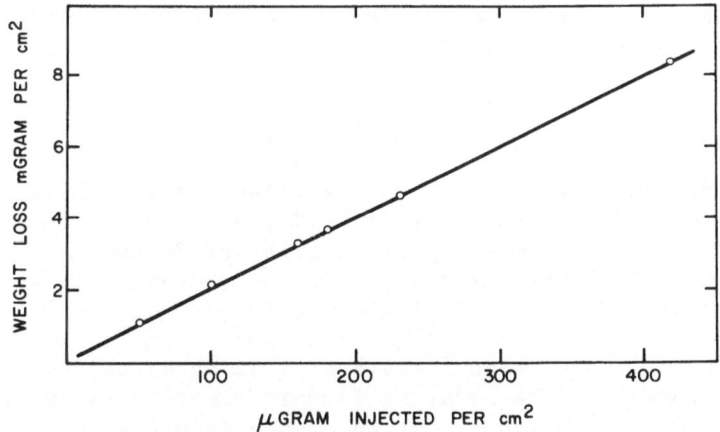

Figure 1. Loss of weight of Zn targets as a result of sputtering by 39 keV [58]Ni ions.

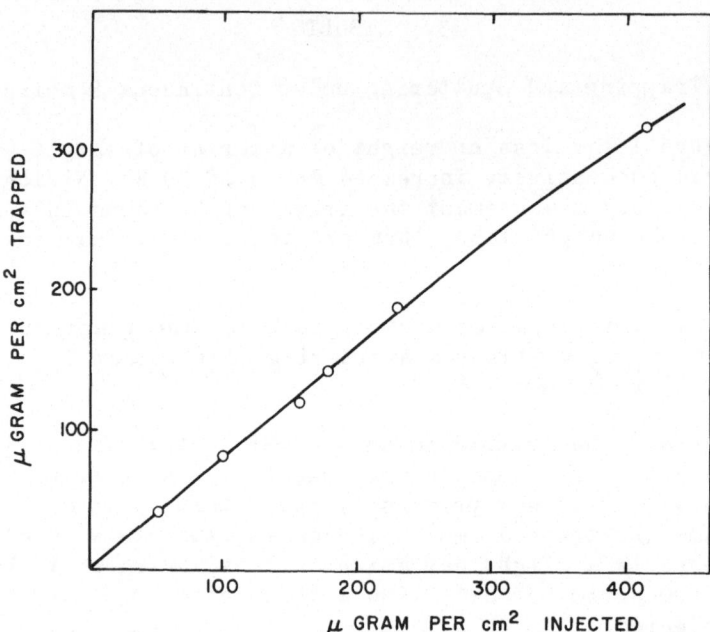

Figure 2. Amount of 39 keV Ni trapped in Zn as a function of dose.

Yet, in spite of the high sputtering rate and the eventual sputtering of 8 mgram per cm^2 of Zn, 80% of all injected Ni is retained in the target.   It must be pointed out that the total amount of trapped Ni is an order of magnitude larger than necessary to completely stop 39 keV Ni ions in pure Ni.   Thus preferential sputtering of Zn atoms to Ni ones alone cannot account for this effect.

This is in accord with what has emerged from previous work[5] viz that the Ni is continuously advancing in front of the surface which itself advances because of sputtering.

3.2.   Thin Zn Targets

In order to investigate what the effect is if limited thickness of Zn is available so that it will be depleted by sputtering, thin targets of Zn were prepared by evaporation of Zn onto Al substrates (A thin layer of Au was first deposited to increase the sticking factor of Zn).

A trapping curve using targets of 1.52 mgram per cm^2 of Zn is shown in Figure 3. (The straight line in Figure 3 is the trapping curve for the usual thickness of Zn, taken from Figure 2).   On the thin targets the trapping curve very soon departs from that for thick

targets, showing that the movement of the Ni is inhibited and that
more Ni is being sputtered.   The trapped amount eventually
decreases and should stabilize at the saturation value of Ni in Al.

        The loss of weight of these targets is shown in Figure 4.
Initially the gradient is the same as that of ordinary thick Zn, as
in Figure 1, but with increasing dose the gradient and thus S,
decreases.   This occurs because the concentration of Ni in the
surface is increasing.   (Ni sputters less readily than Zn).

        In Figure 3 are also shown the trapped amounts in targets of
increasing thickness which have received equal doses of about 50
μgram per cm^2.   The thicker the targets, the nearer to the thick
Zn case the results are lying.   These points are used to construct
Figure 5, which shows the trapped amount of Ni for 50 μgram per cm^2
injected into the various targets as a function of residual thickness of Zn (obtained from the measured weight loss).   For thicker
targets the trapped amount is independent of thickness, but as the
Zn gets thinner, the behaviour is strongly affected by the thickness.
This happens at a thickness of about 2 mgram per cm^2.   This gives
a rough indication of the depth of the diffused Ni distribution.
It is of the same order as previously obtained[5] by measuring the
distribution of trapped Ni in Zn by argon ion sputtering at 77K.

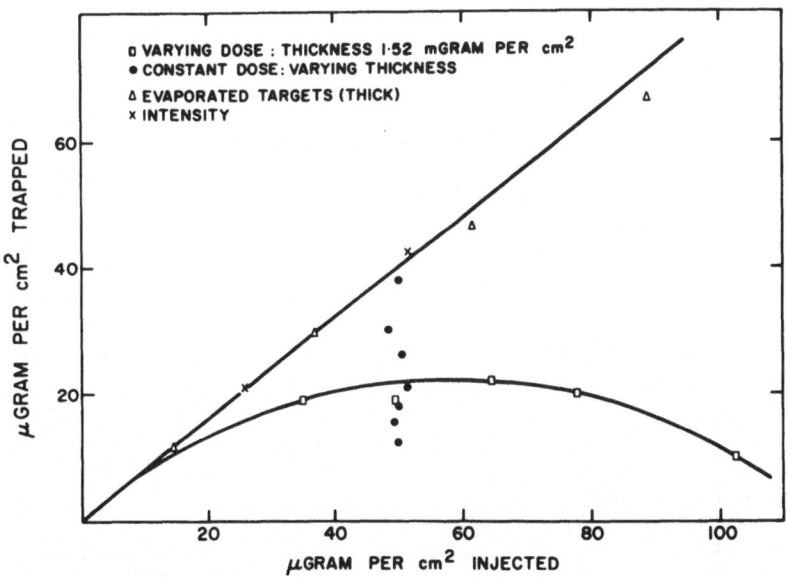

Figure 3.   39 keV Ni trapped in Zn: effect of i) limited Zn thick-
             ness, ii) intensity and   iii) evaporated thick targets.

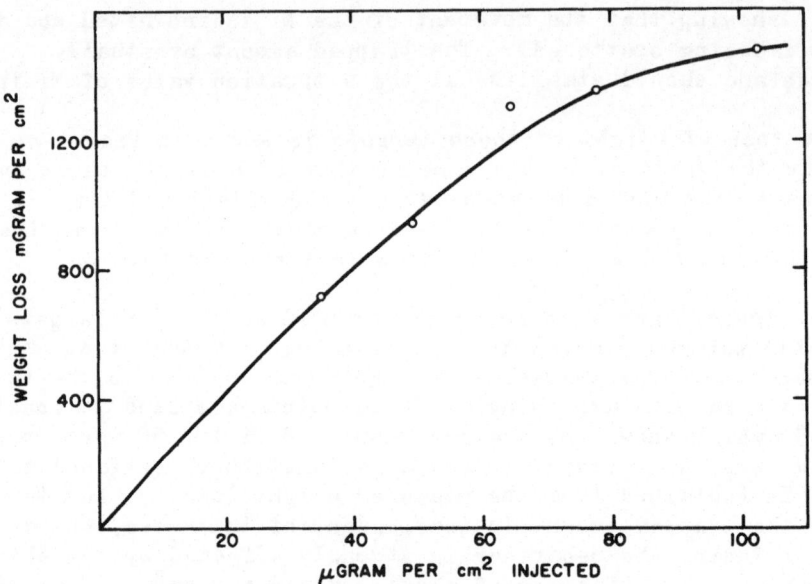

Figure 4.   Weight loss of 1.52 mgram per cm^2 Zn targets as a
            function of dose of 39 keV Ni ions

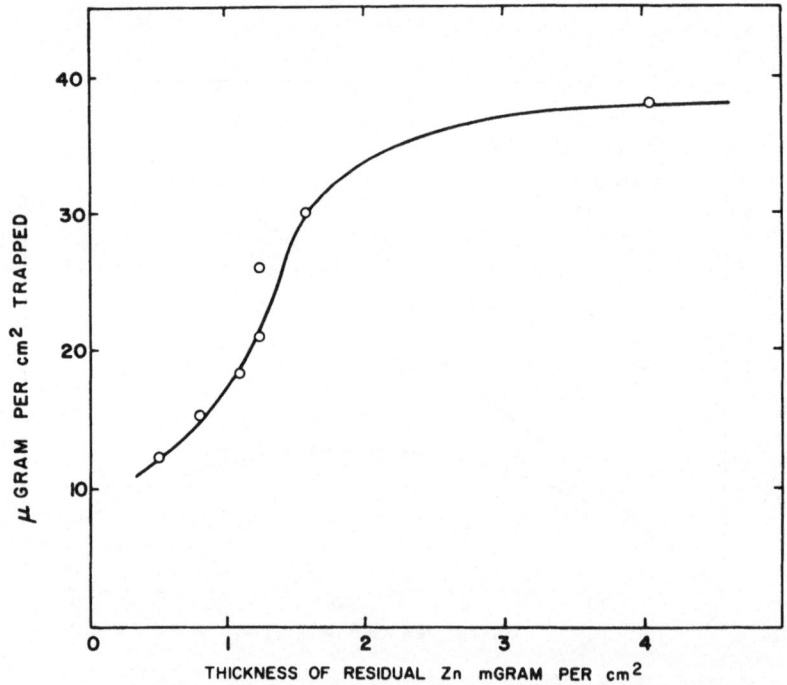

Figure 5.   Amount of Ni trapped as a function of residual Zn
            thickness (injected dose: 50 μgram per cm^3)

3.3. Thick Targets Prepared by Evaporation

The results of trapping measurements on evaporated thick Zn targets are also shown in Figure 3. Although these lie slightly lower than the trapping curve for the normal Zn foil, they show that the effect also occurs on Zn prepared by other methods.

3.4. Ion Beam Intensity

The role of intensity has been checked by measuring points on the trapping curve with lower and higher beam currents than in the normal case. The results are summarized in Table I, and are also included in Figure 3, the lower point being the lower intensity measurement. It appears that for the range of intensities covered, the behaviour is not influenced by intensity.

Table 1

Fraction of 39 keV Ni ions trapped in Zn for different ion beam intensities (I being the usual ion current density described in section 2)

Current density	μgram per cm^2 injected	μgram per cm^2 trapped	Fraction trapped
1.8 I	51.3	42.5	0.83
I			0.80
.3 I	25.7	21.0	0.82

3.5. Ion Energy

The dependence of the trapping and sputtering behaviour on ion energy was investigated by injecting the same dose (50 μgram per cm^2) of Ni ions into a series of targets at different energies.

In Figure 6 the results of the sputtering measurements are shown. At lower energy the sputtering coefficient decreases as expected.

The trapping results, presented in Figure 7, show that the trapped amount (0.8 of injected dose at 39 keV) remains practically unaffected as the energy is reduced. Around 3 keV it increases to a maximum before it decreases sharply at 1 keV.

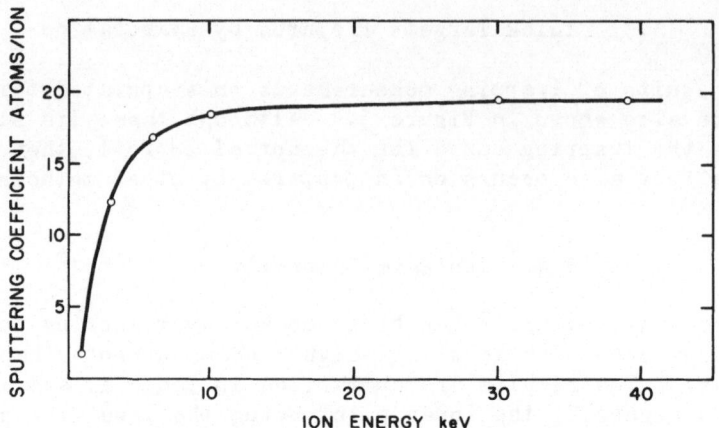

Figure 6.   Sputtering coefficient of Zn bombarded by [58]Ni ions
            as a function of ion energy

        This behaviour conforms to what was noticed before[5] in the
case of argon ion bombardment at room temperature of Zn targets
containing Ni.   In the latter case it was observed that 3 keV
argon ions tended to be more effective in advancing the Ni population than 39 keV argon ions.

        This behaviour is further discussed in the next section.

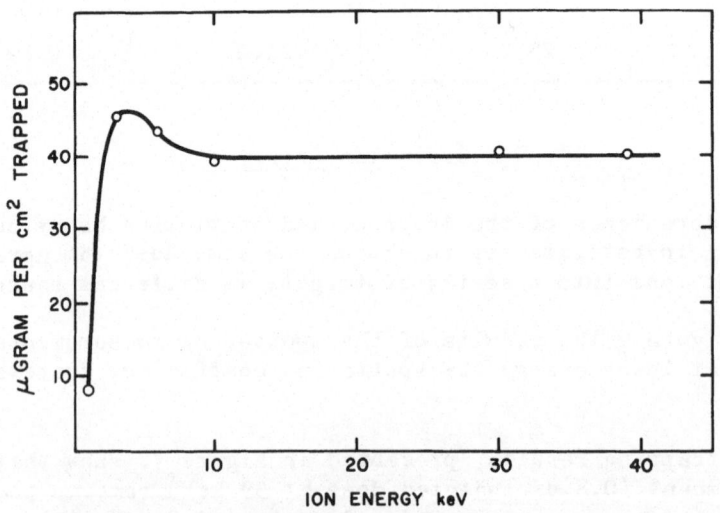

Figure 7.   Amount of Ni trapped as a function of ion energy.
            Injected ion dose: 50 μgram per cm^2.

4. DISCUSSION

4.1. Point Defects

In a previous paper[5] it has been shown that this effect is not due to ordinary diffusion, nor to ordinary diffusion accelerated by the target being heated by the beam, nor to fortuitous channelling. It must be due to point defects resulting from the bombardment.

It has further been indicated that a fast diffusion process accompanied by a trapping mechanism must be involved to account for the disappearance of the diffusion after irradiation has been stopped.

The direct effect of the beam is restricted to the layer in which the ions are being slowed down and stopped.   If the Lindhard Scharff and Schiøtt calculations are applicable, the most probable projected range of 39 keV Ni ions in Zn is 10.5 μgram per cm^2. This implies that nearly all ions should be stopped within about 30 μgram per cm^2.   Taking into account uncertainties in the theory, practically all ions should be stopped within at the utmost 50 μgram per cm^2.   From the deductions about the extent of the Ni distribution in section 3.2 and from the measurements in reference[5] it is clear that diffusion is taking place through a thickness of two orders of magnitude deeper than the layer directly affected by the beam.

The diffusion must thus be due to defects that are created in the first layers but which are mobile enough to move on to greater depths, until they become annihilated or trapped.

The jump frequency ν of a point defect r at temperature T is given approximately by

$$\nu = \nu_o \, e^{- u_m^r / kT} \tag{2}$$

with $\nu_0 \sim 10^{13}$ sec^{-1}, u_m^r the activation energy for migration and k the Boltzman constant (see for instance Thompson[10]).

To be mobile would require at least a jump frequency of the order of 10 to 100 per sec.   From equation (2) this yields at T = 294 K an activation energy u_m^r of 0.70 to 0.64 eV.

This gives a rough upper limit for the activation energy of migration of the defects involved in this diffusion.   In view of the magnitude of the effect and the fact that a trapping mechanism must be involved, a lower activation energy can rather be expected.

Although little information is available about activation

Table II

Activation energies for: selfdiffusion, vacancy formation; vacancy migration and interstitial migration in Zn

	eV	reference
u_{sd}	.95*	11
u_f^v	.54	12
u_m^v	.50	11
u_m^i	.21	11

* Various values have been obtained, for review see the reference article

energies of point defects in Zn, some values from the literature are summarized in Table II.

The value of u_m^i seems rather high in comparison to other metals. From the table it is apparant that vacancies and interstitials (probably also combinations such as divacancies) must be taken into account in the diffusion reported here.   In the diffusion at 77K reported before, vacancies can be excluded and only more mobile defects could be involved.

Under the conditions of the present experiment of continuous bombardment at room temperature the concentration of point defects will be limited by saturation and rapid annihilation because of this high mobility.   It seems possible, however, that some of the interstitials formed on the periphery of the damaged region may move into the bulk of the target, leaving an excess of the less mobile vacancies, behind of which some might also move on. (Models for such condition have been discussed theoretically by Sparks).[13]

The possibility has been suggested[3,6] that as a result of the radiation there could be a high probability of creating Ni interstitials which, if they can persist as such, can diffuse interstitually until a trap is encountered, when it reverts to a substitutional position.   However, from a consideration of ionic radii it seems unlikely that Ni ions can persist as interstitials in Zn.

4.2. The Energy Dependence

The amount of projectile material found trapped in a target
after a given bombardment dose is the result of two competing
processes. The diffusion is advancing the Ni into the target
while the sputtering is advancing the surface.

From the results in section 3.5 it is seen that the sputtering
coefficient is constant over a large energy range, as well as the
fraction of Ni trapped. This indicates that the diffusion is also
independent of energy. At lower energy the sputtering coefficient
decreases so that the diffusion is gaining on sputtering and more
Ni is retained in the target. At still lower energy the diffusion
effect also decreases.

The total number of defects created by an impinging ion is
proportional to its energy. But under conditions of continuous
bombardment and saturation the number of defects surviving and
contributing to diffusion could be independent of energy. As the
energy becomes too low this behaviour could change and the number
of such defects could decrease, until eventually the threshold for
damage is reached.

4.3. Conclusion

As mentioned in the introduction this case of Ni in Zn is an
extreme manifestation of an effect also occuring in other metal
ion-target combinations. More experiments are necessary to clarify
the process. This could provide another way of looking at point
defects in metals.

It could be of importance in the interpretation of measurements
of range distributions in metals. It is of practical importance
for the concentration to be reached and the distribution of
injected ions in metals. The experimental method of removing, by
argon ion bombardment, representative layers from the surface of a
metal containing impurities for investigations must be used with
caution.

ACKNOWLEDGEMENTS

Mr J.J. Selier is thanked for his technical assistance.
Mr F.T. Wybenga and Mr K. Kröger are thanked for x-ray fluorescence
and atomic absorption measurements respectively.

One of us (G.N. vW.) would like to thank the NPRL of the Council
for Scientific and Industrial Research for their hospitality and for
financial support.

REFERENCES

1. O. Almén and G. Bruce, Nucl. Instr. Meth. 11, 257 (1961).
2. G. Carter, J.S. Colligon and J.H. Leck, Proc. Phys. Soc. 79, 299 (1962).
3. H.J. Smith, Thesis, University of Pretoria, 1971.
4. H.J. Smith, Rad. Effects 18, 55 (1973).
5. H.J. Smith, Rad. Effects 18, 65 (1973).
6. H.J. Smith and W.L. Rautenbach, Proceedings of the International Conference on Electromagnetic Isotope Separators and the Technology of their Applications, Marburg, edited by H. Wagner and W. Walcher, p. 197 (1970).
7. H.J. Smith, Phys. Lett. 37A, 289 (1971).
8. H.J. Smith, Rad. Effects 18, 73 (1973).
9. J. Lindhard, M. Scharff and H.E. Schiøtt, Kgl. Danske Videnskab Selsk.   Mat. Fys. Medd. 33 No. 14 (1963).
10. M.W. Thompson, Defects and Radiation Damage in Metals (University Press, Cambridge 1969).
11. D. Schumacher, in Vacancies and Interstitials in Metals, edited by A. Seeger, D. Schumacher, W. Schilling and J. Diehl (North-Holland Publishing Company, Amsterdam, 1970) p. 889.
12. A. Seeger, J. Phys. F : Metal Phys. 3, 248 (1973).
13. M. Sparks, Phys. Rev. 184, 416 (1969).

DISCUSSION

Comment: (J. P. Biersack) I was very interested in learning about your experimental work, as I myself studied the underlying theoretical model last year in a general way. The equations governing implantation under the simultaneous influence of surface sputtering and diffusion have been solved in terms of analytical functions, and the results are being published now in Radiation Effects, 1973.

A: I am interested to hear this. In the present observation the diffusion coefficient is most probably not constant with depth.

Comment: (R. S. Nelson) The implantation region could have a very high stress. In a soft metal like Zn, stress induced vacancies would be produced in large quantities; these would then produce a diffusion of the Ni. A complete temperature dependence even above R.T. would be interesting.

A: Yes, in a final description this will have to be taken into account.

STUDY OF Li-6 IMPLANTED INTO NIOBIUM

J. P. Biersack and D. Fink

Hahn-Meitner Institute

Berlin, Germany

ABSTRACT

The technique applied here utilizes a combination of accelerator and nuclear reactor techniques. The accelerator is used for implantation and for observation of the host lattice (radiation damage), but is difficult to use for depth profile and lattice location studies when the implanted atoms are light compared to the host atoms. Therefore, in order to investigate such systems, a nuclear reactor is used for observing light implants, such as He-3, Li-6, Be-7, or B-10, by means of the (n, p) or (n, α) reactions.

The present paper deals with Li-6 implanted at 220 keV into niobium. Depth profiles are measured and compared with theoretical distributions. At low temperatures, diffusion of a small fraction of Li is observed, while the main part remains immobile up to about 900°C. By heating the Nb(Li) sample up to 990°C, the diffusion of the remaining Li-6 is measured.

Heating in air at atmospheric pressure to 700°C shows particular corrosion patterns: abundant oxidation to white Nb_2O_5 at damaged sites, whereas unirradiated areas turn black due to lower states of oxidation.

INTRODUCTION

Depth profiles of Li-6 are studied after implantation and subsequent diffusion processes by means of the energy spectrum of emitted α-particles following the $^6Li(n, \alpha)T$ reaction of cross

section 950 barns. Furthermore, the lattice location of Li-6 is
studied by observing channeling patterns of the emitted particles.[1]

The decay products of the nuclear reaction, α and T, have
energies of 2.0 and 2.7 MeV, respectively, and can easily be
detected by a surface barrier detector or acetate cellulose foil.
The channeling or blocking patterns are directly related to the
original lattice sites of Li atoms, as the nucleus is not displaced by the nuclear reaction (momentum transferred by thermal
neutrons is far below the displacement threshold).

Other experiments on light atoms in heavy host lattices have
been carried out by S. T. Picraux and F. L. Vook,[2] who studied
lattice locations of D and ^{3}He in W by means of the nuclear
reaction ^{3}He(d, p)^{4}He. In this case, a slight disadvantage may be
seen in the fact that the ratio of damage cross section to nuclear
reaction cross section is not as favorable as in our method; this
may become important in long-time exposures, i.e., in observing
low impurity concentrations. R. S. Blewer[3] successfully observed
p backscattering spectra using thin Cu foils with high implanted
concentrations of He, thus obtaining depth profiles but no lattice
locations.

IMPLANTATION AND RADIATION DAMAGE

Niobium single crystals and polycrystalline niobium foils
were used in the experiments. The single crystals were spark-cut
from a rod of 5.8 mm diameter, and the surface was electropolished.
The surfaces of the polycrystalline samples were cleaned but not
prepared otherwise.

220 keV ^{6}Li$^+$ was implanted at room temperature up to a maximum
concentration of a few atom.-%. The ion beam was directed nearly
perpendicular to the crystal surface, avoiding, however, channeling
directions.

The radiation damage in the Nb crystal was checked by comparing the $\langle 111 \rangle$ minimum yield in proton backscattering from
irradiated and unirradiated crystal areas. Radiation damage was
apparent at room temperature (23% increase in minimum yield), but
disappeared after half an hour anneal at 1000°C.

DEPTH PROFILES AND DIFFUSION

The Li doped Nb targets were irradiated in a beam of thermal
neutrons of flux $2.10^4/cm^2$ sec at the FMRB reactor facility,
Braunschweig, Germany. The emitted T and α particles were observed
at an angle of 45° from the specimen surface by a silicon surface
barrier detector which was placed outside the neutron beam at a
distance of about 4 cm from the target.

Fig. 1 shows a typical measured energy spectrum. All peaks except the left one arise from the ^{6}Li(n, α)T reaction, the right one being the triton peak. The structure at the center depicts the α spectrum which yields a well-resolved depth profile including a surface peak. The left peak at 1.4 MeV is presumably due to an oxygen surface contamination, detected by the ^{17}O(n, α) reaction.

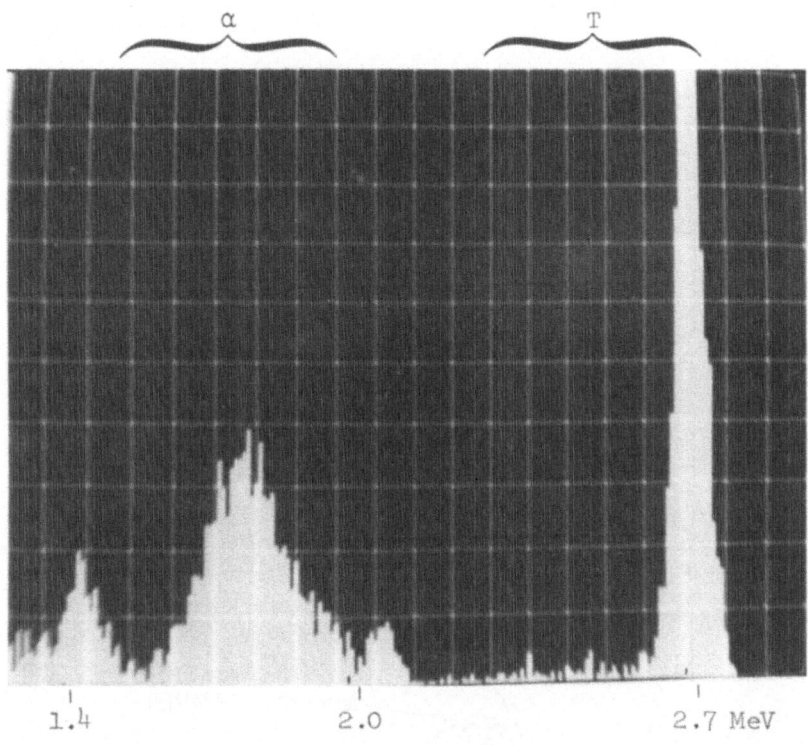

Figure 1.   Energy spectrum of particles emitted from Nb(Li) sample. The central peak is due to α particles from ^{6}Li(n, α)T reactions, the peak to the right corresponds to tritons from the same reaction. The background is zero, as can be seen at the right end of the spectrum. The left tail of the T peak results from interstitially mobile Li which penetrates a few μm in depth. The small α peak at 2.0 MeV also occurs during low temperature diffusion due to Li which gets trapped at the surface.

The mean projected range of the 220 keV Li$^+$ ions is found to be 0.48 μm and agrees well with the theoretical value of 0.5 μm.[4] The measured range distribution, however, yields a smaller range straggling (σ = 0.15 μm) than is predicted by theory (σ = 0.2 μm).[4]

At low temperatures, a small fraction of about 10% of the implanted atoms become mobile and collect at the nearby surface, Fig. 2a. This anomalous diffusion sets in at room temperature, and is completed after 20 min. at about 750°C. This diffusion is not connected to any broadening of the main profile, but just slightly lowers its amplitude.

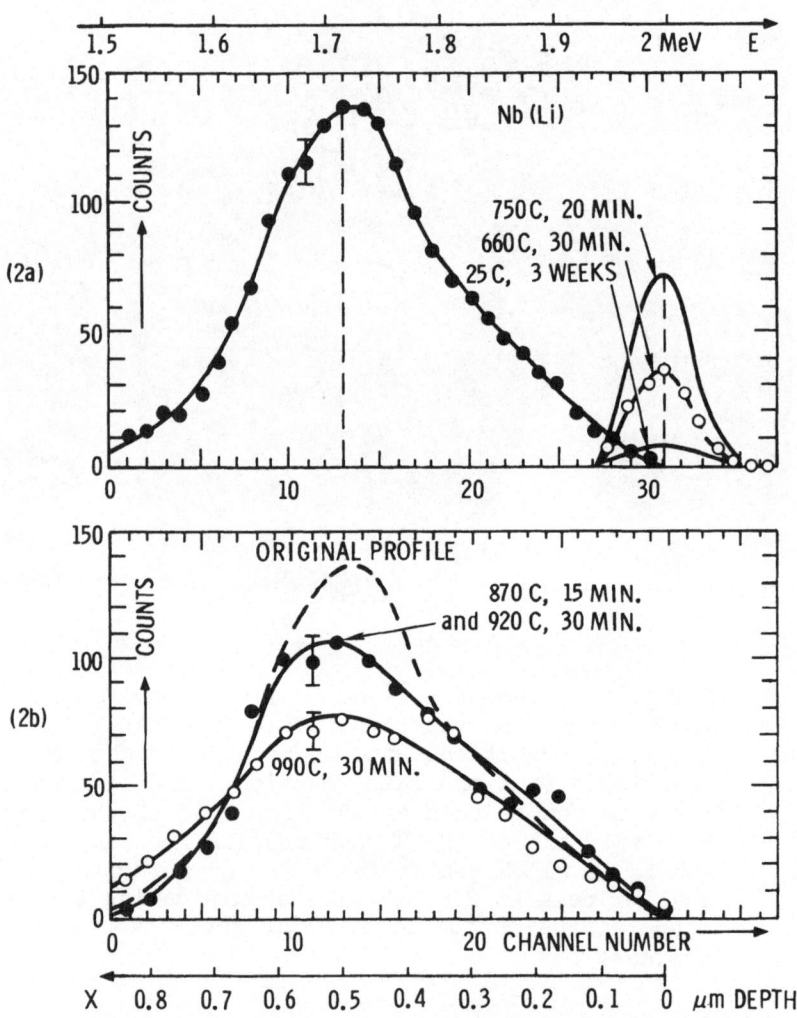

Figure 2. Depth distribution of 220 keV Li implanted in Nb for annealing a) below 800°C and b) above 800°C.

A similar behavior was reported for Ni in Zn at room temperature by H. J. Smith and G. N. van Wyk,[5] and at somewhat higher temperatures of 225 to 400°C for He in Cu by R. S. Blewer.[3] This behavior probably is related to interstitially migrating atoms which, during implantation, come to rest at interstitial sites, or at traps with low binding energy.

The appearance of the surface peak facilitates the depth calibration and the determination of the resolution of the equipment.

Above 750°C the surface peak of Li disappears rapidly due to evaporation. At 860°C, 920°C, and 990°C the Li profile broadens and decreases in amplitude, as shown in Fig. 2b. The observed changes agree with the calculated diffusion profiles under the boundary condition $c(0) = 0$. From these measurements the diffusion coefficient for 990°C, $D = (6 \pm 2) \times 10^{-14} \, cm^2 \, sec^{-1}$, and the activation energy of $\Delta E = (2.1 \pm 0.4)$ eV are obtained. No substantial difference could be detected between single crystals and foils. This was expected, as the mean size of grains in the polycrystalline material was found to be 21 μm which is large compared to the depth of observed profiles.

The low temperature diffusion corresponds to an activation energy of roughly 0.5 to 0.8 eV which compares well with activation energies of interstitially diffusing He,[6] Na, and K[7] in niobium. The activation energy of 2.1 eV, observed at higher temperatures, indicates that the diffusion is slowed down by trapping/detrapping processes, or that the Li atoms in a vacancy cluster migrate as an entity, i.e., microscopic lithium precipitate may become mobile.

LATTICE LOCATION OF Li IN Nb

The lattice location of Li in Nb crystals was studied immediately after implantation without anneal, in order to find both the interstitial and trapped locations. A channeling experiment was set up inside a reactor beam tube, with the Nb crystal positioned near the graphite block at a thermal neutron flux of $5 \times 10^9 \, cm^{-2} \, sec^{-1}$.

A cellulose acetate foil at a distance of 80 cm from the crystal served as a detector with low sensitivity to the reactor background radiation. The T and α particle tracks in the foil were chemically etched, and observed and counted with a Quantimet B of Metals Research, G.B.

The track distribution, as depicted in Fig. 3, exhibits an axial and planar channeling pattern around the ⟨111⟩ direction. This is consistent with interstitial Li atoms, either on tetrahedral or octahedral sites (which cannot be distinguished at present).

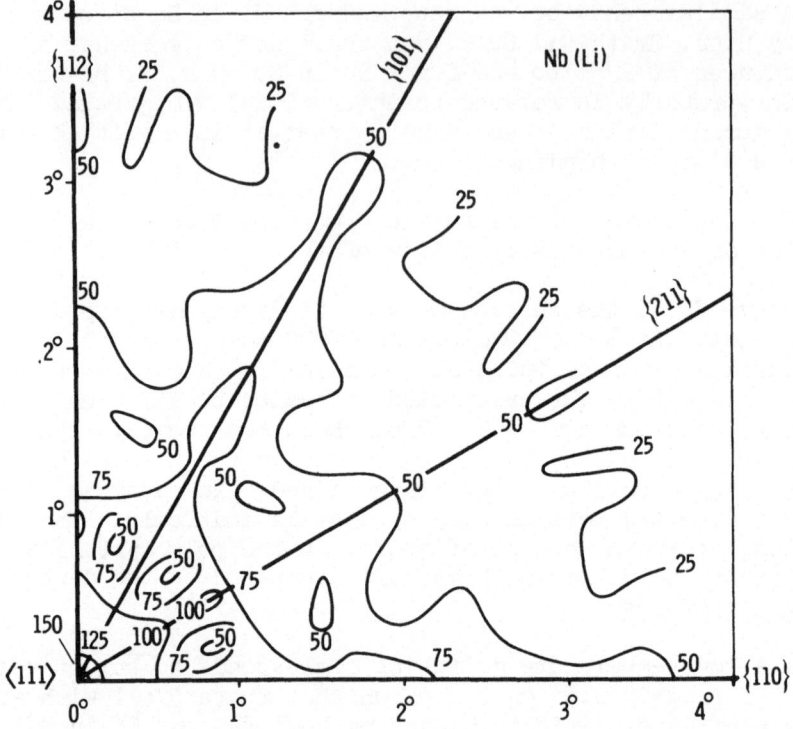

Figure 3.   Channeling emission (blocking) pattern for Li in Nb
around the ⟨111⟩ direction.

There is also some indication of Li atoms trapped in random positions, insofar as the channeling pattern of the Nb(Li) sample is
much less pronounced than the channeling pattern of Ge(Li) which
was observed before with the same experimental technique[8] (it is
known that all the lithium occupies interstitial lattice sites in
Germanium, as the Li is diffused into the crystal rather than
implanted).

    In order to obtain more precise results, the channeling/
blocking experiments have to be repeated with better counting
statistics, and in different crystallographic directions.

ABNORMAL CORROSION EFFECTS

A niobium sample which was implanted with Li to a few atom.-%
at depths between 0 and 0.8 μm, showed "catastrophic" oxidation
at the site of implantation, Fig. 4. The sample was heated for
one hour to 700°C in air at atmospheric pressure with the result
of abundant oxidation to white Nb_2O_5 (more than 0.1 mg/cm²), while
unirradiated areas turned slightly darker due to lower states of
oxidation. This behavior is not clearly understood, as impurity
atoms of electronegativity 1 are expected to inhibit oxidation
rather than enhance it.[9]

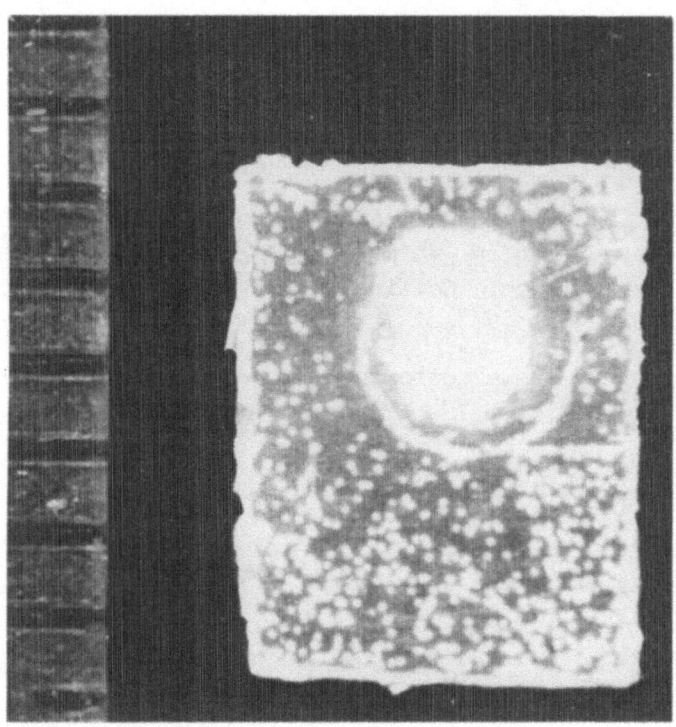

Figure 4.   Corrosion pattern of a Nb foil after one hour at 700°C
            in air. Oxidation to white Nb_2O_5 occurs at damaged
            surface regions; largest amounts of pentoxide being
            observed at site of implantation. (Scale divisions
            are 1 mm.)

ACKNOWLEDGEMENT

We are indebted to the reactor division of the Physikalisch-
Technische Bundesanstalt, Braunschweig, Germany, for making
available to us the FMRB reactor facility. We are very much

obliged to Dr. Adam, Technische Universität, Berlin, for his kind
help in preparing the single crystal samples.

REFERENCES

(1) J. P. Biersack, D. Fink, Nucl. Instr. Meth. 108, 397 (1973).

(2) S. T. Picraux, F. L. Vook, these conference proceedings.

(3) R. S. Blewer, these conference proceedings.

(4) H. Schiøtt, Mat. Fys. Medd., Dan. Vid. Selsk. 35, #9 (1966).

(5) H. J. Smith, G. N. van Wyk, these conference proceedings.

(6) S. Blow, AERE - R 6845, Harwell (1973).

(7) Dubnin, Denediktova, Karpman, Shcherbedinskii, cf. Diff.
     Data 4, 437 (1970).

(8) J. P. Biersack, D. Fink, Proceedings of the International
     Conference on Atomic Collissions in Solids, Gatlinburg,
     September 25-29, 1973.

(9) G. Dearnaley, these conference proceedings.

CHAPTER V

IMPLANTED ATOM LOCATION

LATTICE LOCATION OF IMPURITIES IMPLANTED INTO METALS

H. de Waard

Bell Laboratories, Murray Hill, New Jersey

University of Groningen, The Netherlands

L. C. Feldman

Bell Laboratories, Murray Hill, New Jersey

I. INTRODUCTION

The method of ion implantation has made it possible to substitute impurities in lattice sites of metals in a number of cases where more usual methods such as diffusion or alloying fail, due to the insolubility of the impurity in the host metal or due to its high chemical reactivity with the host.

As implantation becomes more widely used in the process of material modification, and as a tool in all branches of physics, it becomes important to determine and understand the new atomic system that is formed. One item of basic and fundamental interest is the atomic position of the implanted impurity in the lattice. Along with the radiation damage created, this property will determine the final microscopic and macroscopic properties of the implanted material. Ideally, a simple set of rules is desired, similar to the Hume-Rothery rules in alloying theory, to specify the lattice location. Although such a goal is far from realization, certain trends have been established from which we can gain insight. In this paper we shall discuss what has been learned on the lattice location question.

Channeling and impurity hyperfine interactions (HFI), i.e. the interactions between the electromagnetic moments of the implanted nuclei and the electromagnetic field of the surrounding charges, constitute two important means of obtaining information on

the location of the implanted ions. Hyperfine interactions are
largely determined by a few neighbor shells of atoms and therefore
are sensitive to local damage of the lattice. Their measurement
does not directly yield a unique determination of the lattice sites.
In some cases, however, channeling measurements do yield clear and
straightforward results for the lattice location and thus complement the hyperfine interaction results. In many cases the two
techniques feed upon one another to construct a model of the microscopic surroundings of the implanted ions. Channeling measurements
to determine the lattice location of implanted impurities have now
been carried out for a wide range of dopants, primarily in iron,
and to a lesser degree in a variety of other metals. These results
will be reviewed in order to show trends and other characteristics
of implanted systems. Methods using radioactive implants for the
study of HFI in dilute systems are of particular importance because
of their high sensitivity. Since each implanted radioactive atom
in due time emits a signal that conveys information about the interaction, only very small concentrations (generally less than 0.1%)
in a narrow depth region (of the order of 100 Å) are needed, implying
low radiation damage. On the other hand, the total dose of a radioactive implant can easily be increased to that used for channeling
measurements, so that both methods can be compared for equal doses.

In section 2, the hyperfine interactions relevant to our purpose are introduced, and the principles of nuclear spectroscopic
methods successfully used for studying the HFI of impurity configurations are discussed and illustrated by some examples. We will
restrict our discussion to cases where the unique character of the
interaction can be definitely established or where the existence of
two or more inequivalent sites can be shown. In section 3 the
channeling technique will be critically reviewed. We shall consider
some recent channeling measurements of the lattice location of the
series of elements Sb, Te, I and Xe implanted into Fe, present a
review of all relevant channeling measurements known to the authors
and discuss some trends revealed by this body of data. In section 4,
HFI and channeling results will be confronted and their apparent
agreement and discrepancies discussed.

II. THE HYPERFINE INTERACTIONS

The hyperfine interactions that can occur between nuclear
moments and the electromagnetic field of the surrounding charges
are the following:

(a) the (scalar) Coulomb interaction of the nuclear charge, Ze,
and the electrons around the nucleus. For s-electrons, that exhibit
a finite charge density at the nucleus, the interaction energy
depends on this charge density and on the (finite) nuclear dimension. This causes a very small shift of the energies of gamma

lines that can be observed in Mössbauer experiments, where it is called the isomer shift.

(b) the (vector) interaction between the nuclear magnetic dipole moment and the effective magnetic field at the nucleus due to orbital and spin currents of the surrounding electrons and due to external sources.

(c) the (tensor) interaction between the nuclear electric quadrupole moment and the gradient of the electric field set up by the surrounding electronic and nuclear charges.

From measurements of these interactions, many interesting results, both in nuclear physics and in solid state physics have been derived. Since about 1965 they have also been used to varying extent for studying the location of impurities implanted in crystalline solids. The circumstance that the strength of the interaction often depends very strongly on the exact configuration of the host atoms around the impurity has contributed to the success of these studies.

Interaction (a), that only gives rise to small shifts of nuclear level energies, can be treated separately from interactions (b) and (c), that give rise to a splitting of the levels. In a general treatment of the latter two interactions, a Hamiltonian must be constructed that contains terms representing the interactions of the nuclear moments with external fields, crystal fields and the orbital and spin moments of the atomic electrons as well as their charge distribution. We will give theoretical results for the simple cases which result when either interaction (b) or (c) dominates, in direct connection with the methods of measurement to be discussed in the next sections.

II.1 Low Temperature Nuclear Orientation

In the discussion of this method, we will limit ourselves to cases where the hyperfine interaction can simply be described in terms of an effective magnetic field aligned parallel to an externally applied field which provides a polarization axis for the system and where the quadrupole interaction can be neglected. In these cases, a nuclear state of spin I and magnetic dipole moment μ is split in an effective magnetic field H into $2I+1$ equally spaced magnetic substates $|Im\rangle$ with energy differences $\mu H/I$. Radiation emitted in the decay of any particular substate $|Im\rangle$ has a well defined directional distribution with respect to the field axis, but if all $2I+1$ substates are equally populated, the total radiation pattern of the decaying nuclei is isotropic. An anisotropy of the emitted radiation can, therefore, only be observed if somehow unequal populations of the substates can be produced. In such cases the nuclei have acquired a certain degree of orientation. In nuclear orientation experiments this is achieved by cooling down the source to a temperature T, where the equilibrium occupations of the substates, given by

$$W(\beta,m) = \exp(-\beta m)/ \sum_{m=-I}^{m=+I} \exp(-\beta m), \text{ with } \beta = \mu H/kTI \ , \qquad (1)$$

become noticeably different. This happens at temperatures where T $\sim$ $\mu H/kI$, which always lie well below 1°K (e.g. for $\mu/I=1$ nuclear magneton and H=1 Megagauss, $\mu H/kI \sim 0.04$°K). In most cases, the anisotropy of gamma radiation emitted in the decay of the oriented nuclei is studied, though for some particular cases, beta, alpha and fission fragment anisotropies of oriented nuclei have been observed. We will only discuss gamma anisotropy, for which case the directional distribution can be written as (1)

$$W(\theta,\beta) = \sum_{k} f_k(\beta)G_k A_k P_k(\cos \theta) \qquad (2)$$

where the $f_k(\beta)$, called degrees of orientation of order k, are simple functions of β, defined, for instance, in ref. (1) and the G_k are parameters that take into account changes in orientation by nuclear decays occurring before the observed gamma decay. The factors $A_k P_k(\cos \theta)$ give the directional distribution of gamma rays emitted by completely oriented nuclei. The P_k's are Legendre polynomials of order k and the coefficients A_k are determined by the spins of the states between which the gamma transitions take place and by the multipole character of the emitted radiation. Only even values of k are non-zero and almost always only k=0, k=2 and k=4 occur.

The gamma anisotropy is usually determined by placing gamma detectors at angles $\theta=0$° and $\theta=90$° with respect to the magnetic field axis, measuring the gamma intensities at these angles as a function of temperature T. A plot of these intensities (or their ratio) vs $1/T$ is then fitted to expression (2), calculated using known values of the nuclear parameters A_k and G_k. The fit yields a value for the hyperfine interaction strength μH. It is not always possible to obtain a good fit of the data for a model involving only one component of the hyperfine field. In such cases, a fit involving components with two (or more) different hyperfine fields may be tried, choosing their relative contributions to the measured anisotropy as further parameters to be fitted. Following this approach, Niesen (2), Pattijn et al. (3) and Keene and Postma (4)

TABLE I

Nuclear orientation measurements on sources implanted in iron from which information about lattice location can be obtained

Implanted Element	Isotope	Dose (at/cm^2)	Energy (keV)	Field $\|H\|$ (MG)	Percentage in field	Remarks	Ref.
Sb	122	4×10^{12}	80	0.234±0.020	>95[b]	most favourable case	6
	124	8×10^{13}	40	0.227±0.025	>90[b]		5
I	131	10^{13}-10^{14}	110	1.13[a]	62±5	two component model	4
Xe	131m	4×10^{15}	75	1.51±0.16	45±5	two component model	3
	133m	4×10^{13}	75		65±5	two component model	3
	133	10^{13}-10^{14}	50-100	1.2-1.6	50-75	two component model	2
Ce	137	5×10^{12}	100	-	-	no unique site	7
Tb	160	10^{14}	100	3.8	30±5	two component model	2,8
Au	198	4×10^{13}	100	1.35±0.1	>95[b]	most favourable case	6

[a] Taken from Mössbauer effect data.

[b] Concluded from comparison with diffused source.

were able to establish the existence of at least two inequivalent
fractions for implants of Xe-133(2), Xe-131m and Xe-133m(3) and
I-131(4) in iron. As an example, the improvement obtained by
Pattijn et al. for a two component fit compared with a one component
fit for the case of Xe-131m is shown in Fig. 1. So far, the method
has not allowed a differentiation between more than two components.
By combination with the results to be discussed in sec. 4 the high
field fraction could be associated with substitutionally implanted
atoms for all cases investigated so far. Nuclear orientation
results with relevance to lattice location studies are summarized
in Table 1.

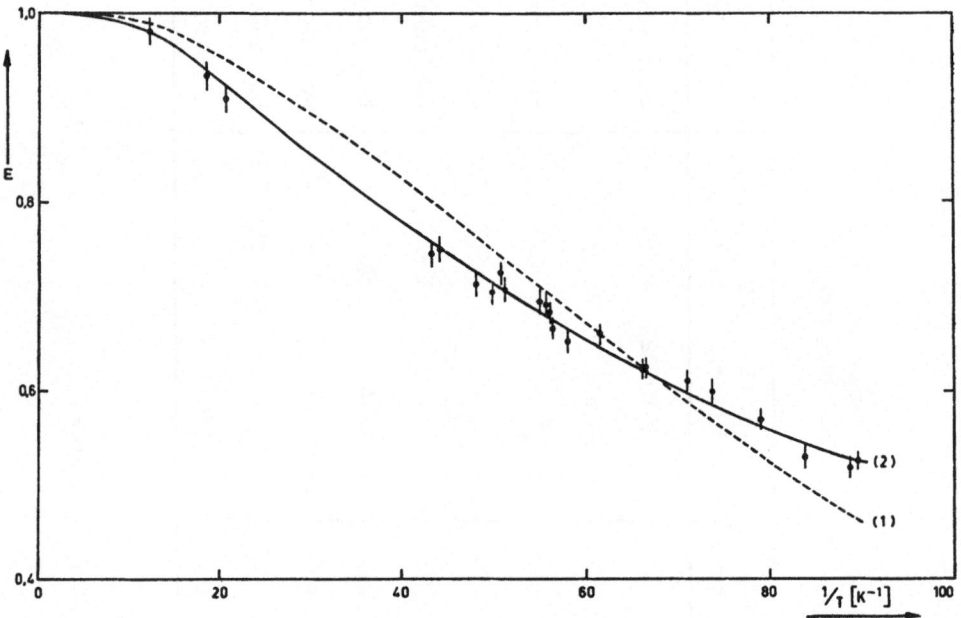

Fig. 1   Degree of gamma anisotropy $\varepsilon = W(0)/W(\pi/2)$ vs $1/T$ for the
164 keV gamma ray of Xe-131m. The solid line represents the best
fit for a substitutional fraction a = 0.60, whereas the broken line
is the best fit for all xenon nuclei in a unique site. Taken from
the work of Pattijn et al. (3).

II.2 Perturbed Angular Correlations

Nuclear spins may become polarized by other processes than the
low temperature nuclear orientation discussed in the previous section. A nuclear reaction or decay process, for instance, will in

general give rise to polarization of the product nuclei with respect to the direction of incidence or emission of the particles involved in the process. As in nuclear orientation experiments, the degree of nuclear orientation can be deduced from the angular distribution of radiation emitted after the process. The most familiar case is that of the angular correlation between gamma quanta emitted in cascade. The first of these (γ_1) defines the process leading to the nuclear orientation and the second (γ_2) is, in general, emitted anisotropically with respect to the direction of γ_1. The gamma quanta γ_1 and γ_2 are detected in coincidence by two detectors 1 and 2 (see Fig. 2) and the coincidences are counted as a function of the angle between the axes of the two detectors to find the directional anisotropy. The directional distribution is given by an expression resembling that for nuclear orientation (9):

$$W(\theta) = \sum_k A_{kk} P_k(\cos \theta) \quad , \tag{3}$$

where the A_{kk} are coefficients determined by the spins I_i, I^* and I_f of the states involved (see Fig. 2) and by the multipole character of γ_1 and γ_2. Usually, only terms with k=0, k=2 and k=4 occur. If the nuclei experience a hyperfine interaction while they are in the intermediate energy state, expression (3) must be modified, because due to this interaction the populations of the magnetic substates (defined with respect to the direction of emission of γ_1 as the quantization axis) change as a function of time and therefore also the angular distribution of the radiation pattern of γ_2. Semiclassically, one may also say that the spin I^* of the intermediate state precesses around the symmetry axis of the interaction with a frequency determined by the interaction strength. The directional distribution of γ_2 precesses together with the spin (Fig. 2). We will discuss the important cases of pure magnetic dipole and pure electric quadrupole interaction separately, restricting ourselves to geometries for which the theory acquires it simplest form.

A. <u>Magnetic interaction</u>. If the hyperfine interaction can be ascribed entirely to an effective magnetic field perpendicular to the detector plane, the nuclear spins in the intermediate state precess with a Larmor frequency

$$\omega_L = -\mu H/\hbar I \tag{4}$$

and, if a time t elapses between observation of γ_1 and γ_2 in a cascade, the observed angular distribution will now be

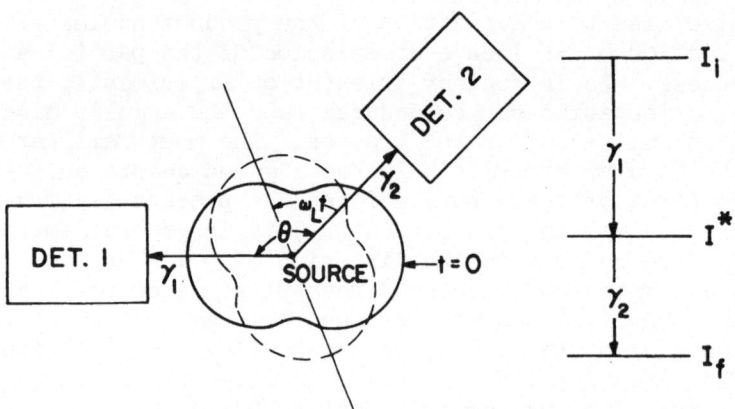

Fig. 2 Principle of perturbed angular correlation measurement for magnetic hyperfine interaction. The rays γ_1 and γ_2 are detected in coincidence by deterctors 1 and 2. If γ_2 is emitted a time t later than γ_1 the correlation pattern has rotated over an angle $\omega_L t = (-\mu H/\hbar I)t$ in a magnetic field H applied perpendicular to the detector plane.

$$W(\theta,t) = \sum_k A_{kk} P_k [\cos(\theta - \omega_L t)] \ , \tag{5}$$

which can also be written as

$$W(\theta,t) = \sum_k b_{kk} \cos k(\theta - \omega_L t) \ . \tag{5'}$$

Since the intermediate state decays while it precesses, the coincidence counting rate will be given by

$$N(\theta,t) = N_o \exp(-t/t^*) \, W(\theta,t) \ , \tag{6}$$

where t^* is the lifetime of the intermediate nuclear state. Depending on the relative magnitude of t^*, the time resolution of the angular correlation equipment t_o and the Larmor period $t_L = 2\pi/\omega_L$, several cases may be discerned. The most important cases for practical application are the following two:

(1) $t^* < t_o$, $t_L \gg t^*$. In these cases, the measurement yields the time average of Eq. (6) and a (small) average precession $\Delta\theta \sim \omega_L t^*$ of the angular distribution of γ_2 is observed. This is called a time integral perturbed angular correlation measurement (IPAC). Such measurements do not yield direct information about lattice location for implanted sources.

(2) $t^* \gg t_o$, $t_o < t_L < t^*$. In these cases, the coincidence rate between γ_1 and γ_2 can be measured as a function of the delay of emission of γ_2 relative to χ. A plot of this rate vs delay (delayed coincidence plot) shows the usual exponential decay of the intermediate state, modulated by "wiggles" representing the periodic behaviour of the factors b_{kk} cos $k(\theta - \omega_L t)$ in Eq. (5'). The inter- action strength can now accurately be determined from a least squares fit of the data to Eq. (6) or some appropriate function closely related to Eq. (6). A representative result is shown in Fig. 3. This is the time differential perturbed angular correla- tion method (TDPAC), which has important applications for lattice location experiments. In principle, the Fourier transform of the oscillating part of the delayed coincidence plot may yield a number of different frequencies if impurity atoms in different sites with different hyperfine interaction strengths are present. In practice, however, experimental conditions do not always allow a very detailed analysis, due to insufficient statistical accuracy or due to the fact that the condition $t_o < t_L < t^*$ is not fulfilled for all field components. In such cases, sometimes only one clearly defined modu- lation frequency is observed, the amplitude of which, however, is reduced relative to that derived from Eq. (5) or (5') by inserting the appropriate nuclear parameters A_{kk} or b_{kk}, respectively. These constants can often be determined accurately from a "normal" angular correlation experiment. Other complications of the method occur if the modulation amplitude decreases with time. A Fourier transform of such a damped modulation shows a frequency distribution of width inversely proportional to the damping time constant. It is, however, not always certain that such a frequency distribution corresponds to a distribution of slightly inequivalent sites, because a damping of the modulation amplitude can also occur as a result of relaxation phenomena that gradually destroy the nuclear orientation. It is often difficult to discern between such time dependent relaxation processes and a distribution of slightly different static fields, even when further experimental data are available. The discussion just given about the PAC method for the case of γ-γ cascades, where the polarizing event was the emission of the first gamma quantum, is equally valid when the polarizing event is a nuclear reaction or excitation induced by a charged particle beam. An important development in this field is the method called "recoil implantation

perturbed angular correlation" (IMPAC) (10) in which nuclei in an
excited state are recoiled into a substrate by virtue of the
exciting nuclear reaction. Like the γ-γ PAC method, the IMPAC
method is useful for lattice location studies primarily in the time
differential mode.

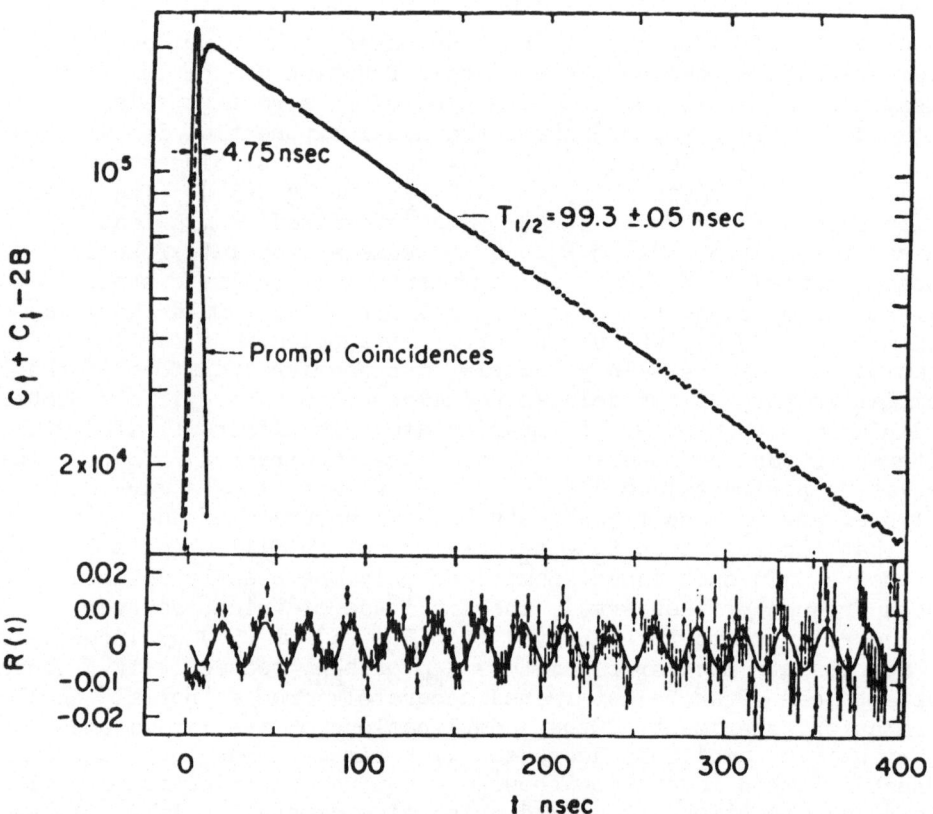

Fig. 3 Example of an angular correlation measurement perturbed by
magnetic hyperfine interaction. Upper graph: background corrected
decay curve of the 14.4 keV state of Fe-57, obtained by summing data
for magnetization up and down and thus removing the precession
oscillations. Lower curve: sinusoidal oscillations of a plot
obtained by forming the (background corrected) relative difference of the data for field up and field down
$(R(t) = (C_\uparrow(t)-C_\downarrow(t))/(C_\uparrow(t)+C_\downarrow(t)-2B))$. Solid line: least
squares fit of theoretical curve to $R(t)$. Example taken from the
work of Hohenemser et al. (62).

The number of results on TDPAC and TDIMPAC that have relevance with respect to lattice location is, as yet, rather limited.
Cases presently known to the authors have been summarized in
Table II. (The results on F-19 implanted into Ni of Stokstad et al.
(11) and Klepper and Spehl (13) cast doubt on the validity of the
older result of Braunsfurth et al. (12).)

B. Quadrupole interaction. Impurities in metals may experience an electric quadrupole interaction for various reasons. For
instance, the crystal field states of incompletely filled electron
shells of the impurity atom may produce a field gradient at the
nucleus. For substitutional sites in cubic metals, this mechanism
produces sizeable field gradients only in the case of the rare
earths which, though of considerable importance, we must leave
outside discussion here. A more direct quadrupole coupling occurs
for substitutional impurities in non-cubic metals and may occur
for non-substitutional impurities in cubic metals. Impurity
associated defects also give rise to quadrupole coupling which
may yield important information about the nature of these defects.
In these cases, a field gradient is produced at the nucleus
directly by the electric charges of the surrounding atoms. Often
this gradient is considerably enhanced by the polarization it
induces in the electron shells of the impurity (by the so-called
Sternheimer factor).

In many cases, the electric field gradient at the impurity has
axial symmetry and, especially if we restrict ourselves to non-
magnetic metals, the theoretical treatment is simplified. Unlike
the case of purely magnetic interaction, where the field axis can
be given the same direction throughout the sample by proper magnetization, the field gradient axes in a polycrystalline sample are
randomly oriented. A unique orientation of these axes can only be
obtained by using one single crystal or a mosaic of properly
oriented small single crystals.

If we choose the axis of symmetry of the field gradient along
the z-axis of a rectangular coordinate system, the field gradient
tensor has only one independent element: $\partial^2 V / \partial z^2 = eq$; the Laplace
equation requires $\partial^2 V / \partial x^2 = \partial^2 V / \partial y^2 = -\frac{1}{2} eq$. A nuclear level of
spin I is split by the electric quadrupole interaction into sub-
states $| Im \rangle$ with energies

$$W = \frac{e^2 qQ[3m^2 - I(I+1)]}{4I(2I-1)} \quad . \tag{7}$$

Apparently, substates with opposite magnetic quantum numbers m
and -m have the same energy for quadrupole interaction, whereas
for magnetic dipole interaction they had opposite energies (+ and
$-m\mu H / I$). This implies that in the presence of quadrupole

TABLE II

Time differential perturbed angular correlation measurements (TDPAC and TDIMPAC) on sources implanted in Fe, Co or Ni from which information on lattice location can be obtained.

Implanted element and host	HFI observed in	Implantation energy (keV)	Field H(kG)	Percentage in field	Remarks	Ref.
IMPAC						
F Ni	F^{19}	2300^a	17.8±0.4	~30%	} about 50% in irregular sites	11
			90.2±1.4	~20%		
		840^a	-21.8±0.4		satellite fields observed	12
		840^a	17.6±0.5	19±4	} rest in undetectable field	13
			91±3	11±3		
F Co	F^{19}	2700^a	31.4±0.5	18±6	rest in undetectable field	13
		840^a	59.5±1.5		satellite fields observed	12
F Fe	F^{19}	840^a	95.7±0.5		satellite fields observed	12
		2300	92		preliminary result	11
A Ni	A^{37}	2250^a	280±23	25	75% assumed in low field sites	15
K Ni	K^{41}	400^a	-197±1.2	60	40% assumed in low field sites	16
Ca Fe	Ca^{42}	~3000^a	-100±9	21±4	rest in undetectable fields	17
PAC						
Hf Ni	Ta^{181}	60	~90	30-70	remaining fraction sees no field	18
Tl Fe	Hg^{199}	60-90	|H| = 670±65	65-80	} at most 20% in irregular sites	19
			|H| = 455±85	15-20		
Pb Fe	Pb^{204}		~300	>90	^{a}Maximum recoil energy	20

interaction states $|Im\rangle$ and $|-Im\rangle$ precess in opposite direction, while for magnetic interaction they precess in the same direction. Therefore, no net precession of the angular distribution of the gamma rays results in quadrupole interaction (as long as the states $|Im\rangle$ and $|-Im\rangle$ are equally populated), but instead, a periodic intensity modulation of the anisotropic distribution is observed. This behaviour is expressed in the perturbed angular correlation formula:

$$W(\theta,t) = \sum_k G_k(t) \, A_{kk} P_k(\cos \theta) \qquad (8)$$

where the G_k's are periodic functions of time for $k > 0$, $G_o = 1$. A simple example: for a state with spin I=3/2 and a randomly oriented field gradient, only $G_o = 1$ and $G_2(t) = (1+4 \cos \omega_L t)/5$, with frequency $\omega_L = [W_Q(m=3/2) - W_Q(m=1/2)]/\hbar$, are non-zero. General expressions for $G_k(t)$ can be found in the extensive review paper of Frauenfelder and Steffen (9). A time differential perturbed angular correlation for the case of electric quadrupole interaction is measured in the same manner as described for magnetic interaction. Again, those cases where the lifetime of the intermediate state is long enough to allow a time differential measurement are most suitable for obtaining information relevant to lattice location. The number of isotopes suitable for quadrupole interaction studies by TDPAC is rather small. For many investigations, the 247 keV state of Cd-111 has been used. This has a very convenient lifetime, $t^*=121$ nsec, and a convenient parent activity, 2.8 day In-111. An example of a TDPAC measurement with quadrupole interaction is shown in Fig. 4. In the contributed paper of Kaufmann et al. (22), location results obtained with implanted In-111 sources in various non-cubic metals are discussed. The fraction of implanted atoms in unique sites is derived by comparing the experimental value of the modulation amplitude of the TDPAC plots with the maximum theoretically possible value. Krien et al. (23) implanted 4 day Pd-100 in Zn and concluded from a TDPAC measurement on the 75 keV state of the Rh-100 daughter ($t^*=348$ nsec) that the palladium lands in unique sites. Kaufmann et al. (24) deduce a more than 95% unique site for a 70 keV implant of Hf[181] in a Ti single crystal from a TDPAC measurement.

A few cases have been reported of TDIMPAC, using nuclear reactions to recoil implant atoms of the host element in non-cubic metals. Haas (25) observed a unique frequency for Zn in Zn if implanted above room temperature. Recoil implants of Cd in Cd were studied by McDonald et al. (26) at 3 different temperatures (T=80, 180 and 300°K). They conclude from the modulation amplitude that within a 30% uncertainty the Cd-recoils occupy substitutional sites at all temperatures. Bleck et al. (27), who also carried out TDIMPAC experiments on Cd recoiled into Cd, observed a modulation

amplitude in agreement with a substitutional implant in the range
200-400° K but at 80°K their modulation amplitude is reduced. They
interpret this result as the effect of frozen-in radiation damage
that causes a spread of electric field gradients at part of the
implanted Cd atoms. Brenn et al. (28) measured the quadrupole interaction of polarized F^{19} nuclei recoil implanted into zinc single
crystals. They conclude from the fact that their modulation amplitude is only 45% of the theoretical value, that about 45% of the
fluorine ions probably occupy substitutional sites while the
remainder has landed in sites with a field gradient too high to be
observed.

Finally, we should mention the interesting work of Minamisono
et al. (29), who recoil implanted the short lived beta-emitters
B-12 and N-12, produced by the nuclear reactions B-11(d,p)B-12
and B-10(He-3,n)N-12, into various cubic metals. The implanted
nuclei are oriented by the reaction process and retain this orientation during their lifetime. Therefore, the directional distribution of the beta-particles emitted in their decay is anisotropic;
the anisotropy is detected with two beta detectors placed at
angles $\theta=0°$ and $\theta=180°$ with respect to the nuclear orientation axis.
The beta anisotropy can be destroyed by applying an r.f. field of
the proper frequency to the implanted samples, that are placed in
an external magnetic field. The r.f. field again induces transitions between the magnetic substates of the impurity nuclei. A
plot of the reduction of the anisotropy as a function of the r.f.
frequency shows the presence of a quadrupole interaction for the
b.c.c. metals Nb, Mo, Ta and W. From this fact, it must be concluded that the implanted B and N atoms occupy interstitial sites
in these metals. For the particular case of B in W it has been
shown by channeling that the interstitial site is the one of octahedral symmetry (30). In the f.c.c. metals Al, Cu, Pt and Au, on
the other hand, a single resonance line is found, which indicates
that the impurities occupy the octahedral interstitial sites which
in these metals have cubic symmetry. It is interesting to note
that channeling, in these cases, yield no evidence of a definite
interstitial site (30).

II.3 THE MÖSSBAUER EFFECT

Gamma rays can be emitted without suffering any appreciable
energy loss due to the nuclear recoil when the emitting nuclei are
incorporated in a solid material. In this case, the recoil momentum is not absorbed by the emitting nucleus, but by the solid as a
whole. The "recoilless" gamma lines, if due to a nuclear transition
to the ground state, in principle, have a very narrow width
$\Gamma=\hbar/\tau^*$, where τ^* is the lifetime of the gamma emitting state. The
emitted gamma rays can be resonantly absorbed by nuclei of the same
isotope, incorporated in some suitable solid absorber. The resonance

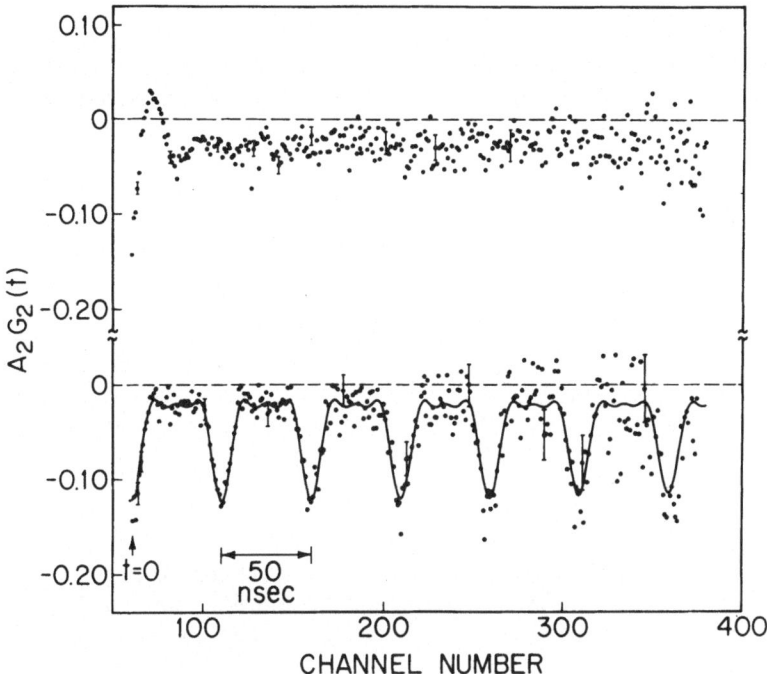

Fig. 4 Time differential angular correlation measurement perturbed by quadrupole interaction. In-111 implanted into Zn; (above) pre-anneal, (below) after 300°C - 1/2 hr. anneal.

can be observed by imparting a velocity to source or absorber that gives the line a sufficient Doppler shift to move it off resonance. These interesting facts were discovered by R. L. Mössbauer in 1958 (31) and the recoilless gamma resonance phenomenon just described was named after him. Typical velocities required to move a gamma line off resonance range from a few mm/sec to several cm/sec. The gamma intensity transmitted through the absorber as a function of the source or absorber velocity, yields a Mössbauer spectrum. The present status of Mössbauer spectroscopy is well presented in the book of Greenwood and Gibb (32).

The extreme sharpness of the resonance makes it possible to measure in many cases the hyperfine splitting of the gamma line due to magnetic dipole and/or electric quadrupole interaction. From this splitting the strength of the interaction can be derived. Moreover, the Mössbauer effect makes it possible to observe the very small shifts of gamma line energies caused by the interaction between nuclear and s-electron charges mentioned under (a) in section 2. Finally, the fraction f, of the gamma rays that are recoillessly emitted (or absorbed) can be determined from the absorption depth of the different components of a Mössbauer spectrum. The hyperfine interaction parameters just mentioned as well as f may depend strongly on the immediate surroundings of the emitting (or absorbing) nuclei, making the Mössbauer effect a particularly useful tool for studying lattice location. In fact, the location information obtained from Mössbauer measurements is often more detailed than that reported for PAC and NO measurements.

Requirements for suitable Mössbauer isotopes are a gamma transition with energy E_γ lower than about 100 keV to a stable (or very long lived) ground state and a lifetime of the excited state in the range 1-100 nsec. Moreover, for practical implantation experiments the lifetime of the parent activity should roughly lie within the range of about 5 hours to one year. There are about 25 isotopes that fulfill these requirements.

The reason for requiring a low transition energy is that otherwise the recoilless fraction would become too small to observe the effect. The recoilless fraction is given by

$$f = \exp\left[-(E_\gamma/\hbar c)^2 \langle x^2 \rangle\right] \tag{9}$$

where $\langle x^2 \rangle$ is the mean square vibration amplitude of the emitting atom. This quantity is an increasing function of temperature, but does not become zero for T=0. It is determined by the Debye temperature of the solid. Typically, for a medium heavy nucleus (A=100) and a normal Debye temperature (θ_D=200°K), $\langle x^2 \rangle \sim (0.04 \text{ Å})^2$ at T=0. Then, f=0.007 for E_γ = 100 keV, making the effect hard to observe, which explains the requirement of a low gamma energy. The quantity $\langle x^2 \rangle$ depends on the elastic forces exerted on the impurity by the surrounding host atoms. If their configuration is altered, $\langle x^2 \rangle$ will change. In particular, if there are vacancies close to the impurity, we must expect $\langle x^2 \rangle$ to be increased, leading to a reduction and a more rapid decrease with temperature of the recoilless fraction. This effect can be used to differentiate between regular substitutional sites and damage sites (see ref. (33) and (8)).

Many of the Mössbauer measurements that have proved to be relevant for lattice location, were performed with sources implanted into ferromagnetic metals, mainly iron. In these cases, the magnetic hyperfine interaction leads to easily identifiable patterns, allowing a unique decomposition of the measured spectrum into a number of fractions corresponding to different fields, i.e. different lattice sites.

Results on lattice location and on annealing of damage obtained from Mössbauer measurements up to 1971 were recently reviewed by one of the authors (H. de Waard (34)). For a summary of all results known at present we refer to Table III. Some recent investigations will now be briefly discussed.

(1) Sb-125 and I-125 → Te-125/Fe. Mössbauer spectra of the 35.6 keV transition in Te-125 were measured using both a diffused source of 1% Sb-125 in Fe and an implanted source of I-125 in Fe, in each case in combination with a single line ZnTe-125 absorber (35). The spectrum obtained with the Sb-125/Fe source is shown in Fig. 5A. It can be fitted very well with the symmetric 6-line pattern expected in this case for a single magnetic field component. This is in agreement with the fact, that Sb dissolves substitutionally into iron for concentrations up to a few percent. The fit of the single component Sb-125/Fe spectrum facilitates the analysis of the more complex spectra obtained with I-125/Fe sources. These spectra, taken at $4°K$ and at $78°K$ are shown in Fig. 5B. They cannot be fitted by less than three components. The result of a three component fit is shown in the figure, together with the shapes of the individual components. The intensity ratios of the Zeeman lines of each of these components were taken equal to those found from the Sb-125/Fe spectrum. The component with the largest splitting yields the same field as the Sb-125/Fe source and therefore is obviously due to substitutionally implanted I-125 atoms. The component with intermediate field (i) is interpreted as due to association of an iodine atom with one vacancy, that with low field (l) to association with two or more vacancies. It can also be observed in Fig. 5B that, when the source temperature is raised from $4°K$ to $78°K$, the intensities of the lower field components are reduced more rapidly than that of the high field component. This indicates that the Debye temperature for the lower field components is lower than for the high field component, in accordance with the hypothesis of vacancy association. From the estimated Debye temperatures, low temperature recoilless fractions can be derived for each of the three components of the Mössbauer spectrum. By combining these with the relative intensities of the components, the fractional occupations of the sites, also given in Table III, are found.

(2) Te-129m → I-129/Metals. The earlier work on this case (34) is being continued by Reintsema (39) who investigates sources of Te-129m implanted into many different transition metals, mainly with the purpose of studying the isomer shift vs the atomic number of the host metal. In some cases, fractions with different isomer shifts, corresponding to different sites, are observed.

Coussement et al. (40) measured Mössbauer spectra of Te-129m/Fe as a function of the dose of implanted tellurium. They find that

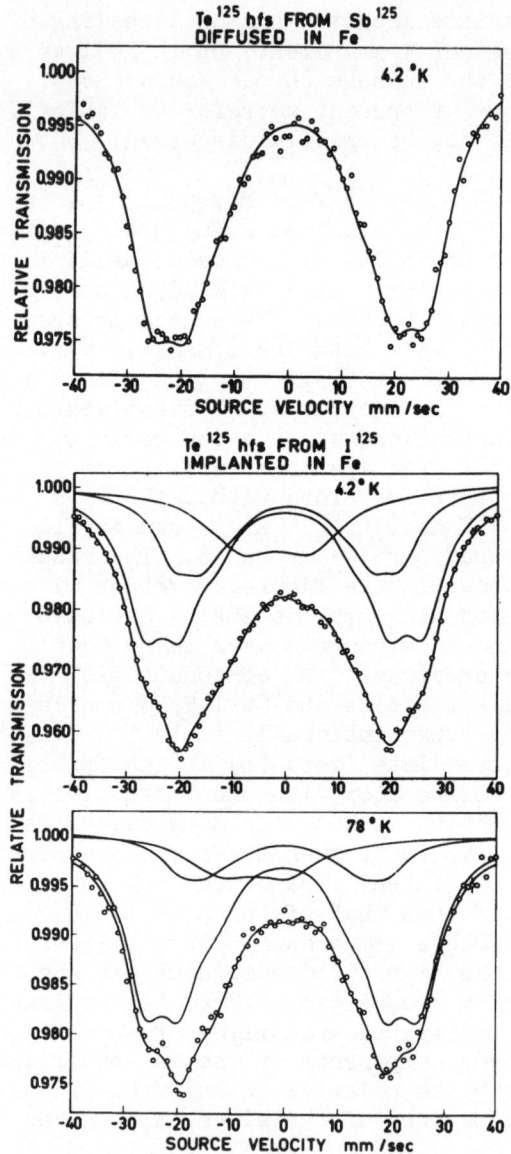

Fig. 5A Mössbauer spectrum obtained at 4.2°K with a diffused source of Sb-125 in Fe and a ZnTe-125 absorber. The drawn line through the measured points represents a least squares fit to a 6 line pattern, obtained for one unique value of the magnetic hyperfine field: H = 678 kG.

Fig. 5B Mössbauer spectra obtained with implanted I-125/Fe source and ZnTe-125 absorber at 4.2 and 78°K. The lines drawn through the measured points represent least squares fits to a 3 component (high, intermediate and low field site) model.

at an implantation energy of 75 keV and at doses above $3\times10^{15}/cm^2$, Te clusters with small hyperfine fields start to appear.

(3) I-131 → Xe-131/Fe. Mössbauer spectra of the 80 keV transition in Xe-131, measured with implanted sources of I-131 in Fe against a Na_4XeO_6 absorber, again exhibit a structure corresponding to at least three different components (35). The general appearance of the spectrum, shown in Fig. 6, is very similar to that obtained for the case of I-125/Fe. Again a good fit can be obtained with

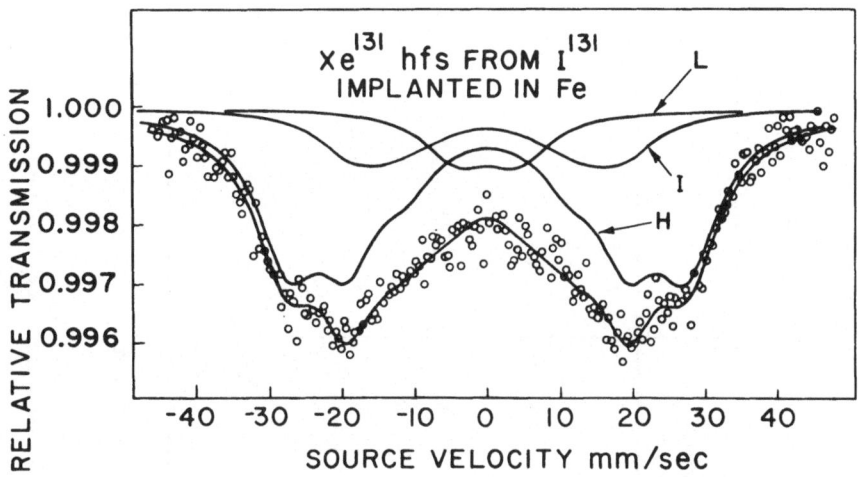

Fig. 6 Mössbauer spectrum obtained with implanted I-131/Fe source and Na_4XeO_6 absorber at 8°K. The line drawn through the measured points represents a least squares fit to a 3 component model. Separate components are given by lines marked H (high field), I (intermediate field) and L (low field).

three magnetic components, corresponding to high, intermediate and low field sites. The h-site is again interpreted as substitutional, the i and l sites as vacancy associated.

(4) Xe-133 → Cs-133/Fe. After the work already reported in ref. (34), research on this case was continued in Groningen. Recently, implantations were carried out above and below room

TABLE III

Results on hyperfine fields and lattice site occupation obtained from Mössbauer spectra of sources implanted in iron.

Activity	Implantation Dose at/cm²	Energy keV	Temp °K	Mössbauer line keV	in	Fields (kG) h	i	l	Site occupation (%) h	i	l	Ref.
Sb^{125}	1%[a]					678±5	-	-	100	-	-	35
I^{125}	10^{15}	120	RT	35.6	Te^{125}	683±5	550±10	190±60	46±5	32±4	-	35
Te^{129m}	10^{14}-10^{15}	50-120	RT	27.7	I^{129}	1130±30	-	-	100	-	-	34
	2×10^{14}	140	90			1130±30	-	-	100	-	-	36
I^{131}	3×10^{14}	120	RT	80	Xe^{131}	1510±100	1050±100	~300	(40)[b]	(20)[b]	(40)[b]	35
Xe^{133}	10^{13}-10^{14}	50-140	90	81	Cs^{133}	280±5	135±10	-	70±10	30±10	-	36
			RT			280±5	135±10	~300	45±10	20±5	35±10	34,36
	2×10^{14}		380			280±5	135±10	~300	35±10	65±10		36
	2×10^{14}		480			280±5	135±10	~300	15±10	85±10		36
Gd^{151}	4×10^{14}	50	RT	21.6	Eu^{151}	1500	-	<50 kG	50	-	50	37
Tb^{161}	10^{14}	50-140	RT	25.6	Dy^{161}	6700	-	-	>90	-	<10	

[a]Diffused source; diffused and implanted Sb-activities yield the same hyperfine field, as was shown by Andrews et al. (5) and by Reid et al. (6).

[b]Estimates based on measurements with I^{125}Fe and Xe^{133}Fe sources.

temperature; at liquid nitrogen temperature, a substantial increase
of the substitutional fraction was found, indicating that fewer Xe
atoms are associated with more than one vacancy. The results are
further discussed in the paper contributed by Drentje et al. (36)
to this conference.

(5) Gd-151 → Eu-151/Fe. As will be reported by Cohen et al. (37)
at this conference, the Mössbauer spectrum of the 21.6 keV line of
Eu-151 emitted in the decay of a 50 keV, 4×10^{14} at/cm^2 implant of
Gd-151 in Fe exhibits a high and a low field component of comparable
intensity. It is concluded from this, that about 50% of the
implanted atoms are substitutional, while the remainder are in
irregular sites of a single but not yet clearly established type.

(6) Tb-161 → Dy-161/Fe. The earlier interpretation of the
Mössbauer spectra of this case, given in Refs. (8) and (41), where
the existence of two components was postulated, has to be revised
in view of a more recent analysis of the old data by Wit and
Niesen (38). They could fit the spectra taken at different temperatures rather well with a theoretical spectrum containing only
one component by incorporating relaxation effects. It must be
remarked, however, that for the rare earths, where the hyperfine
interaction strength is often determined to a large extent by the
4f electron configuration of the impurity itself, impurities in
different sites may exhibit almost the same spectrum. In other
words, in the case of rare earth impurities the hyperfine interaction strength is not always a very sensitive probe of the microscopic surroundings.

III. CHANNELING

III.1. The Channeling Technique

The mechanics of the channeling technique for the lattice
location of implanted impurities has been described a number of
times; one of the most recent and complete descriptions is in the
work of Alexander et al. (42). Basically, the technique depends
on the fact that close encounter processes such as Rutherford
scattering or nuclear reactions from atoms on regular lattice sites
are strongly suppressed relative to those off lattice sites when
the beam is in a channeling direction. When scanning about such a
direction, substitutional impurities will produce the same angular
dependence as the host lattice, while non-substitutional impurities
produce an angular yield different than the host. Although the
method is straightforward, underlying problems in the technique
have prevented its full and confident use. For the problems of
concern in metal implantation, there seem to be three major

uncertainties or deficiencies in the technique: 1) Impurity concentration limitations, 2) effects due to the measurement itself
and 3) interpretation of the results in the case of mixed site
configurations.

The minimum impurity concentration that is possible to investigate by this method depends on the system and the probing conditions. However, there has not been any experiment performed at
less than an approximate ∿0.1% level and almost all have been in
the range of 0.1% - 1% impurity concentration. Since the systems
already studied represent a wide range of situations, it appears
that 0.1% is close to a practical lower limit. One main disadvantage that results from this limitation is the difficulty of comparing the implanted case with the as-grown case because of low
solubilities.

A second category of problems centers about the interaction
of the energetic probing beam with the implanted system. The probing beam creates significant damage in the implanted region and
especially at the end of its range. For example, in the experiments reported here approximately 5% of the iron atoms in the
implanted region are displaced due to the radiation damage of the
probing beam for a dose required for one angular scan. The radiation damage effect of the probing beam has been clearly observed
in some semiconductor studies but has not yet been clearly demonstrated in the case of the metals at room temperature. There is
no change observed in the impurity site in most cases during the
course of the experiment. This has been most carefully explored
in a number of situations in the Fe-impurity system at room temperature. The correlation of the substitutional percentage with similar information extracted from hyperfine measurements also supports
the assertion of no effect from the damage created by the probing
beam. However, this assertion cannot be safely extrapolated to
systems other than Fe or even the Fe-impurity system at low
temperature.

An additional type of interaction with the probing beam has
been reported in the investigation on Br implanted in iron (Ref. 42).
In that case ^{14}N was used as the probe and measurements over a
time period comparable to the time required for the N to diffuse
from its end of range to the vicinity of the implanted impurity
(∿ 300 hrs) indicated a change in the impurity site location. An
effect on such a time scale would not affect most reported location results because of the manner in which the experiment is done.

Perhaps the most frustrating aspect of location experiments
is in the interpretation of the raw data. At the time of this
writing, there is no implanted case in metals which has yielded a
100% substitutional situation, while a large number of cases are

in the 80-100% region. Many situations show dips of one-half the host dip and the simplest interpretation is that $\sim$ 50% of the impurities are substitutional and 50% randomly distributed throughout the lattice. This interpretation is physically quite unreasonable and the understanding of such data remains an important problem to be resolved. One suggestion is that the region about the impurity is highly strained, thus producing a different channeling pattern for impurity and host (43). However, this hypothesis does not bear up under some cases in which additional detailed information from hyperfine interactions is available.

Other aspects of interpretation that remain to be resolved center about the flux peaking phenomena and the quantitative interpretation of it. For pure interstitial impurities one can often find symmetry directions which do not depend on the flux peaking effect for the site determination; however, for cases of mixed sites a detailed understanding is necessary (30).

Although this section has centered on the difficulties in the channeling technique, it is a fact that most of our knowledge about the lattice location of implanted impurities arises from this technique.

III.2 The Lattice Location of Sb, Te, I and Xe in Fe

Using the channeling technique with 2.0 MeV He backscattering, we have investigated the lattice location of ^{123}Sb, ^{128}Te, ^{127}I and ^{132}Xe in iron. These systems lead to interesting comparisons for a number of reasons:

(1) The primary implantation damage will be comparable since the experiments were done with implantation doses of 5.5-8x10^{14} atoms/cm^2 all at 100 keV and at room temperature.

(2) As a control all implants were in the same crystal, the experiments being done two at a time on the back and front of the same sample.

(3) The chemical and physical properties of the dopants are very different. Sb has a solubility in Fe in excess of the implanted dose, while Te is expected to be insoluble, I is highly reactive and Xe is an inert gas.

(4) Measurements of the hyperfine fields which result from the implantation of these elements is available, supplying additional location information.

(5) The I in Fe case might give a result analogous to the previously reported Br case, in a system which is simpler to study (42).

Figure 7 shows the angular yield patterns for these four cases in the $\langle 111 \rangle$ axial direction and Fig. 8 indicates a similar set for the {100} plane. Table IV summarizes the characteristic ratios for these cases. From this information the following points should be noted:

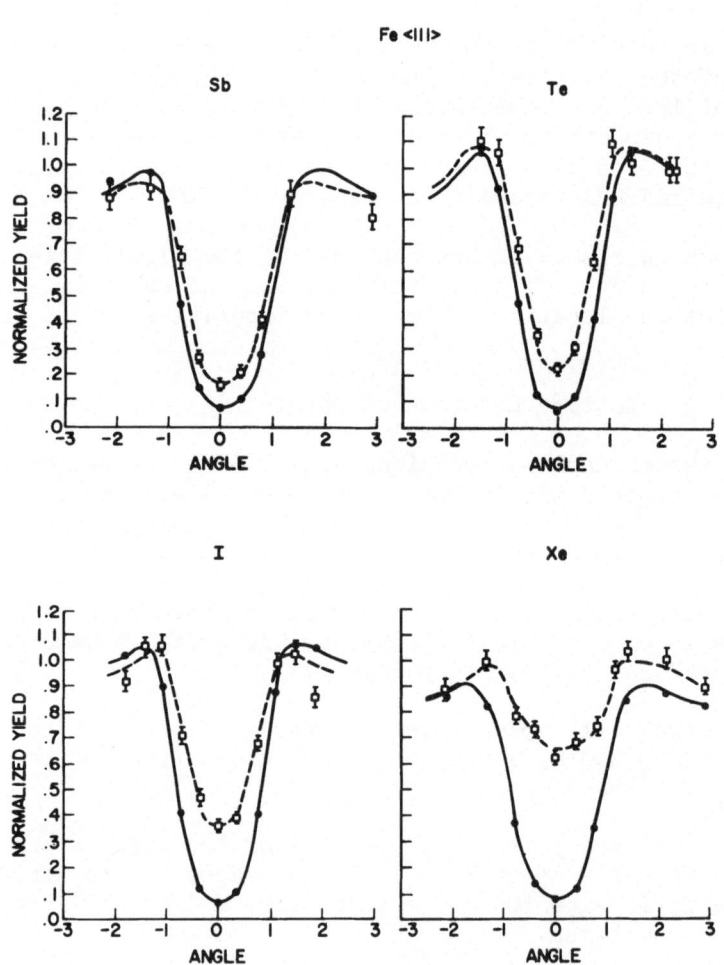

Fig. 7 $\langle 111 \rangle$ angular scans for the iron host (●) and the indicated impurities (□). All the implantations were at room temperature with 100 keV beams and for doses varying from (5.5–8)×10^{14} atoms/cm^2. The probing beam was 1.8 or 2.0 MeV He$^+$.

TABLE IV

⟨111⟩ Axis

imp	χ_{host}	χ_{imp}	$\dfrac{1-\chi_{imp}}{1-\chi_{host}}$	ψ_{imp}/ψ_{host}
Sb	.075	.165	.90±.09	.96±.06
Te	.075	.225	.84±.08	.85±.06
I	.070	.360	.69±.07	.87±.06
Xe	.080	.660	.40±.04	.88±.06

{100} Plane

imp	χ_{host}	χ_{imp}	$\dfrac{1-\chi_{imp}}{1-\chi_{host}}$	ψ_{imp}/ψ_{host}
Sb	.54	.59	.89±.09	1.02±.06
Te	.60	.65	.87±.09	.89±.06
I	.60	.62	.95±.09	.87±.06
Xe	.59	.81	.46±.05	1.10±.06

TABLE V

Fe	Ref.	Ni	Ref.
B – Int.	30	B – Random	30
C – Int.	46	Hf – Mixed	56
F – Random	47	Tl – Mixed	57
Ca – Mixed	48	Bi – Mixed	57
Cu – Sub.	49	Pb – Mixed	57
Br – Mixed	42	_Cu_	
Sb – Sub.	45	D – Int.	55a
Te – Sub.	– –	B – Random	30
I – Mixed	– –	_Al_	
Xe – Mixed	44	B – Random	30
Tb – Mixed	50	_Au_	
Yb – Mixed	43,51	B – Random	30
Au – Sub	45	_Tl_	
Tl – Sub	52	O – Int.	58
Pb – Sub	52		
Bi – Sub	52		
_W_			
D – Int.	53		
He^3 – Int.	53		
B – Int.	30		
Rn – Sub.	54		
_Nb_			
D – Int.	55		

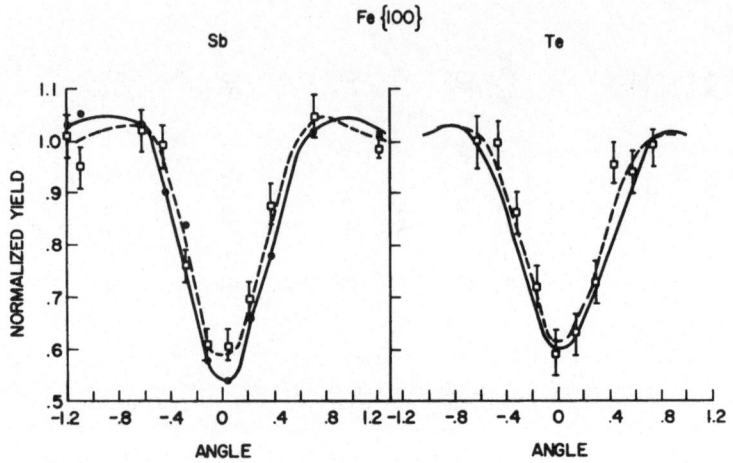

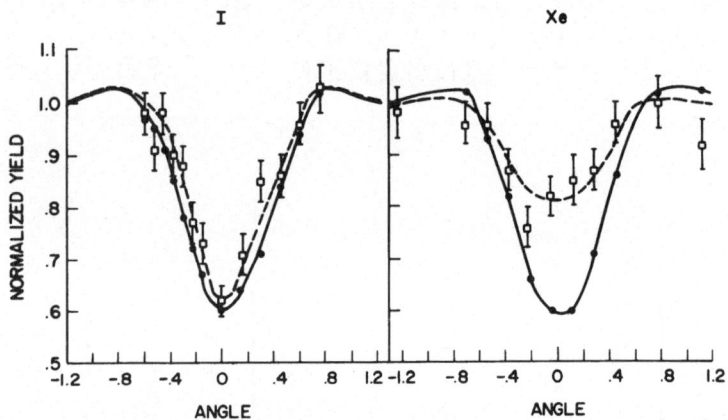

Fig. 8 {100} angular scans for the iron host (●) and the indicated impurities (□). The experimental conditions are the same as
in Fig. 7.

(1) The minimum yield, χ_{host}, in the axial cases is .070-.080, more
than a factor of two larger than the pre-implant minima
(.025-.030) and indicates that some damage has occurred. The
minimum yield for the planes has also increased by a factor of ~2.

(2) The ratio, $(1-\chi_{imp})/(1-\chi_{host})$, indicative of the substitutional
percentage decreases monotonically from Sb to Xe for the $\langle 111 \rangle$
axial case. It is consistent with the similar ratio for the
planar case except for the case of I.

(3) The angular widths of the impurity scans, χ_{imp}, are usually somewhat smaller than those of the host, χ_{host}.

(4) The I {100} scan does not show the structure reported in the case of Br.

It should be noted that the Xe result is in agreement with the work reported in Ref. (44) and the Sb result is in agreement with the work reported in Ref. (45). Furthermore the data affirms the intuitive feeling that high solubility implies high substitutionality. The rather high substitutional values for I and Xe are surprising considering the radii of these impurities. In section IV it will be shown that a part of these impurities are associated with vacancies. The {100} planar anomaly for I is similar to that observed by Abel et al. for the case of Yb in Fe (43), in which it is suggested that the impurity is trapped in a defect with a {100} orientation.

III.3. Lattice Location Measurements

Table V summarizes all channeling-lattice location measurements in implanted metals known to the authors. All of these cases correspond to room temperature implants in the dose range of $2 \times 10^{14}/cm^2$ - $5 \times 10^{15}/cm^2$. In this table "sub" implies impurity substitutional values of >80%, "mixed" is ~40%-80%, "random" is less than 40% and "Int." implies definite proof of an interstitial site. Most of these cases, particularly the ones in Fe have been motivated by H.F. measurements. In many cases they have shed some light on difficulties associated with these measurements, particularly in source preparation effects.

From this body of information it is interesting to note that there is not a good substitutional case in the F.C.C. crystals. In addition, B, a clear interstitial case in two B.C.C. lattices, W and Fe, has shown no structure in the angular distribution for four different F.C.C. lattices. Since Ni and Fe have so many similar atomic properties, one is tempted to ascribe this difference to the crystal structure of these materials. Radiation damage studies have shown similarities in the annealing steps for metals of the same crystal structure and from the hyperfine data already described, it is clear that damage plays an important role in the post-implant situation.

It is also interesting to consider the systematics for the many implants in Fe. In Fig. 9 we have indicated the substitutionality of these impurities on a Darken-Gurry plot (59). These plots, which show each impurity as a point on a graph of radius vs

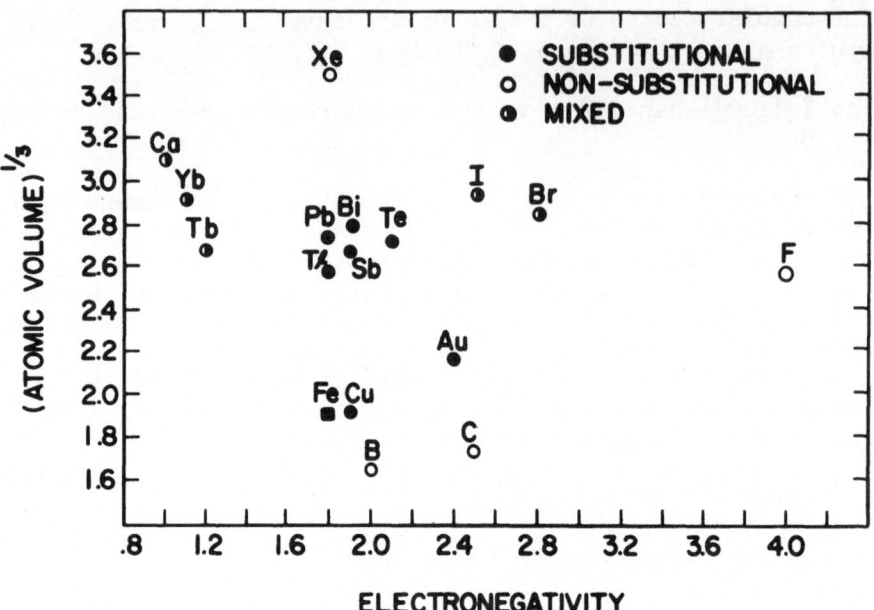

Fig. 9 Darken-Gurry plot summarizing the channeling data for various impurities in Fe. For this figure all the cases correspond to implanted impurities for doses ranging from $2\times10^{14}/cm^2$ to $2\times10^{15}/cm^2$ with the Fe host at room temperature. The symbols are designed as follows: (●), 80-100% substitutional; (◐), 50%-80% substitutional; (○) < 50% substitutional. The atomic volume is in arbitrary units and the electronegativity follows Pauling's scale.

electronegativity, are used in alloying theory to indicate those materials which are expected to substitionally alloy with a given host. Such materials fall within an ellipse on the Darken-Gurry plot; the center of the ellipse is the host of interest and the major and minor axes are defined by various rules, for example the Hume-Rothery rules (60). The qualitative picture this presents is that substitutionality depends on both electronegativity and atomic radius. Following this thought we have summarized the channeling data in the same manner. The grouping of the different cases is apparent although the radius difference between some of the substitutional impurities i.e. Tl and Fe is far in excess of the ~15% given by Hume-Rothery. Hyperfine data suggest that at most 60% of the implanted Tl is substitutional (19) while the channeling indicates >80%. This indicates that some large percentage are in damage associated sites that appear substitutional. Probably, for elements of very different electronegativity than Fe

ionic bonding may occur so that these elements yield low or no
substitutional value. It seems that by implantation fairly large
atoms can occupy substitutional sites. Possibly the vacancies
created by the implantation tend to stabilize such sites. The
microscopic role of these vacancies is perhaps best understood
when comparing the hyperfine data and the channeling results.

IV. DISCUSSION OF HYPERFINE INTERACTION RESULTS AND COMPARISON WITH CHANNELING RESULTS

We will restrict this discussion to the series Sb, Te, I, Xe
implanted in iron that has been extensively investigated both by
channeling and by Mössbauer spectroscopy.

It is seen in Table III that Sb and Te implants yield unique
hyperfine fields, but I and Xe implants do not. This striking
difference can be attributed to the size of the impurities.
According to Flynn (61) Sb and Te are imbedded as neutral atoms
in a high conduction electron density metal like iron. The highly
electronegative iodine, however, forms I^- ions. If this is the
case, the atomic (ionic) radius jumps from about 1.6 Å for Sb and
Te to 2.2 Å for I^- and Xe. The process of site formation may now
be described as follows: when an impurity comes to rest at a
substitutional position in the lattice it has just created several
vacancies and interstitials by collisions near the end of its track.
Some of the vacancies are trapped by the impurities, while others
may be annihilated or form stable clusters. If the implanted atom
is large, the trapped vacancies cannot be annihilated by interstitials because the "binding energy" of the vacancy to the impurity
exceeds the annihilation energy. For the smaller impurities,
however, the annihilation energy may exceed the binding energy so
that vacancies initially present near the impurity are annealed,
leaving practically all impurity atoms in pure substitutional
sites.

We may compare the high field fractions of the Mössbauer
spectra given in Table III with the substitutional fractions
determined from the axial and planar channeling scans given in
Table IV. Clearly, the results for Sb, Te and Xe are consistent
with the assumption that the high field sites correspond with
substitutional impurities. For I, however, the results disagree:
the axial channeling measurement yields a substitutional fraction
0.69±0.07, the planar scan even 0.9±0.1, much higher than the high
field fraction 0.46±0.05 for the I-125 implant or about 0.4 for
the I-131 implant. The channeling results indicate that the I
atoms are localized in the lowest order planes, many of them very
close to substitutional sites. A very tentative explanation of
this discrepancy is the following: Assuming the lower field sites

to be associated with vacancies and also assuming that the beta recoil energy is too small to lead to a change of the vacancy configuration, one could imagine that a large part of the vacancy associated iodine ions is kept close to a substitutional position due to its negative electric charge. Further hyperfine interaction and channeling experiments, especially at low implantation temperatures, would be of importance to verify this tentative interpretation.

We gratefully acknowledge many stimulating and useful discussions with our colleagues R. B. Alexander, R. A. Boie, R. L. Cohen, S. A. Drentje, D. E. Murnick, J. R. MacDonald, R. S. Raghavan, P. Raghavan, and J. M. Poate. E. N. Kaufmann has been particularly close to this work and we have benefited from his involvement and advice. We also are happy to acknowledge the aide and technical expertise of W. F. Flood and J. W. Rodgers.

REFERENCES

1. The general theory is given in S. R. de Groot, H. A. Tolhoek and W. J. Huiskamp, Alpha, Beta and Gamma-ray Spectroscopy (Ed. K. Siegbahn, North Holland, Amsterdam 1968) p 1199. The notation used in Eq. (1) is close to that of R. J. Blin Stoyle and M. A. Grace, Handbuch der Physik 42 (Springer, Munchen) 555.

2. L. Niesen, Thesis, Leiden, Netherlands, 1971.

3. H. Pattijn, R. Coussement, G. Dumont, E. Schoeters, R. E. Silverans and L. Vanneste (Louvain) to be published.

4. B. K. S. Koene and H. Postma, to be published in Nuclear Physics.

5. H. R. Andrews, T. F. Knott, F. M. Pipkin and D. Santry, Physics Letters 26A, 58 (1967).

6. P. G. E. Reid, M. Scott and N. J. Stone, Hyperfine Structure and Nuclear Radiations (Ed. F. Matthias and D. A. Shirley, North Holland, Amsterdam, 1968) p 719.

7. N. J. Stone in Ref. 6, p 240.

8. H. de Waard, P. Schurer, P. Inia, L. Niesen and Y. K. Agarwal, Hyperfine Interactions in Excited Nuclei (Ed. G. Goldring and R. Kalish, Gordon and Breach, London, 1971) p 89.

9. H. Frauenfelder and R. M. Steffen, Alpha, Beta and Gamma-ray Spectroscopy (Ed. K. Siegbahn, North Holland, Amsterdam 1968) p 997.

10. L. Grodzins, in Ref. 6, p 607.

11. R. G. Stokstad, R. A. Moline, C. A. Barnes, F. Boehm and
     A. Winther, in Ref. 6, p 699.

12. J. Braunsfurth, J. Morgenstern, H. Schmidt and H. J. Korner,
     Z. Physik $\underline{202}$, 321 (1967).

13. O. Klepper and H. Spehl, Z. Physik $\underline{215}$, 17 (1968).

14. F. Bosch and O. Klepper, Z. Physik $\underline{240}$, 153 (1970).

15. H. G. Devare and H. de Waard, Phys. Rev. $\underline{5B}$, 134 (1972).

16. H. G. Devare and H. de Waard, Physica Status Solidi $\underline{52}$, 134
     (1972).

17. M. Marmor, S. Cochavi and D. B. Fossan, Phys. Rev. Letters $\underline{25}$,
     1033 (1970).

18. All references to this case have been collected in Ref. 56.

19. R. S. Raghavan, P. Raghavan, E. N. Kaufmann, K. Krien and
     R. A. Naumann, Phys. Rev. $\underline{7}$, 4132 (1973).

20. E. Bodenstedt, private communication to E. N. Kaufmann, see
     Ref. 2 in Ref. 52.

21. M. Behar and R. M. Steffen, Phys. Rev. $\underline{C1}$, 788 (1973).

22. E. N. Kaufmann, P. Raghavan, R. S. Raghavan, K. Krien,
     E. Ansaldo and R. A. Naumann, Proceedings of this conference.

23. K. Krien, to be published.

24. E. N. Kaufmann, R. S. Raghavan, P. R. Raghavan, K. Krien and
     R. A. Naumann, to be published.

25. H. Haas, private communication to R. S. Raghavan.

26. J. M. McDonald, P. M. S. Lesser and D. B. Fossan, Phys. Rev.
     Letters $\underline{28}$, 1057 (1972).

27. J. Bleck, R. Butt, H. Haas, W. Ribbe and W. Seitz, Phys. Rev.
     Letters $\underline{29}$, 1371 (1972).

28. R. Brenn, G. D. Spronse and O. Klepper, J. Phys. Soc. Japan,
     supplement $\underline{34}$, 175 (1973).

29. T. Minamisono, K. Matuda, A. Mizobuchi and K. Sugimoto, J. Phys.
     Soc. Japan 30, 311 (1971).

30. J. U. Andersen, E. Laegsgaard and L. C. Feldman, Rad. Eff. 12,
     219 (1972).

31. R. L. Mössbauer, Z. Physik 151, 124 (1958); Naturwissenschaften
     45, 538 (1958).

32. N. N. Greenwood and T. C. Gibb, Mössbauer Spectroscopy
     (Chapman and Hall, London 1971).

33. W. Mansel, G. Vogl, W. Vogl, H. Wenzl and D. Barb, Physica
     Status Solidi 40, 461 (1970).

34. H. de Waard, "Mössbauer spectroscopy and its applications"
     (Intern. Atomic Energy Agency, Vienna, 1972), p 123.

35. H. de Waard, R. L. Cohen and S. R. Reintsema, to be published.

36. S. A. Drentje, S. R. Reintsema and A. N. Kalkman, Proceedings
     of this conference.

37. R. L. Cohen, G. Beyer and B. I. Deutch, Proceedings of this
     conference.

38. H. P. Wit and L. Niesen (Groningen) private communication.

39. S. R. Reintsema (Groningen), private communication.

40. R. Coussement et al., communicated at the hyperfine interactions
     conference held at Liege Belgium from Sept. 4-7, 1973.

41. P. Inia, Thesis, Groningen 1971.

42. R. B. Alexander and J. M. Poate, "Proceedings of the
     International Conference on Ion Implantation into Semiconductors
     and Other Materials", (B. Crowder ed.) Plenum Press 1973 and
     Phys. Rev. (to be published).

43. F. Abel, M. Bruneaux, C. Cohen, H. Bernas, J. Chaumont and
     L. Thome, Solid State Communications 13, 113 (1973).

44. L. C. Feldman and D. Murnick, Phys. Rev. B5, 1 (1972).

45. R. B. Alexander, Harwell Report No. AERE R-6849 (1971).

46. L. C. Feldman, E. N. Kaufmann, J. M. Poate and W. M. Augustyniak, "Proceedings of the International Conference on Ion Implantation in Semiconductors and Other Materials", Plenum Press 1973.

47. E. N. Kaufmann and Jack R. MacDonald, private communication.

48. R. A. Boie, W. Darcy, R. Hensler and Jack R. MacDonald, to be published.

49. R. A. Boie, Jack R. MacDonald and J. M. Poate, private communication.

50. R. B. Alexander and L. C. Feldman, unpublished work.

51. R. B. Alexander, E. J. Ansaldo, B. I. Deutch, L. C. Feldman and J. Gellert (proceedings of this conference).

52. L. C. Feldman, E. N. Kaufmann, D. W. Mingay and W. M. Augustyniak, Phys. Rev. Lett. $\underline{27}$, 1145 (1971).

53. S. T. Picraux and F. L. Vook (proceedings of this conference).

54. D. Domeij, Arkiv Fysik $\underline{32}$, 179 (1966).

55. H. D. Carstanjen and R. Sizmann, Physics Lett. $\underline{40A}$, 93 (1972).

55a. H. Fischer, R. Sizmann and F. Bell, Z. Physik $\underline{224}$, 135 (1969).

56. E. N. Kaufmann, J. Poate and W. M. Augustyniak, Phys. Rev. $\underline{B7}$, 951 (1973).

57. L. C. Feldman and E. N. Kaufmann (unpublished results).

58. R. B. Alexander and R. J. Petty, to be published.

59. L. S. Darken and R. W. Gurry, "Physical Chemistry of Metals", McGraw-Hill Book Co., N. Y. (1953).

60. W. Hume-Rothery, R. E. Smallman and C. W. Haworth, "The Structure of Metals and Alloys (Institute of Metals, London, 5th edition) 1969.

61. C. P. Flynn, Physics Letters $\underline{41A}$, 45 (1972).

62. C. Hohenemser, R. Reno, H. C. Benski and J. Lehr, Phys. Rev. $\underline{184}$, 298 (1969).

DISCUSSION

Q: (W. K. Chu) Since many laboratories are not able to implant radioactive ions, I would like you to comment on the practicality and limitation of the following modification of the experiment: implanting non-radioactive ions and using a radioactive source outside the sample under investigation.

A: In principle this is possible in some favorable cases and we have, in fact, done it for a study of the Mössbauer effect in Kr-83 in Aℓ. It is necessary, however, to stack a large number of very thin foils that have each been implanted at a number of different energies to provide a homogeneous concentration of the impurity. Even so, the effect will usually be quite small because only a small effective absorber thickness can be obtained without absorbing too much of the source gamma rays. This, however, may be compensated by the use of a very strong source, much stronger than can be obtained by radioactive implantation.

Comment: (H. Bernas) [1] Most of the work covered by your talk has involved studies of the magnetic dipole hyperfine interaction (i.e., spin density distribution). Although the results are interesting from the radiation damage viewpoint, the mechanisms involved in spin density changes due to differing lattice sites are very complicated. Perhaps more effort should be devoted to the study of change density modifications (i.e., quadruple interaction), which are better understood and often more sensitive. [2] In connection with radiation damage and lattice site studies of interest to many participants of this Conference, I believe the Mössbauer effect study of low temperature implantation of ^{57}Co in Aℓ (Mangel and Vogel, Phys. Rev. Lett.) should be mentioned; information was obtained on the evolution of Aℓ interstitials, and on their interaction with Co impurities.

Q: (F. Abel) Please comment on the influence of the disorder created in the surroundings of a large nonsoluble impurity on the χ_{min} and half-width of the impurity angular scans.

A: From basic channeling considerations it is clear that disorder in the immediate surroundings of an impurity should affect (increase) χ_{min} for the impurity and decrease its angular width. The difficulty in many experiments to date has been the observation of large differences between the minimum yield of the host and impurity without an accompanying difference in the angular widths. This is not the expected result for most disorder configurations.

Q: (J. A. Davies) Can you correlate your planar (100) channeling
data for I in Fe with the hyperfine results and with the axial
channeling data by assuming that some of the iodine atoms are displaced along the $\langle 100 \rangle$ rows (i.e., towards the octahedral site)?

A: Yes, but I would prefer them to be vacancy associated and not
interstitial in the normal sense, because of their large radius.
Possibly the negative charge on the I-ions keeps them close to substitutional positions in the lowest order planes but as soon as
they decay to Xe (neutral) they relax out of those planes, thus
explaining the lower high field fraction ($\sim 40\%$) found for the Xe-
daughter as compared with the higher substitutional fraction ($\sim 70\%$)
derived from the channeling data on I in Fe.

Comment: (R. S. Nelson) Your results on Ni could give information
for the binding of substitutional impurities to interstitials. It
is thought that in some cases interstitial-substitutional pairs
will take up a "dumbell" configuration with a binding energy of
~ 1 eV. If you implant at increasing temperatures then you may be
able to see substitutional atoms and get a measure of the binding
energies. Alternatively your room temperature results could be
understood in terms of interstitial loop formation. It is well
known that the addition of substitutional impurities causes an enhanced interstitial loop formation which arises from the growth of
interstitial-impurity pairs. In this case your impurities could
all be at the core of dislocation loops and would not appear substitutional. In Ni these would anneal at self-diffusion
temperatures.

HIGH SUBSTITUTIONAL FRACTIONS IN COLD IMPLANTATIONS OF Xe

AND Te IN IRON AS SHOWN BY MÖSSBAUER EFFECT MEASUREMENTS

S.A. Drentje, S.R. Reintsema and G.N. Kalkman

Laboratorium voor Algemene Natuurkunde

University of Groningen, Netherlands

ABSTRACT

Mössbauer effect measurements show that high substitutional fractions (70 ± 10%) can be obtained in implantations of Xe in Fe held at liquid nitrogen temperature (90 K). Implantations at higher temperatures show that the substitutional fraction is a decreasing function of implantation temperature. For Te in Fe pure substitutional implants can be obtained at 90 K and room temperature.

1. *Introduction*

The Mössbauer effect has proved to be especially suitable for studying the lattice location of implanted impurities[1,2,3]. Earlier results on sources of ^{129m}Te and ^{133}Xe implanted in iron have previously been published[4,5,6]. In those experiments the sources were all prepared by ion implantation at room temperature (RT), at implantation energies ranging from 50 to 140 keV. Here we present Mössbauer spectra of gamma transitions excited in the decay of implanted sources of $^{129m}Te(\rightarrow^{129}I)$ and $^{133}Xe(\rightarrow^{133}Cs)$ in iron. Different implantation temperatures are used, viz. 90 K and RT in the case of ^{129m}Te and 90 K, RT, 380 K and 480 K in the case of ^{133}Xe. All spectra are taken at 4.2 K. Sources of 50 to 300 μCie were prepared with the Groningen isotope separator[7] using neutron-irradiated enriched ^{128}Te metal and fission ^{133}Xe diluted with natural Xe as charge material, respectively. The implantation energy was 140 keV. The concentration of implanted impurity atoms was kept low (< 0.1%, typical doses < 2×10^{14} atoms/cm^2) by beam sweeping. Special precautions were taken to maintain low pressure in the collector box in order to avoid the formation of condensed layers on top of the iron foils (especially important at 90 K implantation temperature).

Under the assumption that the beta decay of the implanted radioactive atom does not change its location in the iron lattice (recoil energies are 6.1 eV and 1.4 eV resp.), the Mössbauer spectra
of the decay products give us two kinds of information: since
inequivalent sites generally produce different hyperfine spectra,
the number of spectral components observed is an index of the
number of inequivalent sites occupied by the implanted ion. By
varying the source temperature, it is also possible to determine
the relative occupation of the different sites.

2. *Results on Tellurium*

The sources of ^{129m}Te in iron have been studied by observing
Mössbauer spectra of the 27.7 keV, $5/2^+ \rightarrow 7/2^+$ transition of the
^{129}I daughter.
The spectra of both 90 K and RT implanted sources exhibit a symmetric magnetic splitting (see fig. 1). They can be fitted with
a theoretical spectrum consisting of one Zeeman-split component
For both spectra a unique magnetic hyperfine field of $|H|$ =
1130 ± 30 kOe is found, suggesting that the Te atoms are situated
at one type of lattice position (however, see section 4). We may
assume this position to be substitutional as justified by
channeling experiments[8].

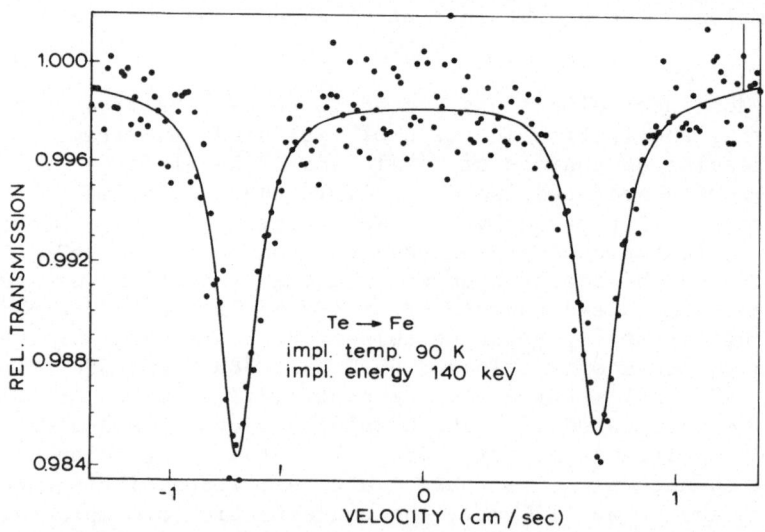

FIG. 1. Mössbauer spectrum of ^{129}I excited in the decay
of ^{129m}Te implanted at 90 K in an Fe foil. Implantations
at RT yield identical spectra.

3. Results on Xenon

Mössbauer spectra of the 81.0 keV, $5/2^+ \to 7/2^+$ transition of the ^{133}Cs daughter are presented in fig. 2. The effect of using different implantation temperatures is clearly visible in the spectra.

It is possible to fit all spectra with two components: a Zeeman-splitted component corresponding to one high-field site of $\sim$280 kOe with large isomer shift and a broad single line corresponding to a set of low-field sites with small isomer shifts. In the fits the excited to ground state magnetic moment ratio μ^*/μ = 1.335[9]) is used. However, the relative total absorption shows large and unsystematic variations in these fits. Decomposition into two Zeeman-split components and one single line (as in ref. 1, pg. 134) gives more consistent results. The components correspond to lattice sites denoted as *high-field sites* (h-sites, hyperfine-field value $|H_h|$ = 280 ± 5 kOe, isomer shift δ_h = -1.10 ± 0.02 mm/sec), *intermediate field sites* (i-sites, $|H_i|$ = 135 ± 10 kOe, δ_i = - 0.80 ± 0.10 mm/sec) and a set of *low-field sites* (l-sites, $|H_l|$ < 30 kOe, δ_l = - 0.25 ± 0.05 mm/sec), respectively. It is mentioned that in the fits of the spectra of the 90 K implantations the single line drowns in the statistical noise. In this case fits with two Zeeman-split components give more accurate results.

The relative absorption of each component and the total relative absorption as a function of the implantation temperature are given in table I. Using results on f-values ("recoilless fractions") obtained by De Waard[1]) we conclude that 70 ± 10% of the Xe atoms implanted at 90 K are located at h-sites. For the RT implantation this amount is 45 ± 10%. We assume this h-site to correspond to substitutional Xe atoms. This assumption is justified by channeling experiments of Feldman and Murnick[10]), who found a substitutional fraction of 50 ± 10% in sources prepared at RT. (for some remarks on "substitutional lattice sites" see sec. 4).

TABLE I

site	relative absorption (in %) as a function of implantation temperature			
	90 K (average of 3 runs	RT (average of 2 runs)	380 K (1 run only)	480 K (1 run only)
h	3.5 ± 0.2	3.3 ± 0.3	2.5 ± 0.3	1.6 ± 0.4
i	0.9 ± 0.1	0.6 ± 0.1	0.9 ± 0.1	1.5 ± 0.5
l	-	0.5 ± 0.1	0.6 ± 0.1	1.4 ± 0.4
total	4.4 ± 0.3	4.4 ± 0.5	4.0 ± 0.5	4.5 ± 1.3

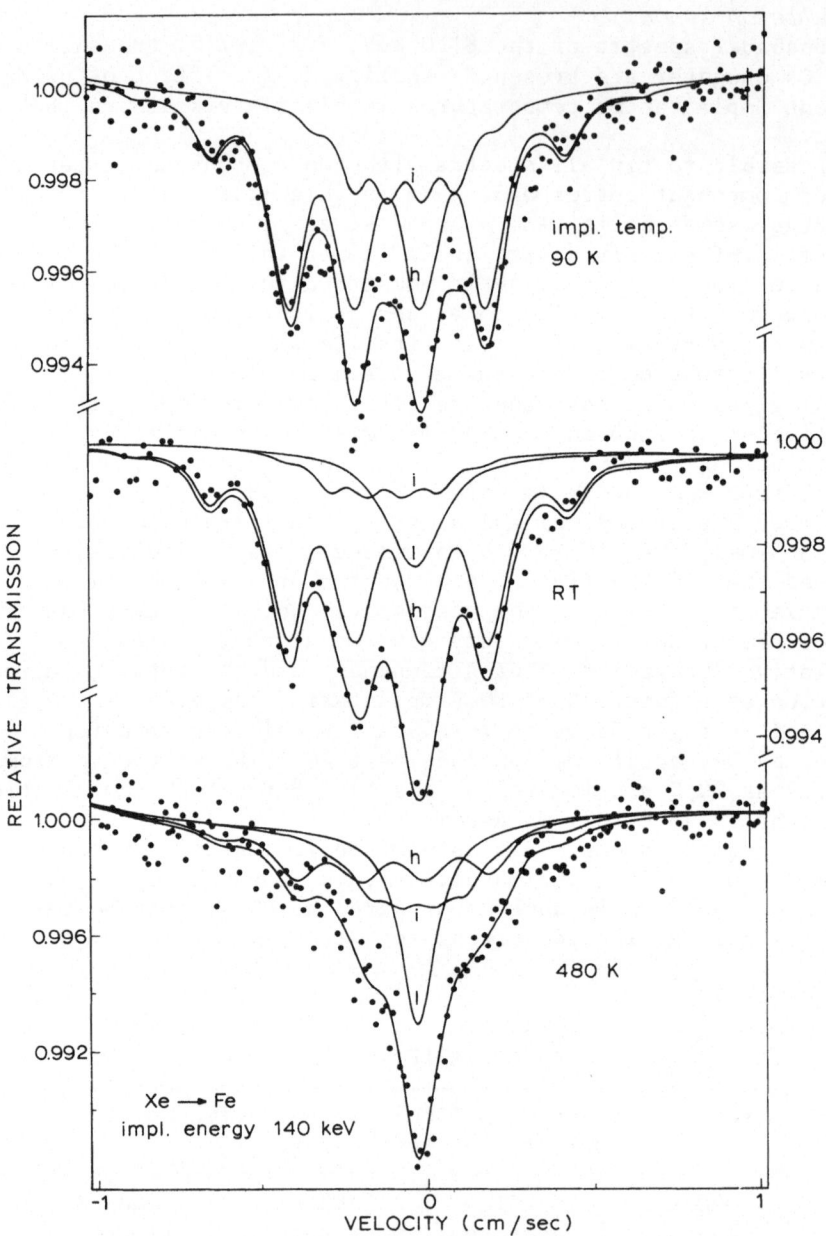

FIG. 2. Mössbauer spectra of ^{133}Cs excited in the decay
of ^{133}Xe implanted at different temperatures in Fe. The
drawn line through the measured points represents a least
squares fit to a two component model (at 90 K implantation temperature), and to a three component model (at RT
and 480 K). Separate components are given by lines h(high
field), i(intermediate field) and l(low field).

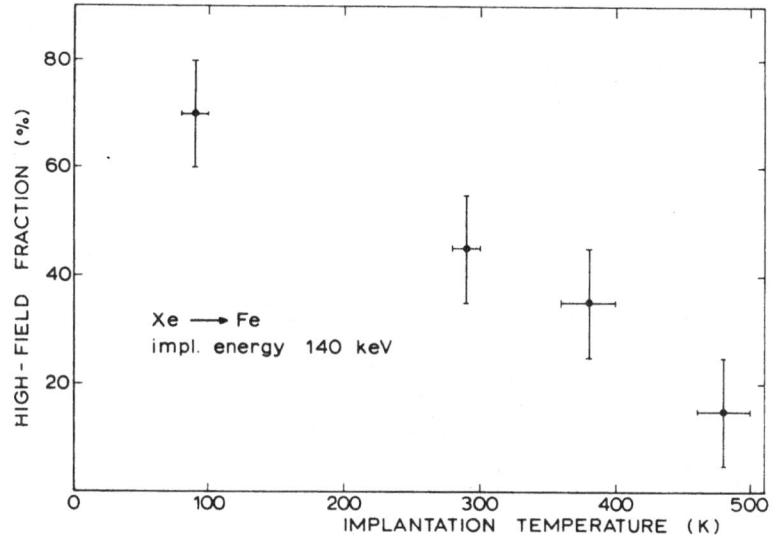

FIG. 3. The high-field fraction of Xe (identified as substitutional fraction) as a function of implantation temperature. f-Values of ref. 1 were used.

At the higher implantation temperatures even lower high-field fractions were found (see fig. 3).

4. *Discussion of results*

We may assume that both Te and Xe atoms on coming to rest
after implantation initially are frequently associated with one
or more vacancies[11,12].
From our experimental results and the channeling results we conclude that in the case of Te the lattice is recovered in the
immediate surrounding of the Te atoms (say, within one lattice
distance). So we may conclude that the vacancies initially associated with the Te atoms can be filled by interstitials that are
able to move at the implantation temperatures considered (thus,
even at 90 K).
The occurence of more than one component in the Xe spectra proves
that such a type of recovering does not take place in the Xe case.
The solution of this problem probably lies in the different size
of the implanted atoms. Whereas Te, with an atomic radius of 1.43Å
does not fit badly in the iron lattice, Xe, with a radius of 2.2Å
is much too large. So, the vacancies associated with the Xe atom
remain trapped because too much energy would be needed to fill

them with migrant interstitials. This can also be concluded from
static calculations on lattice relaxation carried out by Drentje
and Ekster[13]. They calculated binding energies of various xenon +
vacancy clusters in iron and lattice positions of the constituents.
Results for some of the damage clusters investigated are given in
table II. The binding energies show that the energy needed for
filling the trapped vacancies in clusters A and C cannot be provided by migrant interstitials. It should be mentioned that
channeling experiments cannot distinguish cluster B from substitutional implanted Xe atoms (S). This can be concluded from
table II: the displacement of the Xe atom from a lattice site is
about equal to the minimum distance within which atoms are detected
as substitutional atoms in a channeling experiment. We believe that
the hyperfine field associated with the Xe atom in cluster B and
that, associated with a substitutional Xe atom is effectively the
same. Apparently, neither the Mössbauer effect nor the channeling
effect are able to detect the presence of one vacancy in the
vicinity of a substitutional Xe atom down to next-nearest neighbour distance.
We tentatively identify Xe atoms in clusters of type A with the
intermediate-field component. This is pointed out in ref. 14.

TABLE II

Configuration	Binding Energy (eV)	Final position of Xe atom	Final distance of Xe atom to nearest surrounding Fe atoms (in lattice units of Fe=1.43Å)
substitutional Xe atom (S)	–	lattice position	8 Fe atoms at 1.89
subst. Xe atom + 1 vacancy at 1st nd[*] (A)	1.39	0.61Å shifted in vacancy direction	3 Fe atoms at 1.85 3 " " 1.94 3 " " 1.99
subst. Xe atom + 1 vacancy at 2nd nd[*] (B)	0.26	0.10Å shifted in vacancy direction	4 Fe atoms at 1.87 4 " " 1.91
subst. Xe atom + at 1st and 2nd nd 2 vacancies mutually at 1st nd (C)	3.1	in centre of vacancy triangle	2 Fe atoms at 1.88 2 " " 1.93 2 " " 1.96 2 " " 2.12

[*] 1st nd = nearest neighbour distance

    2nd nd = next nearest neighbour distance

Xe-atoms in clusters of type C or Xe atoms associated with even more vacancies we identify with the low-field component. The decrease of the high-field fraction (see fig. 3) from 90 K to RT implantation temperature can be explained by assuming a difference in vacancy mobility at these temperatures during the implantation process. The combined action of vacancy mobility and lattice strain close to the Xe-atoms might then be such, that vacancies can be trapped by the Xe-atoms in the RT implantation but not in the LN temperature implantation. This explanation is inforced by measurements with annealed sources, which will be published later, see also ref. 1, pg. 137, and ref. 10.
Other effects may play a role to explain the decreasing of the high-field fraction for implantations at temperatures above RT.

Acknowledgements
    We thank Mr. Tj. Klootsema for taking care of the Te and Xe implantations. The "Rekencentrum" of the "Rijksuniversiteit" is gratefully acknowledged for a teletype-connection of the Cyber-computer with our institute. Valuable discussions at the Summer-school "Radiation damage processes in materials" (Corsica, 1973) gave us more insight in thermal spikes. This investigation was partly financed by the "Stichting voor Fundamenteel Onderzoek der Materie (FOM)" subsidized by the "Nederlandse Organisatie voor Zuiver Wetenschappelijk Onderzoek (ZWO)".

References
1. H. de Waard, in Mössbauer Spectroscopy and its Applications, I.A.E.A., Vienna 1972. pg. 123-142.
2. H. de Waard et al., Hyperfine Interactions in Excited Nuclei, Gordon and Breach, London (1971), pg. 89.
3. N.S. Wolmarans, H. de Waard, S.R. Reintsema, Intern. Conference on Applications of the Mössbauer effect, Israël 1972, paper F 12.
4. H. de Waard, S.A. Drentje, Phys. Lett. 20 (1966) 38.
5. H. de Waard et al., Hyperfine Structure and Nuclear Radiation, North Holland, Amsterdam (1968), pg. 331.
6. H. de Waard, S.A. Drentje, Proc. Roy. Soc. A311 (1969) 139.
7. S.A. Drentje, Nucl. Instr. & Meth. 59 (1968) 64.
8. L.C. Feldman and H. de Waard, (these conference proceedings).
9. L.E. Campbell and G.J. Perlow, Nucl. Phys. A109 (1968) 59.
10. L.C. Feldman and D.E. Murnick, Phys. Rev. B5 (1972) 1.
11. J.B. Gibson et al., Phys. Rev. 120 (1960) 1229,
    C. Erginsoy et al., Phys. Rev. 133 (1964) A595,
    C. Erginsoy et al., Phys. Rev. 139 (1965) A118.
12. W. Mansel et al., Physica Stat. Sol. 40 (1970) 461.
13. S.A. Drentje and J. Ekster, submitted to Physica Stat. Sol.
14. H. de Waard, R.L. Cohen and S.R. Reintsema, to be published.

DISCUSSION

Q: (F. L. Vook) Have you done annealing measurements or damaging bombardments prior to implantation? What differences are there between a low temperature implant annealed to higher temperatures and an implant at higher temperatures?

A: We did annealing experiments with a RT implant. It turns out that the high field fraction remains constant until $\sim 300°C$ and then drops down to a value of $\sim 10\%$ at $450°C$. An anneal experiment with a 80 K implant is not yet finished, but we have the impression that it will show the same behavior.

Q: (E. N. Kaufmann) Of what significance are the apparently different isomer shifts of the intermediate field component for different temperatures which were evident in your least-squares fit results?

A: The statistical error in the isomer shift of the i-site component is so large that the differences you mentioned are not significant.

VALENCE DETERMINATION AND LATTICE LOCATION VIA MÖSSBAUER

SPECTROSCOPY OF Gd^{151} IMPLANTED INTO IRON*

R. L. Cohen

Bell Laboratories, Murray Hill, New Jersey

G. Beyer

Joint Institute for Nuclear Research, Dubna, U.S.S.R.

B. I. Deutch

University of Aarhus, Aarhus, Denmark

Measurements of hyperfine interactions and isomer shift (a measure of the electronic density around the nucleus of the ion being studied) by Mössbauer spectroscopy (MS) can provide useful information for the study of the configuration and bonding of implanted impurity ions. MS has two advantages over many competing techniques: First, it is microscopic; it measures the properties of discrete ions and hence is rather independent of the presence of inequivalent sites or "inactive" centers. Second, the parameters observed are almost entirely determined from the nearest atomic neighbors. Hence, well-defined spectra can be obtained even with heavily damaged "amorphous" samples.

We have implanted radioactive Gd^{151} into iron, and studied the hfs of the Eu^{151}:Fe decay product via MS. The implantation was carried out at room temperature, with 50 keV Gd ions, to a total dose of 4×10^{14} at/cm^2. A sample spectrum is shown in Fig. 1. The Eu^{151} resonance has been extensively studied by MS. The complex hyperfine pattern shown here arises from two sites of roughly the same intensities. One site is split by magnetic and electric

---

* Extended Abstract

quadrupole interactions. The second site is essentially unsplit.
Both these sites have distinct hf fields and isomer shifts, and
thus can be considered as well-defined. The isomer shift for both
sites corresponds to that for trivalent Eu although the normal
valence of Eu is 2^+ in metallic environments. This observation can

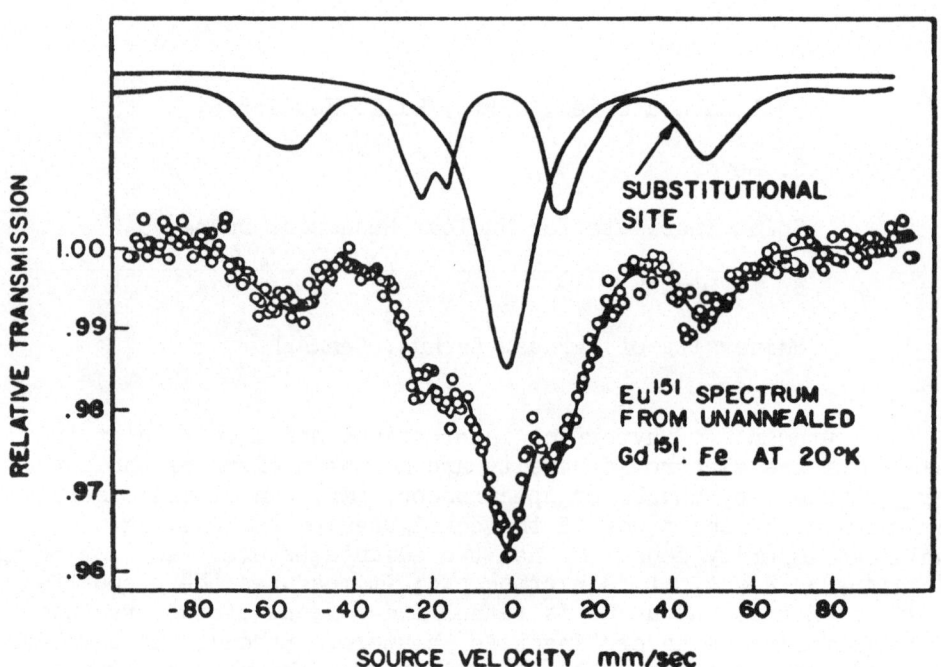

Figure 1. Mössbauer spectrum of the Eu^{151} daughter produced via
the radioactive decay of Gd^{151} implanted into iron. The spectrum
has been decomposed under the assumption of two inequivalent sites
for the Eu ions. The hyperfine spectra arising from each of the
two sites are shown above the experimental data, and the line
through the data points is the sum of the two components. The
split spectrum has been identified with the substitutional site
(for the implanted Gd parent) primarily because its anneal behavior
is similar to that observed by ion channeling measurements for
the substitutional site of Yb implanted into iron.

be explained on the basis that the Eu^{3+} ion with its 1.77 Å radius fits better than Eu^{2+} (r = 2.03 Å) in the iron (r = 1.27 Å) lattice. Recent location measurements on another rare earth, Yb implanted into Fe (R. B. Alexander, et.al.[1]), show that about 50% of the implanted ions embed into substitutional sites. If Gd implants in the same way as Yb, then this present study shows that the remaining ions are very likely all in a single type of second site. The hf field at the high-field site is 1.5 MG, far larger than has ever been observed before in Eu. It arises from the conduction electron polarization of the iron and the local magnetization of the 4f electrons of the Eu ions. The hf field at the second Eu site is close to zero. Time integral perturbed angular correlation (TIPAC) determinations would yield a value of 0.75 MG for the hf field since TIPAC measures the average effective magnetic field, not the specific field at different sites.

REFERENCE

1. R. B. Alexander, E. J. Ansaldo, B. I. Deutch, L. C. Feldman, and J. Gellert (these conference proceedings).

DISCUSSION

Q: (E. V. Kornelsen) There is some evidence from helium injection experiments that large inert gas atoms in both tungsten and nickel trap vacancies in more than one binding state, and that rather distinct conversion temperatures from one to another occur. If true in iron, this might be an alternative mechanism for the changes observed on annealing.

A: Yes. This kind of mechanism could play a role in explaining why the oxidation of the substitutional Gd ions appears to occur during the 500°C anneal.

Q: (W. L. Brown) Can you rule out the possibility that the Ge is in the surface oxide as annealing proceeds?

A: A recent paper by Bernas and co-workers (Solid-St. Commun. 13, 113 (1973)) has established decisively that for Yb implanted into Fe, the Yb depth distribution does not change with annealing.

Q: (H. deWaard) If the low field component in your unannealed Eu-Fe spectrum is due to oxides, wouldn't you expect the high field fraction to depend strongly on source preparation? For Dy-Fe Niesen and Wit now find a high unique field fraction (> 90%). If there were appreciable oxide formation in that case one would not expect this result. Vacancy association, however, might not affect the h-f-field value very strongly in their case. Can you include vacancy association in the unannealed sample as the cause of a low field component also in your case?

A: Yes, different source preparation techniques would be expected to yield different high-field fractions. In the Aarus channeling location experiments, the source preparation technique was very similar to that used for the Gd-Fe source. Also, some possible mechanisms for enhanced oxygen diffusion could be proportional to fluence, which would eliminate one of the major variables in source preparation. We cannot exclude some kind of vacancy association as a cause of the unsplit line, but it would seem unlikely for such a site to have such a small hf field. In the Eu case (in contrast to Xe), the Eu ion is not very much larger than the Fe host and would not be expected to make a large hole for itself. The failure of high temperature anneals (to 1000°C) to change the spectrum strongly supports the oxidation, rather than the defect-association model.

Comment: (H. Bernas) With reference to possible complications in interpretation of the data, we studied the implantation energy and dose dependence of the magnetic hfi for ^{169}Yb in Fe. If you take our results (Nucl. Inst. Met. 107, 423 (1973)) for 50 keV and 4 x 10^{14} at./cm^2, you will find that the experimental average hf field is a factor of ~ 0.6 lower than the hf field measured for a higher-energy ($\geq$ 80 keV), lower-dose (< 10^{14} at./cm^2) implant.

COMBINED LATTICE LOCATION AND HYPERFINE FIELD STUDY OF Yb

IMPLANTED INTO Fe

R. B. Alexander[*], E. J. Ansaldo[+], B. I. Deutch, J. Gellert

Institute of Physics, University of Aarhus, DK 8000,
Aarhus C, Denmark

L. C. Feldman

Aarhus University and Bell Telephone Laboratories
Murray Hill, New Jersey 07974

1. INTRODUCTION

Ion implantation has been extensively used to prepare radio-active sources of impurities in metals for hyperfine interaction studies. However, it has been found that the hyperfine field (hf) at implanted impurities may vary according to the implantation conditions as well as the processing of the host (see, for example, Refs. 1-3). Factors which can influence the measured hf include radiation damage, the concentration and location of the impurity atoms in the host lattice, and their migration or precipitation on annealing. The impurity lattice location (as well as certain types of radiation damage) can be investigated by the technique of ion channeling. In order to understand better the various factors on which the hf may depend, it was decided to make a combined channeling-hf study on a suitable system.

Earlier investigations [1] indicated that implanted Yb in Fe was a reasonable choice for both types of experiment. The large mass difference between stable ^{172}Yb and Fe (atomic weight 56) makes the system especially suitable for the use of backscattering in channeling measurements. At doses $\sim 2 \times 10^{14}$ ions/cm^2, radioactive sources of ^{169}Yb (decaying to excited states in ^{169}Tm) in Fe are sufficiently strong for convenient perturbed angular correlation

[*]Work partly carried out while at the Clarendon Laboratory, Oxford, U.K. and Nuclear Physics Division, A.E.R.E., Harwell, U.K.
[+]Present address: Department of Physics, Princeton University, Princeton, New Jersey, U.S.A.

(PAC) measurements of the hf. Both the rotation and attenuation of the correlation pattern can be measured for γ-γ cascades through <u>two</u> nuclear states, the $5/2^+$ (τ= 0.090nsec) and the $7/2^+$ (τ=0.45nsec). In this way the average hf is overdetermined. Preliminary experiments showed that, upon annealing, the lattice location of the Yb changed; furthermore, the average hf decreased. To determine whether the change in location was correlated with the decrease in the hf, both the location and hf were measured as a function of annealing tempera- ture in the present study.

2. SAMPLE PREPARATION

In all of these experiments single crystals of Fe were used. The crystals were obtained as thin discs or slabs cut perpendicular to either a⟨100⟩ or ⟨110⟩ direction, and one surface was electro- polished. For the channeling experiments, stable ^{172}Yb was implanted at 60 or 80 keV on the Harwell Mk. I electromagnetic separator. For the PAC experiments, radioactive ^{169}Yb was implanted at 60 keV on the Aarhus separator I. The implantations were performed at room temperature, with the ions incident along non-channeling directions. The implantation dose was 2×10^{14} ions/cm^2, which corresponds to an average local concentration of 0.3 at.% Yb in the implanted region (for an energy of 60 keV).

Annealing of the implanted crystals was carried out under vacuum (pressure $\lesssim 10^{-6}$ torr) for 30 min, at temperatures up to 600°C. Back- scattering measurements showed that the surface oxide layer on the crystals was $\lesssim$ 40Å thick in all cases.

3. CHANNELING EXPERIMENTS

3.1 Experimental

Channeling-backscattering measurements on the implanted ^{172}Yb in Fe crystals were performed with ^{12}C$^+$ and ^{14}N$^+$ ions of several MeV; the experiments employed both the 5MV Van de Graaff accelerator at Harwell and the 2 MV Van de Graaff at Aarhus. For ^{12}C and ^{14}N ions, the separation between the Yb peak and the Fe edge in the back- scattered energy spectrum is sufficiently large that the background under the Yb peak due to pulse pile-up is negligible. The incident beam was collimated to an angular divergence of ± 0.03°. Crystal alignment was accomplished using a goniometer which allowed rotation about three independent axes in steps as small as 0.01°. Back- scattered particles were detected with a Si surface-barrier counter, and the ion dose was measured by integration of the target current.

3.2 Results and Analysis

The use of the channeling technique to determine the location of impurity atoms in a crystal lattice is described elsewhere [4,5].

The technique involves measuring the angular dependence of the yield (for example, from backscattering) about the major crystal channeling directions, for both the impurity and host atoms. The impurity location is determined from the features observed in the impurity angular yield curves.

For most samples used in these experiments, angular scans were carried out across at least two of the three major axes in the bcc Fe lattice. The normalized aligned yield χ_0 was always measured for all three axes. The backscattering yields from the implanted Yb and host Fe atoms were measured simultaneously, the Fe yield being determined at the same (calculated) depth from the crystal surface as the Yb. Typical results of the angular scans are shown in Fig. 1. Except for the crystal annealed at 600°C, similar results were obtained in all the scans. The Yb yield showed a dip, of the same width as the Fe dip and a magnitude which depended on the annealing temperature.

The interpretation of these results is straightforward. Whenever there is a dip in the impurity yield of the same width as the host dip for a particular axial direction, it follows that a certain fraction of the impurity atoms is located in sites lying along that

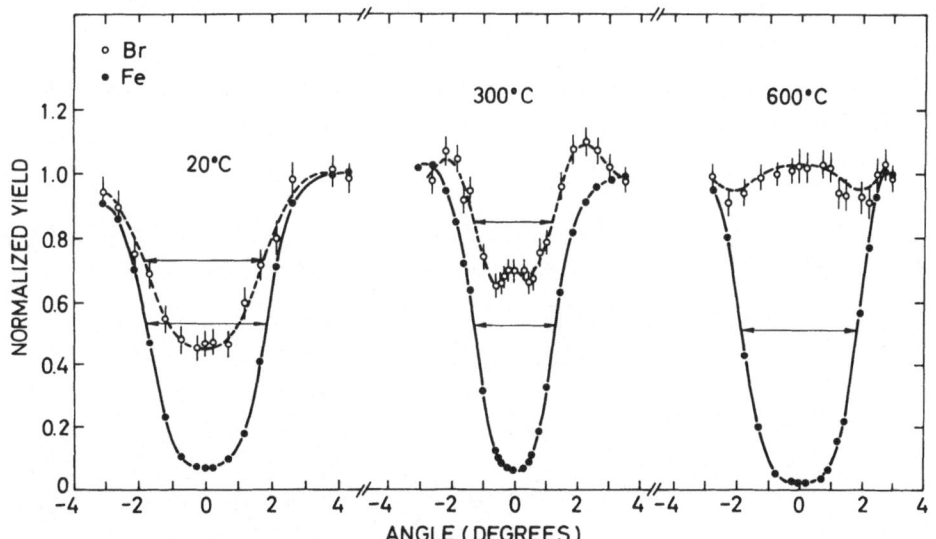

Fig. 1. Channeling angular scans across a ⟨111⟩ axis in ^{172}Yb-implanted Fe single crystals, before annealing (20°C) and after annealing at 300°C and 600°C. The scan on the 300°C-annealed crystal was carried out with 3.5 MeV ^{14}N$^+$ ions, the other two scans with 1.5 MeV ^{12}C$^+$ ions. The smooth curves through the Yb and Fe points are only to guide the eye.

set of axial rows. This fraction is given by

$$S = \frac{1 - \chi_o(i)}{1 - \chi_o(h)} ,$$ (1)

where $\chi_o(i)$ and $\chi_o(h)$ are the normalized yields in the aligned
direction, for the impurity and host respectively. If the same
result is obtained for two different axial directions, S can be
identified as the fraction of atoms in substitutional sites in the
lattice. Furthermore, the lack of any pronounced structure in the
Yb yield curves indicates that the remainder of the Yb is not in any
specific crystallographic site. In Fig. 1, a small, broad peak can
be seen near the center of the Yb yield curves for the crystals
annealed at 300°C and 600°C. This arises from flux-peaking for
channeled ions scattered from those non-substitutional Yb atoms
lying near the center of $\langle 111 \rangle$ channels.

The results of all the measurements are summarized in Table I.
It is seen that approximately 60% of the Yb atoms occupy substitutional sites in Fe following implantation. The substitutional
fraction decreases with increasing annealing temperature, no atoms
remaining substitutional after a 600°C anneal (as indicated by the
lack of any dip in the Yb yield, cf. Fig. 1). Thus there is a
movement of Yb from substitutional to non-substitutional sites in
the Fe lattice upon annealing. Comparison of the energy spectra
for annealed and unannealed crystals showed that the annealing was
not accompanied by any migration of the Yb towards the surface of
the Fe.

Table I. Fraction of implanted ^{172}Yb atoms in substitutional
sites in Fe, as a function of annealing temperature.

Annealing temp. (°C)	Substitutional fraction
20	0.58 ± 0.04
200	0.36 ± 0.04
300	0.31 ± 0.04
350	0.31 ± 0.04
400	0.19 ± 0.04
450	0.26 ± 0.035
500	0.24 ± 0.04
600	0

The non-substitutional atoms most probably reside in impurity clusters, or in a chemical phase or precipitate.

4. PAC EXPERIMENTS

4.1 Experimental

The implanted ^{169}Yb in Fe sources were polarized by means of a small electromagnet; the anisotropy and precession were measured for both cascades in a standard way. A 20 cc Ge(Li) counter in coincidence with a movable 3.81 × 5.08 cm integral-line NaI (Tl) detector was utilized, with conventional fast-slow electronic modules. Measurements were undertaken with the angle between the detectors varying from 90° to 180° in 15° steps. The magnetic field was reversed automatically every 400 sec at each angular position; the γ spectra were stored in different quadrants of a 1024-channel analyzer. The measurements were made with the source at both room temperature and 140°K, after each anneal, to check on effects from electronic relaxation.

4.2 Theory

The non-specialist should consult Ref. 6 for a summary of PAC theory. The method often employs a γ-γ cascade but any two step nuclear decay can be utilized. The first γ-ray, by selecting one unit of angular momentum in its emitted direction, unequally populates the m substates of the intermediate nuclear state. When γ-ray 1 is observed in coincidence with γ-ray 2 at an angle θ in the second counter, there results a change of the counting rate with angle; this is denoted by $W(\theta) = \Sigma A_{kk} P_k(\cos \theta)$, the integral angular correlation (see later definition of terms). Perturbations of the nuclei by atomic, crystal or external fields can change or destroy the angular correlation. For example, an external field may cause the nuclei to Larmor precess due to the interaction between the magnetic moments of the nuclei and the external field (a static magnetic interaction), resulting in a rotation of the correlation pattern. If the field is large enough, the correlation will be attenuated as well.

In general when rare earths are embedded in ferromagnetic material, the integral rotation is accompanied by an appreciable attenuation. The attenuation can be caused by four possible interactions: the static magnetic, the static-electric quadrupole and the two time-dependent interactions (electric and magnetic). Fortunately, ^{169}Yb in Fe can be described by a combined static and time-dependent magnetic interaction and hence the correlation can be written in closed form. For a correlation (relaxation) time τ_c short compared to the nuclear mean life τ, Abragam and Pound [7] have derived an expression for the angular correlation which is appropriate:

$$W(\theta) = 1 + \frac{1}{4}A_{22}G_{22}\left\{1 + \frac{3\cos 2(\theta - \Delta\theta_{22})}{[1 + (2\omega_L \tau G_{22})^2]^{1/2}}\right\}$$

$$\text{(2)}$$

$$+ \frac{1}{64}A_{44}G_{44}\left\{9 + \frac{20\cos 2(\theta - \Delta\theta_{24})}{[1 + (2\omega_L \tau G_{44})^2]^{1/2}} + \frac{35\cos 4(\theta - \Delta\theta_{44})}{[1 + (4\omega_L \tau G_{44})^2]^{1/2}}\right\},$$

where

$$\tan\Delta\theta_{Nk} = \frac{N\omega_L \tau}{1 + \lambda_k \tau}, \qquad G_{kk} = \frac{1}{1 + \lambda_k \tau}. \qquad (3)$$

Here $\omega_L = -(g\mu_n/\hbar)H_{eff}^z$, λ_k is the relaxation parameter and the A_{kk} are the expansion coefficients defined in Ref. 6.

As discussed by Caspari et al. [8], the application of this formula can be extended to describe the case of a partially ordered system showing fast relaxation provided that $\langle H_{eff}^z \rangle \ll \langle H_{int}^2 \rangle^{1/2}$; $\langle H_{eff}^z \rangle$ is the average static magnetic field along the z-axis and $\langle H_{int}^2 \rangle^{1/2}$ is the mean ionic field, the brackets denoting sample averages. The frequencies ω_L and ω_m can be defined such that $\omega_L = -g(\mu_n/\hbar)\langle H_{eff}^z \rangle$, which describes the precession, and $\omega_m = -g(\mu_n/\hbar)\langle H_{int}^2 \rangle^{1/2}$, which accounts for the attenuation of the correlation. The conditions for Eq. (2) to be valid are that $\omega_m \tau_c \ll 1$, and $\tau_c \ll \tau$. To first order in the exchange field, the explicit expression for the λ_k for a time-dependent magnetic interaction is

$$\lambda_k = \frac{1}{3}k(k+1)\omega_m^2 \tau_c. \qquad (4)$$

In the absence of quadrupole effects the correlation pattern for [169]Yb in Fe would be expected to follow Eqs. (2)-(4) at temperatures above 120°K, corresponding to a combined static magnetic interaction (from the effect of the magnetized Fe) and a magnetic time-dependent interaction (from the relaxation of the 4f electrons with the Fe lattice). Furthermore, for [169]Yb in Fe, the relaxation parameters for the 7/2 and 5/2 states would be related by their g factors. From Eq. (4), the ratio $\lambda(7/2)/\lambda(5/2) = g^2(7/2)/g^2(5/2) = 1.66 \pm 0.28$ (a number known from experiment [9]). Thus with the assumption of a combined static magnetic and time-dependent magnetic interaction, the measured PAC of the 7/2 state predicts the PAC of the 5/2 state. From the redundancy of measuring both correlations (as well as the measurements at two different temperatures for both states after each anneal), the behaviour of Yb in physically different surroundings can be studied via the measured hyperfine interactions. For example, if the Yb atoms after anneal were composed of two population groups, one in a magnetic environment obeying Eqs. (2)-(4) with a hf H_{eff} (the z in the superscript has been omitted) and a relaxation parameter λ, the second in a non-magnetic environment (precipitation center, oxide, intermetallic compound, or near some defect center) obeying Eqs. (2)-(4)

but with $H_{eff} = 0$ and $\lambda = \lambda$, then the correlation pattern would be

$$W(\theta) = fW(\theta, H_{eff}, \lambda) + (1-f)W(\theta, 0, \lambda') \quad . \qquad (5)$$

If H_{eff}, λ and λ' are known equation (5) can be fit to the data to extract f. In general, these quantities are not known and a correlation of the form

$$W(\theta) = W(\theta, H_{eff}, \lambda) \qquad (6)$$

may be fit where H_{eff} and λ are fitting parameters. H_{eff} and λ extracted in this manner represent some type of average values. We also define a quantity

$$F = \frac{H_{eff}(T)}{H_{eff}(o)} \qquad (7)$$

where $H_{eff}(T)$ is the effective magnetic field after annealing at temperature T and $H_{eff}(o)$ the field before annealing (i.e. after implantation). In a two site model and under the condition $\omega\tau < 1$, F can represent the fraction of atoms at high field sites after anneal at temperature T relative to those prior to anneal.

4.3 Results and Analysis

Equations (5), (6) and (7) are based on the assumption that the Yb atoms lie in two quite different environments, a magnetic and non-magnetic site. Mössbauer studies of two other implanted rare earths in Fe, [151]Gd [10] and [161]Dy [11], show that the implanted atoms occupy two main sites, a magnetic and non-magnetic one, in roughly equal proportions. Furthermore, nuclear orientation experiments with implanted [169]Yb in Au [12] indicate two sites, in one of which the nuclei do not align in the applied field.

The results of the integral PAC measurements with both the 198-110 keV γ-cascade (through the 5/2 state) and the 177-131 keV γ-cascade (through the 7/2 state) were least squares fitted with Eq. (5) or Eq. (6). Figure 2 displays some typical correlations and indicates that a better fit is obtained with Eq. (6), in which the average effective field H_{eff} and the average relaxation parameter λ are determined. It is seen from Table II that the ratio of the λ's for the two states, calculated from Eq. (4) on the basis of a combined static magnetic and magnetic time-dependent interaction, agrees with the theoretical prediction at both measurement temperatures. The assumption of an additional small quadrupole interaction results in a poorer fit to the observed correlations.

Table III lists the values of H_{eff} and λ after annealing at the different temperatures, as calculated from Eq. (6). Note that after the anneal at 600°C, the value of λ approaches 1/3 of the original

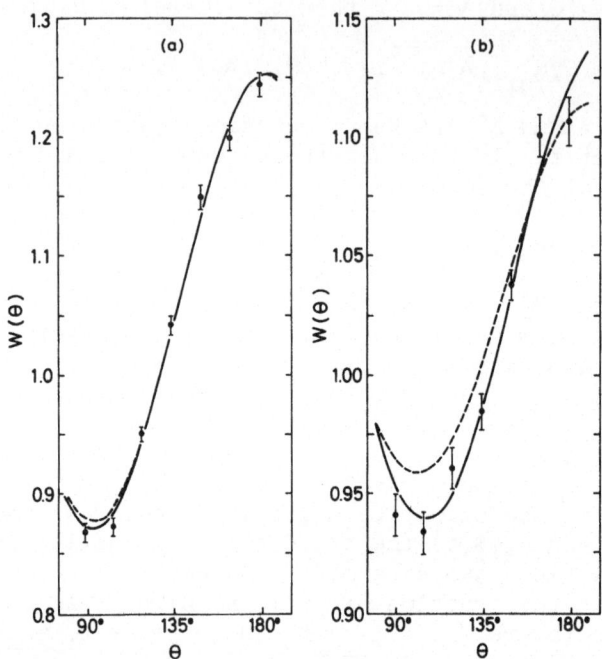

Fig. 2. Typical γ–γ correlations obtained with ^{169}Yb-implanted Fe single crystals, magnetized in a direction normal to the plane of the detectors. Parts (a) and (b) are for the 198–110 keV γ-cascade (5/2 state) and the 177–131 keV γ-cascade (7/2 state) respectively in ^{169}Tm. The dashed curves are fits to the data using Eq. (5), the solid curves using Eq. (6).

Table II. The experimental and theoretical ratio of the relaxation parameters λ for the 7/2 and 5/2 states in ^{169}Tm (see Eq. (4)), at two different temperatures. The experimental ratio is an average of the values obtained for all annealing temperatures. The theoretical ratio has been calculated using g(7/2) = 0.38 and g(5/2) = 0.30 [9].

Measurement temp.	Experimental $\lambda(7/2)/\lambda(5/2)$	Theoretical $(g(7/2)/g(5/2))^2$
300°K	1.47 ± 0.18	1.66 ± 0.28
140°K	1.33 ± 0.28	1.66 ± 0.28

Table III. Values of H_{eff} and λ, both measured at 300°K, as a function of annealing temperature. The values have been determined for both the 5/2 and 7/2 states independently from Eq. (6).

Annealing	H_{eff}(MOe)		$\lambda(10^8 \text{ sec}^{-1})$	
temp. (°C)	5/2 state	7/2 state	5/2 state	7/2 state
30	1.65 ± 0.15	1.60 ± 0.15	11.1 ± 2.5	18.2 ± 2.8
200	0.86 ± 0.08	0.70 ± 0.08	8.8 ± 2.5	11.6 ± 1.3
300	0.75 ± 0.08	0.68 ± 0.10	7.8 ± 2.5	12.2 ± 1.3
350	0.70 ± 0.08	0.69 ± 0.15	8.8 ± 3.0	13.0 ± 2.0
400	0.56 ± 0.08	0.49 ± 0.08	7.8 ± 2.5	10.4 ± 1.3
450	0.69 ± 0.15	0.57 ± 0.10	7.5 ± 3.0	12.0 ± 2.0
500	0.71 ± 0.08	0.64 ± 0.08	6.6 ± 2.5	11.8 ± 1.3
600	0.05 ± 0.07	0.01 ± 0.02	6.6 ± 2.5	6.6 ± 1.0

value after implantation; the effective field disappears. The results for both states are consistent and yield the same H_{eff} and F (the fraction of atoms at the high field site after a given anneal, relative to that after implantation) and consistent values of λ.

5. DISCUSSION

The fraction of Yb atoms in substitutional sites, normalized to the fraction substitutional before annealing, is plotted as a function of annealing temperature in Fig. 3. This fraction is derived from the channeling measurements (Table I). Also plotted is the function F which represents the normalized fraction of Yb atoms in high magnetic field sites, determined independently for both states studied in the PAC measurements (Table III). It is seen that there is a definite correlation between the channeling and hf results, the high magnetic field being associated with the substitutional site. The value of F decreases monotonically with annealing temperature, except for an apparent kink at 400°C. Nevertheless, it is clear that Yb atoms move from magnetic to non-magnetic sites during the annealing procedure. Mössbauer measurements [10] indicates a very similar annealing behaviour for implanted [151]Gd in Fe; the fraction of atoms in high field sites after a 450°C anneal is the same as in the present case of [169]Yb in Fe (F $\sim$ 0.4). This might be expected as Gd and Yb atoms in the same charge state have almost the same radius.

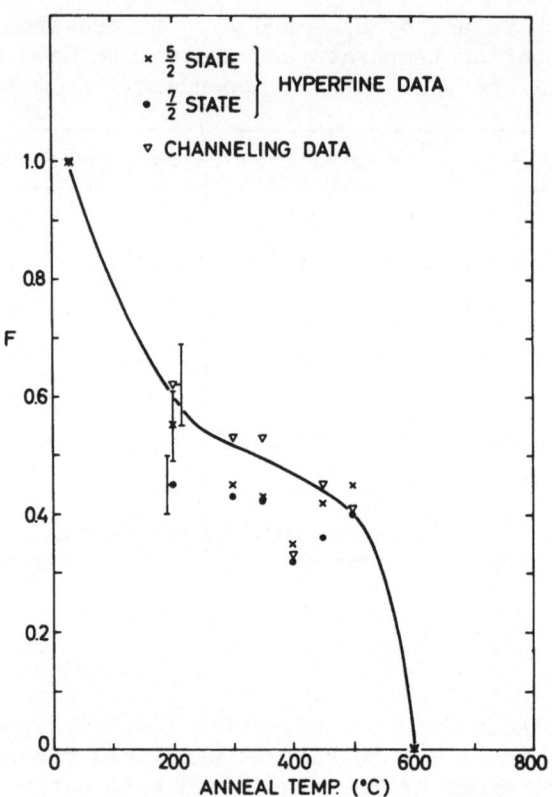

Fig. 3. The parameter F as a function of annealing temperature. For the channeling measurements, F is the fraction of [172]Yb atoms in substitutional sites normalized to the fraction substitutional before annealing. For the PAC measurements, F is the normalized fraction of [169]Yb atoms in high magnetic field sites (determined from Eqs. (6) and (7), for both the 5/2 and 7/2 states). The smooth curve is only to guide the eye.

       For a negligible hf at the non-substitutional Yb sites, the hf for [169]Tm (the radioactive daughter of [169]Yb) at substitutional sites in Fe should be $\dfrac{1}{0.58}$ of the value of -1.63 ± 0.11 MOe (Table III) measured in this study. Here 0.58 is the fraction of Yb atoms in substitutional sites following implantation (before annealing). Thus we expect H_{eff} (Tm) = -2.8 ± 0.2 MOe. A similar analysis can be applied to the closed 4f shell rare earth [175]Lu (the daughter of

[175]Yb) in Fe, giving H_{eff} (Lu) = -0.83 ± 0.10 MOe from the measured value for [175]Yb in Fe of -0.48 ± 0.06 MOe [13,14]. Hence the contribution of the unpaired 4f electron in Tm can be determined as H_{eff} (4f Tm) = H_{eff} (Tm) - H_{eff} (Lu) = -2.0 ± 0.2 MOe.

A similar channeling-hf study of Yb implanted in Fe has recently been reported by Abel et al. [15]. The results of these authors for the substitutional fraction are in agreement with our results in the case of an unannealed crystal, but show a different depencence on annealing temperature. It is thought that this difference is due to the implantation energy in their work being 400 keV, compared with 60-80 keV in the present work; the proximity of the crystal surface may be an important factor in annealing. From an investigation of quadrupole effects for the 36 nsec state in [169]Tm, Abel et al. [15] conclude that implanted Yb in Fe forms precipitates after a 600°C anneal. However, in the Mössbauer measurements [10] on [151]Gd in Fe mentioned above, the isomer shift following a 600°C anneal is found to correspond to Gd_2O_3. This suggests that implanted Yb in Fe may also reside in an oxide after annealing at 600°C. PAC measurements of the relaxation parameter λ for sources of [169]Yb_2O_3 and [169]Yb implanted in Yb metal are in progress to elucidate this point.

In conclusion, the present study should illustrate the need to determine the lattice location of implanted impurity atoms for a proper interpretation of hf measurements on implanted sources. Conversely, detailed hf measurements may provide useful information on the location and local environment of implanted atoms.

6. ACKNOWLEDGMENTS

We are grateful to Dr. F. Abel and his colleagues for providing data and information before publication.

REFERENCES

1. B. I. Deutch and G. M. Heestand, "Angular Correlations in Nuclear Disintegration", eds. H. van Krugten and B. van Nooijen (Rotterdam University Press, 1971), p. 487.

2. L. C. Feldman, E. N. Kaufmann, D. W. Mingay and W. M. Augustyniak, Phys. Rev. Lett. 27, 1145 (1971).

3. H. de Waard, "Mössbauer Spectroscopy and its Applications" (IAEA, Vienna, 1972), p. 123.

4. J. A. Davies, "European Conference on Ion Implantation" (Peregrinus, Stevenage, England, 1970), p. 172.

5. R. B. Alexander, P.T. Callaghan and J.M. Poate, to be published
    in Phys. Rev. B.

6. H. Frauenfelder and R.M. Steffen, "Alpha-, Beta- and Gamma-Ray
    Spectroscopy", ed. K. Siegbahn (North-Holland, 1965), Chap. XlXA.

7. A. Abragam and R.V. Pound, Phys. Rev. $\underline{92}$, 943 (1953).

8. M.E. Caspari, S. Frankel and G.T. Wood, Phys. Rev. 127, 1519
    (1962).

9. V.S. Shirley, "Hyperfine Interactions in Excited Nuclei", eds.
    G. Goldring and R. Kalish (Gordon and Breach, 1971), Vol. 4,
    p. 1255.

10. R.L. Cohen, G.Beyer and B.I. Deutch, (these conference pro-
    ceedings).

11. P. Inia and H. de Waard, "Angular Correlations in Nuclear
    Disintegration", eds. H. van Krugten and B. van Nooijen.
    (Rotterdam University Press, 1971), p. 519.

12. A. Benoit, J. Flouquet and J. Sanchez, to be published.

13. K. Bonde Nielsen and B.I. Deutch, Phys. Lett. $\underline{25B}$, 208 (1967).

14. L. Thomé, H.C. Benski and H. Bernas, Phys. Lett. $\underline{42A}$, 327 (1972).

15. F. Abel, M. Bruneaux, C. Cohen, H. Bernas, J. Chaumont and
    L. Thomé, Solid State Comm. $\underline{13}$, 113 (1973) and these conference
    proceedings.

EFFECT OF RADIATION DAMAGE ON LATTICE LOCATION AND HYPERFINE INTERACTIONS OF IMPURITIES IMPLANTED IN IRON[*]

F. Abel, H. Bernas[+], M. Bruneaux, J. Chaumont[+], C. Cohen, and L. Thome[+]

Groupe de Physique des Solides, Université Paris VII, 75005 - Paris
[+]Institut de Physique Nucléaire et Centre de Spectrométrie de Masse, Université Paris-Sud, 91406 - Orsay

Previously, we studied radiation damage effects via electron microscopy, hyperfine interaction (hfi) and lattice location experiments[1] in Yb-implanted Fe, before and after annealing. Some form of ytterbium precipitation was shown to take place during a well defined annealing stage around 450° C.

Here we present new hfi results confirming this mechanism: a differential perturbed angular correlation experiment in the 36 ns (379 keV)--level of ^{169}Tm after annealing of the implanted $\underline{Fe}$ ^{169}Yb alloy above 450°C, shows a pure static quadrupole interaction with a distribution (width 15%) of quadrupole frequencies around an average value ($\omega_0 = (2.60 \pm 0.8) \times 10^8$ rad s^{-1}). This would be expected after precipitation.

We report also new lattice location experiments performed to study whether: a) impurity radiation damage interactions do appear in annealing experiments on metals implanted with more soluble impurities than Yb; b) if the impurity damage interactions are affected when the damage mechanisms are modified.

a) Lattice-location experiments in Au-implanted Fe show that while the corrected extinction ratio on the impurity is $\tau = 0.85 \pm 0.03$ before annealing (in agreement with previously published results), it rises to unity after annealing at 450°C. This is probably related to annealing of radiation damage near the Au atoms.

---

[*] Full paper to be published.

b) Lattice-location results on Yb implanted Fe at higher-than-room temperatures (up to 500° C) indicate a lowering of the transition reported above. Similar hfi experiments will be reported at the conference.

The relation between the so-called "substitutional fraction" and the value of τ in lattice-location experiments will be discussed in the light of these results and those of Ref. 1.

REFERENCES

1. F. Abel, M. Bruneaux, C. Cohen, H. Bernas, J. Chaumont and L. Thomé, Solid State Comm. <u>13</u>, 113 (1973).

DISCUSSION

Q: (E. N. Kaufmann) It seems clear from the plot of your time differential data that you cannot exclude the possibility that your spectrum represents a mixed site situation where some ^{169}Tm is in an unperturbed environment and the rest is in magnetically hard precipitates such that Tm sees a weak polycrystalline (nonalignable) somewhat inhomogeneous magnetic interaction. How can you exclude this?

A: [1] If a significant fraction of the ^{169}Tm nuclei were in an unperturbed environment, the anisotropy of the TDPAC curve would be reduced. Taking into account the experimental time resolution, our measured anisotropy at time zero agrees with the value calculated from the nuclear parameters. So essentially all the Tm nuclei are in perturbed environments. [2] We have not done the calculation for a distribution of nonaligned magnetic sites. It is possible that a fit could be found to both the IPAC and TDPAC results at 550°C; we will try it. This would not change our main point, however; above 550°C, the Yb ions are no longer in a dilute solid solution, and this is due to interactions with the radiation damage.

Q: (L. C. Feldman) Can you imagine any single site for the Yb which would yield a large χ_{min} impurity (~ 0.5) and not have a large effect on the angular width?

A: Not easily; can you?

DETERMINATION OF UNIQUE SITE POPULATION IN VARIOUS In IMPLANTED NON-CUBIC METALS USING ANGULAR CORRELATIONS AND THE NUCLEAR ELECTRIC QUADRUPOLE INTERACTION

E. N. Kaufmann, P. Raghavan and R. S. Raghavan

Bell Laboratories, Murray Hill, N. J.

and

K. Krien, E. J. Ansaldo and R. A. Naumann

Princeton University, Princeton, N. J.

INTRODUCTION

As part of an investigation of the electric quadrupole interaction (EQI) at impurity nuclei in non-cubic metals[1], we have applied the time-differential perturbed-angular correlation (TDPAC) technique[2] to the measurement of the EQI at a ^{111}Cd impurity in several hosts. Since in most of the cases studied the ^{111}In parent radioactivity is difficult to introduce in the host lattice by metallurgical means, we prepared all sources by implantation of the ^{111}In activity. As a by-product of the investigation, therefore, we obtained information on the efficiency of the implantation process in providing dilute alloy systems with a large fraction of In impurity atoms at lattice sites which appear unique insofar as the EQI is concerned. This definition of uniqueness implies not only that the impurity site be a specific crystallographic position but also that it sees identical surroundings in the first few shells of neighbors.

Following the electron-capture decay of the ^{111}In, the 247 keV $I^{\pi} = 5/2^{+}$ state of ^{111}Cd is populated and depopulated in succession by a gamma-ray cascade.[3] The angular correlation of the two gamma-rays was measured as a function of the time spent by the excited nucleus in the intermediate state ($T_{\frac{1}{2}} = 84$ nsec) using standard detection and coincidence electronic processing methods

which have been described in the literature.[2] By combining data
taken for different angles of emission of the second radiation
relative to the first, the trivial nuclear decay factor can be
removed and the angular correlation anisotropy can be plotted as a
function of time. For nuclei not influenced by perturbing fields,
the anisotropy is a constant whose value depends only on the known
nuclear state spins and radiation multipolarities and on geometrical
factors which account for finite detector solid angles and absorption and scattering of radiation in the source.[4] On the other hand,
the amplitude of the anisotropy will be modulated as a function of
time if the nuclei precess while in the intermediate state under the
influence of a static extranuclear field. For the case of a unique
static axially symmetric electric quadrupole interaction, the modulation function is simply a sum of cosine functions of the fundamental interaction frequency and two harmonics plus a small constant
term. In the event that the nuclei are distributed among sites of
different EQI strengths and orientations, the resultant modulation
function is simply the linear superposition of functions for each
site weighted by the fraction of nuclei in each site. The superposition of several sites with, in general, incommensurate EQI frequencies will result in a modulation function which is rapidly damped
from its initial maximum (unperturbed) value at time zero to a
small constant value referred to as the polycrystalline "hard core"
anisotropy. Thus the fraction of impurity atoms in unique sites
may be determined by comparing the amplitude of modulation at unique
frequencies with the unperturbed amplitude at t = 0.

SOURCE PREPARATION

The ^{111}In isotope was produced in the Princeton cyclotron
through the reaction ^{109}Ag$(\alpha,2n)$ ^{111}In on a natural Ag target. The
^{111}In (plus some natural In carrier) was chemically separated from
the target material and subsequently implanted into the various host
lattices at room temperature with an energy of 90 keV using the
Princeton isotope separator. The source strengths were of the order
of 10 μCi which implies an ^{111}In dose of $\sim 10^{12}/cm^2$ over an area of
~ 0.2 cm^2. Because the nearest stable isotope is two mass units away
from the mass 111 position, any additional stable dose was at or
below levels comparable to that of the ^{111}In. This is also indicated by the absence of ^{109}In activity in the implanted sources.

The host materials studied so far include foils of Be, Bi, Re,
Sc, Zn and Zr and single crystals of Te, Ti and Hf. In each case
the sample was chemically etched to provide a clean surface for
implantation.

ANALYSIS

The amplitude vs. time spectra for the various samples fall into two groups: those which displayed readily discernible unique frequencies (Be, Re, Te, Ti) and those where no significant amplitude in a unique frequency was seen (Bi, Sc, Zn, Zr, Hf). For the first case, the analysis was carried out by assuming that a fraction f_u of impurity atoms was in the unique site and the remainder $(1-f_u)$ was in a distribution of sites and surroundings such that the modulation function for this group was of a strongly damped character. A least-squares fit of the theoretical unique EQI modulation function[2] was made to that portion of the data where it was reasonable to assume that the "non-unique" function was completely damped to its hard core value. It is easily shown for the case considered here that the hard core value is 1/5 of the initial amplitude of the non-unique function.[2] Thus, if the total available amplitude as read from the spectra at t = 0 is denoted A and the amplitude resulting from the least-squares fit beyond t = 0 is denoted A_F, the relation

$$A_F = f_u A + \frac{1}{5}(1-f_u)A$$

holds and the unique site fraction can be computed from

$$f_u = (5A_F/A-1)/4 \ .$$

Since the primary objective of these experiments had been to determine the EQI frequencies which are easily extracted from the data without precise knowledge of the modulation amplitudes, no special precautions were taken to standardize the detection geometry or to correct the data for scattering and absorption in the samples which had varying dimensions and mass absorption coefficients. The values of A were therefore estimated from each spectrum separately and can be in error by as much as 15%.

The interpretation of the spectra for which no significant unique frequencies could be observed is not so straightforward. The main difficulties arise from our ignorance in most cases with regard to what the EQI frequency might be at a unique site and from the limited statistical accuracy of the data. From known EQI frequencies for [111]Cd in many materials we can say that it is very unlikely that a unique (or at least substitutional) site frequency is too large to be resolved with the time resolution of our electronics. In addition, if a large fraction of nuclei were to give such a high frequency, a severe attenuation of the t = 0 amplitude would be evident in the data. On the other hand, a rather low frequency, i.e. less than one full precession period within the time range available for study, could be present and may not be distinguishable from a damped modulation. Measurements with greater statistical accuracy, precise knowledge of geometrical effects and "time-zero" stabilization would probably overcome this difficulty. We have been able to extract some information from the present data by observing the shape of the spectra around t = 0, by setting a limit on the amplitude of any unique components and in one case (Zn) by measuring the EQI frequency in an annealed sample.

RESULTS

In what follows we give a brief description of the spectra obtained for each system studied and quote results of the analysis as described above and of some corollary experiments.

Titanium (Fig. 1). The unique frequency fit (A_F = 0.105) accounts for nearly the full t = 0 amplitude (A = 0.11) yielding a unique site fraction f_u = 0.95. As seen in the figure, the spectrum is predominantly a single cosine of the fundamental EQI frequency as expected for a single crystal sample with symmetry axis perpendicular to the detection plane. A channeling experiment was performed on inactive In implanted Ti crystals ($5 \times 10^{14}/cm^2$ at 100 keV) using a 2.0 MeV He^+ beam. Backscattering yields for impurity and host were measured on and off two major crystal axes. The substitutional fractions obtained were $(88 \pm 5)\%$ for the $\langle 0001 \rangle$ axis and $(93 \pm 5)\%$ for the $\langle 11\bar{2}0 \rangle$ axis which strongly suggest that the unique site is substitutional.

Rhenium (Fig. 2). The unique site fit yielded an amplitude extrapolated to t = 0 of A_F = 0.038. Combining this with the unperturbed amplitude at t = 0 of A = 0.075 gives a unique site fraction f_u = 0.39. This value of f_u must be regarded as a lower limit because an alternate fit, which allowed a slight frequency spread (i.e. a slow damping) for the unique site corresponding to lattice imperfection beyond the first few neighbor shells, yielded a value f_u = 0.62 which may be regarded as an upper limit on f_u. A more accurate determination of f_u would require greater statistical accuracy in data in order to better define the amplitude at large times.

Beryllium (Fig. 3). The unique site fit in this case is very well determined because three full modulation periods fall within the spectrum. The resulting value of A_F = 0.029 taken together with the unperturbed amplitude A = 0.062 gives a unique site fraction f_u = 0.33.

Tellurium (Fig. 4). The highest symmetry element in the tellurium crystal structure is a diad axis. A unique site EQI will therefore correspond in general to an axially asymmetric interaction. This additional feature results in a relaxation of the restriction that the frequencies in the modulation function form a harmonic series.[5] It can be seen from the figure that this results in an aperiodic oscillation. The analysis for our purposes however is the

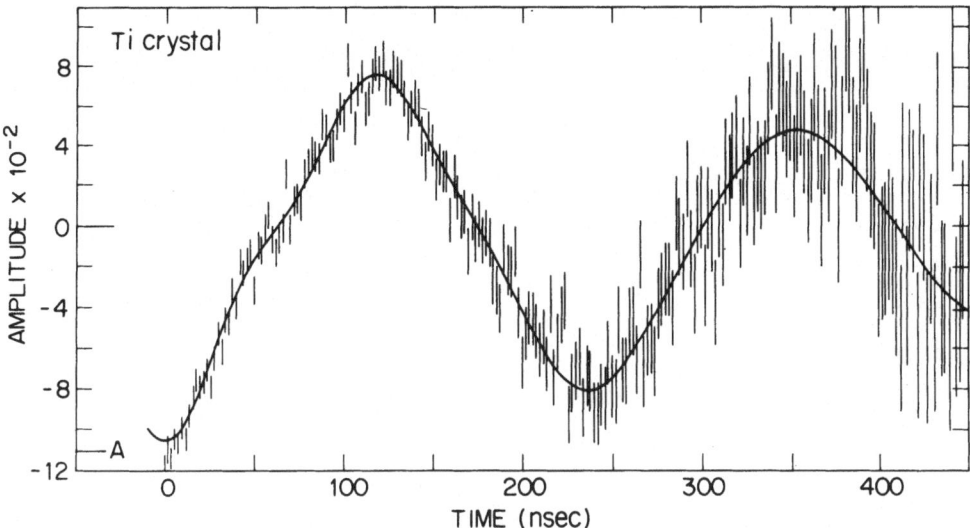

Figure 1. TDPAC spectrum for In implanted in a titanium single crystal. (The amplitude scale is not comparable to any of the following figures because a different combination of the raw data was used to compute this spectrum.) The solid curve resulted from a least-squares fit to the data.

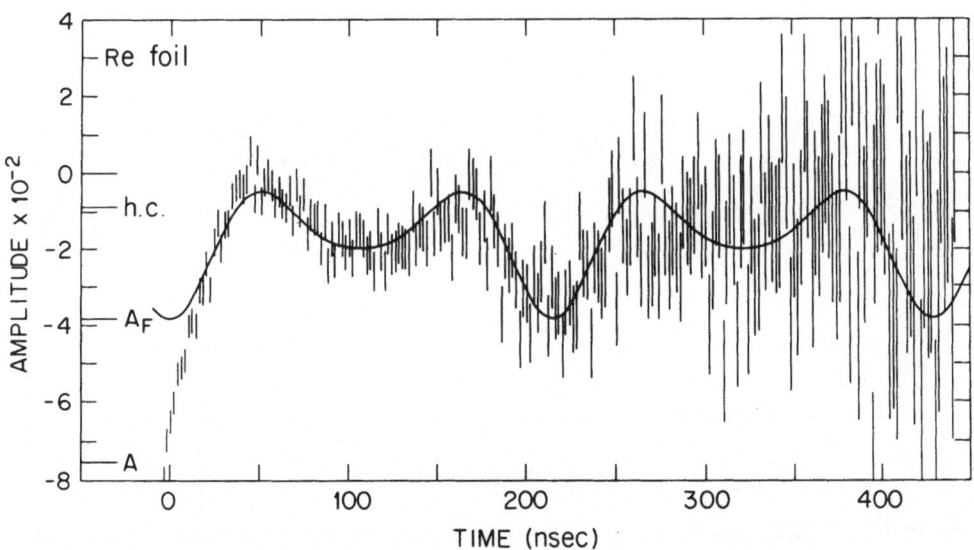

Figure 2. TDPAC spectrum for In implanted in a rhenium foil. The solid curve resulted from a least-squares fit to the data.

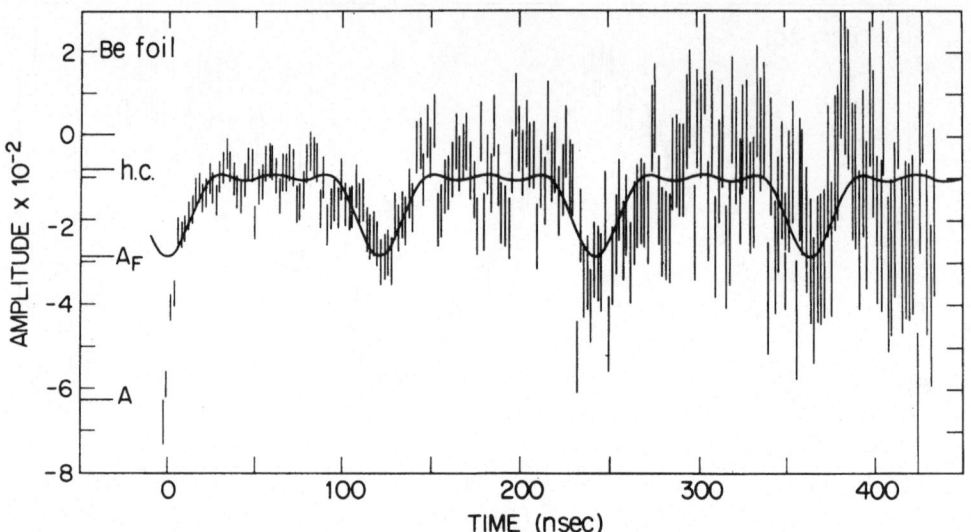

Figure 3.   TDPAC spectrum for In implanted in a beryllium foil.   The solid curve resulted from a least-squares fit to the data.

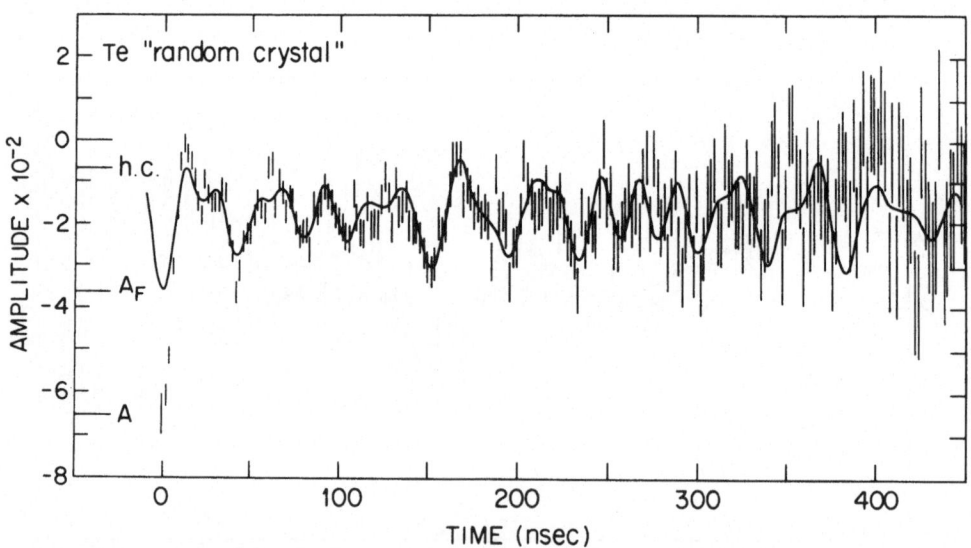

Figure 4.   TDPAC spectrum for In implanted in a tellurium single crystal which was set in motion to simulate a random polycrystalline sample.   The solid curve resulted from a least-squares fit to the data assuming an axially asymmetric EQI.

same. A value of A_F = 0.036 obtained from the fit compared to the
unperturbed amplitude A = 0.065 gives a unique site fraction
f_u = 0.44. Although the physical sample used was a single crystal of
Te metal, it was mounted on a two axis motorized drive and set into
rotation in order to simulate a randomly oriented sample. This was
necessary because the three crystallographically equivalent atom
sites in the Te unit cell display inequivalent orientations of the
principal axes of the EQI. The "randomization" eliminates the effect of this inequivalence on the data and allows a simpler analysis.
Fourier analysis of Te spectra have indicated a secondary unique site
may be populated to the extent of ∼15% of the primary site, but this
possibility has not yet been verified.

Zinc (Figs. 5 and 6). No unique frequency was clearly evident
in the spectrum for zinc as shown in Fig. 5. The dominant feature
is the rapid decay of the amplitude at t = 0 to the hard core value.
However in this spectrum, unlike most others reported here, there is
a pronounced overshoot before the hard core value is reached. Based
on computer simulations for various widths of EQI frequency distributions, we believe this underdamping is characteristic of a somewhat
narrower spread in the frequencies at the impurity sites as compared
to the other cases studied here. From the data of Fig. 5, an upper
limit can be estimated for the parameter A_F of 0.02 since any unique
oscillation corresponding to a larger value would be visible unless
the frequency were too low. Thus from the unperturbed amplitude
A = 0.065 we can set the limit $f_u \leq$ 0.13 for the unique site fraction.

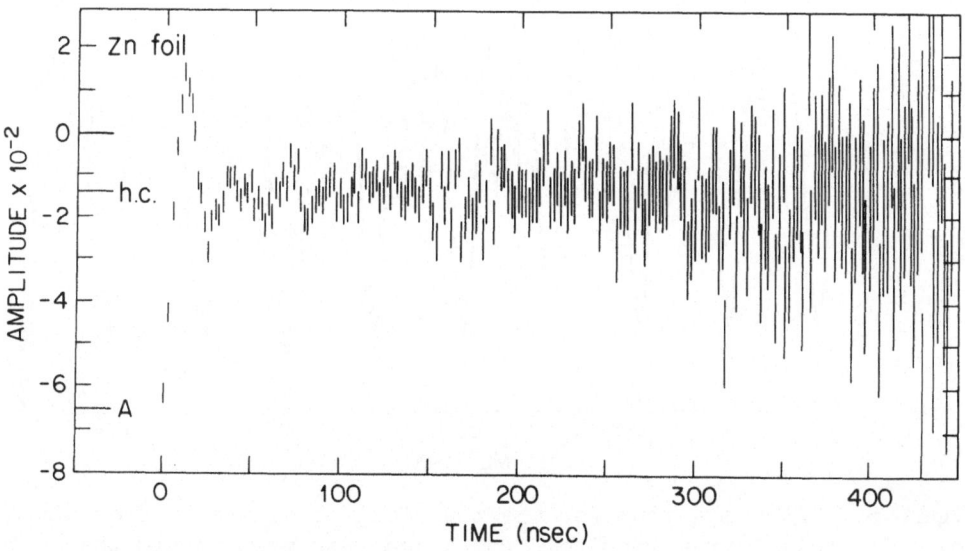

Figure 5. TDPAC spectrum for In implanted in a zinc foil.

In a subsequent experiment the implanted Zn foil was annealed at
300°C for one hour before measurement and gave the spectrum shown in
Fig. 6 where it is evident that the unique (presumably substitutional)
site frequency is indeed fast enough to have been seen in Fig. 5 if
it were present to a significant extent. A channeling experiment was
performed on an In implanted Zn single crystal (3×10^{14}/cm^2 at 100 keV)
using a 2.0 MeV He$^+$ beam. The "on axis" to "off axis" scattering
yields from host and impurity gave an apparent substitutional fraction along the $\langle 0001 \rangle$ axis of (7 ± 15)%. This result and the TDPAC
spectrum both suggest that the In is trapped in regions of lattice
damage.

Scandium (Fig. 7). This data shows no apparent oscillatory
features. The initial amplitude of A = 0.07 decays to about one-
fifth of its value at large times. This observation alone would suggest a unique site fraction consistent with zero. However, an additional feature in the spectrum is evident on closer inspection. There
is a distinct break in the slope of the decay from t = 0 at an amplitude of ~0.04. If the less steep portion were actually part of the
t = 0 maximum in a unique site modulation function, its shape would
predict that the second maximum occurs beyond the time range of the
measurement. Such a low frequency would not be surprising in view of
NMR data on pure Sc metal which has found a very small EQI for ^{45}Sc.[6]
If this interpretation is correct and a value of $A_F = 0.04$ is assumed, the unique site fraction $f_u = 0.46$ would follow. This

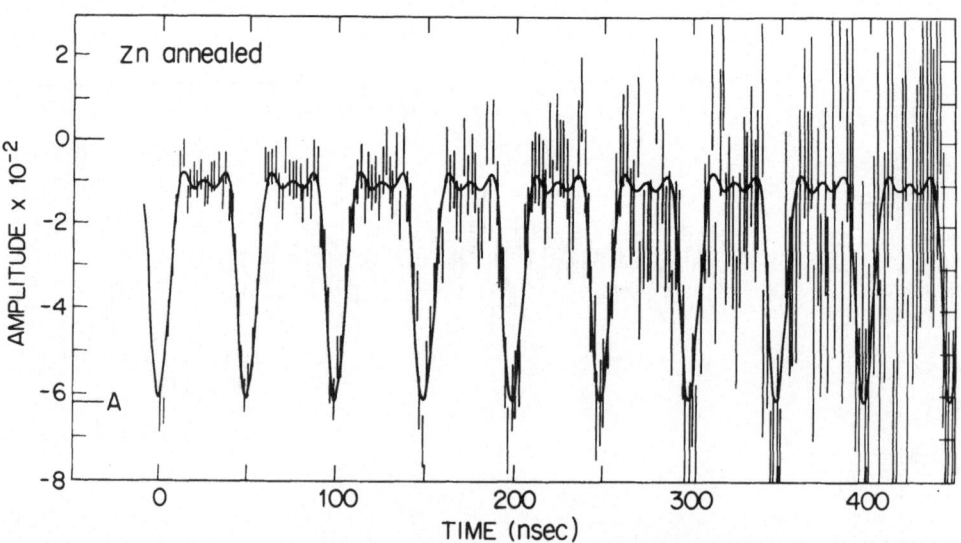

Figure 6. TDPAC spectrum for the zinc foil of figure 5 after annealing. The solid curve resulted from a least-squares fit to the data.

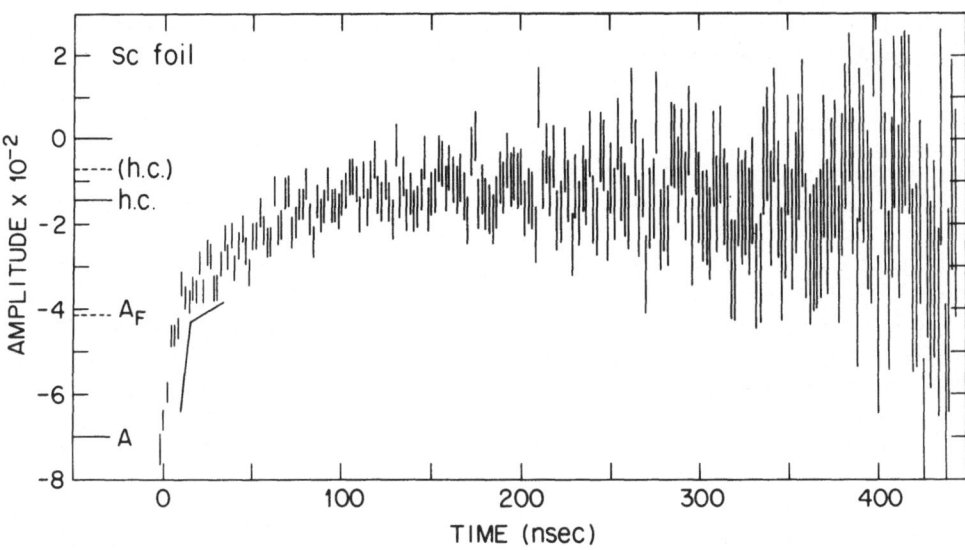

Figure 7. TDPAC spectrum for In implanted in a scandium foil. A break in slope near t = 0 is indicated by the solid line segments.

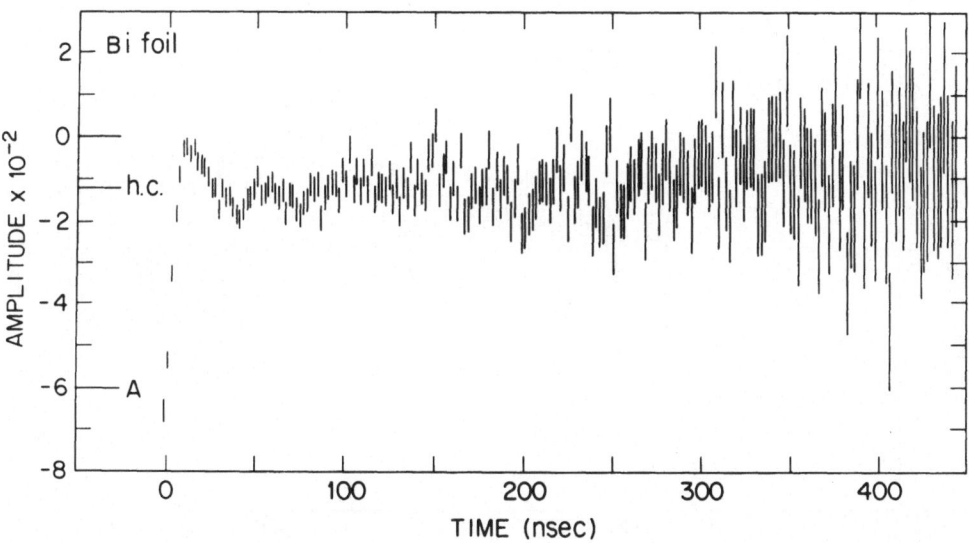

Figure 8. TDPAC spectrum for In implanted in a bismuth foil.

experiment must be repeated with better statistical accuracy and control of time-zero instabilities before the above evaluation of the data can be verified.

Bismuth (Fig. 8). This data strongly resembles that for the case of Zn aside from the less pronounced overshoot in the spectrum. An estimation of $A_F \leq 0.02$ from the largest spectral feature beyond t = 0 combined with the unperturbed value A = 0.06 yields a limit on the unique site fraction of $f_u \leq 0.17$. Annealing of this sample did not provide a unique site for comparison, however a recent measurement[7] on the In in Bi system performed on a rapidly quenched alloy gave a frequency about twice that seen in Zn and would have been quite apparent in our data if the corresponding site were significantly populated.

Hafnium (Fig. 9). In this case also no clear evidence for a unique site is present in the data. An estimation as in the previous case gives $A_F \leq 0.02$ and A = 0.069 which implies $f_u \leq 0.11$. Since the Hf sample was a single crystal with symmetry axis perpendicular to the detection plane, a unique site EQI which has the crystal symmetry would have given an oscillatory pattern like that of the titanium host (cf. Fig. 1). This type of pattern is not visible in the data although a hint of some minor oscillation is present and might be revealed when greater statistical accuracy is obtained.

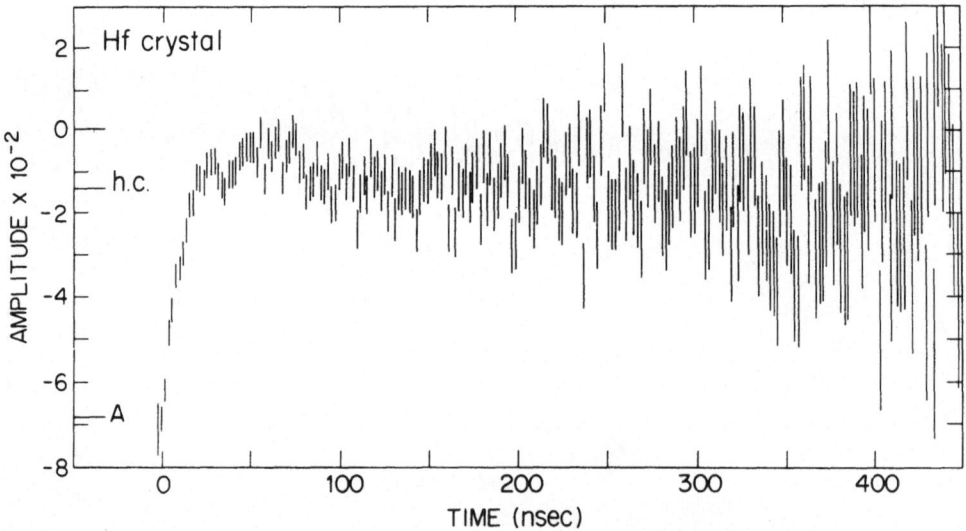

Figure 9. TDPAC spectrum for In implanted in a hafnium single crystal. The crystal symmetry axis was perpendicular to the detection plane.

Zirconium (Fig. 10). It is unfortunately apparent from the
error bars in the figure that insufficient data was collected to
make a definitive interpretation. This occurred because of a relatively weaker source than usually available in this instance. It
would not be surprising if the unique EQI frequency in Zr were quite
low and the hint of a break in slope in the decay near t = 0 may
indicate a situation similar to that encountered in the scandium
host.

CONCLUSION

The above description of the currently available results has
demonstrated the utility of the TDPAC technique in studying low dose
implanted impurity systems and has also pointed out some of the
problems encountered in interpreting the data. Most of the ambiguities that arise however can be resolved by performing suitable additional experiments and by increasing statistical accuracy. For
example, a recent measurement by Budtz-Jorgensen et al.[8] by this
technique on In implanted In, InBi and In$_2$Bi has demonstrated that,
with suitable time-zero stabilization and background subtraction
techniques, the presence of rather low unique frequencies can be
verified and that, by measuring with an expanded time scale, the mean
EQI frequency and relative frequency spread for non-unique sites can
be extracted from data near t = 0. Another recent measurement by
Behar and Steffen[9] on In implanted Ag metal targets, which had been
irradiated by α-particles to produce and recoil-implant the ^{111}In,

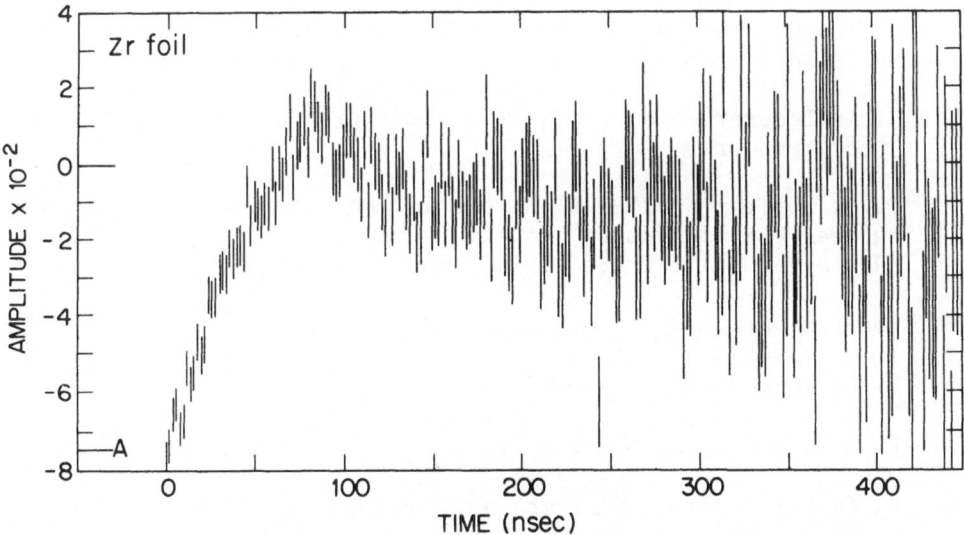

Figure 10. TDPAC spectrum for In implanted in a zirconium foil.

has shown this technique to be useful in the study of damage in cubic materials. The statistical accuracy achieved by them was sufficient to expose very low amplitude high frequency components in the data corresponding to unique damage configurations in the cubic lattice. We therefore have good reason to anticipate that further experiments with the hosts described here will clarify the interpretations of the data and perhaps reveal new weakly populated unique sites which we have so far been unable to identify.

The unique site fraction or limits thereon which we have quoted here do not seem to follow any systematic trend when compared to such variables as In solid solubility, host electronegativity, relative atomic volumes, and so forth. The study of additional host-impurity systems and the effects of such variables as implantation temperature are needed to aid in identifying the relevant parameter of the problem. A recent measurement[10] of the EQI at ^{100}Rh in Zn from a ^{100}Pd implanted source which has shown nearly complete unique site occupancy, when compared to our opposite result for In in Zn, points to this need quite vividly.

We would like to thank F. Loeser for help with the implantation and W. Flood for sample preparation.

REFERENCES

1. P. Raghavan, R. S. Raghavan, E. N. Kaufmann, K. Krien and R. A. Naumann, Bull. Am. Phys. Soc. 18, 134 (1973) and to be published.

2. H. Frauenfelder and R. M. Steffen in "Alpha-, Beta- and Gamma-Ray Spectroscopy" ed. by K. Siegbahn (North Holland, 1965) 997.

3. C. M. Lederer et al., "Table of Isotopes" (John Wiley & Sons, 1968, New York).

4. P. C. Lopiparo and R. L. Rasera in "Angular Correlations in Nuclear Disintegration" ed. by H. van Krugten and B. van Nooijen (Rotterdam Univ. Press, 1971, Netherlands) p. 66.

5. See for example E. Gerdau et al., Proc. Roy. Soc. A311, 197 (1965).

6. R. G. Barnes et al., Phys. Rev. 137, A1828 (1965).

7. H. Haas and D. A. Shirley, J. Chem. Phys. 58, 3339 (1973).

8. C. Budtz-Jorgensen (private communication, 1973).

9. M. Behar and R. M. Steffen, Phys. Rev. C 7, 788 (1973).

10. K. Krien (private communication, 1973).

DISCUSSION

Q: (W. D. Wilson) You seem to have neglected the Sternheimer antishielding factor in your analysis. This could make measurement of the actual quadrupole moment and quadrupole resonance shift rather difficult. Furthermore, if this factor (which admittedly is difficult to calculate) varies with the position of the nucleus in the crystal, it may make the assignment of position rather difficult also. Do you have any estimates of this effect?

A: The Sternheimer atomic antishielding factor as well as more subtle (and less calculable) enhancement factors due to metallic conduction electrons of nonspherical orbital character are implicit in determining the total electric field gradient at the impurity nucleus and are of great importance when trying to understand the particular magnitudes of the impurity-host interaction. In this contribution, however, we are concerned only with the distribution of the impurities among unique and nonunique surroundings. The various electronic enhancement factors are, by definition, unique for a unique site. For the nonunique sites (presumably defect associated), an inhomogeneously distributed range of interaction strengths are seen and the electronic enhancement is very likely also nonunique for such impurity sites.

Q: (R. L. Cohen) What would you see if you did similar measurements in cubic hosts?

A: This has been done in the one system of In implanted in Ag in the presence of strong α-beam irradiation (see Ref. 9). Very encouraging results were reported in which evidence of unique defect interactions were seen. The TDPAC technique is particularly well suited to this technique.

THE LOCATION OF DISPLACED MANGANESE AND SILVER ATOMS IN IRRADIATED

ALUMINUM CRYSTALS BY BACKSCATTERING

M.L. Swanson, F. Maury* and A.F. Quenneville

Atomic Energy of Canada Limited, Chalk River Nuclear

Laboratories, Chalk River, Ontario, Canada

ABSTRACT

By measuring the energy spectra of backscattered 1 MeV He$^+$ ions from Al-0.09 at.% Mn and Al-0.08 at.% Ag crystals, it was found that substitutional Mn and Ag atoms were displaced from lattice sites by irradiation with 0.3-1.0 MeV He$^+$ ions at 35°-70°K. From detailed channeling and annealing results, it was demonstrated that the Mn and Ag atoms were displaced by trapping irradiation-induced interstitial Al atoms, forming Al-Mn and Al-Ag atom pairs in the <100> dumbbell configuration.

INTRODUCTION

The positions of foreign atoms in good single crystals of metals can be determined by analyzing the energy spectra of back-scattered high energy ions in various crystallographic directions (1-10). This technique is based on the channeling phenomenon. Under suitably favourable conditions, the impurity atom positions in the host lattice can be found to within 0.2 Å, for impurity concentrations as low as 10^{-4} atomic fraction(4).

Since the displacements of impurity atoms toward other point defects in metals are often presumed to be at least 0.2 Å(11), the channeling technique can be used to investigate impurity-point defect interactions, including impurity-impurity, vacancy-impurity

---

* On leave from Université Paris-sud, Bt350, Orsay, France.

and interstitial-impurity interactions. As the displacement of impurity atoms should be greatest when they are associated with self-interstitial atoms, we have investigated these interactions first. It has been shown from electrical resistivity measurements after low temperature irradiations of metals that self-interstitial atoms are trapped by impurity atoms(12-16). These interstitials are apparently released during stage II or stage III recovery. The form of the interstitial-impurity pair has not been specified, but it might be expected that the configuration would be the <100> dumbbell (or split interstitial) configuration, which is the calculated stable form of the self-interstitial in f.c.c. metals (11,16,17). This configuration would give a large displacement of impurity atoms ($\gtrsim$ 1Å) in < 100 > directions, which could easily be studied by the channeling technique.

In order to investigate self-interstitial-impurity atom interactions, it is desirable to begin with an annealed alloy, in which the impurity atoms are all in normal substitutional sites. This is perhaps best accomplished by doping a metal uniformly from the melt, and then growing a good single crystal of the dilute alloy. We have chosen the alloys Al-0.09 at.% Mn and Al-0.08 at.% Ag for our backscattering studies because Mn and Ag are much heavier than Al (permitting easy identification by measurement of backscattered ion energies), and because the trapping of self-interstitial atoms by these impurities has been studied by electrical resistivity measurements(12-14,16,18). Impurity concentrations near 0.1 at.% have been chosen because that value is close to the saturation point defect concentration which can be produced by low temperature irradiation(19).

EXPERIMENTAL PROCEDURE

Single crystals of Al-0.09 at.% Mn, Al-0.08 at.% Ag, and Al-0.2 at.% Ag were grown from the melt by the Bridgman technique at a speed of 1 cm/hr. Slices were spark cut, polished on successively finer emery paper and diamond grit wheels, vibratory polished for 16 hr. with 0.05 μm alumina powder, and electro-polished 100 s. in a solution of 20% perchloric acid plus 80% ethyl alcohol.

The samples were examined in the low temperature arm of the 2.5 MV Chalk River Van de Graaff accelerator, using backscattering of 1 MeV He$^+$ ions(20,21). The analyzing He$^+$ beam was 1 mm in diameter, with divergence < 0.1°. The beam current was 2-20 nA. The energies of backscattered ions were measured using a surface barrier detector at a scattering angle of 150°, with single and multichannel analyzers. The energy resolution was 20 keV full width at half maximum. Each sample, mounted on a goniometer(20),

was cooled to ∿35°K by means of a self-contained refrigeration
unit(21). A shield at the same temperature surrounded the sample,
thus avoiding surface contamination from hydrocarbons. A heater
attached to the sample holder was used to vary the temperature
from 35°K to 300°K. The temperature was measured with a copper-
constantan thermocouple. The crystals were damaged at 35-70°K by
irradiating them in a random crystallographic direction with a
0.3-1 MeV He$^+$ beam which was swept uniformly over a 5 mm diameter
area.

RESULTS

For annealed crystals, the minimum yield of backscattered
1 MeV He$^+$ ions (χ_{min} = [aligned yield]/[random yield]) from Al
atoms at 35°K in a <110> direction was $(\chi_{min})_{Al} \simeq 0.02$, while
the minimum yield from Mn or Ag atoms was $\chi_{min} \leq 0.05$. These
values show that >96% of Mn or Ag atoms had replaced Al atoms at
f.c.c. lattice sites.

After irradiation of these crystals with 0.3-1.0 MeV He$^+$ ions
in a random direction at 35-70°K, the minimum yield from Mn atoms
$(\chi_{min})_{Mn}$ or Ag atoms $(\chi_{min})_{Ag}$ had increased greatly compared with
the increase in $(\chi_{min})_{Al}$, indicating that the impurity atoms had
been displaced from lattice sites. This effect of irradiation is
shown in Figures 1(a) and 1(b), and in Table 1. Extrapolated <110>
minimum yields from Mn and Ag atoms did not reach the random
levels even for very large irradiations, but saturated at values
up to 0.6 (depending on the temperature of irradiation). It will
be noticed that surface peaks appear for Al and Mn disordered atoms
in the surface oxide film. An oxygen peak is also seen for this
layer, but the usual C peak caused by surface contamination from
the ion beam is not present. The appearance of a large surface
peak for Ag, occurring even in the random spectrum, is due to Ag
enrichment at the surface by electropolishing, and corresponds to
about a monolayer of Ag. The Cl peak is caused by a residual
surface layer from the electropolishing solution. This impurity
proved difficult to remove without introducing some more
undesirable impurities, such as K. The channels over which Mn and
Ag yields were measured are indicated in Figure 1.

The fraction of displaced impurity atoms can be estimated
from the χ_{min} values. If a simple geometric model is used for the
backscattering from impurity atoms in a channel, where the ion
flux is assumed uniform, the fraction of shadowed impurity atoms,

$$C = \left[1 - (\chi_{min})_i\right] / \left[1 - (\chi_{min})_{Al}\right] \qquad \text{....(1)}$$

where $(\chi_{min})_i$ refers to the minimum yield from the impurity atoms.
It is seen from Table 1 that C decreased with increasing irradiation

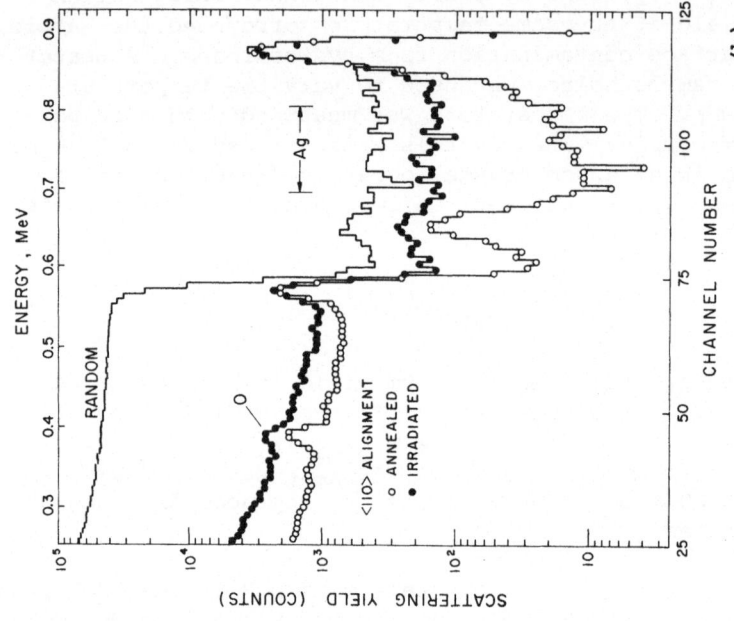

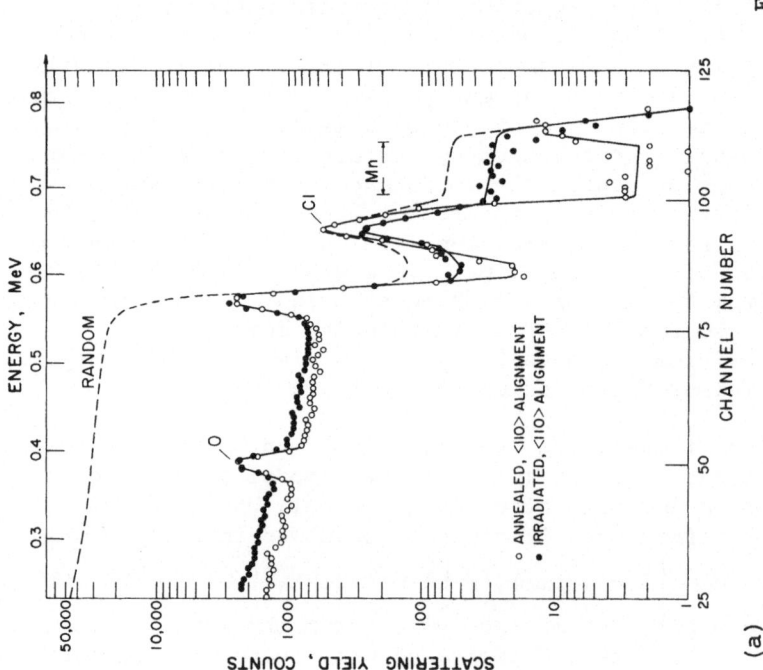

Figure 1

Typical energy spectra of backscattered 1 MeV He$^+$ ions from Al crystals at 35°K, showing the <110> aligned spectra for annealed and irradiated crystals, as well as random spectra. Fig. 1(a): Al–0.09 at.% Mn, irradiation dose 10^{16} 1MeV He$^+$ ions/cm^2 at 35°K plus a 10 min. anneal at 60°K. The irradiation increased (χ_{min})Al from 0.024 to 0.029 and (χ_{min})Mn from 0.039 to 0.48. Fig. 1(b): Al–0.08 at.% Ag, irradiation dose 3×10^{15} 0.3 MeV He$^+$ ions/cm^2 at 60°K. The irradiation increased (χ_{min})Al from 0.020 to 0.036 and (χ_{min})Ag from 0.034 to 0.39.

TABLE 1

Backscattering Yields of 1 MeV He$^+$ Ions in a <110> Direction
at 35°K for Annealed and Irradiated Al-0.09% Mn and
Al-0.08%Ag Crystals

Irrad. Dose (10^{15} ions/cm^2)	Irrad. Energy (MeV)	Irrad. Temp. (°K)	$(\chi_{min})_{Al}$ <110>	$(\chi_{min})_i$ <110>	C from Eq. (1) (%)
Al-0.09% Mn					
0			0.024	0.039	99
2.8	1.0	35	0.025	~0.1	~90
6.1	1.0	35	0.031	~0.3	~75
0			0.022	0.068	95
7.8	0.3	65	0.049	0.59	43
Al-0.08% Ag					
0			0.020	0.027	99.5
0.5	1.0	40	0.021	0.050	97
1.5	1.0	40	0.021	0.079	94
2.5	1.0	40	0.023	0.104	91.7
3.5	1.0	40	0.026	0.140	88.3
4.5	1.0	40	0.029	0.152	87.3
0			0.020	0.024	99.5
0.5	1.0	60	0.020	0.103	91.5
1.5	1.0	60	0.020	0.19	82.5
2.5	1.0	60	0.025	0.26	76
3.5	1.0	60	0.025	0.33	69
0			0.020	0.034	99
3.0	0.3	60	0.036	0.39	63

with 0.3–1 MeV He$^+$ ions at 35–65°K. If impurity atoms are displaced far enough into the channels to give the random yield, then
C corresponds approximately to the concentration of substitutional
impurity atoms, and 1 − C equals the concentration of displaced
impurity atoms. This assumption is shown to be valid from the
measurements of channel widths which will be discussed later.

The recovery of ΔC, the irradiation-induced change in C,
during 10 min. isochronal annealing is shown in Figure 2 for both
Mn and Ag atom displacements. The concentration of displaced Mn
and Ag (not shown in Figure 2) atoms increased during annealing
from 40–60°K. This is strong evidence that the impurity atoms
were displaced by trapping irradiation-induced Al interstitial
atoms, since self-interstitial Al atoms migrate freely from 40–50°K
(16). It was observed also that an irradiation at 60°K produced
more displaced Mn or Ag atoms than an equal irradiation at 40°K,
followed by an anneal at 60°K, since irradiation-induced defects
at 40°K provided competitive sinks for trapping of self-interstitial
Al atoms. The trapping of Al interstitials by impurities during
irradiation at 35–40°K is assumed to be largely due to Al atom
migration in Stage I$_D$, occurring near 35°K(16). The direct

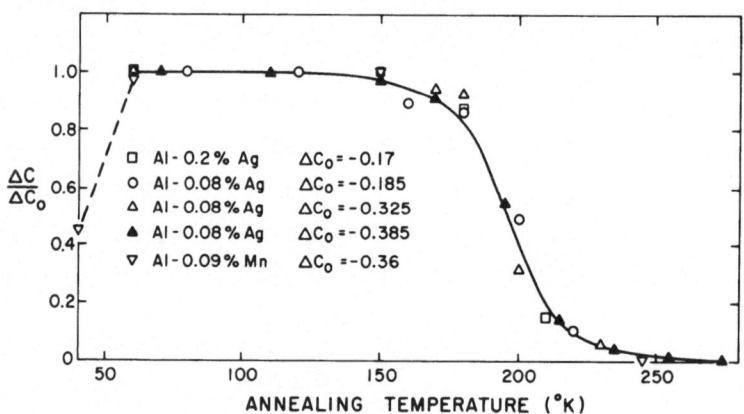

Figure 2
Recovery of the concentration of displaced impurity atoms during
isochronal annealing (10 min. pulses). All measurements were taken
at ≈35°K and refer to backscattering of 1 MeV He$^+$ ions in a <110>
direction for Al-0.2 at.% Ag and for Al-0.08 at.% Ag, and in a <111>
direction for Al-0.09 at.% Mn. The recovery is expressed in terms of
ΔC/ΔC$_o$, where ΔC$_o$ is the initial irradiation-induced change in the
fraction C of shadowed impurity atoms, and ΔC is that part of ΔC$_o$
which remains after annealing at a given temperature. ΔC/ΔC$_o$ is
normalized to the 60°K value.

displacement of impurity atoms during irradiation was very
unlikely since <0.1% of lattice atoms were directly displaced
during the irradiation.

Almost complete recovery of the concentration of displaced
Mn and Ag atoms occurred during annealing from 180-220°K, as shown
in Figure 2. Also, after these anneals, the <110> aligned
spectra were identical to the pre-irradiation spectra. These
results indicate that self-interstitial Al atoms were released
from the impurity traps during this recovery stage, and annihilated
vacancies created by the irradiation. (Alternatively, the
annihilation of trapped interstitials by migrating vacancies
cannot be excluded.) Thus almost all point defects created by the
irradiation were removed, and dechanneling resulting from these
defects was eliminated. In this case, it thus appears that most
of the Al dechanneling was due to simple irradiation-induced
defects. The concentration of such defects was large, approximately 0.1%, and relaxation of their near Al neighbours apparently
caused the observed dechanneling from Al atoms.

In the case of Al-0.09% Mn, measurement of the angular
dependence of backscattering yields from Al and Mn atoms near
<110> and <111> axes, as well as the measurement of {100} planar
channeling, indicated strongly that the displacement of Mn atoms
by irradiation was in <100> directions. Thus the Mn-interstitial
Al atom configuration was the <100> dumbbell.

In Figure 3(a) an angular scan through a <110> direction is
shown for Al-0.09% Mn. The width of the dip in backscattering
yield from Mn atoms is the same as the width of the Al dip. This
result indicates that the Mn atoms responsible for the dip were
in perfect lattice sites, because a narrowing of the channeling
dip would be observed if all the Mn atoms were displaced a small
amount. The presence of a narrow peak in the Mn yield (having a
peak to valley ratio of 1.5) at exactly the <110> direction shows
conclusively that some Mn atoms were displaced far enough into the
channel to reach the flux peaking region. If each displaced Mn
atom was associated with an Al self-interstitial atom to form a
<100> dumbbell configuration, the displacement of the Mn atoms
would be ∿1.3 Å (assuming equal atom spacing in <100> directions).
Such a displacement would give a calculated ratio of 1.5 for peak
to valley scattering yields from Mn atoms in the <110> channel,
which is in agreement with our results. This calculated value is
based on an approximation for the flux distribution in a channel(5),

$$F_i = \ln (A_o/A_i). \qquad\qquad ....(2)$$

Here A_o is the channel area, and F_i is the normalized flux at the
edge of an area A_i enclosed by an equipotential contour. In the

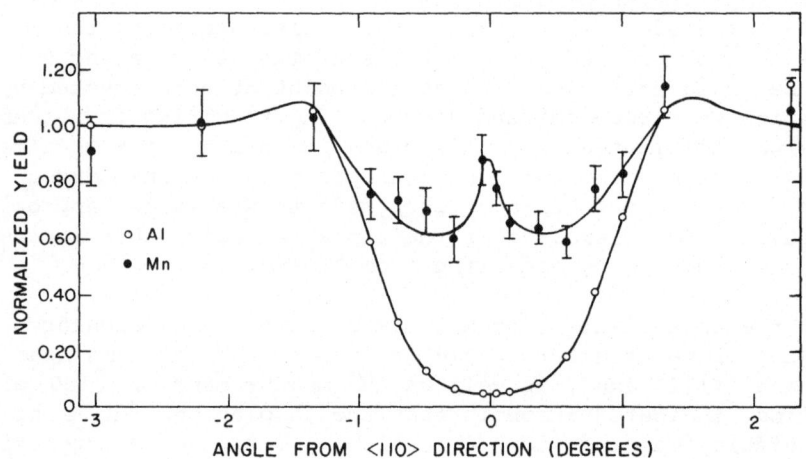

Figure 3(a)

Angular dependence of the normalized yield of backscattered He+
ions (incident energy 1 MeV) from Al and Mn atoms in an Al-0.09
at.% Mn crystal after irradiation at 65°K with ~10^16 0.3 MeV
He+ ions/cm^2. The yield is plotted as a function of the angle
from the [110] direction near the (112) plane. The backscattering
yield was measured from a depth of 1000 Å (for Mn this was an
average from 500-1500 Å or 12 channels).

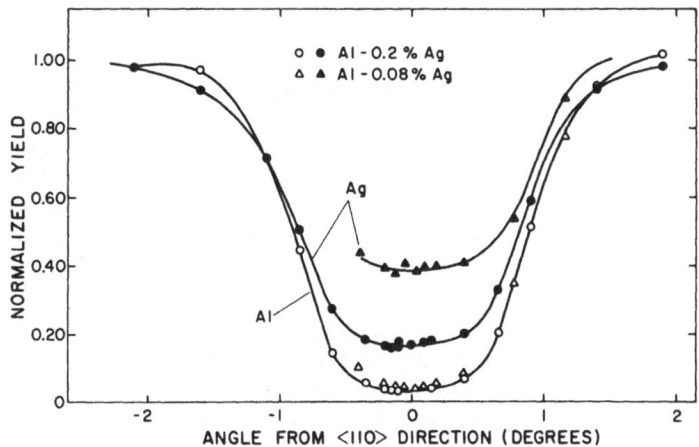

Figure 3(b)

Angular dependence of the normalized yield of backscattered He[+]
ions (incident energy 1 MeV) from Al and Ag atoms in an Al-0.2
at.% Ag crystal after irradiation with ~4 x 10[15] 1 MeV He[+] ions/
cm[2] at 60°K, and in an Al-0.08 at.% Ag crystal after irradiation
with ~4 x 10[15] 0.3 MeV He[+] ions/cm[2] at 60°K. The yield is
plotted as a function of the angle from the $[\overline{1}1\overline{0}]$ direction near
the $(2\overline{2}\overline{1})$ plane. The backscattering yield was measured from a
depth of 2000 Å (for Ag this was an average from 1200-2800 Å or
16 channels).

derivation of Eqn. (2), multiple scattering is neglected(22).   In
addition, it is assumed that statistical equilibrium of ion positions in a channel has been reached, which is not accurate for
small depths(22,23).   Calculations have shown that there is a depth
dependence of the flux at the center of a channel which is most
pronounced for depths less than 1000 Å(22).   Since our values for
Mn yields are an average from ∿500-1500 Å, this depth dependence
will be a minor effect.   In any case, it is expected that the
experimental error in determining the peak height of the Mn yield
in the <110> channel is greater than the error in calculation of
the height by this means.   The important point to be noted is that
the presence of this peak in Mn yield demonstrates that the Mn
atom displacement is large, almost certainly greater than 1 Å.

     As shown in Figure 3(b), an angular scan through the <110>
direction showed no peak in the Ag yield.   The width of the Ag dip
was the same as that of the Al dip, indicating that a certain
fraction of Ag atoms were displaced a large amount (∿1 Å), rather
than 100% being displaced a smaller amount (<0.5 Å).

     In order to demonstrate the existence of irradiation-induced
Al-Mn and Al-Ag <100> dumbbells, it remains to be shown that the
displacement of the Mn and Ag atoms is in <100> directions, but that
it is not sufficient to displace the atoms into body-centered
positions.

     If impurity atoms were displaced into b.c. positions, they
would be entirely shadowed in <111> directions.   As shown in
Figure 4 for Al-0.09 at.% Mn, the observed shadowing of Mn atoms
was only 63% for a dose of ∿10^{16} 0.3 MeV He^+ ions/cm^2.   This dose
gave 43% shadowing in a <110> direction, for the extrapolated
minimum yield of Figure 3(a).   The larger amount of shadowing in a
<111> direction could be explained by 20% of the Mn atoms being
in body-centered positions, but is more simply explained by a
unique Mn atom displacement of ∿1.3 Å in <100> directions, which
moves Mn atoms from one <111> string close to a neighbouring
string (within 0.6 Å).   The expected narrowing of the <111>
channel for Mn atoms which are displaced in this way could not be
verified from our results (Figure 4), because of large statistical
errors in counting.   In the case of Al-0.08 at.% Ag, the shadowing
of Ag atoms in <111> directions was almost the same as in <110>
directions, which is consistent with Ag atom displacements of
∿1 Å in <100> directions.

     In a f.c.c. lattice, the existence of impurity displacements
in <100> directions can be demonstrated by measurement of {100}
planar channeling.   Since no more than 1/3 of impurity atoms which
are displaced in <100> directions are in a given {100} planar
channel (taking into account the possibility of displacements into

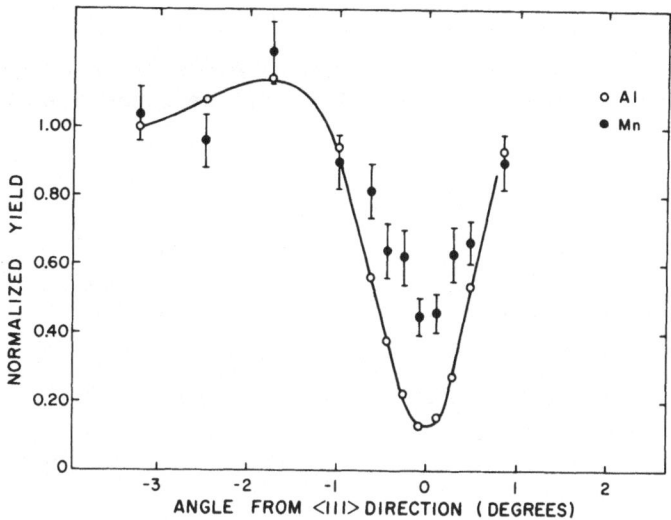

Figure 4

Angular dependence of the normalized yield of backscattered He$^+$ ions (incident energy 1 MeV) from Al and Mn atoms in an Al-0.09 at.% Mn crystal after irradiation at 65°K with ~10^{16} 0.3 MeV He$^+$ ions/cm^2. The yield is plotted as a function of the angle from the $[\overline{11}]$ direction near the $(\overline{13}\overline{2})$ plane. The backscattering yield was measured from a depth of 1000 Å (for Mn this was an average from 500-1500 Å or 12 channels).

b.c. positions), the shadowing along {100} planes will be greater than or equal to the concentration of substitutional impurity atoms plus 2/3 of the concentration of atoms displaced in <100> directions. For the Al-0.09 at.% Mn alloy, an irradiation which gave 69% Mn atom shadowing in a <110> direction showed 92% shadowing in a {100} plane (at an angle of 5° from a <110> direction), in agreement with the above formula (69% + 2/3 x 31%). This result shows that the Mn atom displacements were in <100> directions. It will be noted that 2/3 of <110> displacements lie in a given {100} channel, and that all <111> displacements lie in a given {100} channel. A similar result was obtained for Ag displacements, thus indicating that both Mn and Ag atoms are displaced to form half of a <100> dumbbell interstitial when they trap self-interstitial Al atoms.

ACKNOWLEDGEMENTS

The encouragement and advice of the entire channeling group at CRNL, and especially J.A. Davies, I.V. Mitchell and J.L. Whitton, is greatly appreciated.

REFERENCES

1. E. Bøgh, in "Interaction of Radiation with Solids" (Ed. A. Bishop) Plenum Press, New York, 1967, p. 361.

2. L. Eriksson, J.A. Davies, J. Denhartog, H.-J. Matzke and J.L. Whitton, Can. Nucl. Tech. $\underline{5}$, 40 (1966).

3. J.W. Mayer, L. Eriksson, J.A. Davies, "Ion Implantation in Semiconductors", Ch. 4, Academic Press, New York (1970).

4. J.A. Davies in "Channeling in Solids:" Ch. 11 (Ed. D.V. Morgan) John Wiley and Sons, New York (1973).

5. J.U. Andersen, O. Andreasen, J.A. Davies and E. Uggerhøj, Radiation Effects $\underline{7}$, 25 (1971).

6. B. Domeij, G. Fladda and N.G.E. Johansson, Radiation Effects $\underline{6}$, 155 (1970).

7. E.N. Kaufmann, J.M. Poate, and W.M. Augustyniak, Phys. Rev. B $\underline{7}$, 951 (1973).

8. J.U. Andersen, E. Laegsgaard and L.C. Feldman, Radiation Effects $\underline{12}$, 219 (1972).

9. L.C. Feldman and D.E. Murnick, Phys. Rev. B $\underline{5}$, 1 (1972).

10. R.B. Alexander and J.M. Poate, Radiation Effects $\underline{12}$, 211 (1972).

11. A.C. Damask and G.J. Dienes, "Point Defects in Metals", Gordon and Breach, New York 1963.

12. C. Ceresara, T. Federighi and F. Pieragostini, Phys. Letters $\underline{6}$, 152 (1963).

13. C. Dimitrov-Frois and O. Dimitrov, Mem. Sci. Rev. Met. $\underline{65}$, 425 (1968); C.R. Acad. Sc. Paris $\underline{266}$, 304 (1968).

14. A. Sosin in "Lattice Defects and Their Interactions" (Ed. R.R. Hasiguti), Gordon and Breach, New York, 1967 (p. 235).

15. R. Brugière and P. Lucasson, Radiation Effects <u>11</u>, 55 (1971).

16. W. Schilling, G. Burger, K. Isebeck and H. Wenzl in
     "Vacancies and Interstitials in Metals" (Eds. A Seeger,
     D. Schumacher, W. Schilling, J. Diehl), North-Holland,
     Amsterdam 1970 (p. 255).

17. J.B. Gibson, A.N. Goland, M. Milgram and G.H. Vineyard,
     Phys. Rev. <u>120</u>, 1229 (1960).

18. K.R. Garr and A. Sosin, Phys. Rev. <u>162</u>, 669 (1967).

19. H.J. Wollenberger in "Vacancies and Interstitials in Metals"
     (Eds. A. Seeger, D. Schumacher, W. Schilling, J. Diehl),
     North-Holland, Amsterdam 1970 (p. 215).

20. J.A. Davies, J. Denhartog and J.L. Whitton, Phys. Rev. <u>165</u>,
     345 (1967).

21. J. Bøttiger, J.A. Davies, J. Lori and J.L. Whitton, Nucl.
     Instr. & Methods <u>109</u>, 579 (1973).

22. D. Van Vliet, Radiation Effects <u>10</u>, 137 (1971).

23. F.H. Eisen and E. Uggerhøj, Radiation Effects <u>12</u>, 233 (1972).

DISCUSSION

Q: (W. L. Brown) What fraction of the interstitials created by
the He radiation are trapped at Mn sites?

A: The fraction of self-interstitials trapped by impurities depends on the irradiation dose and temperature; i.e., it depends on
the ratio of impurities to vacancies or other sinks. At our low
doses, most of the free interstitials created by irradiation at
$\geq$ 60 K are trapped, whereas at our higher doses, perhaps half of
the free interstitials or one-fifth of all interstitials are
trapped.

LATTICE LOCATION STUDIES OF ^{2}D and ^{3}He in W[*]

S. T. Picraux and F. L. Vook

Sandia Laboratories

Albuquerque, New Mexico 87115

ABSTRACT

Direct determinations of the lattice locations of implanted hydrogen and helium isotopes in tungsten single crystals have been made for the first time by means of ion channeling and ion induced nuclear reactions. Channeling angular distribution measurements along the ⟨100⟩ axial and {100} and {110} planar directions indicate that the implanted ^{2}D occupies the tetrahedral interstitial site in W. Similar measurements indicate that the implanted ^{3}He atoms are trapped at lattice vacancies in configurations consisting primarily of three helium atoms trapped at a vacancy with a possible smaller component consisting of two helium atoms trapped at a vacancy. The channeling data agree with independent calculations by Bisson and Wilson of the lattice locations of the He atoms in a vacancy.

I. INTRODUCTION

The behavior of hydrogen and helium in metals has both important scientific and technological consequences. The nature of hydrogen in metals is an important problem in solid state research, and theoretical work is in progress to obtain a unifying understanding of the stability and electronic and atomic structure of hydrogen in metals and metal hydrides over a wide concentration range.[1]

[*]This work was supported by the U. S. Atomic Energy Commission.

407

Particularly relevant are transition metals for which very little
information exists. The technological importance of hydrogen in
metals includes hydrogen embittlement at low concentrations and
hydrogen storage in hydrides at high concentrations; both problems
are related to energy usage in the transportation and storage of
hydrogen. In addition, the first wall structural material for
controlled thermonuclear reactors and the fuel cladding for fast
breeder reactors will include transition metals that will be subject
to large fluences of hydrogen and helium. It is for these reasons
that the behavior of hydrogen and helium in transition metals is
important.

Very little data exist concerning the solubility and diffusivity
of hydrogen and helium in tungsten. However, values for hydrogen are
known to be much lower in W than for most metals. Frauenfelder[2] has
investigated the solubility and diffusivity of hydrogen in tungsten
between 1100 and 2400 K. He found the diffusion constant to be
$D = 4.1 \times 10^{-3} \exp(-9000/RT) cm^2/sec$. The diffusion constant was
previously found to be orders of magnitude smaller by Moore and
Unterwald[3] who reported $D = 7.25 \times 10^{-4} \exp(-41,500/RT) cm^2/sec$ for
temperatures between 1200 and 2500 K. Extrapolations of these
values to room temperature give diffusion constants of 10^{-9} and
$10^{-34} cm^2/sec$, respectively. The solubility constant[2] extrapolated
to room temperature and one atmosphere is approximately 3×10^{-20}
atom fraction. Recent measurements[4] above 1170 K give a much different extrapolated value of 10^{-7} atom fraction. Because of the low
solubility at room temperature there are no data on the lattice location of H in W. Insufficient H can be dissolved in W to use either
NMR or neutron diffraction techniques for lattice location. In
addition, no data exist on the solubility of He in W. The only
diffusivity data for He in W are the results of Kornelson[5] which
indicate that He implanted into W undergoes rapid interstitial
diffusion unless it is trapped by defects postulated to be vacancies.

Relatively few ion beam analysis studies of the behavior of
hydrogen and helium in solids have been reported. Recently, the
$D(d,p)T$ nuclear reaction has been used for channeling lattice location studies of D in niobium, and the results were interpreted to
indicate tetrahedral interstitial site occupation.[6] However, for
cases where the crystal is not uniformly loaded with D this technique
suffers from D buildup at the end of range which results in an
additional orientation dependent signal due to reduced energy loss
rates for channeled ions and the energy dependence of the reaction
cross section. More recently the possibility of profiling D and ^{3}He
in solids using the $^3He(d,p)^4He$ reaction has been investigated.[7] We
find this latter nuclear reaction to be well suited for lattice location studies of both D and ^{3}He in solids.

II. EXPERIMENTAL TECHNIQUE

Electropolished tungsten (W) of 99.999% purity, obtained from Materials Research, was used for these experiments. The ^{2}D or ^{3}He was introduced by ion implantation along nonchanneling directions 7° from the ⟨100⟩ axis. Thirty keV D was introduced by accelerating the molecular species D_3^+ to 90 keV; ^{3}He$^+$ implants were done at 60 keV. Except where otherwise noted, implants were performed at 296 K to fluences of 3 x 10^{15} D atoms and 1 x 10^{15} He atoms/cm^2. The projected range for 30 keV D$^+$ is 1270 Å and for 60 keV ^{3}He$^+$ is 1100 Å.[8] If all the implanted ^{2}D and ^{3}He were retained, range distribution calculations would predict concentrations of ≈ 1.5 x 10^{20} and 5 x 10^{19} atoms/cm^3, respectively. Details of the in situ ion implantation and ion beam analysis system have been given previously.[9]

Ion channeling analysis was performed by monitoring the ion backscattering from the W and by using the nuclear reaction ^{3}He(d,p)^{4}He to monitor the implanted D or ^{3}He. This nuclear reaction has a Q value of 18.3 MeV and a peak cross section ~ 70 mb/sr at a D beam energy of 430 keV.[10] For D analysis we used a 750 keV ^{3}He$^+$ beam, and for ^{3}He analysis we used a 500 keV D$^+$ beam; these energies correspond to equivalent projectile momentum at energies slightly above the peak in the reaction cross section. A 300 mm^2 surface barrier detector with 500 μm depletion depth was located at a scattering angle ≈ 135° with a solid angle ≈ 0.1 sr. The detector was covered with a 12 μm aluminized mylar foil to absorb the backscattered ions while allowing the more energetic nuclear reaction products to pass through the foil to the detector. Several small holes were introduced into the foil to allow simultaneous monitoring of the backscattered yield from the W crystal. Single channel analyzers were set so as to accept the W signal from a depth corresponding to the range of the implanted impurity. In the case of the ^{3}He location studies, the ^{4}He recoils as well as the proton nuclear reaction products were detected.

An analysis beam spot of 0.9 mm diameter was used, and the beam collimation had a maximum full angle divergence of less than 0.06°. All analyses were performed at 296 K, except in one case where the ^{3}He location measurement was made at 105 K. The analysis beam was moved laterally to different spots on the implanted W crystal in all samples, and the beam was not observed to influence the impurity lattice location or to induce migration of the implanted impurities. After sufficiently long analysis on a given spot, however, the flux peaking structure was observed to wash out, and the W channeling yield increased slightly. This was observed for both ^{3}He and D analysis beams and is attributed to blistering as indicated by the surface appearance under the microscope. Typical threshold beam fluences for this to happen were of the order of $10^{17} - 10^{18}/cm^2$, consistent with blistering studies of the similar transition metal. Mo.[11] The blistering in effect simply washes out the single crystal alignment, similar to our observations for a crystal with mosaic spread.

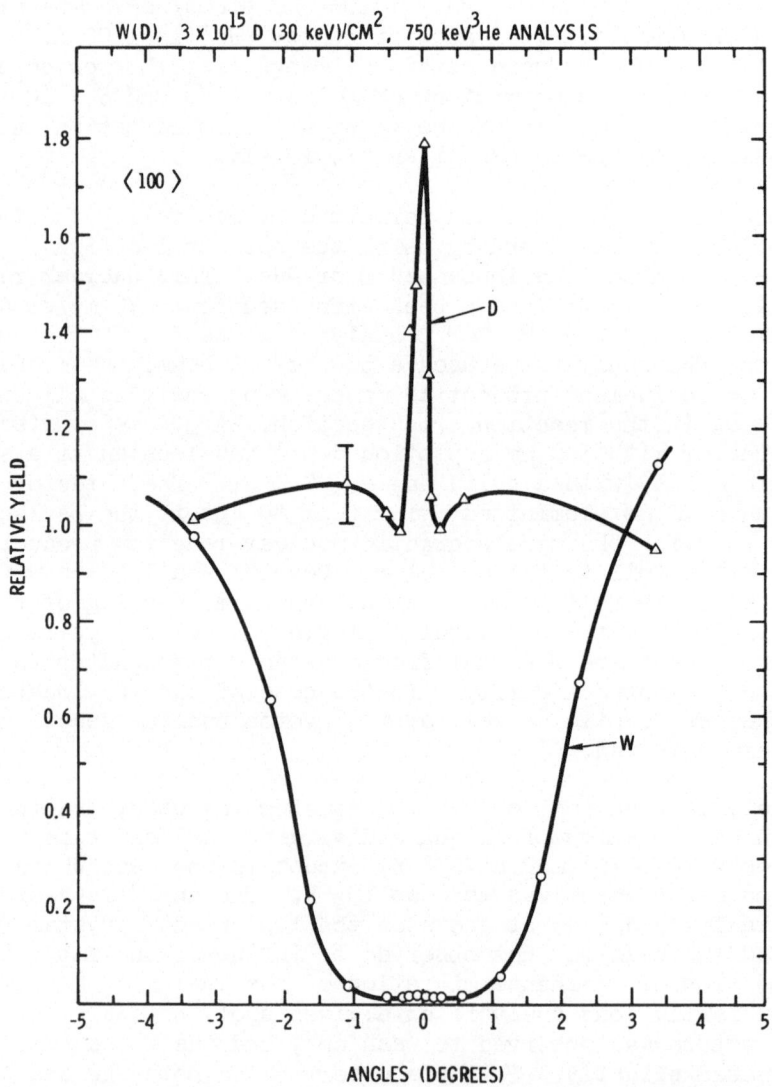

Fig. 1. Angular scan through the $\langle 100 \rangle$ axis for W with $3 \times 10^{15}/cm^2$, 30 keV D implant at 296 K using a 750 keV ^{3}He analysis beam: O corresponds to backscattered yield from W and Δ corresponds to proton nuclear reaction yield from D.

III. RESULTS AND DISCUSSION

A. D Location

Results for a ⟨100⟩ axial scan for D implanted into W are shown in Fig. 1. The D signal is given by the open triangles and the W signal by the open circles. The random W level was established by other data points (not shown) and the normalized yield of one corresponds to 14,500 counts for W and 180 counts for D. The W minimum yield observed of 1.5% is typical for all of these measurements. The strong enhancement in the yield (flux peak) from the D is seen in Fig. 1 with a full angular width of 0.2° and a maximum relative yield of 1.8. The W critical angle is observed to be 2.0°. Planar channeling data are shown for the same system for the {100} plane in Fig. 2a and for the {110} plane in Fig. 2b. For the {100} plane a dip of the order of 25% is observed with an indication of a flux peak of the order of 0.2° angular width in the center of the dip. The angular width of the dip due to those D atoms that lie in the {100} planes is narrower than for the W host. The reason for this is not quantitatively understood but some narrowing would be anticipated due to the larger vibrational amplitude of the D relative to the W. For the {110} planar data there is no obvious structure in the D signal, although there is an indication of a small enhancement of the yield along this planar direction.

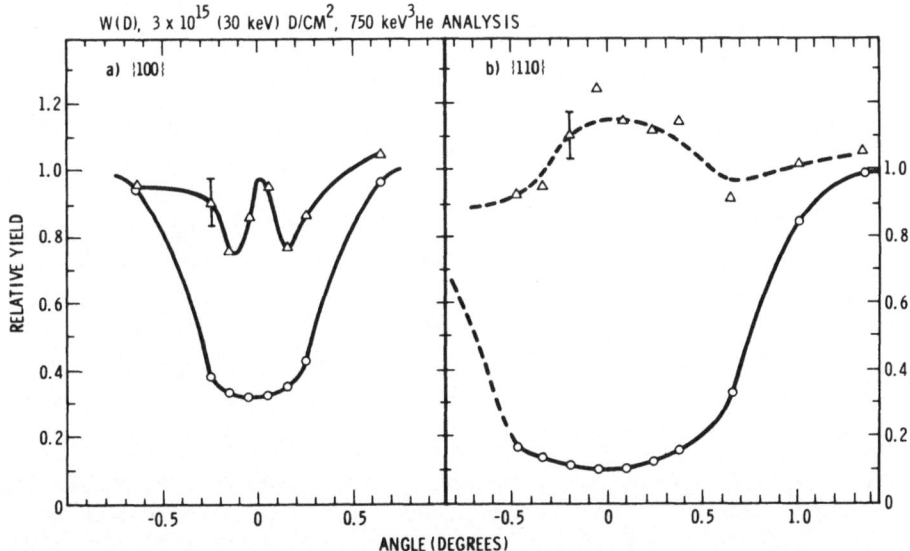

Fig. 2. Planar angular scans for a) {100} and b) {110} with same
        conditions as for Fig. 1.

    The large flux peak seen in Fig. 1 indicates that the D atoms
are located in an interstitial site between the lattice rows,[12] as
might be expected for a hydrogen isotope in a metal.[13] The two most
likely interstitial sites for the bcc lattice of W are the tetrahedral site located at (1/2, 1/4, 0) in the unit cell and the octahedral site located at (1/2, 1/2, 0) or equivalently at (0, 0, 1/2).
The relative positions with respect to the lattice rows and planes
for the three channeling directions measured in Figs. 1 and 2 are
shown in Fig. 3. The interstitial position is shown by the square
boxes and the relative probability of occurrence by the number in
the box. The greatest distinction between these two interstitial
sites is given by the {100} planar channeling direction. The octahedral site is always contained within a {100} plane for all equivalent positions, whereas the tetrahedral site lies within {100}
planes two-thirds of the time and is located centrally between the
{100} one-third of the time. Thus, for octahedral site occupation
the D signal should show the same channeling dip as that for the W
lattice, whereas for the tetrahedral site the D dip should be ~
two-thirds the W dip but with a flux peak superimposed within the
center of that dip.

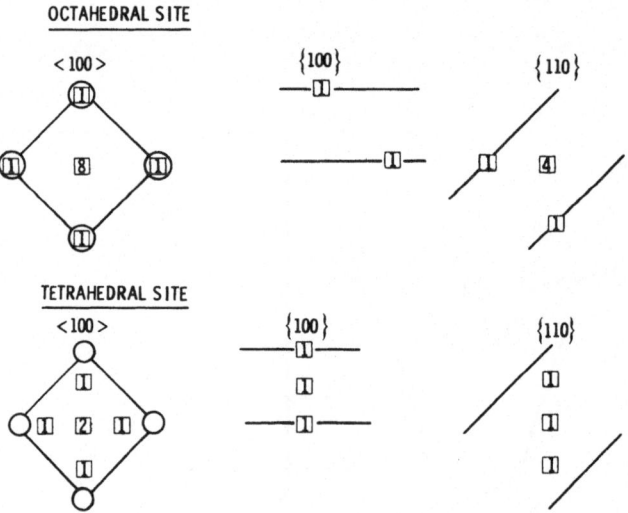

Fig. 3. Relative position of octahedral and tetrahedral sites in
         bcc lattice. Circles represent crystal rows normal to
         plane of the figure, lines represent crystal planes and
         squares represent interstitial positions with the numbers
         giving the relative probability of occurrence for the
         portion of the unit cell shown.

As seen in Fig. 2a, the experimental data are clearly inconsistent with the octahedral site but consistent with the D located in the
tetrahedral interstitial site. The ⟨100⟩ axial data are also consistent with the tetrahedral site since for this site there should be a
strong central flux peak from one-third of the D atoms which would be
located in the center of the channel. Also, there would be no dip
except perhaps a very narrow one near the center of the axis, since
the remaining equivalent sites are displaced 0.8 Å from lattice rows.
Whereas, for the octahedral site one third of the equivalent sites
are directly along the ⟨100⟩ lattice rows; and thus there would be
a 33% dip with angular width similar to that for the W lattice, in
addition to the central flux peak. The {110} planar results are
also consistent with the tetrahedral location, although channeling
information from this planar direction is less definitive. Thus,
we conclude from these implantation and lattice location measurements
that dilute D is located in the tetrahedral interstitial site in bcc
tungsten.

Another result observed for D implants in W is a greater D concentration at the beam spot where the W crystal was prealigned by
the 750 keV ^{3}He beam before D implantation. A 50% larger nonchan-
neled yield was observed (not shown) over that from the surrounding
implanted region. This enhanced fraction of the yield did not exhibit ⟨100⟩ channeling structure but rather appeared as a constant
fraction added to the D signal. While this effect is not entirely
understood, it suggests that prebombardment damage by the helium beam
resulted in trapping of D during implantation (≤ 33% of the implanted
fluence) which otherwise escapes from the surface of the crystal
during room temperature implantation. The additional trapped component does not exhibit either a simple substitutional or well-defined
interstitial location. The D signal from the prebombardment spot
was consistent with absolute calculations based on the reaction
cross-section and the measured implant fluence, although the accuracy
(±30%) of this determination was not sufficient to be certain that
the prebombarded spot represents 100% of the implant fluence.

These results suggest this technique may also be a sensitive
way to study the stabilizing influence of damage on the microscopic
behavior of D in metals in the low concentration regime. In addition, such lattice location measurements should be quite sensitive
to the microscopic nucleation of D gas bubbles at higher concentrations.

B. He Location

The results of ⟨100⟩ axial channeling measurements for 60 keV
^{3}He implanted to a fluence of 1 x 10^{15}/cm^2 are shown in Fig. 4a.
The W yield is given by the open circles and the ^{3}He yield by the
open triangles. For the ^{3}He yield the zero has been offset and a

S. T. PICRAUX AND F. L. VOOK

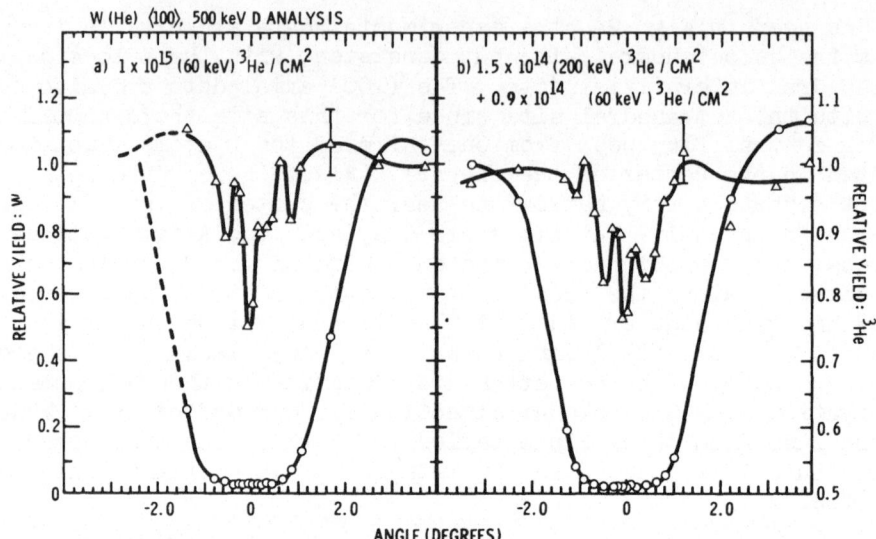

Fig. 4.   Axial ⟨100⟩ angular scans using 500 keV D analysis beam
          for W implanted with ³He under conditions indicated at
          top of curves.

x2 magnification has been used so that the detailed structure may
be better seen in the figures. The angular scan gives an indication
of fine structure in the ³He signal which we will refer to as flux
peaks superimposed on a narrow dip. To gain some confidence that
this structure is indeed present, we have repeated the ⟨100⟩ axial
scan experiments seven different times under four different experimental conditions. In Fig. 4b we show the scan for a broader and
lower fluence implant at multiple energies. The 200 keV implant at
$1.5 \times 10^{14}/cm^2$ and 60 keV implant at $0.9 \times 10^{14}/cm^2$ correspond to a
factor of 10 lower helium concentration in the implanted region
according to LSS range distribution calculations.[8]

       Additional results are shown in Fig. 5a where, after a room
temperature implant to a fluence of $1 \times 10^{15}/cm^2$ at 60 keV, the
angular scans were measured at 105 K. By cooling from 296 to
105 K, the W atom rms vibrational amplitude transverse to the ⟨100⟩
row, ρ_{xy}, is reduced from ≈ 0.07 to $\simeq 0.05$ Å for a Debye characteristic
temperature of $\Theta_W \simeq 300$ K. This cooling results in an increased W
critical angle. For He atoms bound with the same force constant in
the harmonic well approximation $\Theta_{He} \simeq \Theta_W (m_W/m_{He})^{\frac{1}{2}}$ where m is the
mass; only a slightly larger rms ³He vibrational amplitude $\rho_{xy} \simeq$
0.11 Å is predicted at 296 K compared to the W lattice. This value
does not change appreciably on cooling to 105 K ($\rho_{xy} \simeq 0.10$ Å).

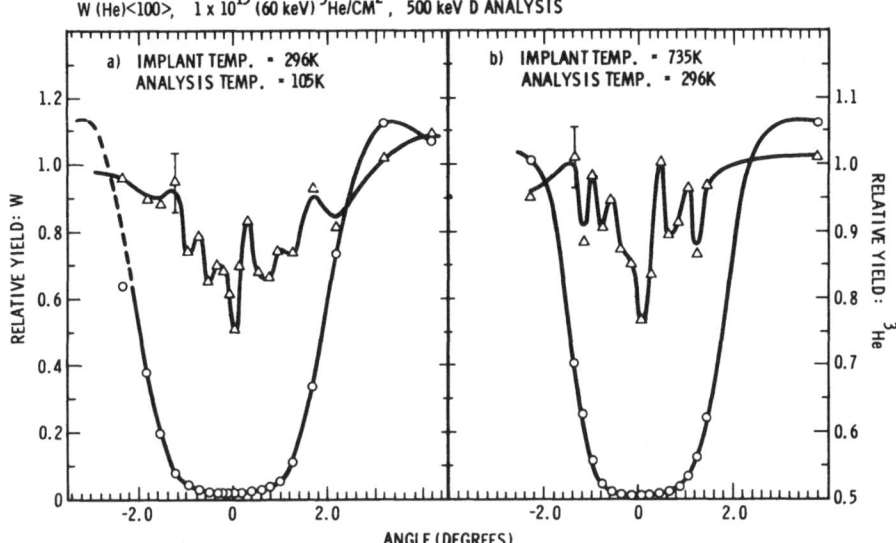

Fig. 5.   Axial ⟨100⟩ angular scans as in Fig. 4 for indicated
          conditions.

However, if hopping of the He atoms to equivalent sites over a moderately high energy barrier were a limiting factor on the He
channeling structure, then improvement in resolution might be
expected upon cooling. As seen in Fig. 5a, the observed angular
scan remains similar to that for 296 K measurements. In Fig. 5b
we show a ⟨100⟩ angular scan at 296 K after a 60 keV, 1 x 10^{15}/cm^2
implant at 735 K. This hot implant corresponds to a position in
temperature approximately half way down stage III annealing in
tungsten. Again, the angular scan results are similar.

    Planar angular scans (not shown) did not reveal any well-defined
structure for the ^{3}He signal. The {100} planar scan frequently
gave an indication of a shallow dip of magnitude the order of 10%
and width the order of the W dip. The {110} planar angular scans
did not reveal any resolvable structure.

    For all the ⟨100⟩ axial scans the following features are ev-
ident: (1) a narrow central dip of 23 to 25%, with a full-width
of the central region ≈ 0.2°; (2) a flux peak to either side of
this central region located on the average $\psi_{m1} \approx 0.27°$ from the
center; and (3) a weaker second flux peak spaced $\psi_{m2} \approx 0.8°$ from
the center of the dip. These results are summarized in Table I
for the four ⟨100⟩ scans shown.

TABLE I

Data from $\langle 100 \rangle$ angular scans

Shown in Fig.	Implant Fluence cm^{-2}	Implant Temp. K	Analysis Temp. K	W $\psi_{1/2}$ (deg)	X_{min}	^{3}He ψ_{m1} (deg)	ψ_{m2} (deg)
4a	1 x 10^{15}	296	296	1.75	0.75	0.25	0.70
4b	2.4 x 10^{14}	296	296	1.65	0.77	0.22	0.82
5a	1 x 10^{15}	296	105	1.98	0.75	0.30	0.85
5b	1 x 10^{15}	735	296	1.60	0.77	0.30	0.80

The central narrow dip and the wider envelope of a dip as seen in Figs. 4 and 5 indicate some component of the He is located along or near the $\langle 100 \rangle$ rows. However, in contrast to the D location data, the He location is neither a simple interstitial nor a pure substitutional location. Flux peaks on either side of the central axial direction indicate some component of the helium is located at a position displaced from the center of the channel.

A simple first order way to relate the angular position of the peak to the displacement position from the center of the channel is given by

$$E\psi_m^2 = U(x) \quad , \tag{1}$$

where ψ_m is the angle of the peak position with respect to the $\langle 100 \rangle$ direction; E is the ion energy; and $U(x)$ is the continuum potential of the atom rows, normalized to a value of zero at the potential minimum which here corresponds to the center of the channel. This balance of the transverse kinetic energy and the potential energy can be easily calculated using a summation of Lindhard continuum potentials for the nearest neighbor rows:

$$U(x) = \sum_i \frac{Z_1 Z_2 e^2}{d} \ln [(Ca/r_i)^2 + 1] \quad , \tag{2}$$

where the distance from the center of the channel x must be related to distances from the rows $r_i = f_i(x)$, $Z_1 e$ and $Z_2 e$ are the atomic charge of projectile and target atoms, d the atom spacing along the row, a is the Thomas-Fermi screening distance and $C \approx \sqrt{3}$.

The lattice constant for tungsten is 3.16 Å and the distance from the center of the $\langle 100 \rangle$ channel to each of the four nearest neighbor rows (see Fig. 3) is 1.58 Å. Using Eq. (1), one can estimate that for [001] continuum rows a displacement distance from the center of the channel along the [010] direction (towards the nearest neighbor row) of 0.83 Å would correspond to the average observed flux peak displacement of $\psi_{ml} \approx 0.27°$, and a displacement of 1.28 Å would correspond to the second flux peak located at $\psi_{m2} \approx 0.8°$ from center. Similar estimates for the [110] direction (a direction half-way between two nearest neighbor rows forming the channel) indicate that the displacement cannot be along this direction since the potential U(x) is nowhere large enough to give the observed ψ_m values. These general constraints on the ³He location are implied directly by these data.

Additional information is available from independent calculations of He locations in W by Bisson and Wilson.[14] They calculate that a purely interstitial He atom would migrate rapidly through W, whereas vacancy trapping of single and multiple atoms results in highly stable structures. Kornelson has performed extensive detrapping measurements of implanted He from W after prebombardment damage with Kr ions.[5] His experimental results indicate that interstitial He migrates very rapidly at room temperature, and that detrapping peaks can be identified with highly stable defects composed of 1-, 2-, and 3-He atoms trapped in a vacancy. Quantitative comparisons of the energetics between the experimental measurements of Kornelson and the calculations of Bisson and Wilson show good agreement.[15]

For a single He atom trapped by a vacancy, Bisson and Wilson calculate the He to be located in the substitutional (0,0,0) position.[14] For 2-He and 3-He atoms trapped at a vacancy, stable structures are found as shown in Fig. 6. The central intersection of the

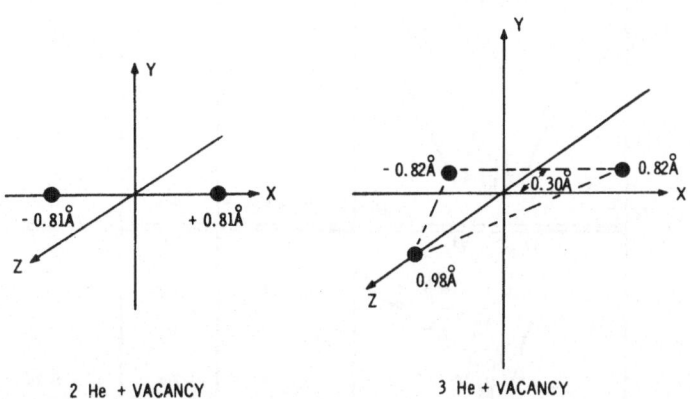

2 He + VACANCY   3 He + VACANCY

Fig. 6.   Calculated He defect positions in W for 2-He atoms trapped at vacancy and for 3-He atoms trapped at vacancy, as given in Ref. 14.

xyz axes corresponds to the substitutional position and the axes are the [100], [010], and [001] directions. For 2-He atoms trapped in a vacancy, displacements of ±0.81 Å are calculated with both He atoms lying along the same ⟨100⟩ direction. For 3-He atoms in a vacancy, the calculated positions in Å units are (0, 0, 0.98), (-0.82, 0,-0.30), and (0.82, 0,-0.30).

Semiquantative calculations of the anticipated ⟨100⟩ channeling structure expected for these multiple He vacancy trap locations were made and are given in Table II. For a given ⟨100⟩ direction the various projected displacements from the ⟨100⟩ for equivalent defect orientations are given together with the relative probability of occurrence in columns 1 and 2, respectively. In column 3 the calculated flux peak positions using Eq. (1) are given and the resulting structure is indicated qualitatively in the sketches. The He atoms which are displaced only 0.3 Å from the row in the case of the (3He + vacancy) center would give the flux peak at wide angles ≈ 0.8°, but also are sufficiently close to the rows to give a narrow central dip to the structure. This 2/9th component plus the 1/9th component which is located exactly along the ⟨100⟩ directions would result in a dip somewhat less than a 33% atom fraction, but significantly more than the 11% contribution from the 1/9th component exactly along ⟨100⟩ rows. Reference 14 also calculates a second (3He + vacancy) orientation slightly rotated from that shown in

TABLE II

Predicted ⟨100⟩ channeling structure for the
calculated He locations of Ref. 14

	Proj. Disp. from ⟨100⟩	Rel. Prob.	ψ_m
(2 He + V)	0	2/6	dip
	0.81 Å	4/6	0.24°
(3 He + V)	0	1/9	dip
	0.30 Å	2/9	0.81° + dip
	0.82 Å	2/9	0.23°
	0.87 Å	2/9	0.20°
	0.98 Å	2/9	0.16°

Fig. 6 that is of only 0.01 eV higher energy. This second orientation would not give significantly different channeling results and
so is not discussed further. However, rotation between the two
orientations could account for the observed channeling structure
not being stronger even for measurement at 105 K.

As seen from Table II, the (3He + vacancy) calculated positions
would give a channeling angular distribution structure in good
agreement with our observations (Table I and Figs. 4 and 5). Some
fractional component of the (2He + vacancy) defect, probably less
than 50%, could be present without drastically changing the angular
distribution. In addition, for these He locations the He atoms
displaced $\sim$ 0.8 to 0.9 Å from the row appear for the {100} planar
direction to be essentially in the center of the channel (1/3
component) with the remaining 2/3's component to be along or only
0.3 Å from the {100} planes. Thus, the {100} planar structure
would have a superposition of a channeling dip plus a flux peak,
which for these complex structures would be difficult to resolve
experimentally due to the narrowness of the {100} critical angle
and the greatly varying flux distribution with depth for planes.
Thus, we conclude that our data are consistent with a majority of
the implanted He atoms being trapped as (3He + vacancy) defects
with some smaller component being (2He + vacancy) defects.

Calculations of the nuclear reaction yield based on reaction
cross sections for a given analysis beam fluence are in agreement
with our implant fluences, suggesting that all the implanted ^{3}He
atoms are trapped within the implanted region near the W surface.
In order to confirm this, a calibration run was made implanting the
same ^{3}He fluence of 1 x 10^{15}/cm^2 into Pd at 100 K and analyzing the
nuclear reaction yield without warmup. This was done since the He
re-emission studies of Bauer and Thomas indicate that in excess of
80% of the He is trapped in Pd at these temperatures.[16] We observed
the same signal in this case as we did for the 296 K and the 735 K
W implants, again indicating that in all cases the He is 100% trapped
within the resolution ($\pm$30%) of these measurements.

Calculations by Brice[8] indicate that $\approx$ 2.3 keV goes into atomic
displacement processes for a 60 keV incident ^{3}He ion in W. While
defect recombination factors are not precisely known, one may estimate $\approx$ 50 interstitial-vacancy pairs produced/ion and, based on
experiments of Kornelson,[17] a recombination factor somewhere between
4 and 10. This gives $\approx$ 12 to 5 vacancies generated per incident He
ion, which would certainly be sufficient to provide complete trapping
of the He ions. It is surprising in fact that we do not observe a
larger fraction of the He atoms located substitutionally (He + vacancy) or as (2He + vacancy) defects. Even when the He concentration
was lowered from $\approx$ 5 x 10^{19}/cm^3 to $\approx$ 5 x 10^{18}/cm^3, no strong shift to
substitutional He could be distinguished. Comparing our 60 keV,

$1 \times 10^{15}/cm^2$ implants to the low energy detrapping data of Kornelson[5] in terms of calculated He range profile distributions, we would estimate that our concentrations correspond to those obtained by Kornelson for implant fluences $\approx 2 \times 10^{13}/cm^2$. At this fluence Kornelson begins to observe the dominance of the (2He + vacancy) center over the substitutional He center. However, there is some indication that with increasing He energy there is a tendency to populate more heavily (multiple He + vacancy) centers over substitutional He centers.[17] Apparently lattice location measurements at significantly lower implanted He concentrations than those currently measurable by this experimental technique are required to observe predominately substitutional He after implantation.

In summary, for the case of deuterium implanted into W, we report the first experimental determination of the lattice location of a hydrogen isotope in W, with the D found to be situated in the tetrahedral interstitial site. For the case of He implanted into W, the lattice location was consistent with the He being predominately located as a (3He + vacancy) center. This study indicates how structural information can be obtained for dilute hydrogen and helium in refractory metals and also indicates for the case of He the possibility of directly observing the structural form of He bubble nucleation.

ACKNOWLEDGMENTS

We acknowledge valuable technical discussions with W. D. Wilson, E. V. Kornelson, L. C. Feldman, and expert technical assistance by R. G. Swier.

REFERENCES

1. A. C. Switendick, Ber. Bunsenges. Physik. Chem. 76, 535 (1972).

2. R. Frauenfelder, J. Vac. Sci. Technol. 6, 388 (1969).

3. G. E. Moore and F. C. Unterwald, J. Chem. Phys. 40, 2639 (1964).

4. W. J. Arnoult and R. B. McLellan (to be published).

5. E. V. Kornelson, Rad. Effects 13, 227 (1972).

6. H. D. Carstanjen and R. Sizmann, Ber. Bunsenges. Physik. Chem. 76, 1223 (1972) and Phys. Lett. 40A, 93 (1972).

7. P. P. Pronko and J. G. Pronko, Thin Solid Films (to be published October 1973).

8. D. K. Brice (private communication), and Rad. Effects 6, 77 (1970).

9. S. T. Picraux, Rad. Effects 17, 261 (1973).

10. J. L. Yarnell, R. H. Lovberg, and W. R. Stratton, Phys. Rev. 90, 292 (1953).

11. S. K. Erents and G. M. McCracken, Rad. Effects 18, 191 (1973) and G. M. McCracken (private communication).

12. J. U. Andersen, O. Andreasen, J. A. Davies, and E. Uggerhøj, Rad. Effects 7, 25 (1971).

13. See, for example, T. Ebisuzaki and M. O'Keeffe, Prog. Solid State Chem. 4, 187 (1967).

14. C. L. Bisson and W. D. Wilson (this conference, following paper).

15. W. D. Wilson and C. L. Bisson, Rad. Effects (to be published).

16. G. J. Thomas and W. Bauer, Rad. Effects 17, 221 (1973) and private communication.

17. E. V. Kornelson (private communication).

DISCUSSION

Q: (E. N. Kaufmann) Did you see any evidence of gettering of the implanted impurity at the end-of-range damage region of the probing beam and how did your observed reaction yield compare with the dose implanted?

A: The nuclear reaction yields did not decrease with increasing analysis time. Since the reaction cross-section was peaked at the approximate implant range this indicates that the implanted impurity did not migrate to the end-of-range of the analysis beam. Preimplantation damage by the analysis beam did give rise to ~ 50% additional deuterium retention (see text) and this measured yield corresponded to that calculated for the implanted dose within absolute accuracies (± 30%).

Q: (P. P. Pronko) Your results are quite impressive and clarify some of the characteristics of helium and hydrogen implantation in metals. However, I think it is important to point out that, in

most metals, hydrogen isotopes migrate very rapidly at room temperature and may not be observable by this method. For example, in our
work on helium and deuterium profiling in metals using this same
reaction, we found that deuterium implanted in niobium was not observable by the resonance reaction. This was probably due to the
rapid redistribution of deuterium throughout the material with a
resultant volume concentration below the sensitivity limit. My
question is therefore, what characteristics of binding can you infer from the result that your deuterium remains interstitial? Do
you suppose a deuteride compound is formed?

A: Diffusion coefficients of hydrogen isotopes in Nb are 10^{-6} to
10^{-5} cm^2/sec whereas from our studies we know that $D \leqslant 10^{-15}$ cm^2/sec
in W. So naturally D migration into the bulk of Nb would occur be-
fore room temperature observation could be carried out. Such limitations can be overcome, however, by carrying out experiments in situ
at low temperatures. Deuterium also can be introduced by means other
than ion implantation. Therefore these techniques should be applicable to a wide range of H and He studies in metals. The low diffusivity for deuterium in W is consistent with bulk high temperature
measurements and the observed tetrahedral interstitial position is
consistent with thermodynamic data for dilute deuterium in the W
lattice. Electronic energy level information can be obtained by
combining crystal structure information with band theory calculations. As far as is known W hydrides do not form and hydrogen solubilities in W are low. We see no evidence for deuteride formation
or other phase changes in W.

LOCATION OF He ATOMS IN A METAL VACANCY[*]

C. L. Bisson and W. D. Wilson

Sandia Laboratories

Livermore, California 94550

Recent lattice location experiments of ^{3}He in tungsten[1] show that the equilibrium position of the implanted He atom is neither substitutional nor simple interstitial. We have previously reported binding energies for up to four He atoms in a vacancy.[2] Here we discuss the minimum energy configurations resulting from multiple-He-atoms clustering in the hope of aiding in the interpretation of these experimental results.

We used previously determined interatomic potentials[3] and allowed 643 atoms surrounding the defect complex to relax in each case. The He-He potential was taken from Beck[4] and the He atoms in the vacancy were free to relax. Firstly, we verified that a single substitutional He atom occupies the center of the vacancy and not an off-center position. Secondly, we found that two He atoms "share" the vacancy and lie along a $\langle 100 \rangle$ direction, each a distance of 0.51 r_o (r_o is the half-lattice constant = 1.58 Å in tungsten) from the central vacancy. We searched for other possible configurations by initially placing the He atoms along $\langle 110 \rangle$ and $\langle 111 \rangle$ directions, but in all cases the $\langle 100 \rangle$ configuration was obtained. In Cu, the $\langle 100 \rangle$ direction is also energetically preferred and each He lies at 0.46 r_o (r_o = 1.81 Å) from the central vacancy.

If three He atoms occupy a vacancy in a tungsten lattice, they will lie in a $\{100\}$ plane forming a nearly equilateral triangle of side $\sim r_o$ (to within 5%). Again we initialized the positions

---

[*] Work supported by the U. S. Atomic Energy Commission.

423

of the He in several widely differing ways (including along a
⟨100⟩ direction with one He atom in the vacancy) in order to estimate the uniqueness of the final configuration. A nearly equilateral triangle is always found but its exact orientation is difficult to determine with great certainty. Indeed, in Fig. 1, two configurations are shown which differ in energy by less than 0.01 eV.
We, therefore, fixed one He atom at $(0.5\ r_0,\ 0.25\ r_0,\ 0.0)$ while
allowing the other two to relax in an effort to determine the
possibility of "rotation" or rearrangement of the atoms comprising
this triangle. This position was found to lie only ∼ .01 eV above
those shown in the figure indicating that the defect complex can
quite freely rotate at liquid helium temperatures. Such a barrier
height is also suggestive of a tunneling process but the magnitude
of the barrier may well be potential dependent.

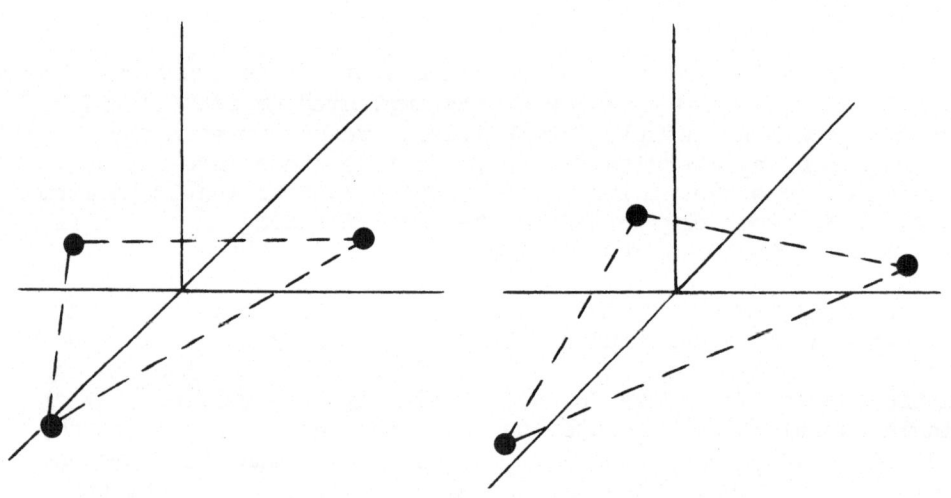

CONFIGURATION A   CONFIGURATION B

Fig. 1. Configurations which result independent of the
initial positions of the three He atoms in a Cu or tungsten
vacancy. In A, the He atoms are at (0.52, 0.19, 0.0), (-0.52, 0.19,
0.0) and (0.0, -0.62, 0.0) in tungsten and at (0.48, 0.12, 0.0),
(-0.48, 0.12, 0.0) and (0.0, -0.72, 0.0) in Cu. In B, the He atoms
are at (-0.35, 0.35, 0.0), (0.64, 0.08, 0.0) and (-0.08, -0.64, 0.0)
in tungsten and at (-0.28, 0.30, 0.0), (0.66, 0.07, 0.0) and (-0.03,
-0.62, 0.0) in Cu. These positions are in units of the half-lattice
constant, r_0. The rotational-tunneling barrier between the configurations is less than 0.01 eV.

Remarkably, the same situation occurs in copper. The three helium atoms lie in a {100} plane forming an equilateral triangle of side $\sim r_0$ and two configurations were found which are oriented precisely as in Fig. 1. The energy difference between these orientations was again found to be less than 0.01 eV. We have not yet determined the "rotational" energy of the ^{3}He in copper but it is not unreasonable to expect that it is low.

It was felt that a He atom might also lie in a divacancy in tungsten away from an exactly substitutional site to explain the experimental data. Actual calculations ruled out this possibility, however, as the He atom was found to reside within 0.015 r_0 of one of the vacancies for second neighbor or more distance divacancies. A first neighbor divacancy will force a He atom to lie 0.26 r_0 from one of the vacancies along a $\langle 111 \rangle$ line connecting them. The second neighbor divacancy containing a He atom is energetically preferred (by 0.14 eV relative to the first neighbor case).

From these calculations it seems that the simplest defect configuration consistent with the experimental results is the three helium atoms in a vacancy (He_3V) complex.

Acknowledgement

The authors are grateful to Drs. Picraux and Vook for communicating their experimental results to us prior to publication.

References

1. S. T. Picraux and F. L. Vook, private communication.
2. C. L. Bisson and W. D. Wilson, Bull.Am.Phys.Soc. 18, 447 (1973).
3. W. D. Wilson and C. L. Bisson, Rad.Effects (to be published) and references therein.
4. D. E. Beck, Mol.Phys. 14, 311 (1968).

SIMULATION OF INERT GAS INTERSTITIAL ATOMS IN TUNGSTEN

Don E. Harrison, Jr., G.L. Vine, J.A. Tankovich, and
R.D. Williams, III
Naval Postgraduate School
Monterey, California 93940

ABSTRACT

Computer simulation was used to investigate the equilibrium
positions assumed by interstitial inert gas atoms in a tungsten
crystal. Light atoms, He and Ne, occupy positions near the octahedral void. As the foreign atom mass increases the equilibrium
position moves along the $\langle 110 \rangle$ direction from the void toward the
split-interstitial position which has previously been associated
with the self-interstitial in the bcc crystal. The preliminary
studies were carried out in a perfect (cold) lattice. Subsequent
studies of Ar in a warmed lattice indicate that the interstitial
sites are not well localized. The foreign atom occupies a shallow
trough-well lying along the $\langle 110 \rangle$ axis near the tetrahedral void
position. The lattice does not force the interstitial atom toward
a predetermined site, rather the atom moves, but the surrounding
lattice atoms also shift their positions to accommodate to its
final position. The result is almost a local liquifaction effect;
as contrasted to a distorted lattice which retains some symmetry.
Presumably this behavior occurs because the bcc lattice is not
close packed. The presence of a nearby surface affects the stability of the foreign atom in the crystal. Interstitial atoms
placed in the first two (010) atomic layers did not reach stable
equilibrium positions. Reproducible sites were defined for all of
the inert gas atoms in the third atomic layer. In the third
atomic layer of a cold lattice the interstitial Ar atom binding
energy lies between 0.04 and 0.02 eV.

427

COMPUTATIONAL GENERALITIES

These computer simulation results are derived from a model which is computationally very similar to the sputtering simulations previously reported[1,2]. Although the approach is theoretically very naive, in the past it has given valuable insights into physical processes. The results reported here lead to a picture of foreign atom implantation near the surface of bcc crystals which is somewhat different from what one might predict based upon previous studies. This investigation raises questions; it does not solve problems.

Almost all of the simulations were performed with a single interstitial atom placed in a 250 atom microcrystallite of tungsten. The lattice is oriented to the (010) surface. The upward direction, toward the surface, is $\langle 0\bar{1}0 \rangle$. The binding energy studies were performed in a smaller microcrystallite whose development as a sub-set of the basic crystal will be discussed below.

These simulations are computationally very difficult. As the atom velocities become small, numerical problems associated with the integration of the atom trajectories become very evident. In particular, the numerical integrations are very sensitive to the choice of timestep, the length of the time interval used in the computations. Preliminary computations compared timesteps scaled on the maximum resultant force on any atom with timesteps based on the maximum kinetic energy of any atom. Most of the results reported here were obtained by the kinetic energy method, but a sufficient number of cases were run both ways for comparison to provide assurance that both methods lead to essentially the same results.

It is also difficult to make a definition of equilibrium which is computationally feasible and which does not require a great deal of computer time. In the interest of computer efficiency most computations were terminated after 30 timesteps if the kinetic energy of every atom was below thermal (0.025 eV). Final positions are actually uncertain to about $\pm$ 0.02 L.U. (1 L.U. = 1.58Å for W).

POTENTIALS

The two-body W-W interaction potential function is a composite constructed from a Born-Mayer repulsive potential function matched (with a cubic function) to a Morse potential based on Girifalco and Weizer parameters[3]. The parameters of the repulsive potential were developed during channeling simulations[4]. A similar potential construction is discussed in detail in ref. 2. The matching procedure places constraints upon the parameters of the repulsive function, but does not uniquely determine them. The potential function parameters are listed in Table I.

TABLE I. Tungsten Composite Potential

R < 0.9494 L.U. (1.5 Å)
    Born-Mayer Potential: $V(R) = A \exp(-BR)$
        A = 80.8 keV
        B = 7.50 L.U.$^{-1}$

0.9494 ≤ R ≤ 1.2658 L.U. (2.0 Å)
    Cubic Potential: $V(R) = C_0 + C_1 R + C_2 R^2 + C_3 R^3$
        C_0 = 3717.3
        C_1 = -9013.5
        C_2 = 7348.9
        C_3 = -2000.3

1.2658 < R ≤ 3.40 L.U.
    Morse Potential: $V(R) = D_1 \exp(-B_1 R) + D_2 \exp(-B_2 R)$
        D_1 = 5.17 keV
        D_2 = -0.143 keV
        B_1 = 4.46 L.U.$^{-1}$
        B_2 = 2.23 L.U.$^{-1}$

        The interstitial atom - W potential functions were derived by
matching Born-Mayer functions to Wedepohl potentials[5]. The match
is excellent for He and Ne with W, but the procedure becomes un-
workable for two heavy atoms. This approach leads to an Ar - Cu
potential function which agrees very closely with the function
used in the sputtering work[2], so that, subject to the usual caveats
about this whole class of potential functions, we have some confidence that these repulsive potentials are not grossly in error.
The matching procedure becomes ambiguous for Xe - W; so, as in
ref. 3, the Xe - W repulsive function is the same as the W- W repulsive function. We shall use this fact to good advantage later.
The coefficients of the Born-Mayer potential functions are listed
in Table II.

TABLE II. Repulsive Interatomic Potentials

Born-Mayer Potential: $V(R) = A \exp (-BR)$

Atoms	A(keV)	B(L.U.$^{-1}$)
He - W	3.64	7.17
Ne - W	15.1	7.42
Ar - W	11.3	5.60
Kr - W	13.0	5.43
Xe - W	80.8	7.50
W - W	80.8	7.50

EQUILIBRIUM POSITIONS

All program development and the exploratory investigation was made with Ar. Systemetic energy reduction during the computation was a major problem. A foreign atom placed in an undistorted lattice acquires a great deal of potential energy which must be dissipated before equilibrium can be established. To facilitate the dissipation process, 'friction' was added by modifying the kinetic energy of every atom at the end of every timestep. In one approach all kinetic energy is removed; so that during each timestep each atom moves in the direction of the force acting upon it. This technique can be used, but it converges very slowly. A better method is to discard only a fraction of the kinetic energy; then the atom 'remembers' where it was going, and the convergence is much more rapid. Various fractions were tested, and a fifty percent reduction was finally accepted as the best compromise between computational stability and convergence efficiency.

Because of the separations involved, the nearest neighbor interactions between the foreign atom and any lattice atom is clearly repulsive; so the interstitial atoms are always in a positive energy state, compared to infinite separation. By contrast, the attractive tail of the W-W interaction is included to the fourth-nearest-neighbor; so the microcrystallite is held together by its own interatomic forces even when the foreign atom is present. When the attractive potential is truncated in this way the binding energy per atom is only approximately correct. Although they exist at a positive energy level, the interstitial atom sites are found to be metastable for small displacements.

Previous investigators[6],[7] found that a self-interstitial atom assumes a split-interstitial position oriented along a $\langle 110 \rangle$ axis. The host-interstitial pair is centered on the host atom site. The possible pair axes from a single host site are illustrated in Fig. 1, which is an exploded drawing a single bcc cell. The central atom of this figure is the host atom, and its location in the undistorted lattice is the host site.

The equilibration process for Ar was studied in great detail. The interstitial atom was always found to move to the (100) plane which contains the interstitial axis, and thence to its final position on the axis. When the Ar atom was inserted into an otherwise undisturbed lattice it assumed the position indicated in Fig. 2, and the host atom withdrew to the indicated position. The positions may be in error by a few hundredths of a L.U., but the interstitial Ar atom and host atom are no longer sharing the host site symmetrically. The uncertainty of final position for the

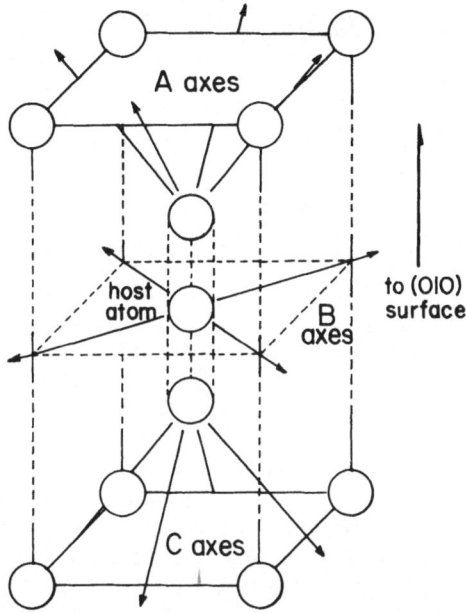

Figure 1. Exploded view of a single cell of the bcc lattice with ⟨110⟩ axes drawn from the host site.

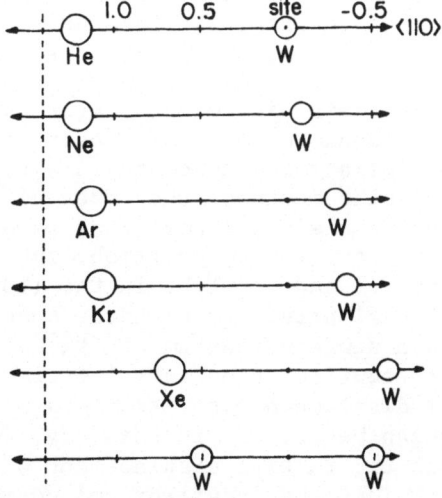

Figure 2. The equilibrium interstitial atom and host locations. Distances are measured in L.U.

Ar-W pair is relatively lower than for the other interstitial atoms, because the same final positions were obtained from a number of Ar atom starting positions.

The uncertainty is higher for the other interstitial atom positions because they were not subjected to so rigorous and detailed computational testing. In particular, it appears that Kr may be slightly too far from the host site, but this also may be a potential function problem.

The final position of an interstitial W atom was calculated with both the composite potential described above and with the repulsive term from that function. The interstitial atom is sufficiently crowded into the lattice that the total potential energy of the interstitial atom is positive for either function, and the final positions agree within the usual uncertainty.

One should keep these results in perspective. The model is crude; so the actual positions may be in error, but the trend is unmistakable. Light atoms tend to stabilize in positions which are close to the octahedral void. As the interstitial atom mass increases the equilibrium position approaches the shared host site position of the self-interstitial atom. There is a strong presumption that this is truly a mass effect, as opposed to a potential function effect, because both the Xe - W and W - W calculations were made with the same potential function. The evidence points toward a mechanism in which the equilibrium position of the host atom depends strongly upon its ability to move out of the way of the interstitial atom. A massive host acquires a smaller acceleration from the host-interstitial interaction than does the less massive interstitial atom; so it does not recoil so far.

As part of the analysis of the equilibration process, the Ar equilibrium site were found to be stable under displacements of the interstitial atom of a few hundredths of a L.U., but a new effect appears for displacements of the order of 0.1 L.U. With the lattice atoms relaxed to the final positions assumed with the interstitial atom in its equilibrium position, the interstitial Ar atom was displaced by 0.1 and 0.2 L.U. in the (100) plane which contains the interstitial axis. Under these conditions the interstitial atom assumes a new equilibrium position! In some sense this is equivalent to equilibration in a lattice which has been adjusted for thermal displacement of its atoms. For Ar a unique, new, final position can be defined in this way, for Xe, the final position depends upon the initial displacement of the Xe atom in the relaxed lattice. Thus the 'equilibrium' position depends upon the initial positions of neighboring lattice atoms, and, for more massive foreign atoms, upon the initial position of the interstitial atom as well.

In the static lattice used in these simulations there is no
single, unique, interstitial atom equilibrium position. Such a
position exists in the fcc Cu lattice for Ar. In a real
crystal, where the atoms are undergoing thermal oscillations the
evidence indicates that the interstitial atom will oscillate in a
shallow trench-like potential well, which may extend for two or
three tenths of a lattice unit along the $\langle 110 \rangle$ axis. For heavy
foreign atoms this well appears to be associated with the tetrahedral void center, but it would be a gross oversimplification to
say that the interstitial atom moves into the tetrahedral void.

Because the lattice atoms, which are located in the stress
field which surrounds the interstitial atom, do not have unique
positions, it is not really fair to continue to describe this
portion of the crystal as a lattice. The short range order is distorted beyond recognition, although the radial distribution
function is probably little modified. The resultant structure has
characteristics more like a liquid than a solid. Certainly calculations based upon symmetry arguments should be rigorously
prescribed.

CRYSTAL SURFACE EFFECTS

When the host site is deep within the crystal the twelve
interstitial axes illustrated in Fig. 1 are crystallographically
equivalent. When the site is near the surface, 'above' and 'below'
become important. Three catagories are obvious: A axes are directed from the site diagonally upward toward the surface, B axes
lie in an (010) plane, parallel to the surface, and C axes are directed from the site diagonally downward into the crystal. Interstitial sites will be designated by the number of the atomic plane
which contains the host site, followed by A,B, or C to designate a
sense with respect to that site. Sites 1C and 2A are almost equivalent for light atoms (they are almost certainly degenerate), because they are near the octahedral void, but they are not
equivalent for heavy interstitial atoms because then the interstitial site lies close to the host site.

Because the initially undistorted lattice produced a unique
interstitial equilibrium site in the preceding investigation, the
fiction of a unique equilibrium site is maintained through this
section. If a foreign atom precedes to that site when placed on
the interstitial axes, but displaced from the site, the site is
said to be stable. If the foreign atom moves toward the surface,
and acquires additional kinetic energy as it moves, the site is
defined to be unstable.

Subject to the definitions as described, all sites 3A, and
below, were found to be stable for all of the interstitial atoms

considered. For He the 3A interstitial site was found to be
slightly above the $\langle 110 \rangle$ axis. The displacement was of the order
of magnitude of the site uncertainty, but was reproducible with
various starting positions. This 'floating' displacement from the
interstitial axis also occurs for Ne, Ar, and Kr, but to a decreasing extent as the mass increases. It is marginally detectable
for He at greater depths, but not for the other atoms.

The 2C site, which is not equivalent to the 3A site for the
heavier foreign atoms, was investigated for Ne, Ar, Kr, and Xe
stability. The tests are ambiguous. For all of these atoms an
interstitial approaches the 'correct' interstitial position as
defined deep in the crystal and the kinetic energy of the foreign
atom decreases, but as the end of the calculation the remaining
kinetic energy is not zero, and it is not truly thermal, because
the velocity is always directed toward the surface. These results suggest that the 2C sites are not truly stable, but that a
minimal variation in the potential function might well lead to
stability.

The same sort of ambiguity exists for the 2B and 2A sites.
The position is reasonable, and the total energy is below thermal,
but the final velocity is always directed toward the surface. All
three of the 2-level sites behave almost exactly the same; so presumably they can be grouped together for analysis purposes. In
fact, the A,B, and C designations lose meaning at this level. The
'final' positions are significantly above the normal equilibrium
positions. For Ar and Ne, the displacement is about 0.1 L.U.
above the 2C site, and 0.3 L.U. above the 2A site.

The simulations lead to a picture of the foreign atom population near the surface. Apparently there are no stable interstitial sites in the first layer or two of the surface. Foreign
atoms may lie on the surface, or replace lattice atoms in the host
atom sites. We have simulated these replacements, but the absence
of any lattice atom-foreign atom attractive force at long range
gives an unrealistic model. In the computations foreign atom replacements in the first two layers are so strongly repelled by
their nearest neighbors that they burst out of the network of
lattice atoms and float away to the surface. A foreign atom replacement in the third layer tends to float away from the lattice
site, but it is constrained to a position near the lattice site by
the attractive forces between the other atoms which lie between it
and the surface. For Ar the replacement energy state is about
+3.5 eV, which may be compared with +16.3 eV energy at the interstitial site. This suggests that Ar replacement may actually be
possible in the real crystal. At similar depths below the surface,
the third layer and below, stable interstitial sites are possible.
The replacement sites were stable to foreign atom displacements as
great as 0.1 L.U. below the fourth atomic layer.

INTERSTITIAL ATOM BINDING ENERGY

Efforts to confirm site stability by displacement of the interstitial atom led to the anomolous site definitions reported above. In short, the lattice did not supply a restoring force in the sense by which one usually defines stability, but the atom did not leave the lattice. The situation is more akin to neutral stability than to the usual concept of a potential well. As it would have been expensive in computer time to attempt to define the volume of neutral stability, the problem was approached in a different way.

The early sputtering simulations determined that when a moving atom has significant energy compared to the other atoms in the lattice, the collision effects spread out from the moving atom, and there is relatively little reaction back onto the energetic atom from distant atoms. Thus it is feasible to perform computations in a microcrystallite which is very much smaller than that required for the equilibration analysis. This reduction in crystal size made a multiple run approach possible where it would have been impractical in the full crystal.

The size reduction was accomplished by storing the locations of those atoms which form the immediate environment of the interstitial atom in the equilibrated crystal, and also the locations of those which lie between it and the crystal surface. Most of these atoms are slightly displaced from their perfect lattice positions. A new microcrystallite constructed from the stored information provides a completely unchanged environment for the interstitial atom. In this way the number of interacting atoms was cut to 60, approximately one fourth the number used in the equilibration studies. Direct comparisons between results obtained from the full and reduced lattices confirmed the validity of this approach for the computations described below. The computer run time was reduced by approximately a factor of four.

In the equilibrated lattice the interstitial Ar atom was given a specified velocity by aiming it toward a point in the plane above the 3A site and assigning it a kinetic energy. The trajectory of the Ar atom was followed to determine whether it left the lattice. A search over a set of points, as defined, determined that the energy at exit is direction sensitive.

These studies were begun with the Ar atom energy at 20 eV, in which case all of the interstitial atoms escaped. The energy was reduced by steps, and as it was reduced, the number of open directions decreased. Rough contours of 'openness' were drawn, and the number of test directions was reduced as the atom energy decreased. Two open directions remained when the atom energy was 0.04 eV, but both had closed at 0.02 eV.

This is a very crude approach to a sensitive computational problem, because angular variations of the order of one degree are known to have large effect upon similar lattice dynamics calculations. Clearly, the interstitial Ar atom is not 'stable' at an energy slightly above thermal, and may be stable at an energy slightly below thermal. These energies are very much below those at which gas burst emission of inert gas atoms has been detected[8]. There is every reason to believe that deep interstitial sites will be somewhat more stable, but the computer cost to perform a complete survey would be prohibitive.

We must conclude that interstitial Ar atoms are effectively unbound in a room temperature W lattice; and strongly suspect that the other inert gases will behave similarly. Thus this study supports the interpretation that the experimental emission peaks should be associated with the release of inert gas atoms from substitutional positions.

These simulations were undertaken to provide background for the interpretation of the experimental results reported by Kornelsen and Sinha[8]. It would be naive to suggest that they comprise a theory of these implantation experiments, but there is sufficient consistency between simulation and experiment to encourage experimental examination of some of the conclusions and insights derived from the computer model.

ACKNOWLEDGEMENT

This research was supported by the Office of Naval Research.

REFERENCES

1a. D.E. Harrison, Jr., N.S. Levy, J.P. Johnson III, and H.M. Effron, J. Appl. Phys. _39_, 3742 (1968).

1b. D.E. Harrison, Jr., J. Appl. Phys. _40_, 3870 (1969).

2. D.E. Harrison, Jr., W.L. Moore, Jr., and H.T. Holcombe, Radiation Effects _17_, 167 (1973).

3. L.A. Girifalco and V.G. Weizer, Phys. Rev. _114_, 687 (1959).

4. D.E. Harrison, Jr., and D.S. Greiling, J. Appl. Phys. _38_, 3200 (1967).

5. P.T. Wedepohl, Proc. Phys. Soc. (London), _92_, 79 (1967); and J. Phys. (B), _1_, 307 (1968).

6. C. Erginsoy, G.H. Vineyard, and A. Englert, Phys. Rev. _133_, A595 (1964).

7. R.A. Johnson, Diffusion in Body-Centered Cubic Metals, American Society for Metals, p 357 (1965).

8. E.V. Kornelsen, and M.K. Sinha, Can. J. Phys. _46_, 613 (1968).

DISCUSSION

Q: (W. D. Wilson) Would you describe how you include the kinetic
energy in your calculation and how well the total energy is therefore conserved? This may have a bearing on your finding several
final configurations depending upon the initial positions and velocities (which we don't find). It may also shed some light on the
mass effect you described for Xe which differed from W in the mass
only.

A: We have taken a fundamentally different approach to simulation
from the model used in your calculations. If I understand your
calculations, you treat the microcrystallite of atoms simulated in
your program as an entity, and require that the energy content of
this system remain constant. We treat the microcrystallite as only
a portion of a much larger system, and allow transfers of energy
and momentum with the larger system, because we feel that these
transfers actually exist in a real crystal. As to the details of
our calculational procedure, at the beginning of each timestep we
calculate a new position and velocity for each atom. When the calculations are completed, the atoms are moved to the new position,
and each velocity component is multiplied by a constant factor,
say q, $0 \leq q \leq 1$, which has the effect of removing kinetic energy
and momentum from each atom. We speak of q as the friction factor,
and allow it to be set as an input in each simulation. Thus, q = 0
corresponds exactly to your computation, while q = 1.0 corresponds
to a full and complete conservation of energy and momentum. We have
examined the equilibration process for Ar in W with a number of q
values, and found that they led to essentially the same final Ar
position.

Effectively then, our condition is a minimization of the total
potential energy of the system once most of the energy created by
the insertion process has been dissipated. The variation in final
position as a function of the initial lattice configuration is at
least an order of magnitude greater than the position uncertainties
of our equilibration process. We are well aware that our procedure
will not produce unique final positions, but configuration sensitivity is an effect which is very much greater than the uncertainty
of our process. So we have every reason to believe that it may also
occur in real crystal in which the energy need not remain localized
in the strain field of the foreign atoms.

Q: (W. D. Wilson) C. L. Bisson and I have done calculations of
He, Ne, Ar, and Kr in a W lattice using a volume dependent potential
for W and potentials for the rare gas-metal determined from first
principles. We find He and Ne to form interstitially, Ar to have a
flat potential and Kr to form a "split" configuration along the
⟨110⟩ direction. This suggests that either your potential or your

method of calculation is causing the off-site position of the He. Johnson finds C in Fe to be off-site but the void has an attractive position which does not exist for He-W. Would you like to comment?

A: If the He atom is placed in the octahedral site, our program retains it in that site, and adjusts the lattice around it to reduce the total strain energy. However, when the He atom is displaced from the octahedral site, the final position is slightly displaced from the site. This suggests that the energy minimum at the site is not sharply defined, i.e., the potential well at the octahedral site is flat bottomed, so that the rms displacement from the mean position would be unexpectedly large. Your comment that Ar has a "flat potential" seems to confirm our result that the Ar atom sits in a trough lying along the $\langle 110 \rangle$ axis.

Q: (H. deWaard) Did you consider putting some vacancies in your lattice in order to see how these will stabilize the impurity positions?

A: Yes, this will almost certainly be our next project.

Comment: (H. deWaard) We have performed a Mössbauer study on Xe^{133} implanted in W (unpublished) and found 3 different sites, with different isomer shifts and Debye temperatures. These are interpreted as (1) substitutional, (2) associated with one vacancy, and (3) associated with two or more vacancies.

CHAPTER VI

ION LATTICE DAMAGE

ION DAMAGE EFFECTS IN METALS AS STUDIED BY TRANSMISSION ELECTRON MICROSCOPY

Manfred Wilkens

Max-Planck-Institut für Metallforschung

Institut für Physik, Stuttgart, Germany

SUMMARY

The paper deals with the application of transmission electron microscopy (TEM) to the study of radiation damage in metals due to ion bombardment. In a first part a brief review is given over the methods which are used for the analysis of the diffraction contrast of small point defect clusters. The following part is concerned with the application of these methods to the TEM study of ion damage. Special emphasis is given to the analysis of point defect clusters which are produced by displacement cascades.

1. INTRODUCTION

Irradiation of crystalline solids with energetic particles (ions, neutrons, electrons) leads to permanent displacements of crystal atoms if the recoil energy which is transferred from the energetic particle to a crystal atom exceeds the displacement energy. Thus an interstitial atom is formed leaving a vacancy behind. At the present time single point defects are not observable by transmission electron microscopy (TEM). However, if point defects of the same kind cluster together forming point defect agglomerates (PDA) these agglomerates may become visible on the TEM images if they exceed a critical size in the order of 15 - 20 Å.

441

Different mechanisms may be responsible for such
an agglomeration: (i) thermally activated or (ii) radiation-induced, i.e., athermal, motion of single point
defects or (iii) collapsing of defect-rich cores of
displacement cascades which are mechanically unstable.
The latter mechanism is especially important for specimens irradiated with energetic heavy ions.

The resultant PDA tend to form energetically stable
(or at least metastable) configurations such as dislocation loops, stacking fault tetrahedra, cavities etc.
Each of these configurations gives rise to more or less
specific contrast effects on the TEM images. If the diameter D of a PDA is comparable with, or larger than,
the effective extinction length ξ of the TEM imaging
mode used, the type and size of the PDA is rather easy
to recognize from the contrast figure (under conventional imaging conditions ξ is of the order of several 100Å).
However, radiation damage in metal crystals leads often
to the formation of PDA with D$\lesssim$100 Å (i.e. D$\ll\xi$). In
these cases the identification of the PDA becomes more
difficult. More sophisticated methods are required for
the interpretation of the contrast figures.

In Sect.2 of the present paper a number of such
methods are briefly discussed. More extended surveys
over this particular field of TEM are given in recent
review articles by Rühle/1/, Eyre/2/, Wilkens et al./3/,
and Wilkens/4/ to which we direct the reader. We summarize in the following sections some recent results,
mainly of the author's group, which may be relevant for
the scope of the conference.

2. SURVEY OVER THE EXPERIMENTAL METHODS

A full description of the damage pattern as produced in a crystalline specimen by particle-irradiation
requires the determination of a number of crystallographic and geometrical parameters of the PDA. The most important examples of such PDA are dislocation loops. This
section is therefore mainly concerned with dislocation
loops. Stacking fault tetrahedra and cavities are only
occasionally mentioned.

For the evaluation of these parameters the following
imaging modes of TEM have been adopted and successfully
used:
(a) Kinematical bright field. This mode means essentially that no low-order reflexion is strongly excited. The

kinematical two-beam mode is a special case. In this
mode one particular low-order reflexion is (moderately)
excited with a sufficiently large excitation error s.

(b) Dynamical bright field or dark field. In this case
one particular low-order reflexion is exactly (or almost
exactly) excited. The excitation of so-called non-sys-
tematic reflexions is avoided as far as possible.

(c) Dark field taken with a weakly excited beam (so-
called weak-beam technique, Cockayne et al./5,6/).

(d) Out-of focus technique/1/.

Each of these imaging modes has its special merits
regarding the study of particular parameters of PDA. In
the following it is briefly outlined how the main parameters can be determined.

(i) Sign of the misfit volume. Dislocation loops of
interstitial type and of vacancy type are characterized
by a misfit volume $\Delta V > 0$ and $\Delta V < 0$, respectively. If the
loop diameter D is sufficiently large $(D \gtrsim \xi)$ the sign of
ΔV can be determined from the "inside-outside" contrast
on kinematical two-beam images, once the inclination of
the loop plane with respect to the imaging plane is known
by stereo microscopy or by tilting experiments (cf.
Hirsch et al./7/, Maher and Eyre/8/).

For loops which are too small to be resolved on the
micrographs $(D \lesssim 150 \text{ Å})$ the sign of ΔV is best determined
by the black-white (BW) contrast technique (Rühle et al.
/9,10/): Small loops located within a distance of $1-1.5 \xi$
to either of the surfaces of the transmission specimen
reveal, due to their strain fields, under dynamical two-
beam conditions, a so-called BW contrast figure which is
characterized by a BW vector $\underline{l}$ pointing from the centre
of the black lobe to the centre of the bright lobe (on
positive prints), cf. Fig.1b,c. The sign of the scalar
product $(\underline{g} \cdot \underline{l})$ ($\underline{g}$ = operating two-beam diffraction vector)
depends on the sign of ΔV and in an oscillatory fashion
on the distance z_o (or $t-z_o$, t = foil thickness) of the
loop centre from the adjacent foil surface. The full
period of these "depth oscillations" is given by the
extinction length. Once z_o is determined, e.g., by the
stereo technique as developed by Diepers and Diehl/11/
the layer of the depth oscillations is known in which
the loop is located. Then the sign of ΔV follows from
the observed sign of $(\underline{g} \cdot \underline{l})$, cf. Rühle and Wilkens/10,12/.

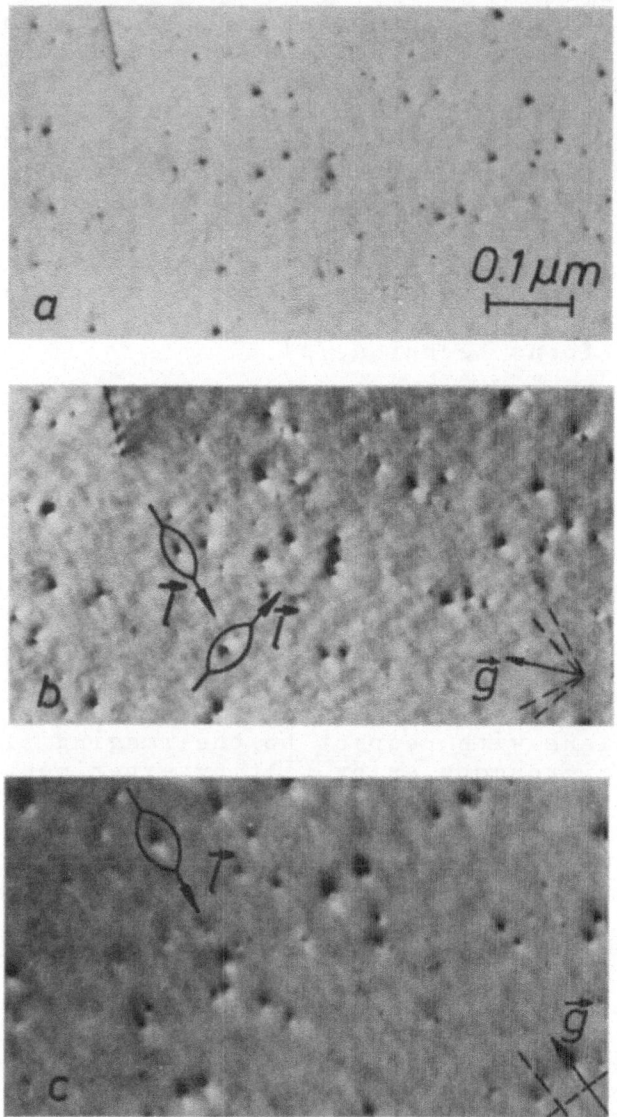

Fig. 1. Transmission micrograph of Cu irradiated with
60 keV Au ions, foil normal ≈ (001), bright field.
(a) kinematical image, (b) and (c) dynamical image with
$\underline{g}$ = (200) and $\underline{g}$ = (220), respectively. In (a) and (b)
the same foil area is shown. Dislocation loops of Frank
type ($\underline{b}$ = 1/3⟨111⟩) are visible as dark dots in (a) and
as BW dots in (b) and (c). The BW vectors $\underline{l}$ of most of
the dots lie parallel to projections of the ⟨111⟩-
directions onto the image plane (dashed lines).

Compared with those around dislocation loops, the strain fields around stacking fault tetrahedra (SFT) are fairly small. Therefore, so far, the application of the BW contrast technique to the study of the sign of ΔV of SFT has only been successful for SFT with edge lengths $\gtrsim 50$ Å (Chik/13/). Similar restrictions are valid for the detection of the strain fields of cavities /1/. In this case $\Delta V > 0$ may be caused by a high internal pressure of gas atoms stored inside the cavity (bubble), whereas $\Delta V < 0$ may be due to surface tension.

(ii) Diameter and shape. For dislocation loops with $D \lesssim 150$ Å the theoretical relation between the width of the contrast figure and the true loop diameter is not very well established, cf., e.g., /4/. Nevertheless, at present the diameters of small loops are best obtained from the black dot contrast figures as revealed under kinematical imaging conditions, cf. Fig.1a. The reason is that in the latter case, as compared with the dynamical two-beam imaging case, the contrast figures of small loops are much less sensitive to the depth position z_0 of the loops inside the transmission foil/10/. For the weak-beam contrast the reverse is true. Pronounced depth oscillations, with a period $1/s$, of the shape and the intensity of the contrast figures of small loops make this method, in spite of its strongly improved resolution, suspicious for counting and size measurements of small PDA (Häussermann et al./14/). (s = excitation error of the weakly excited beam used for the imaging. Usually, $1/s$ is in the order of 50 Å.)

In comparatively thin foils (low influence of the chromatic aberration of the objective lens on the image quality) imaged under kinematical conditions, SFT can be identified by means of the geometrical shape of their contrast figures for edge lengths down to about 25 Å (Bourret and Dautreppe/15/).

If cavities can be imaged under conventional in-focus conditions with a sufficiently high contrast their diameters are easily derived from the contrast figures as calculated by Van Landuyt et al./16/. With decreasing diameter the in-focus contrast of cavities decreases and disappears almost completely if D becomes smaller than about 50 Å depending on the foil thickness. At the same time the contrast is considerably improved by imaging the cavities under appropriate out-of focus conditions. Cavities with D down to ≈ 10 Å can be made visible by this technique /1/. However, in this case the expansion of the contrast figure due to the Fresnel fringes must

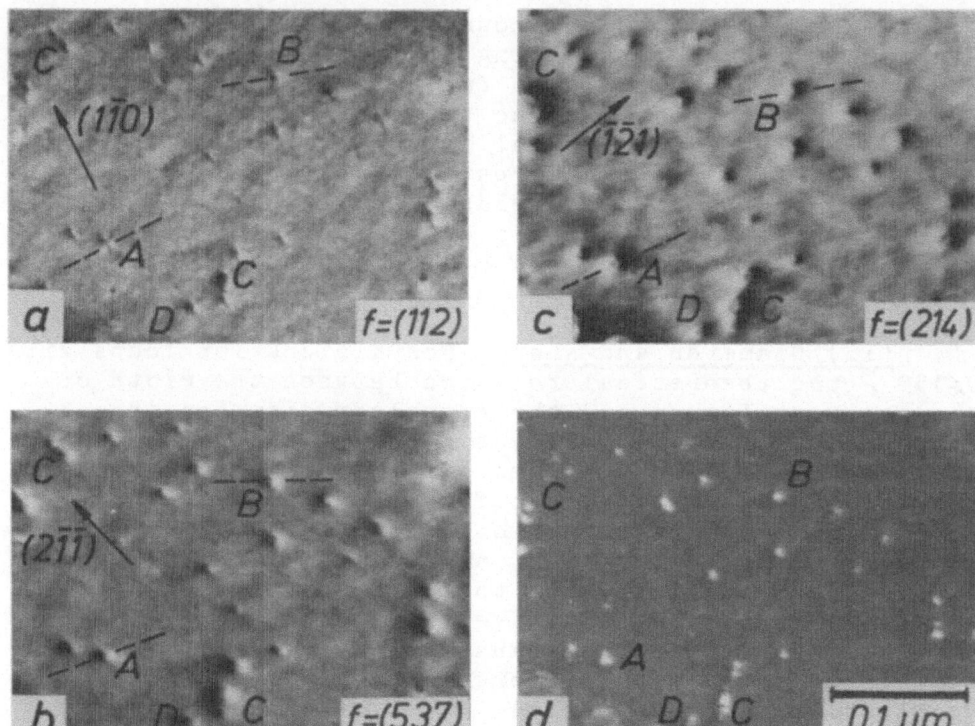

Fig.2. Vacancy loops in W irradiated with 60 keV Au ions, cf. Sect.3.2. (a) to (c): the same foil area is dynamically imaged with different g vectors; (d): weak-beam image. The pictures reveal a number of specific contrast features which depend on the orientation of a vector m with respect to the normal, f, of the image plane (z-axis) and g (x-axis); y-axis perpendicular to f and g (for details of the analysis cf. /2,4,20,43,44/). m = vector which bisects the acute angle between b and n; dashed lines = projected directions of m onto the x-y-plane. For m in the y-z-plane and not too far from the y-axis, a "butterfly" contrast is revealed (A in (a)). BW contrast figures such as B in (a) are typical for m being slightly out of the y-z-plane. With m approaching the x-z-plane the BW contrast dots become more and more symmetric with the direction of l as a mirror line, cf. (b), (c) and also Fig.1b,c. For $|(g \cdot b)| > 1$ the centre of the BW contrast figure may reveal a typical fine structure, e.g., A in (c). (A: b = 1/2[11$\bar{1}$], n = (110), B: b = 1/2[11$\bar{1}$], n = (10$\bar{1}$).) In (d) double dots as A and D are probably due to contrast effects rather than caused by two closely spaced loops, cf./14/. However, the complex contrast of C is certainly due to multiple clusters.

be taken into account properly /1/. Furthermore, the
observation of the out-of focus contrast of small cavities opens up the possibility of studying the mean inner
potential of the surrounding crystal matrix in comparison
to that inside the cavity (Rühle and Wilkens/17/, Ruedl
and Rühle/18/).

(iii) Burgers vector and plane normal of a dislocation loop. If the loop is sufficiently large to be resolved as a loop on the micrographs, the Burgers vector
$\underline{b}$ can be established by searching for the condition
$\underline{g} \cdot \underline{b} = 0$ for contrast extinction. The direction of the
normal, $\underline{n}$, of the loop plane may be derived by stereo
microscopy or by tilting of the specimen until the loop
appears in an "edge on" orientation. For small unresolved
loops the directions of $\underline{b}$ and $\underline{n}$ must be extracted from
the orientation and shape of the BW contrast figures ob-
tained under different dynamical two-beam conditions
(Wilkens and Rühle/19/). Characteristic features of BW
contrast figures of small loops which are useful for such
an analysis are shown in Fig.2. The possible directions
of $\underline{b}$ are restricted by crystallographic arguments.
Therefore in suitable cases the directions of $\underline{b}$ and $\underline{n}$
can be determined fairly unambiguously (Häussermann et
al./20/). For pure edge loops, i.e., $\underline{b}$ parallel to $\underline{n}$
(e.g., Frank loops in f.c.c. metals), the BW vector $\underline{l}$
coincides roughly to the projection of $\underline{b}$ onto the imaging plane, cf. Fig.1b,c, (Eßmann and Wilkens/21/,
McIntyre et al./22/).

(iv) Number density and size distribution function.
From (ii) it is obvious that both parameters are best
obtained under kinematical imaging conditions. However,
if the PDA are accompanied by a weak strain field only,
they may be undetectable unless they are located close
to a surface of the transmission foil and imaged under
dynamical two-beam conditions (Häussermann/23/). In comparatively thick foils the visibility of PDA with $D \lesssim 50$ Å
is noticeably influenced by the chromatic aberration of
the objective lens (Noggle et al./24/).

(v) Depth distribution. Especially for specimens
irradiated with heavy ions of moderately high energy the
depth distribution of the resultant PDA below the irradiated foil surface may be of considerable interest. To
this purpose the technique of stereo microscopy in the
form as developed by /11/ has proved to be most successful.

3. DISPLACEMENT CASCADES IN ION BOMBARDED METALS

3.1 Introduction

In the following sections we report briefly on some experimental results concerning the damage in metals as produced by energetic heavy ions. We restrict ourselves to those PDA which are formed more or less spontaneously (i.e. without long-range diffusion of single point defects) as a consequence of the formation of displacement cascades. For an introduction to the theory of displacement cascades (DC) we refer to the papers of Lindhard et al./25,26/ and Winterbon et al./27/, cf. also Nelson/28/ and Mayer et al./29/. TEM results are reviewed in /2a/.

The area density of PDA as produced by DC is expected to be proportional to the irradiation dose (as long as overlaps of the cascade volumes are negligible). Therefore, TEM investigations of this type of PDA are usually performed at ion doses $\lesssim 10^{11}$ -10^{12} ions/cm . Other mechanisms which may lead to the formation of PDA by means of a statistical accumulation of single point defects require, for their TEM investigation, ion doses which are at least two orders of magnitude higher (Bowden and Brandon/30/, Thomas and Balluffi/31/, Diehl et al./32/, Hertel et al./33/).

Displacement cascades are characterized by a number of parameters such as, e.g., diameter, density of point defects, spatial separation of vacancies and interstitials etc. These parameters depend in a characteristic manner on the mass and the energy of the incident ion, the mass of the target atoms, the crystal structure of the target etc. A direct determination of the DC parameters is difficult if even impossible. However, they can be investigated indirectly by a study of the PDA, into which a DC may be converted, e.g., by thermal vibrations of the lattice.

3.2 Nature of Point Defect Agglomerates

The nature of PDA in ion-bombarded metals was first studied for f.c.c. metals such as Al, Ni, Cu, and Au. Mainly from the features of the BW contrast figures, such as the directions of the BW vector $\underline{l}$, cf. Sect.2(iii), it was concluded that most of the PDA are dislocation loops of Frank type with $\underline{b}$ = 1/3$\langle$111$\rangle$ (Merkle et al./34/,

Howe et al./35/, Rühle and Wilkens/10,36/). In a number
of subsequent papers (e.g., Norris/37/) this result has
been confirmed. By application of the BW contrast technique together with the stereo method it could be shown
that the loops are of vacancy type /10,23,36,37/, Thomas
et al./38/, Wilson/39/.

The dose dependence of the area density of the vacancy loops was measured by /23,38/ and a well defined
linear dose dependence was found. From this result it
could be concluded that the loops are in fact formed by
the DC. The observation of only vacancy loops is most
easily understood in terms of the diluted-zone model of
Seeger/40/. This model postulates that the interstitials
produced by the energetic ion are deposited at the periphery of the DC leaving a vacancy-rich core behind.
Such a separation of the interstitials from the vacancies
may be most easily explained as being due to long-range
collision sequences (Silsbee/41/) or due to channeling
of the knocked-on interstitials.

Structural details of the vacancy clusters in Cu
damaged by Cu or Au ions were investigated by /23/ (Cu
foils bombarded by Cu or Au ions). For ion energies E
between 6 and 10 keV the author found vacancy clusters
with no preferential direction of the strain field (for
similar results cf./38/). In the interpretation of /23/,
these clusters can be considered as DC with vacancy-rich
cores which are still uncollapsed into loops. (For E<6keV
no visible PDA were observed, cf. Sect.3.3). For E>10 keV
the formation of predominantly Frank loops could be confirmed. However, a certain fraction of these vacancy
clusters are found to be in an intermediate stage between
a (planar) Frank loop and a (three-dimensional) stacking
fault tetrahedron (Wilson and Hirsch/42,23/).

The nature of the vacancy clusters in ion-bombarded
W was extensively investigated by Häussermann et al./20,
43/ and Wilkens and Jäger/44/ (cf. also Maher/45/). The
loops were produced by Au ions with E = 60-70 keV. For
the analysis of the BW contrast figures (cf. Fig.2) the
authors used the method of /19/. Most of the loops could
be indexed with $\underline{b}$ = 1/2⟨111⟩ and $\underline{n}$ ≈ {110} (∢($\underline{b}$,$\underline{n}$)≈35°),
$\underline{n}$ = normal of the loop plane, (In some cases the result
was b = 1/2⟨110⟩, $\underline{n}$ ≈ {110}). Some implications of these
results will be discussed later.

Vacancy loops with shear components were also found
in Co(h.c.p.) irradiated by Au ions with E = 40-60 keV
(Föll and Wilkens/46/). Nearly all observed loops could

be indexed with $\underline{b}$ = 1/3$\langle 11\bar{2}0 \rangle$ and $\underline{n} \approx \{1\bar{1}00\}$ $(\not<(\underline{b},\underline{n}) \approx 30°)$.
By application of the method of /19/ all relevant contrast details could be properly explained.

3.3 Yield Factor

The threshold energy E_c for the formation of visible
PDA in Au was first studied by Merkle/47/ using light
ions for the production of energetic primary knock-ons
in the specimen. In more direct experiments the values
for E_c and the yield factor Q in Au were determined by
/38/ (Au ions), Högberg and Norden/48/ (Ar, Xe, and Kr
ions) and Merkle et al./49/ (Xe ions). Specimens irradiated along a channeling direction $\langle 100 \rangle$ revealed
E_c<3 keV /38/ or E_c<10 keV /49/. However, for a "random"
direction of incidence, E_c was found to be between 10
and 20 keV /49/. The increase of E_c for a random direction of incidence was explained by the following assumption /49/: for low ion energies the random penetration
depth becomes so small that the DC are formed directly
beneath, or even at, the specimen surface, which diminishes the chance for the formation of a vacancy loop.
With increasing E>E_c, the yield factor Q increases and
reaches asymptotically the value 1 for E$\approx$50 keV /49/.
(The corresponding value E = 15 keV /38/ was critically
commented by /49/.

In the case of Cu irradiated with Cu or Au ions E_c
was determined to be 6 keV /23/. Using Au ions the asymptotic value Q$\approx$1 was nearly reached for E = 70 keV. Cu
ions lead to about 50% lower values of Q (cf. also /47/).
This is a reasonable result, since the DC in Cu produced
by Cu ions are expected to be fairly extended and divided
into small sub-cascades (von Jahn/50/). On the other hand
the heavy mass of the Au ions facilitates the separation
of the interstitials from the vacancies. The influence
of channeling on the yield factor in Au was clearly demonstrated by Noggle and Oen/51/ who used 51 MeV iodine
ions. For an off-channeling direction of incidence Q was
larger by an order of magnitude than for bombardment along a $\langle 110 \rangle$ channeling direction.

With increasing ion energies the DC tend to broaden
and if energetic primary knock-ons are transfered into
the surrounding lattice, e.g., by quasi-channeling, subcascades are formed which may lead to the formation of
closely spaced vacancy loops. Such multiple defects have
been observed and theoretically discussed by /23,47,52,
53/.

The DC as produced in Al irradiated by Al ions are
even more extended. Consequently, no evidence for the
formation of vacancy loops by means of the DC mechanism
was found (Ruault et al./54/). However, using ions of
higher mass such as Hg /37/ or Au (Gomez-Giraldez et al.
/55/) the formation of (fairly small) vacancy loops could
be detected.

The yield factor Q in W and Mo irradiated with Au
ions (E between 20 and 70 keV) was studied by Häussermann
/56/. The author found that Q is about two orders of
magnitude lower than in f.c.c. metals and, furthermore,
very sensitive to the direction of incidence of the ion
beam. To some extend these observations could be explained in terms of ion-channeling. For another explanation
we refer to Sect. 3.6.

3.4 Size Distribution and Depth Distribution

The size distribution of the vacancy loops was investigated by /47/ and Thomas et al./57/ (Au), by /23/
(Cu) and by Norris/58/ (Au,Ni). Accordingly, the mean
loop diameter $\bar{D}$ increase slightly with increasing ion
energy (cf. however /39/). Calculating the number of
Frenkel pairs which are created per incident ion, e.g.,
by a Kinchin-Pease model a "cascade efficiency" (i.e.
the fractional number of vacancies stored in the loop)
can be calculated from the loop diameter. Reasonable values for the cascade efficiency in the order of 0.3 to 0.5
are obtained if $\bar{D}$ is inserted /23,52/. However, the cascade efficiency becomes too high (in some cases larger
than one) if the extreme loop diameters corresponding to
the tails of the size spectra are used for the calculation /59/, cf. also /55/.

The depth distribution of the vacancy loops below
the irradiated specimen surface was measured and interpreted in terms of the random penetration depth and the
channeling parameters of the incident ions /23,38,57,58/.
Comparing the tails of the depth distribution function
with Whitton's data/60/ reasonably good agreement was
found /57/. (/60/ was concerned with the depth distribution of radioactive Au ions which were implanted into Au
specimens.)

3.5 Ion Channeling

A fairly direct way for the study of characteristic

ion channeling parameters by means of TEM was demonstrated by Häussermann/56/. The author evaluated the damage structure in a Cu specimen which was covered, during
the ion irradiation, with a wedge-shaped W foil, Au ions
of 60 keV energy were used. After the ion irradiation
both foils were inspected by TEM separately: The W foil
was analysed with respect to its local thickness $t_w(x)$
as a function of the distance x from the foil edge. In
the Cu foil the vacancy clusters as observed in the region of the foil shadowed by the W foil were analysed
with respect to their local area density and their local
mean diameter as a function of x, or $t_w(x)$, respectively.
The data were evaluated in two ways:

(i) By means of the yield factor of Au ions in Cu
foils as determined in /23/, the numbers $N(t_w)$ of the
Au ions which had penetrated through the W foil could be
determined. For 250 Å $\leqslant t_w \leqslant$ 1800 Å a well defined exponential decay $exp(-\mu \cdot t_w)$ was found with $\mu \approx 0.8 \cdot 10^{-3}/$
Å ($\mu = 0.43 \cdot 10^{-3}$/Å as given in /56/ is incorrect due to
a calculation mistake). This value fits pretty well to
$\mu \approx 1.1 \cdot 10^{-3}$/Å which can be extracted from the work of
Whitton/61/ who measured the depth distribution of 40 keV
Xe-ions stopped in $\langle 111 \rangle$ channels of W. For $t_w < 250$ Å
the value of $N(t_w)$ was higher than predicted by the exponential law. This effect was attributed to random penetration.

(ii) By comparing the mean diameter $\overline{D} = \overline{D}(t_w)$ with
the mean diameter of the loops in Cu as a function of
the Au ion energy ($\overline{D} = \overline{D}(E)$) /23/ a mean energy loss of
$\Delta E \approx 14$ eV/Å could be estimated which appears reasonable.

3.6 Elastic Interaction of Loops with a Free Surface

As already mentioned in Sect. 3.3 the vacancy loops
in ion-damaged W could be indexed with $\underline{b} = 1/2\langle 111 \rangle$ and
$\underline{n} \approx \{110\}$ /20,43,44/. Such loops are glissile and may
slip out of the foil due to the image forces caused by
the adjacent foil surface, cf. also /56/ (distances loop-
surface in the order of 100 Å or even less). The results
of /43,44/ revealed that among the equivalent configurations which are compatible with the indexing given above,
some were much more populated than the others:

(i) Most of the loops were found with $\underline{b}$ having an
angle $\gamma \gtrsim 65°-70°$ ($\gamma = \sphericalangle$ between $\underline{b}$ and the foil plane nor-

mal). Obviously, for $\gamma < 65°$ the image forces were strong
enough to pull the loops out of the foil. This effect
may strongly influence the effective yield factor, cf.
Sect. 3.3.

(ii) The population of the loop configurations hav-
ing the same $\underline{b}$ but different $\underline{n}$ were found to be significantly different.

The latter observation is hard to understand in
terms of the slipping-out mechanism. However, it could
be interpreted qualitatively if two reasonable assumptions were made: (a) The loops are first nucleated as
Frank loops ($\underline{b}_F$ = 1/2⟨110⟩) on {110} planes. In a second
step $\underline{b}_F$ is converted by a shear ±1/2⟨001⟩ into a perfect
Burgers vector $\underline{b}$ = 1/2⟨111⟩. (b) From the two shear directions the one which causes the stronger reduction of the
elastic energy, due to the adjacent free foil surface, is
preferred. (With respect to (b), the elastic energy of small
loops in an isotropic half-space for arbitrary directions
of $\underline{b}$ and $\underline{n}$ were calculated by Jäger et al./62/.) Thus,
the observation (ii) can be understood as fairly direct
evidence for the two-step model for the formation of
loops of perfect Burgers vectors in b.c.c. metals as
proposed by Eyre and Bullough/63/.

3.7 Volumes of Displacement Cascades

The sizes of the vacancy loops formed by ion-irra-
diation of metal specimens are sometimes used as a measure of the size of the DC, cf., e.g., /57/. However, such
an interpretation is rather ambiguous since the number
of vacancies, which survive in the core of a DC in form
of a dislocation loop, depends sensitively on the spatial
separation of vacancies and interstitials. In order to
make the cascade volume directly visible by TEM, Seiler
and Wilkens/64/ have studied the damage in a well ordered
Cu_3Au-alloy irradiated by 60 keV Au ions. The idea was
that, irrespective of the number of surviving vacancies,
which may, or may not, agglomerate in form of a loop,
the superlattice order should be practically destroyed
inside the cascade volume. As a consequence the ion-ir-
radiation should lead to the formation of small disordered zones embedded in the ordered matrix. If such a
disordered zone is not accompanied by a vacancy loop it
should be invisible by imaging the foil with a fundamental reflexion of the disordered lattice. On the other

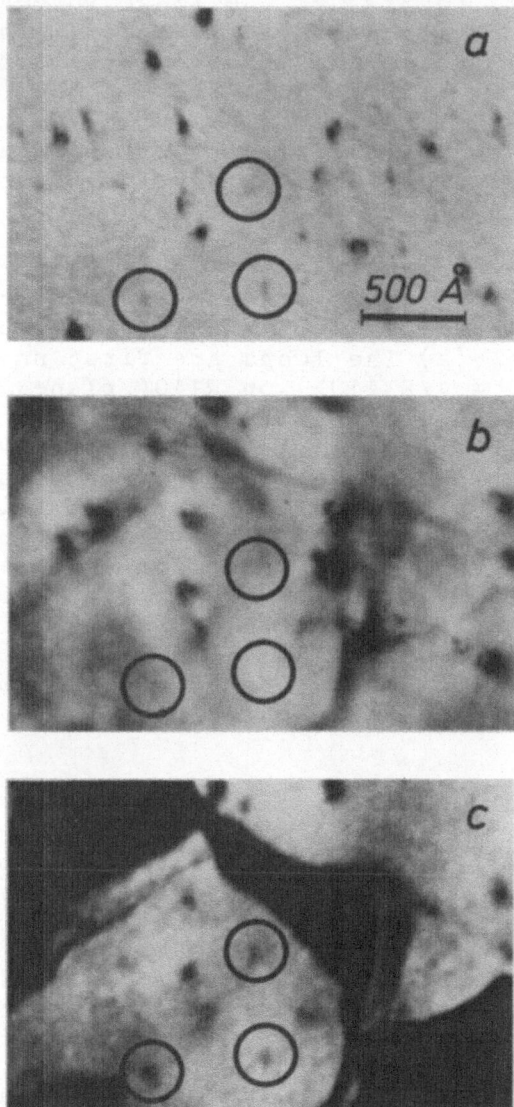

Fig.3. Transmission micrograph of ordered Cu₃Au irradiated with $5 \cdot 10^{10}$ Au ions/cm² (60 keV), foil normal≈(001).
(a) and (b) kinematical bright field images with $\underline{g}$ =
(1$\bar{1}$1) and $\underline{g}$ = ($\bar{2}$20), respectively. (c) dark field image
taken with the superlattice reflexion $\underline{g}$ = (1$\bar{1}$0) fully
excited; the dark areas in (c) are due to antiphase
boundaries. The dark dots which are visible in all the
three images are due to dislocation loops. The dark dots
encircled in (c) are (almost) invisible in (a) and (b).
These dots are assumed to be due to disordered zones.

hand such a disordered zone should become visible on dark-field images taken with a superlattice reflexion. The following experimental results were obtained:

(i) As in pure Cu and Au, dislocation loops and PDA with no preferential directions, both being of vacancy type, were found. Due to their strain fields these defects were visible on bright-field images and on dark-field images irrespective of whether a fundamental reflexion or a superlattice reflexion was strongly excited.

(ii) Additionally, a number of dark contrast dots with diameters in the order of 50 Å could be detected, but which were visible only on dark-field images taken with a well excited superlattice reflexion.

An example is shown in Figs.3a-c. The same foil area was imaged in bright-field using two different (non-parallel) fundamental diffraction vectors (a,b) and in a dark-field mode using a superlattice diffraction vector (c). In the latter case some dark dots are visible which are absent on Figs.3a,b. According to the above considerations we interprete these dark dots as direct images of such displacement cascades for which the separation of the vacancies and interstitials were not effective enough in order to allow the formation of a vacancy loop. If corresponding contrast calculations, which are under way, corroborate the validity of this interpretation, the TEM study of ion damage in ordered alloys may provide an interesting tool for the determination of the true cascade volume.

Acknowledgements

The author thanks Dr. M. Rühle for his continuous collaboration and for his help during the preparation of the present paper. Further, the cooperation with his colleagues, especially Dipl.Phys. H. Föll, Dr. F. Häussermann, Dipl.Phys. W. Jäger, Dipl.Phys. K.-H. Katerbau, Miss M. Rapp, and Dipl.Phys. G. Seiler is gratefully acknowledged.

References

/1/ M.Rühle, in Radiation-Unduced Voids in Metals, eds
    J.W.Corbett and L.C.Ianniello, USAEC p.255 (1972).
/2/ B.L.Eyre, in Defects in Refractory Metals, eds
    R.de Batist, J.Nihoul and L.Stals, S.C.K./C.E.N.
    Mol, p.311 (1972).

/2a/ B.L.Eyre, J.Phys.F: Metal Phys. 3, 422 (1973).

/3/ M.Wilkens, M.Rühle and F.Häussermann, J.de Micro-
     scopie 16, 199 (1973)

/4/ M.Wilkens, in Proc.Intern.School on Electron Micro-
     scopy in Materials Science, Erice 1973, C.I.D.
     Luxemburg, in the press.

/5/ D.J.H.Cockayne, I.L.F.Ray and M.J.Whelan, Phil.Mag.
     20, 1265 (1969).

/6/ D.J.H.Cockayne, Z.Naturf. 27a, 455 (1972).

/7/ P.B.Hirsch, A.Howie, R.B.Nicholson, D.Pashley and
     M.J.Whelan, Electron Microscopy of Thin Crystals,
     Butterworths, London (1965).

/8/ D.M.Maher and B.L.Eyre, Phil.Mag. 23, 409 (1971).

/9/ M.Rühle, M.Wilkens, and U.Eßmann, phys.stat.sol.
     11, 819 (1965).

/10/ M.Rühle, phys.stat.sol. 19, 263, 279 (1967).

/11/ H.Diepers and J.Diehl, phys.stat.sol. 16, K109
     (1966).

/12/ M.Rühle and M.Wilkens, Phil.Mag. 15, 1075 (1967).

/13/ K.P.Chik, phys.stat.sol. 16, 685 (1966).

/14/ F.Häussermann, K.-H.Katerbau, M.Rühle and M.Wilkens,
     J.Microscopy 98, August (1973).

/15/ A.Bourret and D.Dautreppe, phys.stat.sol. 29, 283
     (1968).

/16/ J.Van Landuyt, R.Gevers and S.Amelinckx, phys.stat.
     sol. 10, 319 (1965).

/17/ M.Rühle and M.Wilkens, to be published.

/18/ E.Ruedl and M.Rühle, to be published.

/19/ M.Wilkens and M.Rühle, phys.stat.sol.(b) 49, 749
     (1972).

/20/ F.Häussermann, M.Rühle and M.Wilkens, phys.stat.sol.
     (b) 50, 445 (1972).

/21/ U.Eßmann and M.Wilkens, phys.stat.sol. 4, K53 (1964).

/22/ K.G.McIntyre, L.M.Brown and J.A.Eades, Phil.Mag.
     21, 853 (1970).

/23/ F.Häussermann, Phil.Mag. 25, 537 (1972).

/24/ T.S.Noggle, O.S.Oen and J.C.Crump, Proc.28th
     Annual EMSA Meeting, Houston, Texas, p.406 (1970).

/25/ J.Lindhard, V.Nielsen, M.Scharf and P.V.Thomson,
     Mat.Fys.Medd.Dan.Vid.Selesk. 33, No.10 (1963).

/26/ J.Lindhard, Mat.Fys.Medd.Dan.Vid.Selesk. 34, No.14
     (1965).

/27/ K.B.Winterbon, P.Sigmund and J.B.Sanders, Mat.Fys.
     Medd.Dan.Vid.Selesk. 37, No.14 (1970).

/28/ R.S.Nelson, The Observation of Atomic Collisions
     in Crystalline Solids, North-Holland Publ.Comp.
     Amsterdam (1968).

/29/ J.W.Mayer, L.Eriksson and J.A.Davies, Ion Implanta-
     tion in Semiconductors, Academic Press, New York
     and London (1970).

/30/ P.B.Bouden and D.G.Brandon, Phil.Mag. $\underline{8}$, 935 (1963).

/31/ L.E.Thomas and R.W.Balluffi, Phil.Mag. $\underline{15}$, 1117
     (1967).

/32/ J.Diehl, H.Diepers and B.Hertel, Can.J.Phys. $\underline{46}$, 647
     (1968).

/33/ B.Hertel, J.Diehl, R.Gotthardt and H.Sulze, these
     conference proceedings.

/34/ K.L.Merkle, L.R.Singer and R.K.Hart, J.Appl.Phys.
     $\underline{34}$, 2800 (1963).

/35/ L.M.Howe, J.F.McGurn and R.W.Gilbert, Acta Met. $\underline{14}$,
     801 (1966).

/36/ M.Rühle and M.Wilkens, Proc. 6th Intern. Conf. on
     Electron Microscopy, Kyoto, Vol.I, 379 (1966).

/37/ D.I.R.Norris, Phil.Mag.$\underline{19}$, 527 (1969).

/38/ L.E.Thomas, T.Schober and R.W.Balluffi, Rad.Eff. $\underline{1}$,
     257 (1969).

/39/ M.M.Wilson, Phil.Mag. $\underline{24}$, 1023 (1971).

/40/ A.Seeger, Intern. Conf. Peaceful Uses of Atomic
     Energy, $\underline{6}$, 250 (1958).

/41/ R.H.Silsbee, J.Appl.Phys. $\underline{28}$, 1246 (1957).

/42/ M.M.Wilson and P.B.Hirsch, Phil.Mag. $\underline{25}$, 983 (1972).

/43/ F.Häussermann, Phil.Mag. $\underline{25}$, 561 (1972).

/44/ W.Jäger and M.Wilkens,to be published.

/45/ D.H.Maher, Proc. 7th Intern. Conf. on Electron
     Microscopy, Grenoble, Vol.II, 349 (1970).

/46/ H.Föll and M.Wilkens, to be published.

/47/ K.L.Merkle, phys.stat.sol. $\underline{18}$, 173 (1966).

/48/ G.Högberg.and H.Norden, phys.stat.sol. $\underline{33}$, K71
     (1969).

/49/ K.L.Merkle, L.R.Singer and J.R.Wroble, Appl.Phys.
     Lett. $\underline{17}$, 6 (1970).

/50/ R.von Jahn, phys.stat.sol. $\underline{8}$, 331 (1965).

/51/ T.S.Noggle and O.S.Oen, Phys.Rev.Lett. $\underline{16}$, 395
     (1966).

/52/ K.L.Merkle, Radiation Damage in Reactor Materials,
     I.A.E.A. Vienna, Vol.I, 159 (1969).

/53/ L.E.Thomas, T.Schober and R.W.Balluffi, Rad.Eff. $\underline{1}$,
     279 (1969).

/54/ M.O.Ruault, B.Jouffrey and P.Joyes, Phil.Mag. $\underline{25}$,
     833 (1972).

/55/ C.Gomez-Giraldez, B.Hertel, M.Rühle and M.Wilkens,
     these conference proceedings.

/56/ F.Häussermann, Phil.Mag. $\underline{25}$, 583 (1972).

/57/ L.E.Thomas, T.Schober and R.W.Balluffi, Rad.Eff. 1, 269 (1969).
/58/ D.I.R.Norris, Phil.Mag. 19, 653 (1969).
/59/ M.Rühle, private communication.
/60/ J.L.Whitton, Can.J.Phys. 45, 1947 (1967).
/61/ J.L.Whitton, Can.J.Phys. 46, 581 (1968).
/62/ W.Jäger, M.Rühle and M.Wilkens, to be published.
/63/ B.L.Eyre and R.Bullough, Phil.Mag. 12, 31 (1965).
/64/ G.Seiler and M.Wilkens, to be published.

DISCUSSION

Q: (R. S. Nelson)[1] Is there a denudation of vacancy loops close
to the surface due to image force effects? [2] Can this affect
the visibility of loops as a function of ion energy at low
energies?

A: [1] In our work on ion-bombardment of Cu no denuded layer
which is substantially larger than the loop diameters is observed.
Occasionally in Cu a small number of perfect loops are observed.
These loops sometimes slip out. In bcc metals, the situation is
different as discussed in the paper. [2] The threshold energy
for loop formation in Cu is partly due to the decrease of visibility with decreasing size of the point defect cluster. Also the
surface may play a role.

Q: (G. L. Kulcinski) Has your group done any low temperature
(4-100°K) ion bombardment followed by analysis at low temperature?

A: No, not so far.

Q: (K. L. Merkle) Regarding the imaging of the cascade volume by
means of super-lattice reflections, is the effect that one gets
from the disordering large enough to be separated from the influence at the strain fields?

A: The contrast dots of interest are obviously not caused by
strain contrast since they are invisible by imaging with the
fundamental reflexion. Our interpretation of these as being disordered zones are strongly supported by contrast calculations
which were done in the week before the Conference. So far all
observations are in agreement with the calculations.

Transmission Electron Microscopy Study of

Implantation Induced Defects in Gold

M.O. RUAULT[*], B. JOUFFREY[*], J. CHAUMONT[**],
and H. BERNAS[***]

[*] Laboratoire d'Optique Electronique, CNRS, Toulouse
[**] Centre de Spectrométrie de Masse, CNRS, Orsay
[***] Institut de Physique Nucléaire, Orsay

Theoretical studies of heavy ion stopping in solids have shown
that the energy loss profile generally does not completely overlap
the implanted ion profile [1,2]. A variety of methods has been used
to study the latter [3] ; the former may be determined by transmission electron microscopy (TEM) studies of the implantation-induced
defects. In many cases, the TEM technique also gives information on
the geometry and on the nature of the defects. If the defect's
depth under the sample surface has been measured, as well as its
diffraction vector in the conditions of observation, a study of the
image contrast shows whether the defect is made up of vacancies or
interstitials [4]. This stereoscopic method has been implemented
recently by various groups (see, e.g., [5] [6]). The experiments
reported here on ion-implanted gold had a twofold purpose :
(i) compare the experimental defect depth distribution to the predictions of ref. [1], and (ii) identify the nature of the defects
in the various cases. Also, some results are presented on ion-
implanted samples submitted to high-dose 2.5 MeV electron irradiation.

Annealed polycrystalline Au foils were first thinned down.
Implantations were then carried out on the Orsay ion-implantation
[7] ; doses were chosen so that defect images were well-separated
(image diameters ranged from 50 to 200 Å). A typical case is presented in Fig. 1 (the small dots in the background are due to the
$\sim$ 20 Å - thick evaporated Au layer which identifies the sample
surface for stereoscopic measurements). The strain direction of the
defects was determined from the depth dependence of the image
contrast.

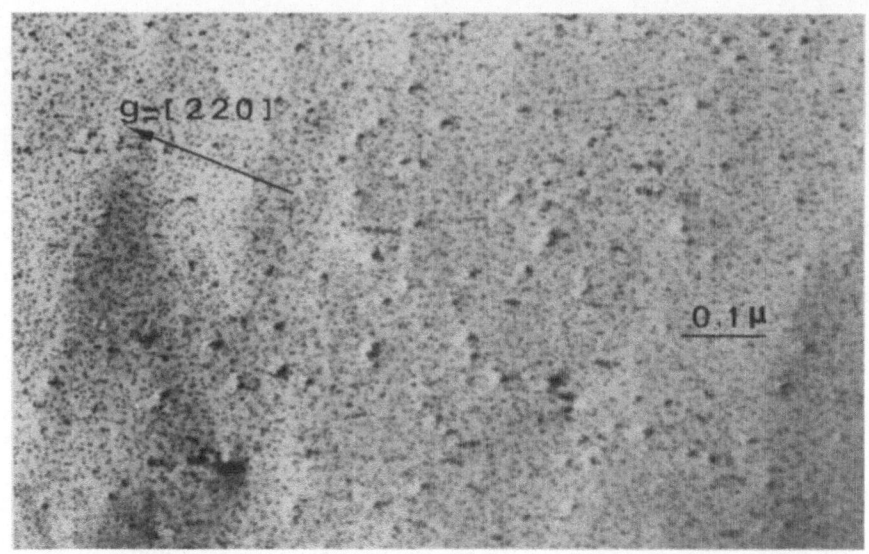

<u>Figure 1</u> : Implantation of 50 keV Kr[+] in gold. Dose 3×10^{11} ions/cm^2. TEM at 100 kV.

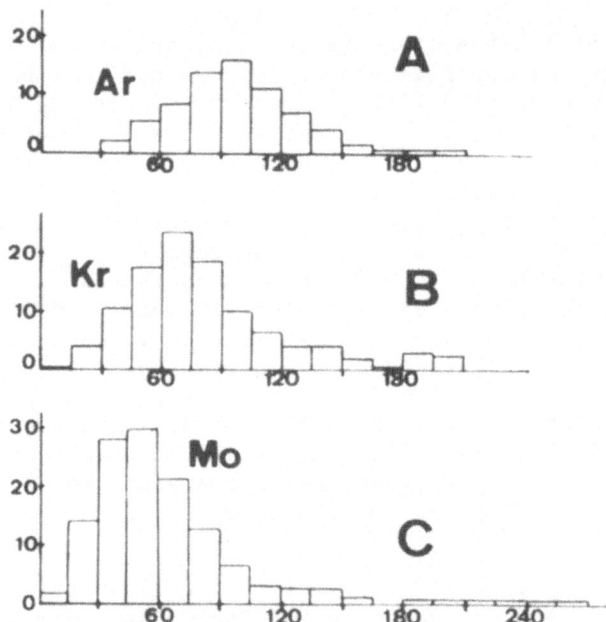

<u>Figure 2</u> : Experimental defect distributions for ion-implanted Au. A = 50 keV Ar[+], dose 3×10^{11} ions/cm^2 ; B = 50 keV Kr[+], dose 2.5×10^{11} ions/cm^2 ; C = 50 keV Mo[+], dose 2.5×10^{11} ions/cm^2.

TABLE 1

Sample	Energy (keV)	Doses (ions/cm²)	Channeling	h_p exp (Å)	$\sqrt{<\Delta h^2>}$ exp (Å)	h_p (theo) in Å m=1/3	h_p (theo) in Å m=1/2	$\sqrt{<\Delta h^2>}$ theo. (Å) m=1/3	$\sqrt{<\Delta h^2>}$ theo. (Å) m=1/2	x_p (Å) m=1/3	x_p (Å) m=1/2	Interstitial loops (%)	Vacancy loops (%)
Au, Yb	130	10^{13}	No	70		65	65	70	60	90	90	10	90
Au, Mo	50	2.5×10^{11}	No	50	40	45	40	35	30	60	65	20	80
Au, Xe	50	3×10^{11}	No	45	30	40	35	50	30	60	50	0	100
	150	* 10^{12}	<100>	130 to 230	130	80	105	100	95	120	150	3	97
Au, Kr	50	* $2 \cdot 10^{10}$	<100>	90	45(30)(1)	50	40	55	55	55	75	80	20
	50	2.5×10^{11}	<110>	90	45(15)(1)	50	40	55	55	55	75	98	2
	50	3×10^{11}	No	65	40	50	40	55	55	55	75	80	20
	150	2.5×10^{12}	<100>	240	60	100	120	115	165	120	220	90	10
	150	2×10^{11}	No	120		100	120	115	165	120	220	90	10
Au, Ar	50	3×10^{11}	No	95	40	55	100	100	140	80	145	100	0
	150	* 10^{12}	<100>				300		420		440	75	25
Au, Fe	50	2.5×10^{11}	No	80	35(20)(1)	60	70	70	55	80	115	80	20

* 6 months after irradiation

(1) When the value of $\sqrt{<\Delta h^2>}$ exp towards the surface is different from its value away from the surface, the former is given in parentheses.

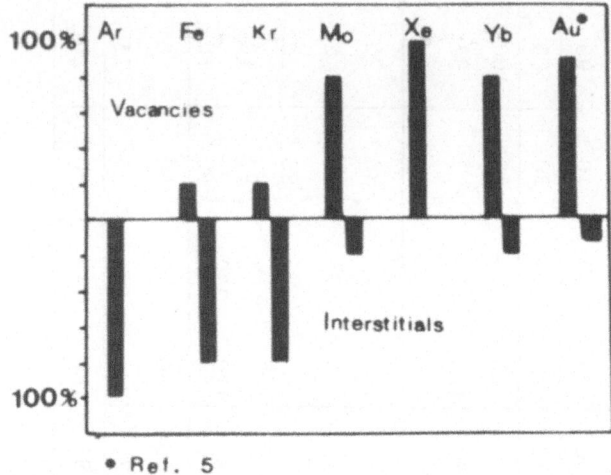

Figure 3 : Proportions of interstitial loops and vacancy loops for 50- and 150- keV ions in gold.

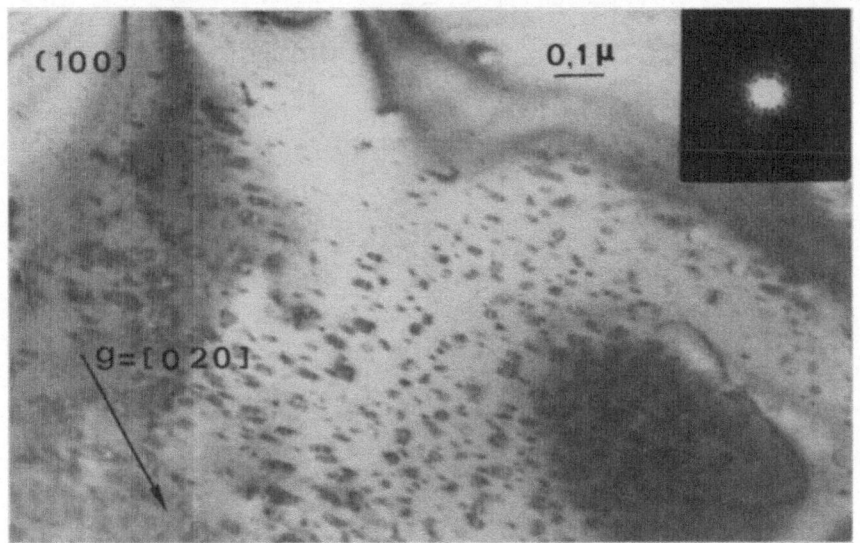

Figure 4 : Gold sample, implanted with 50 keV Kr$^+$ (dose 2×10^{10} ions/cm^2), after irradiation with 2.5 MeV electrons. Picture taken during the electron irradiation . The optical diffraction diagram is shown in the upper right hand corner. Electron dose was 10^{21} e/cm^2.

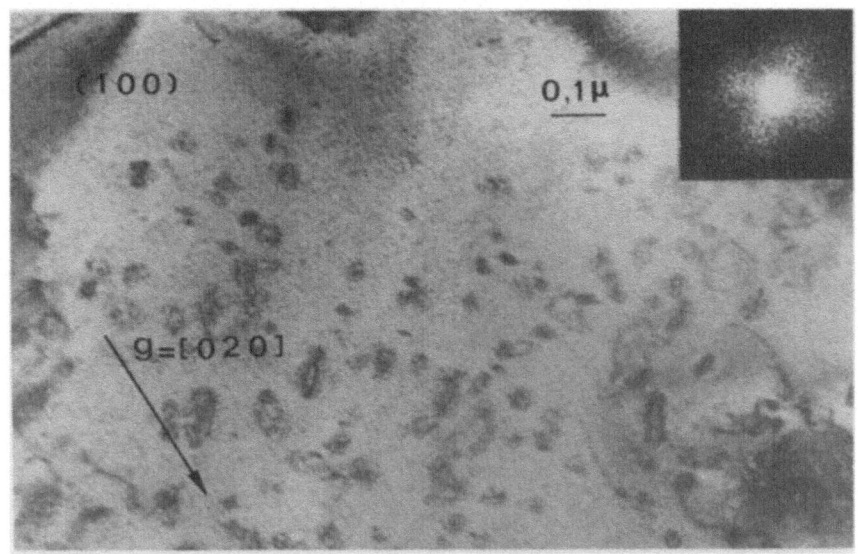

Figure 5 : Same sample after an electron dose of 10^{22} e/cm^2.

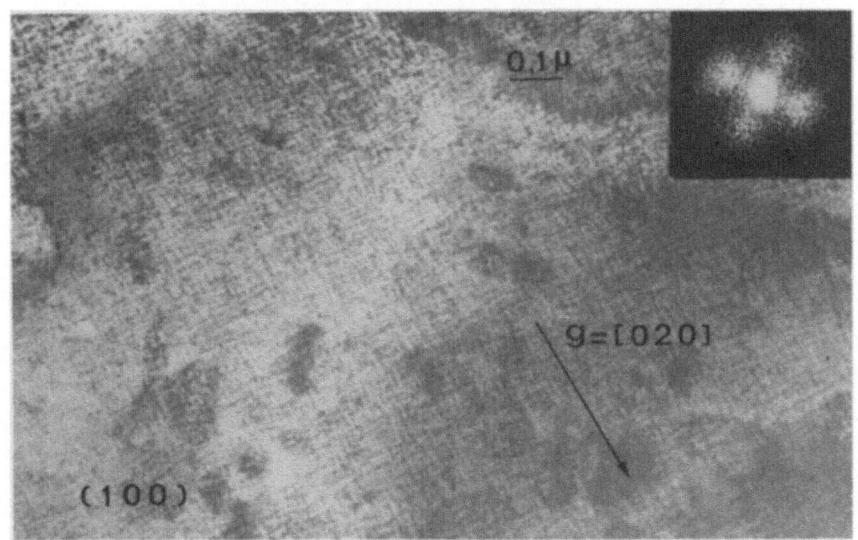

Figure 6 : Same sample after an electron dose of 8×10^{22} e/cm^2. Note ordering.

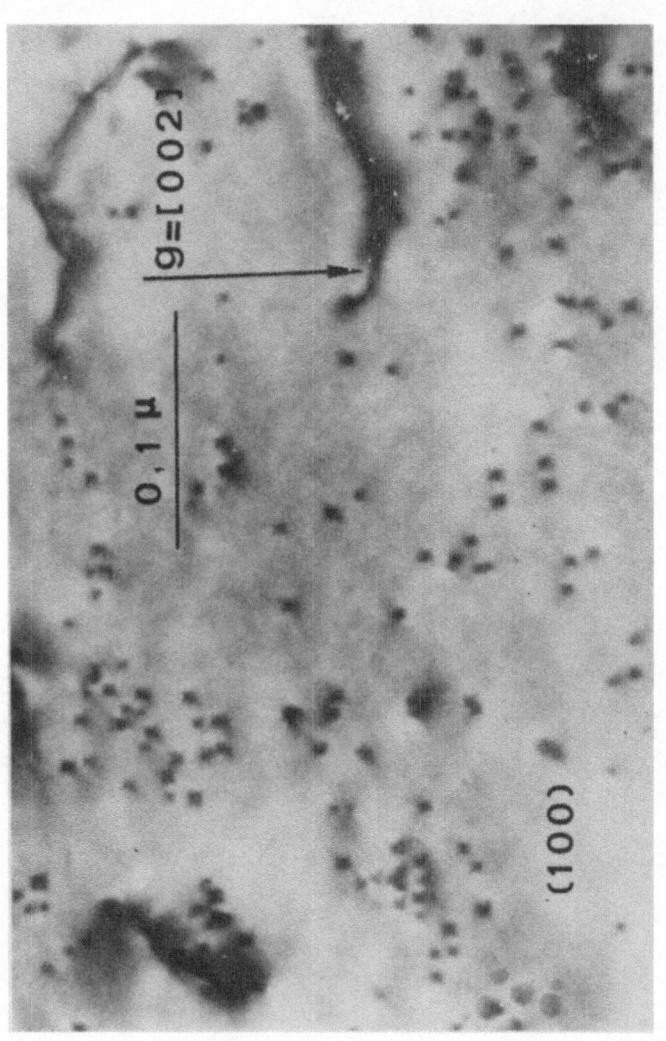

<u>Figure 7</u> : TEM picture (100 kV) of same samples as in Figs 4-6
(total dose $\sim 10^{23}$ e/cm^2), after annealing at 700°C.

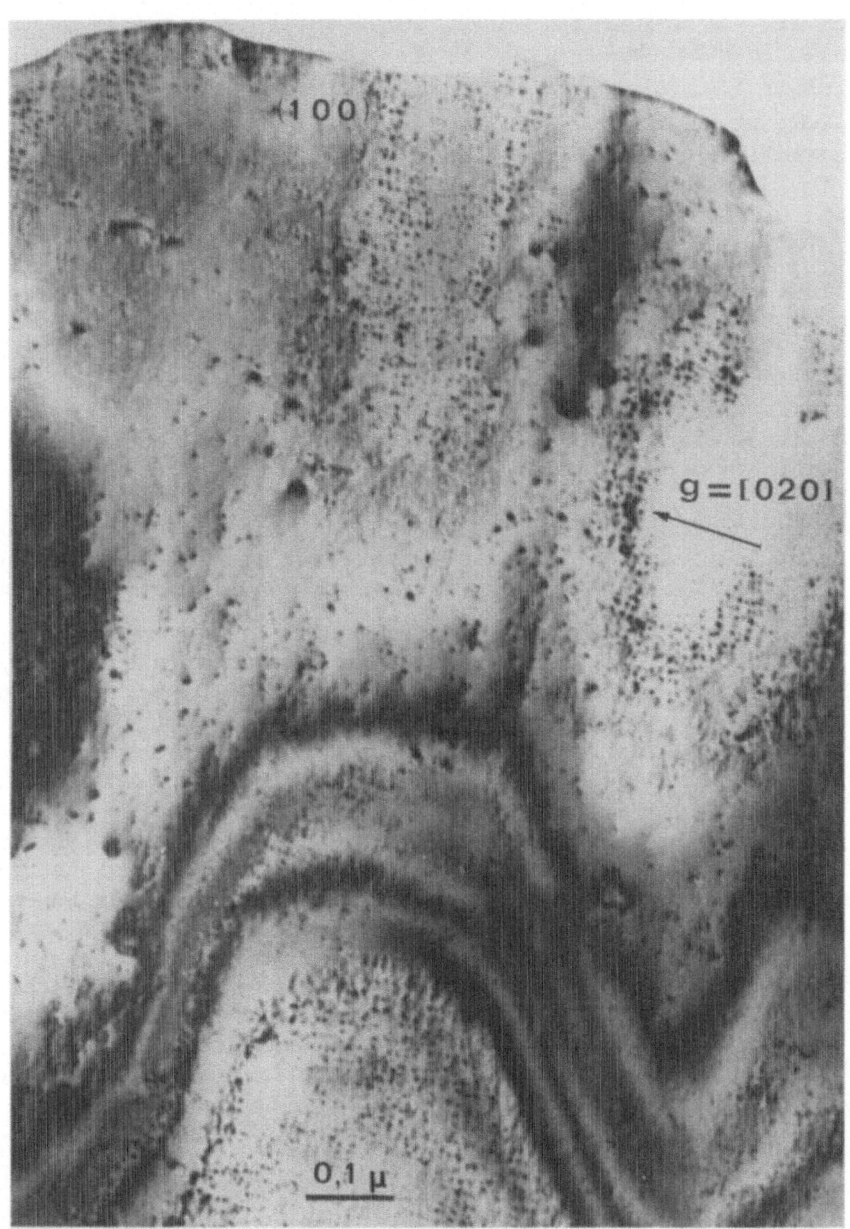

<u>Figure 8</u> : Gold sample (purity 4N) during 2.5 MeV electron irradia-
tion. Flux was $2,3 \times 10^{18}$ e/cm^2/s ; dose was 2.5×10^{22}
e/cm^2.

Typical experimental profiles are presented in Fig. 2. The nature of the defects is also shown. Similar statistics were obtained on all the profiles studied. Our results are summarized in Table 1, in which $h_{p \exp}$ and $\sqrt{<\Delta h^2_{\exp}>}$ are the peak and r.m.s. deviation, respectively, of the experimental depth distribution. These values are compared to the values deduced from ref. [1] : the overall agreement is seen to be quite good. Channeling occurred in a few cases, leading to considerable increases both in the range and the width of the damage distribution, but with apparently little effect on the nature of the observed damage.

An interesting result, displayed in Fig. 3, emerges from our experiments : for a gold host, at least, and at the energies indicated in Table 1, elements lighter than Mo produced mostly interstitial loops, while Mo and heavier elements produce mostly vacancy loops*. This behaviour is not understood at the moment. Possible explanations might involve kinematical effects or condensation of point defects around impurities already present in the host before the implantation ; clearly, it will be interesting to see how this result depends on the implantation energy, and on the nature and purity of the host. If implanted impurities are involved in the effect, hyperfine interaction experiments can also give information [8].

A number of 50 keV Kr-implanted Au samples were submitted to 2.5 MeV electron irradiation, at fluxes of $3-9 \times 10^{18}$ $e/cm^2/s$. Dislocation loops appeared in the thicker portions of the sample, and grew as the irradiation was continued. It was not possible to determine whether these loops developed from preexisting defects. At doses higher that 10^{22} e/cm^2, small defects appeared, which ordered along the <200> directions (Figs. 4-6). Optical diffraction in coherent light showed the ordering process : two-dimensional alignment of broken chains was observed, which remained after annealing at 200°C. After annealing at 700°C, the remaining defects were stacking fault tetrahedra (Fig. 7) which appeared to retain some directional alignment. We have no explanation to offer for these results. It is not clear whether the implanted ions (or impurities) are necessary for the ordering to occur, since similar effects were observed in electron bombardment experiments on unimplanted gold samples (purity 99,99 %) annealed for two hours in air at 1030°C (Fig. 8).

---

* For reasons related to our choice of the diffraction vector in that case, the result on Fe-implanted Au, albeit reasonably well-established, still warrants confirmation.

References

[1] Winterbon, K.B., Sigmund, P., and Sanders, J.B., Klg. Danske
    Vid. Selskab. Mat. Fys. Medd. 37, n° 14 (1970).
    Winterbon, K.B., private communication.

[2] Brice, D.K., Rad. Eff. 11, 51 and 227, (1971).

[3] See for example Mayer, J.W., Eriksson, L., and Davis, J.A.,
    Ion Implantation in Semiconductors, Academic Press, N.Y. (1971)

[4] Rühle, M.R., and Wilkens, M., Phil. Mag. 15, 1075 (1967).

[5] Norris, D.I.R., Phil. Mag. 19, 527-653 (1969).
    Haüssermann, F., Phil. Mag. 25, 561 (1972).
    Thomas, L., Schober, T., and Ballufi , R., Rad. Eff. 1, 257-
    279 (1969).

[6] Ruault, M.O., Jouffrey, B., and Joyes, P., Phil. Mag. 25, 833
    (1972), and Ruault, M.O., Thesis, Orsay 1971 .

[7] Chaumont, J., et al., 3rd Int. Conf. Appl. Vac. Tech., Suppl.
    Le Vide 52, 105 (1971).

[8] Bernas, H., Ruault, M.O., and Jouffrey, B., Phys. Rev. Lett.
    21, 859 (1971), and Bernas, H., Centennial Meeting of the
    French Physical Society, Vittel, May 1973 (to be published).

TRANSMISSION ELECTRON MICROSCOPE STUDIES OF DEFECT

CLUSTERS IN ALUMINIUM IRRADIATED WITH GOLD IONS

C. Gómez-Giráldez, B. Hertel, M. Rühle
and M. Wilkens

Max-Planck-Institut für Metallforschung
Stuttgart

1. INTRODUCTION

After irradiation of gold and copper with self ions defect clusters can be observed by means of transmission electron microscopy (TEM), cf., Merkle/1/, Thomas et al. /2/, Häussermann/3/. The electron microscopical analysis of the radiation-induced defect clusters (cf., Wilkens /4/) showed that after low dose irradiation the defect clusters are of vacancy type. Each observable defect cluster represents the collapsed vacancy rich region of a cascade.

Similar experiments were performed in aluminium /5-9/. However, there were differences in the results obtained by TEM in Al (Ruault et al./9/) compared to Au and Cu:

(i) After irradiation with projectiles (mass M_p) with $M_p \leq M_t$ (M_t = mass of the target Al) defects were only observed after a high dose irradiation.

(ii) The number density N_d of the defects was ≤ 0.01 of the ion dose ϕ. N_d did not increase linearly with ϕ.

(iii) Both vacancy and interstitial type defects were analysed.

For the explanation of the latter results we make use of theoretical evaluations of the sizes and the structures of the cascades (Sigmund and Sanders/10/, Seeger/11/, von Jan/12/). The calculations show that the sizes of the cascades in Al produced by irradiation with self ions are more extended compared to self ion irradiated gold. Therefore, the density of point defects within the Al cascade is very low. As a consequence there is no

noticeable separation of vacancies and interstitials in
the Al cascade. From these arguments it follows that the
observable defect clusters in Al /5-7, 9/ irradiated with
light ions are not spontaneously formed during the irradiation process but by a diffusion process of single point
defects.

However, after irradiation of Al with heavy projectiles ($M_p \gg M_t$) one expects cascades with a small volume
(and high defect concentration) and also a vacancy rich
core in the centre of the cascade. Similar as in gold or
copper this vacancy rich core could collapse to a vacancy
defect cluster observable by TEM. Some experimental studies of Norris/8/ support this presumption. We irradiated Al foils with Au ions. Before the experimental results are described (Section 3) some theoretical results
concerning the sizes of cascades are discussed in the
next section.

2. THEORETICAL CONSIDERATIONS

The range and the straggling of the range of the
projectile as well as the sizes of the cascades can be
calculated theoretically /10-14/. In the theory (Lindhard et al./13/, Sigmund and Sanders/10/, Winterbon et
al./14/) a random distribution of the atoms in the target is assumed, no crystal lattice effects are included.

For the irradiation of Al with Au ions (mass ratio
$M_t/M_p = 0.137$, exponent m in the power law cross section
m = 1/3, if $E_p < 200$ keV) one expects that the projectile
knocks off many Al atoms and leaving behind a vacancy rich
region /14/. The range of the projectile is larger than
the average damage depth. In Tab. 1 the calculated ranges,
the average depth x_D (centre of the cascade) and the radii
R_{cas} of different cascades are summarized.

Table 1: Calculated Ranges and Properties of Cascades

	Energy E_p keV	m	Range Å	x_D Å	R_{cas} Å	T_{max}/T
Au→Al	70	1/3	299	183	127	0.424
	40	1/3	201	123	89	
	20	1/3	130	79	55	
Al→Al	70	1/2	1172	937	746	1
	30	1/2	523	418	309	

The calculations described above were performed for
a random distribution of the atoms. The lattice structure
will influence the size and defect concentration since

interstitial atoms can be transported over long distances far away from the centre of the cascade, e.g., due to focussing collision sequences (Silsbee/15/). The latter effect supports the spatial separation of vacancies and interstitials.

All calculations were performed at 0 K. At elevated temperature one expects that defects within a cascade can rearrange themselves or can migrate. Those effects influence the calculated quantities noted in Table 1.

If we assume (Kinchin and Pease/16/) that

$$\nu_a = \frac{E_p}{2\,E_d}$$

is the (upper) number of formed Frenkel pairs (E_p = energy of the projectile, E_d = displacement threshold energy), than only

$$\nu_a = \eta_a \cdot \nu_o$$

Frenkel pairs will survive after the annealing and only

$$\nu_{cl} = \eta_{cl} \cdot \nu_o$$

point defects will cluster in observable defect clusters. η_a is the displacement efficiency (cf., Robinson/17/) and η_{cl} the clustering efficiency (cf., Torrens and Robinson /18/, Doran and Burnett/19/). η_a was determined by computer simulation of a displacement cascade for a static lattice with thermal vibrations /18/ and also for a cascade where the migration of vacancies and interstitials was included /19/. Accordingly, η_a and $\eta_{cl} < \eta_a$ should be very small. For the irradiation of Al with Au ions at room temperature one expects

$$\eta_{cl} \leq 0.05 \ .$$

3. EXPERIMENTAL DETAILS

Cold rolled aluminium foils (99.999 % purity) were annealed at 550 $^\circ$C and polished electrolytically. The orientation of the foil normal was either close to a {100} or close to a {110} direction. On the surface of the foil an Al_2O_3 layer was observed. The thickness ($\leq$ 50 Å) depends slightly on the orientation of the foil.

The foils were irradiated with Au-ions of energies of 20, 40 and 70 keV. The accelerator (cf., Häussermann /3/) produced an ion flux of $\sim 1.10^9$ ions cm^{-2} s^{-1}, the total dose ranged between $\sim 5.10^{10}$ ions cm^{-2} and $\sim 2.10^{12}$ ions cm^{-2}. The specimens were examined in a JEM 200A electron microscope operated at 150 keV and in a Siemens Elmiskop 102 operated at 125 keV.

4. EXPERIMENTAL RESULTS

The most extensive studies were done at specimens irradiated with 70 keV ions. Few irradiations were performed at 20 and 40 keV.

The irradiated specimens were examined under different dynamical diffraction conditions. Under kinematical diffraction conditions only the largest defects could be seen. Defects were best visible in the very thin part of the foil near the dark part of low order thickness fringes. Fig. 1 shows a foil irradiated with $1.7.10^{12}$ ions.cm^{-2}. At the flanks of the thickness fringes the contrasts of the defects are black and white dots. In the centre of the dark thickness fringes black-white contrasts (BW contrasts) are best visible. In thicker areas of the foil (thickness $t \gtrsim 3\ \xi$; ξ = extinction length) the visibility becomes worse (Noggle et al./20/), often only few very faint contrast figures are observable. It has been proved experimentally that the black and white dots and the BW contrasts are caused by the same defect type (cf., Ruault et al./9/).

4.1 Dose Dependence

The number density N_d of defect clusters was de-

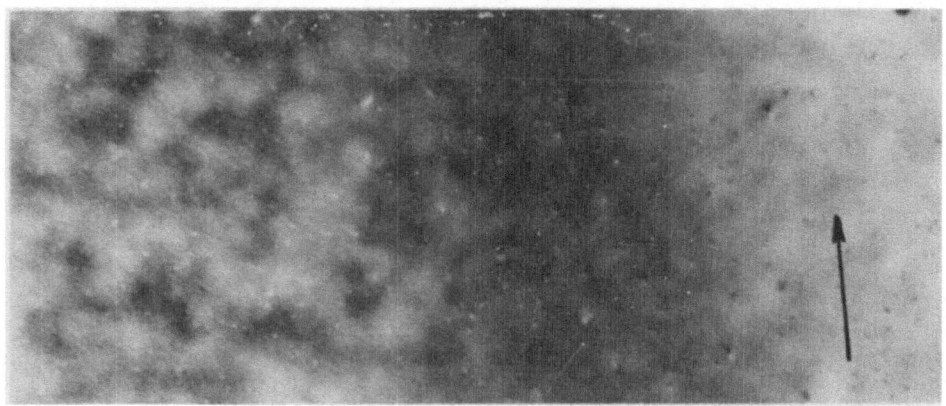

Fig. 1 Aluminium bombarded with gold ions ($\phi = 1.7.10^{12}$ cm^{-2}) electron beam direction parallel to [001] , the operating diffraction vector $\underline{g}$ = (200) is noted in the micrograph; dynamical diffraction conditions. 220000 x. The black and white spots and the BW contrasts (in the central area of the dark thickness fringe) are irradiation induced defects of the same type.

termined as a function of the irradiation dose ϕ. Only those areas were used for a quantitative evaluation where black or white dots were observable (close to the dark thickness fringes). The results are plotted in Fig. 2 which shows that N_d increases linearly with ϕ. The yield factor $Q = N_d/\phi$ for the 3 energies are noted in Table 2.

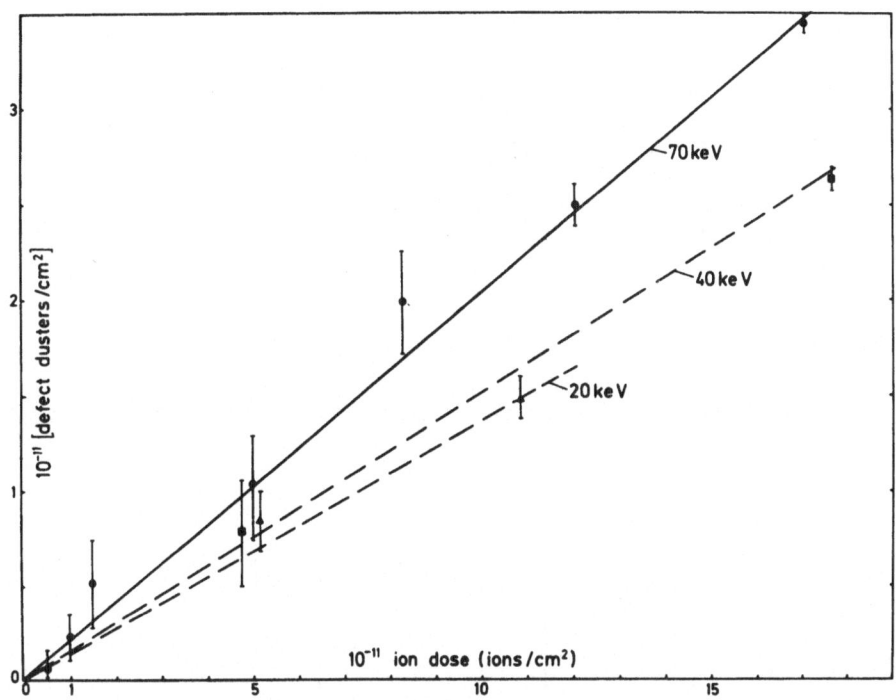

Fig. 2 Number density N_d of observable defect clusters as a function of irradiation dose ϕ after 70, 40, and 20 keV gold ion irradiation.

Table 2: The Yield Factor $Q = N_d/\phi$

Energy (keV)	70	40	20
Q	.204	.15	.14

4.2 Shape and Nature of the Defects

The shape and the nature of the defects was determined by imaging the defects with different diffraction vectors and by applying the stereo technique (cf., Wilkens/4/). The stereo micrographs were also taken under

dynamical diffraction conditions. From the feature of
the BW contrasts and from the results of stereo measurements it could be concluded that nearly all defects are
Frank dislocation loops of vacancy type. This result is
in agreement with observations by Norris/8/.

4.3 Size Distribution

The sizes of the black and white dots observable in
foils irradiated with $\sim 10^{12}$ ions.cm^{-2} of 20, 40, and 70
keV were measured. The normalized size distribution are
represented in Fig. 3. Katerbau/21/ calculated the widths
w of the contrasts of Frank dislocation loops (diameter
d) lying on {111} planes in a very thin {100}-foil for
the determination of the correlation between w and d
(cf., Wilkens/4/). The preliminary result is

$$w = (1.5 \ldots 2.5)\ d \ .$$

The size distribution measurements (Fig. 3) have to be
corrected appropriately.

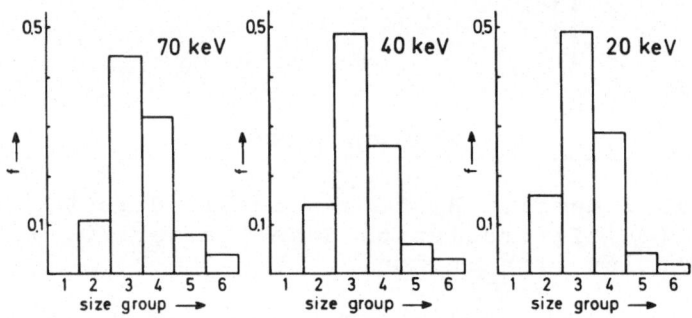

Fig. 3 Size distribution of observable black and white
spots. The fraction f (= number of defects per size
group/total number of measured defects) is plotted as a
function of the size group. The range of each size group
is 12.5 Å. Note that the actual diameters of the defect
clusters are 1.5...2.5 times smaller than the contrast
widths.

4.4 Depth Distribution

Stereo pairs were taken under dynamical diffraction
conditions. Most defects were lying in a layer of about
350 Å thickness close to the irradiated surface. Some

defects lying much deeper inside the foil. They were
produced by ions which penetrated the crystal to high
depths by a channeling process (cf., Häussermann/3/).

4.5 Clustering Efficiency η_{cl}

If we assume that the defect clusters consist of
circular monoatomic layers of vacancies (Frank dislocation loops) we can calculate the clustering efficiency
η_{cl} as introduced in Section 2. The displacement threshold energy was assumed to be 17 eV (Wolfenden/22/). The
actual value of η_{cl} can only be determined with the uncertainty described in Section 4.3. As shown in Table 3
the efficiency increases with decreasing energy.

Table 3: Clustering Efficiency

Energy keV	value	w Å	Diameter Å	η_{cl}
70	mean	35	14...21.2	0.011...0.024
	max	72	29...48	0.045...0.124
40	mean	30	12...20	0.014...0.037
	max	65	26...43.3	0.064...0.176
20	mean	28	11...18.7	0.024...0.066
	max	60	24...40	0.109...0.302

5. DISCUSSION

5.1 Influence of the Al_2O_3 Layer on the Defect Structure

On electrolytically polished aluminium specimens a
layer of Al_2O_3 will always be present. The thickness of
the layer is $\lesssim$ 50 Å. In the oxide layer the projectiles are
slowed down causing a damage of the layer. Jespersgård and
Davies/23/ studied the ranges of noble gas ions in amorphous Al_2O_3. The measurements of the ranges of Xe ions in
Al_2O_3 can be used for the determination of the ranges of
gold ions in Al_2O_3. It is found that the mean free path
of heavy ions in Al_2O_3 is about 5 % smaller than in pure
Al. Therefore, we have to assume that part of the energy
of the incoming ion is already dissipated in the Al_2O_3
layer. However, since (after electropolishing) the oxide
layer has always about the same thickness we conclude that

the influence of the Al_2O_3 layer is the same in all
specimens used for evaluation of the defect structure.
It will reduce the density of visible defects.

5.2 Formation of the Defect Clusters

In this section we will discuss the possibilities
by which the defect clusters can be formed and with which
mechanisms the sizes and concentrations could be changed.
The defect clusters observable in the irradiated
specimens were analysed as Frank dislocation loops of
vacancy type. We assume that the defects are created by
a collapse of the vacancy rich regions in the centres of
the cascades which are produced by the Au ions. The assumption is strongly supported by the observed linear dose dependence (Fig. 2).
During the irradiation of Al with Au ions energetic
aluminium ions are also formed (maximal transferred energy = 0.424 T, cf. Table 1). Those energetic Al ions (for
T_p = 70 keV, T_{max} = 30 keV) create, similiarly as in /9/,
point defects. However, the density of the random distributed defects is so small that the clustering probability
is negligible small. This effect will not be effective in
our irradiation experiments.
Nearly all electron microscopical observations were
carried out at an operating voltage of the electron microscope of 150 keV. Wolfenden/22/ has shown that after a 40
minute irradiation with electrons of that energy defect
clusters are formed if the Al was doped with impurities
prior to the observations. We assurred that this effect is
not effective in our Al after Au ion bombardment by
imaging the defect clusters with microscopes operated at
150 keV (JEM 200A) and at 125 keV (Elmiskop 102), respectively. On the micrographs the same number densities of
defect clusters were observed.

5.3 The Yield Factor

After irradiation with 70 keV gold ions the yield
factor was Q = 0.21 (see Table 2). This value should be
compared with the Q values of Cu irradiated with 70 keV
Au ions Q = 0.4...0.5 (estimated from /3/) and Au irradiated with Au ions Q $\approx$ 1 (Merkle et al./24/). Under this
point of view the rather small yield factor in our experiments seems reasonable since the theoretically calculated concentrations of point defects in the cascade
is smallest for the Au $\rightarrow$ Al irradiation.

On the other hand there may be two experimental
reasons for the explanation of the small value of Q:
(i) The number density in the smallest size group is,
especially in Al, extremely sensitive to the imaging
conditions in TEM.
(ii) In the 50 Å thick Al_2O_3 layer cascades are formed
which are not visible in the electron microscope.

5.4 The Shape of the Defects

The electron microscopical analysis revealed that
nearly all observable defects were Frank dislocation
loops. Also by defocusing experiments (Rühle and Wilkens
/25/) no cavities could be determined. This is in contrast to observations in gold and copper /2, 3/ where
the smallest defects (preferably after low energy irradiations) were three dimensional clusters. From calculations of the stable cluster configurations (Siegler
and Kuhlmann-Wilsdorf/26/) one also expects small cavities.
Our observations can be explained by either assuming
that defects with three dimensional structure are only
formed very close to the surface, i.e., still in the
Al_2O_3 layer, or that those defects are not stable after
irradiation of Al at room temperature.

5.5 The Clustering Efficiency

The experimental determination of the clustering
efficiency resulted in values which were unexpectedly large
(see Table 3). Even larger values than in our experiments
were obtained in ion bombarded Au and Cu (cf., Wilkens
/4/). The large values are in contradiction to computer
simulation experiments /18, 19/.
The difference could be explained by the different
ways of the production of the cascades. In our experiment
both interstitials and vacancies are mobile at the irradiation temperature. In the computer simulation /18, 19/,
however, the point defects are immobile while the cascade
is formed and they become mobile during the annealing
treatment. Under the latter conditions there may be a
higher probability for the interstitials to recombine
with the vacancies of the vacancy rich core of the cascades than under our experimental conditions. This leads
to a smaller clustering efficiency. The model may be
supported by experimental results obtained in neutron-
irradiated copper (Rühle et al./27/). In this experiment
foils were irradiated with about the same neutron doses
at $\sim 80\ ^{\circ}C$ (HTI), and at 4.2 K with subsequent warming up

to room temperature (LTI). Both specimens were investigated by TEM at room temperature. As shown in /27/ the
size distribution of the observable vacancy clusters in
a HTI specimen extends to larger diameters than in a LTI
specimen. Also more interstitial clusters are formed
after HTI than after LTI. However, for a full explanation
of the high values of the clustering efficiency in ion
bombarded metals more experimental work has to be done.

6. SUMMARY

Aluminium foils are irradiated with gold ions of
20, 40, and 70 keV. The damage structure is studied with
the electron microscope. Small defect clusters can be
analysed. The number density of the defect clusters increases linearly with the dose with a yield factor of
0.21 after 70 keV irradiation. The defects are analyzed
as Frank dislocation loops ($\underline{b}$ = 1/3 <111>) of vacancy
type. The loops are formed by a collapse of the vacancy
rich region in the rather small cascade. Small spherical
defects observed in Au and Cu are missed in our experi-
ments. This could be caused by the presence of the 50 $\overset{\circ}{A}$
thick Al_2O_3 layer on the electropolished aluminium. The
unexpected high clustering efficiency is discussed. The
value could be explained by the assumption that during
irradiation at room temperature interstitials escape
with a high probability from the vacancy rich region of
the cascade.

ACKNOWLEDGMENT

We would like to thank Dr. J. Diehl for helpful
discussions, Mr. H. Föll and Mr. R. Schindler for experimental assistance during the irradiation experiments.
    The financial support of the Fundacion Juan March/
Spain is gratefully acknowledged, who gave one of us
(C.G.-G.) a grant for a one year stay at the Max-Planck-
Institut für Metallforschung in Stuttgart.

REFERENCES

/ 1/ K.L. Merkle: phys. stat. sol. <u>18</u>: 173 (1966); in:
      "Radiation Damage in Reactor Materials", Int. Atomic
      Energy Commission, Vienna, Vol. 1: 159 (1969).

/ 2/ L.E. Thomas, T. Schober, and R.W. Balluffi: Radiation
     Effects <u>1</u>: 257 (1969).

/ 3/ F. Häussermann: Phil. Mag. <u>25</u>: 537 (1972).

/ 4/ M. Wilkens: these proceedings.

/ 5/ C.J. Beevers and R.S. Nelson: Phil.Mag. <u>8</u>: 1189 (1963).

/ 6/ L.M. Howe, J.F. McGurn, and R.W. Gilbert: Acta Met.
     <u>14</u>: 801 (1966).

/ 7/ L. Henriksen, A. Johansen, J. Koch, H.H. Andersen,
     and R.M.J. Cotterill: Appl. Phys. Let. <u>11</u>: 136
     (1967).

/ 8/ D.I.R. Norris: Proc. Symp. on the Nature of Small
     Defect Clusters, A.E.R.E. Report R 5269: p. 433
     (1966); Phil. Mag. <u>19</u>: 527 (1969).

/ 9/ M.O. Ruault, B. Jouffrey, and P. Joyes: Phil. Mag.
     <u>25</u>: 833 (1972).

/10/ P. Sigmund and J.B. Sanders: Proc. Int. Conf. on
     Application of Ion Beams to Semiconductor Technolo-
     gy (ed.: Ph. Glotin), Editions Ophrys, p. 215 (1967).

/11/ A. Seeger: Proc. U.N. Intern. Conf. Peaceful Uses
     Atomic Energy, 2nd Geneva <u>6</u>: 250 (1958).

/12/ R. von Jan: phys. stat. sol. <u>6</u>: 925 (1964); <u>7</u>: 299
     (1964); <u>8</u>: 331 (1965).

/13/ J. Lindhard, M. Scharff, and H.E. Schiøtt: Mat.
     Fys. Medd. Dan. Vid. Selesk. <u>33</u>: No. 14 (1963).

/14/ K.B. Winterbon, P. Sigmund, and J.B. Sanders: Mat.
     Fys. Medd. Dan. Vid. Selesk.<u>37</u>: No. 14 (1970).

/15/ R.H. Silsbee: J. Appl. Phys. <u>28</u>: 1246 (1957).

/16/ G.H. Kinchin and R.S. Pease: Rep. Prog. Phys. <u>18</u>: 1
     (1955).

/17/ M.T. Robinson: in "Radiation-Induced Voids in Metals
     (Edts.: J.W. Corbett, L.C. Ianniello), U.S. Atomic
     Energy Commission, p. 397 (1972).

/18/ I.M. Torrens and M.T. Robinson: in "Interatomic
     Potentials and Simulation of Lattice Defects"
     (edts.: P.C. Gehlen, J.R. Beeler, Jr., and
     R.I. Jaffee) Plenum Press New York-Londn (1972) p.423.

/19/ D.G. Doran and R.A. Burnett: cit. 18, p. 403.

/20/ T.S. Noggle, O.S. Oen, and J.C. Crump III: Proc.
     28th Annual EMSA Meeting, Houston, Texas, p. 406
     (1970).

/21/ K.-H. Katerbau: private communication.

/22/ A. Wolfenden: Radiation Effects 14: 225 (1972).

/23/ P. Jespersgård and J.A. Davies: Can. J. Phys. 45:
     2983 (1967).

/24/ K.L. Merkle, L.R. Singer, and J.R. Wrobel: Appl.
     Phys. Let. 17: 6 (1970).

/25/ M. Rühle and M. Wilkens: Proc. of the Vth Europ.
     Congress on Electron Microscopy; The Institute of
     Physics, London and Bristol: p. 416 (1972).

/26/ J.A. Sigler and D. Kuhlmann-Wilsdorf: cit. 8, p.125.

/27/ M. Rühle, F. Häussermann, and M. Rapp: phys. stat.
     sol. 39: 609 (1970).

DECHANNELING FROM DAMAGE CLUSTERS IN HEAVY ION IRRADIATED GOLD

P. P. Pronko and K. L. Merkle

Argonne National Laboratory, Argonne, Illinois   60439

INTRODUCTION

The dechanneling of a well aligned beam of light particles can be initiated through interaction with a variety of crystal defects(1). Examples of such defects are displaced atoms, stacking faults, dislocations and clusters of vacancies or interstitials. Radiation damage in metals bombarded by heavy ions is characterized by very low concentrations of free interstitials and rather high concentrations of defect clusters. In gold, the large clusters are of the vacancy type. Direct backscattering from defect clusters is generally not observable; however, dechanneling of an aligned beam can be induced by such defect clusters and has been observed for a variety of experimental conditions(2,3).

We recently reported on the dechanneling of 2 MeV $^{4}He^{+}$ beams in gold from defect clusters produced by random irradiation with 2 MeV $^{4}He^{+}$(2). The work being presented here extends our previous results to the case of heavy ion Au^{+} bombardment of gold. Dechanneling cross sections for 270 keV $^{4}He^{+}$ and protons have been measured as well as the saturation damage characteristics for high concentrations of discrete clusters. Collision cascade volumes are obtained from such saturation measurements. Variations in the dechanneling cross sections and cascade volumes can thus be determined as a function of the incident Au^{+} energy.

DEFECT CLUSTERS

Electron microscopy observations have shown that, in ion bombarded gold, rather large defect clusters can be directly

481

formed by energetic-displacement cascades(4). These black-spot
defects are produced within individual displacement cascades provided that the dose is small enough so that the individual cascades are well separated in space. Such defects are predominantly
vacancy clusters that are precipitated at the depleted zone.

  Observations by TEM (Transmission Electron Microscopy) make
it possible to quantitatively compare the effects of defect
clusters on dechanneling with the actual size and concentration of
these defects. Figure 1 shows an example of a heavy ion induced
collision cascade in gold as observed by TEM.

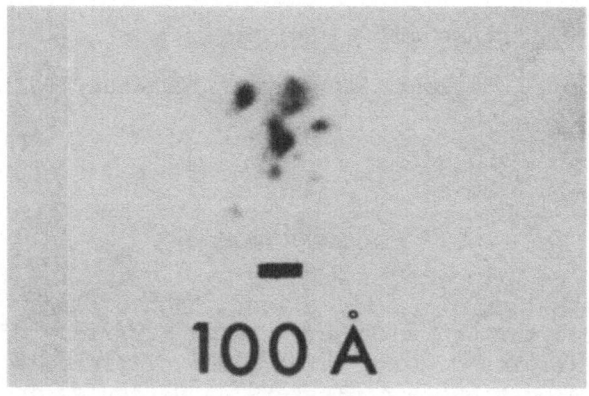

Fig. 1

Example of TEM "black-spot" defect clusters resulting from 300 keV
self ion irradiation of gold. Shown in the figure is a cascade
from one incident ion. The formation of a sub-cascade structure
is clearly visible.

A system of sub-cascades is seen which is formed by the clustering
of vacancies. The structure in Fig. 1 is the result of a single
incoming ion at 300 keV. Many such cascades are formed as the
irradiation proceeds and eventually a condition of saturation will
be generated where adjacent cascades begin to overlap. Our dechanneling experiments are performed through the range where initially
these clusters are well separated and up to the condition where a
significant number of them are beginning to overlap. Should irradiations proceed well beyond the saturation level of single defect
clusters, then a new form of damage is expected which is characterized by the kinetic interaction of a larger number of clusters
resulting in the formation of extended structures such as dislocation loops and networks. Our present study is not concerned
with such extended structures.

TEM Observations of Defect Clusters in Au[+] Bombardment of Au

The depth distribution and size distribution of defect clusters from Au[+] irradiation of gold have been obtained from TEM observations(5). Stereographic methods were used in obtaining the depth distributions. An example of these data is given in Fig. 2a and 2b for a 250 keV Au[+] irradiation of gold.

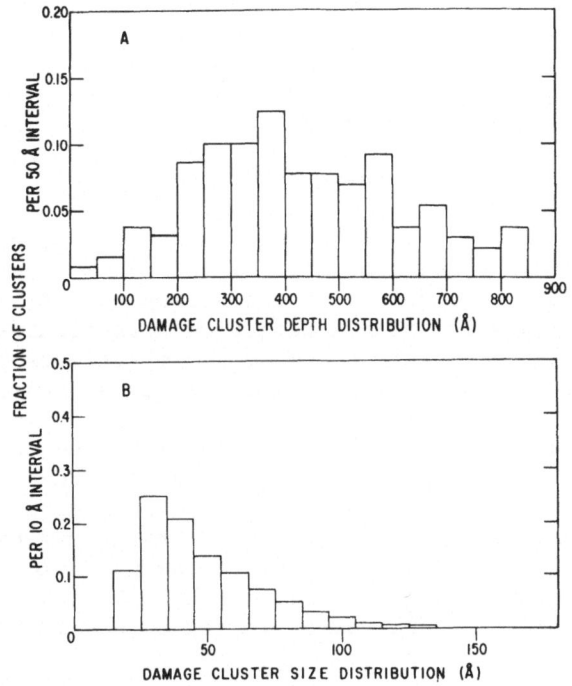

Fig. 2a
Damage cluster depth distribution observed by TEM for 250 keV Au[+] irradiation of gold.

Fig. 2b
Damage cluster size distribution from TEM for 250 keV Au[+] irradiation of gold.

The distribution in depth is spread over a range of about 1000 Å with its maximum concentration in the region of 400 Å. The distribution in size appears to be skewed with a most probable value of 30 Å for the cluster diameter. The skewing in the distribution may be due, in part, to the resolution limitations of the microscope at 10 to 20 Å. The cluster diameters being presented here are the diameters of the sub-cascades as seen in Fig. 1. It is observed that the average size of the sub-cascades is almost

constant over a wide range of incident ion energies. The number
of sub-cascades (i.e. clusters) per cascade increases in a linear
fashion with increasing bombarding energy. It is observed that
on the average, 50 keV in nuclear displacement energy is absorbed
by each cluster, resulting in a relatively constant cluster size
over a range of 50 to 500 keV incident ion energy. Subtracting the
electronic losses from this incident energy gives the damage energy
E_D which goes into nuclear displacements. As the energy goes up
an increase occurs in the total cascade size as well as in the number of defect clusters.

DECHANNELING

Background

Dechanneling of an aligned beam occurs when the transverse
energy of a particle exceeds a critical value and it can no longer
be retained within the confines of the string potentials of the
atomic rows. The way in which a defect imparts transverse energy
to a channeled particle will determine the efficiency with which
it can dechannel that particle. Single displaced atoms, edge
dislocations, and defect clusters will all induce transverse
momentum to a channeled particle as it passes close to such a
structure. However, the particle may or may not dechannel immediately after passing such a defect. In some cases the particle
may dechannel immediately, however, in others it may take a large
number of anharmonic oscillations before the particle escapes the
channeled configuration. In the latter case the location and
character of the responsible defect is not as easily identifiable
as in the former.

Mory and Quéré(6) have presented a picture of dechanneling
where stacking faults and dislocations dechannel according to
their integrated concentration along the path of the channeled
beam. They calculate dechanneling cross sections on the basis of
geometrical details of the defects and compare these, with reasonable success, to measured values. They assume that the channeled
particles either pass through the defects with only a slight
change in transverse energy and oscillatory wavelength or else
they dechannel completely between 0 and $\lambda/4$, where λ is the
oscillatory wavelength and $\lambda/4$ is a turning point of the motion.
Picraux(7) has examined dechanneling from stacking faults in
silicon and used a multiple scattering calculation to determine the
number of such defects. Matsunami and Itoh(8) have used the
concept of a diffusion of trajectories in transverse energy space
to calculate the dechanneling from displaced atoms. They use this
model to determine the shape of the backscattered energy spectrum
and a dechanneling cross section is obtained from the slope of such

a curve. They find that for 1.5 MeV protons the dechanneling cross-section of a displaced atom is on the order of 10^{-18} cm^2. The dechanneling cross sections we obtain for heavy ion defect clusters as reported previously(2) and in the present work are between 4 and 5 orders of magnitude larger than this and clearly represent dechanneling centers much larger than single displaced atoms.

Elementary Model for Cluster Dechanneling

Given a crystal with N_s cascades per unit volume, a channeled beam of intensity I_c will be partially dechanneled by an effective cross section σ_d from each cascade. At a given depth z below the surface the channeled intensity will be reduced by dI_c in going from z to z + dz. To first approximation(2) the change in channeled intensity with depth is:

$$\frac{dI_c}{dz} = -I_c \left(N_s\sigma_d + \xi\text{th}\right) \quad , \tag{1}$$

where σ_d is the total dechanneling cross section per cascade and ξth takes into account dechanneling from thermal vibrations. Integrating this effect from the surface inward results in an expression for the dechanneled fraction of:

$$\chi = 1 - (1 - \chi_{min}) \exp\left[- (N_s\sigma_d + \xi\text{th})z\right] \quad . \tag{2}$$

The dechanneling cross section is obtained from the slope of the dechanneled fraction as a function of depth and is given by:

$$\sigma_d = \frac{1}{N_s} \Delta(\frac{d\chi}{dz}) \frac{1}{1-\chi_{min}} \quad . \tag{3}$$

The slope is calculated by fitting a line to the dechanneled spectrum over the first 1200 Å. Thus σ_d can be extracted from the data since the number of cascades per unit volume may be obtained from the total fluence (ions/cm^2) and the range over which it is distributed. In our calculations we have used 1000 Å as the range over which the damage is distributed. This is a reasonable first approximation. Ideally one should use the exact limits of the damage distribution (or the rms straggling distance). TEM observations indicate that the damage distributions are spread out rather broadly within the first 1200 Å with a wide peak in the distributions centered at about 400 Å for 250 keV Au$^+$ bombardments and 500 Å for 500 keV bombardments.

As the irradiations proceed to higher fluences and the damage clusters begin to overlap, the cluster production rate will be proportional to the undisturbed lattice volume $1-N_sV_o$, where

V_o is the volume of the cascade. Since, for low damage concentrations $N_s = \phi/t$, where t is the depth over which the damage is distributed and ϕ is the damaging ion fluence per cm^2, then:

$$\frac{dN_s\sigma_d}{d\phi} = \frac{1}{t} (\sigma_d - N_s\sigma_d \cdot V_o) \qquad (4)$$

for the case where the number of cascades per unit volume begins to approach $1/V_o$. Thus, a plot of $dN_s\sigma_d/d\phi$ versus $N_s\sigma_d$ should result in a straight line of negative slope with V_o being given by the abscissa intercept. That is, $V_o = \sigma_d/(N_s\sigma_d)_\infty$.

OBSERVATIONS AND RESULTS

Experimental

The single crystal gold specimens were prepared by vapor deposition of gold on a heated (350°C) sodium chloride substrate cleaved parallel to the (001) face. This resulted in a (001) orientation on the gold as well. The film (usually between 1000 and 2000 Å) was left on the substrate throughout the channeling experiments. In order to reduce strain at the interface and also to minimize the dislocation density, an annealing cycle, in air, of 350°C for 30 minutes was performed prior to mounting the specimen in the channeling apparatus. Minimum yields from the gold film of 2.5 to 3% demonstrated that the films were free of strain and contained very few dislocations.

The channeling set up consisted of an ultra high-vacuum target chamber operated in the 10^{-8} torr range. The gold irradiations, as well as the helium or proton channeling were all performed in situ on a 300 keV heavy ion accelerator. Collimation of the beam for channeling measurements was obtained with a pair of precision slits as forward aperture (.04") and a fixed second aperture (1/16") 10 feet down the beam line. The gold irradiations were performed by orienting the crystal in a random direction as determined from backscattering of ^{4}He and then opening the forward collimator. Electrostatic beam sweeping was used to obtain a uniform bombardment. Current integration was obtained from an in-line Faraday Cup that collected an annular portion of the swept beam. For the channeling experiments, the collimation allowed an unswept beam to pass directly through the Faraday Cup and on to the target. In this case, current integration was taken from the target. Crystal orientation could be controlled reproducably to better than 0.05 degrees. Single alignment 150° back scattering was used in all cases.

Detector resolution for ^{4}He at 270 keV incidence was 8.5 keV yielding a depth resolution of 60 Å.   The ^{4}He data was taken with a room temperature detector. The proton spectra were taken with a cryogenically cooled silicon surface barrier detector. The first stage of the preamplifier was also cooled with the detector. This arrangement gave 2.5 keV proton resolution corresponding to about 55 Å depth resolution in gold.

Data

An example of the dechanneling spectra observed for 270 keV Au$^+$ bombardment of gold is presented in Fig. 3.

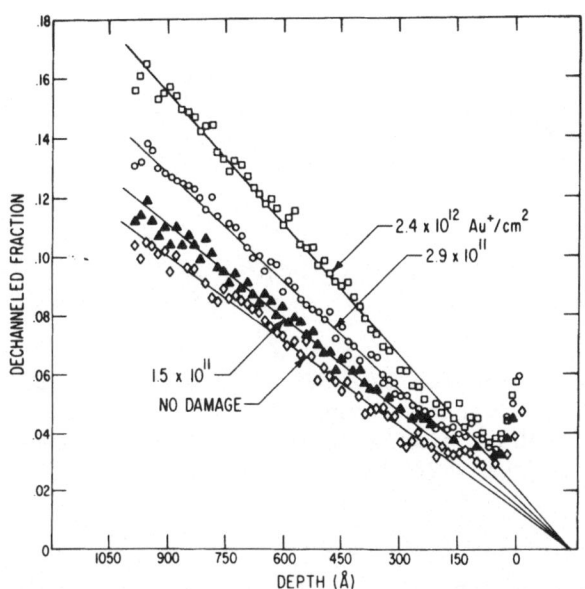

Fig. 3
Dechanneling spectra for 270 keV ^{4}He$^+$ in gold irradiated with 270 keV Au$^+$. Dose levels are indicated in the figure.

These spectra show the dechanneled fraction (ratio of aligned to random spectra) as a function of depth. The depth scale was determined from the known film thickness as determined from gravimetric measurements. As can be seen from Eq. (2), these curves should fit a straight line when $N_s \sigma_d z \ll 1$. In general this will be the case for depths less than one micron. The change in slope for the dechanneling spectrum is shown starting with the case of no damage and going to the case of 2.4 x 10^{12} Au$^+$ ions/cm^2

which is approaching the upper limit of dechanneling observable from defect clusters. The damage cluster concentration is well into the saturated region at this point.

The slopes of the dechanneling spectra of Fig. 3 are used to calculate the dechanneling cross sections according to Eq. (3). The results of this analysis are shown in Fig. 4, for two different gold films, using in both cases 270 keV Au$^+$ irradiations and 270 keV ^{4}He$^+$ dechanneling. In the figure $N_s \sigma_d$ is plotted versus the

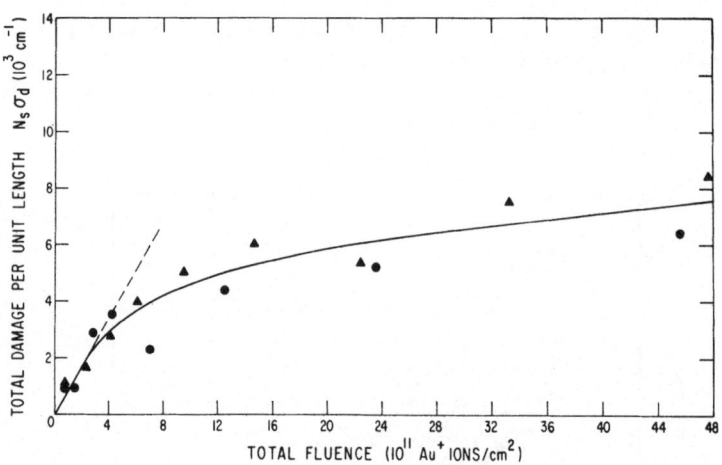

Fig. 4

Damage – fluence curve for 270 keV Au$^+$ bombardment of Au observed with ^{4}He$^+$ (270 keV) dechanneling. The filled circles were obtained from a 1200 Å single crystal Au film on NaCl and the filled triangles from a 2000 Å film.

total fluence of Au$^+$ irradiation. A linear damage-fluence relation can be fitted to the asymptotic data at low fluence. Significant departures from this linearity begin to occur at 5×10^{11} cm^{-2} signaling the onset of saturation effects. The reproducability of the results from the two films is reasonable. A dechanneling cross section for the clusters is obtained from the low dose, straight line portion of the damage fluence curve.

With increasing energy more defect clusters per cascade are formed. Thus, the average dechanneling cross section per cascade should go up with increasing energy. In Fig. 5 are shown the results from 540, 270 and 80 keV Au$^+$ irradiations. The 270 keV curve was transferred from Fig. 4 with the points being left out

for clarity.   The dechanneling cross sections obtained from the
asymptotic low dose regions are 3.25×10^{-13} cm^2, 8.5×10^{-14} cm^2,
and $(2.3 \pm 1.3) \times 10^{-14}$ cm^2 in going from the highest to the
lowest energy.   A discussion of the above cross section will be
given in the following section.

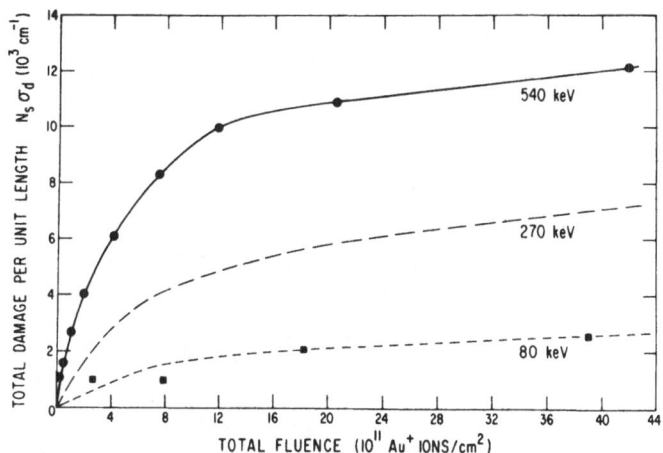

Fig. 5

Comparison of damage-fluence curves for 540, 270, and 80 keV
Au$^+$ bombardment of Au.   The curve for 270 keV was transferred from
Fig. 4.

     The high dose approach to saturation, of the three different
energy curves, is somewhat peculiar in that the curves do not
appear to approach each other as one would expect they eventually
should.   This means that either the saturation characteristics of
the film change with bombarding energy, or the measuring technique
produces an artifact in the high dose region. There is some possibility that the latter is the case since, if dechanneling occurs at
rather long distances from the defects (as previously discussed)
then the response of the dechanneling spectra to changing damage
distributions may be such as to give an aritifical reduction in
observed damage for highly concentrated surface distributions
as compared to distributions that are spread more uniformly through
the film.   Further experiments will be required to determine whether
or not this is the case.

Analysis of the approach to saturation, as described by Eq. (4), has been done for the 540 keV and 270 keV irradiations. The results appear in Fig. 6 where we have plotted the slope of the damage fluence curve versus total damage $N_s \sigma_d$. The results of

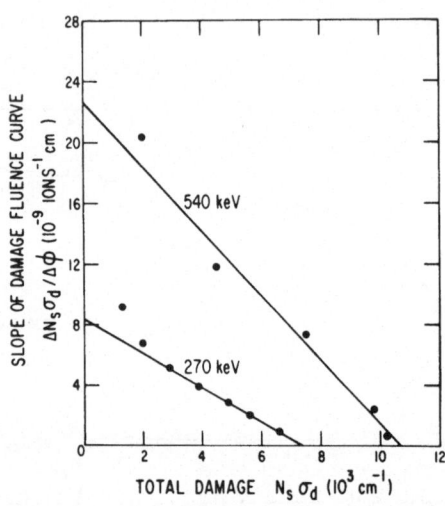

Fig. 6

Slope of the damage-fluence curves versus total damage.

this analysis should yield a straight line with negative slope. This is seen to be the case for damage levels where $N_s \sigma_d$ exceeds 2×10^3 cm^{-1}. The abscissa intercepts $(N_s \sigma_d)_\infty$ are to be used for an evaluation of the cascade volume V_0. The results of that analysis yield V_0 values of 3.04×10^{-17} cm^3 for 540 keV and 1.1×10^{-17} cm^3 for 270 keV which correspond to collision cascade diameters of 388 and 276 Å respectively. These diameters are the range over which the collision cascade originally extends as distinct from the defect clusters that remain after the cascade has come to equilibrium at room temperature.

Charge Dependence of Cross Sections

The observed dechanneling cross sections are determined by the mechanism of interaction of the channeled particles with the defect clusters. A better understanding of this process can be achieved by examining the differences in dechanneling behavior for particles of different nuclear charge and varying incident energy.

In this regard, we have compared cross sections for $^4He^+$ and protons at 270 keV. These data were taken on the same sample after each irradiation so that both particle beams are probing exactly the same damage under identical orientation. In all cases the proton spectrum was taken before the $^4He^+$ spectrum. The results are presented in Fig. 7.

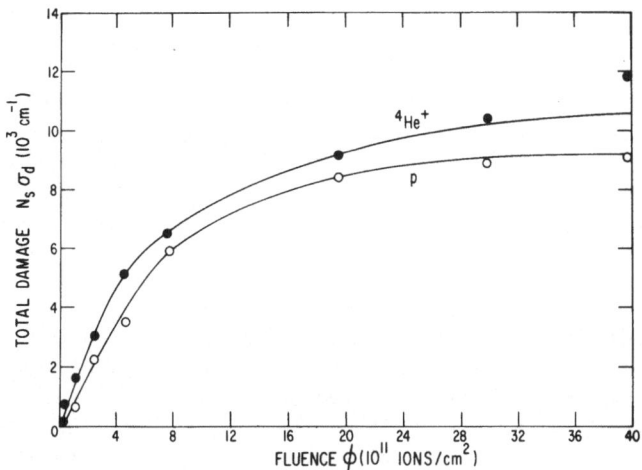

Fig. 7

Comparison of 270 keV $^4He^+$ and proton dechanneling cross sections in 1200 Å single crystal Au film on NaCl after self ion irradiation.

It can be seen in the figure that the proton cross sections are consistently lower than those obtained with $^4He^+$. The critical angle for protons is less than for $^4He^+$ and one would expect that higher proton cross sections should result. However, the charge dependence of the particle collision cross section with the lattice atoms increases with charge and apparently dominate the dechanneling process. In the linear damage-fluence region the ratio of cross section for $^4He^+$ to that of protons is 1.5, suggesting a relationship of $\sqrt{Z}$ dependence on the dechanneling cross section.

DISCUSSION

We have used a simplified model of cluster dechanneling to arrive at dechanneling cross sections from vacancy clusters in self-ion irradiated gold. The model assumes that dechanneling results from encounters with lattice distortions associated with the cluster. The backscattered energy spectrum exhibit good linearity in the first 1000 Å in accordance with the model.

However, as the damage distributions are made narrower by going to
lower energy (e.g. 80 keV Au^+) the dechanneled spectra do not
exhibit changes in shape to reflect the known changes that are
occurring in the damage distributions. This, along with the way
in which different energy implants approach saturation suggests
that cluster dechanneling is probably not occurring until particles
have traveled a significant distance beyond the clusters. The
clusters form in subcascades with energy input of about 50 keV
per cluster. Thus, taking our cross sections for bombardment energies of 540, 270, and 80 keV respectively and dividing by E_D/50 keV,
the number of clusters per cascade, we get 4.6×10^{-14}, 2.3×10^{-14},
and 1.9×10^{-14} cm^2 as the effective dechanneling cross sections for the individual clusters. This roughly agrees with
the results of our previous work with $^4He^+$ damage in gold at 2 MeV
(2), where it was found that the dechanneling cross section of $2.3
\times 10^{-14}$ cm^2 was a factor of 5 smaller than that observed by TEM.
In that case the average cluster size was rather close to the ones
observed in self ion bombardments. The factor of 2 difference
between the 540 and 270 keV run might in part be due to the difficulties in determining the initial slope of the damage vs. fluence
curves accurately. If we take in fact the extrapolated values of
Fig. 6 we get somewhat closer agreement. There is also the possibility that the combined strain field of all the clusters within
a cascade gives rise to enhanced dechanneling for more energetic
cascades.

Measurements of the cascade volume, from saturation experiments yield cascade diameters of 388 and 276 Å for 540 and 270
keV Au^+ irradiations. The random cascade model predicts 460 and
280 Å. respectively(9).

CONCLUSIONS

The results of the present work indicate that dechanneling of
light particle beams is an effective way of studying heavy ion damage clusters in metals. The reasons for drawing this conclusion
are:

(a) the experimental dechanneling cross sections confirm
that defect clusters, observable as black-spots in the electron
microscope, are responsible for dechanneling in the fluence range
up to 5×10^{12} ions/cm^2.

(b) analysis of the saturation damage behavior yields
dimensions for the collision cascades that are consistent with
the random cascade model.

(c) the cluster dechanneling cross section for protons at 270 keV is consistently lower than for $^4He^+$ at the same energy for all fluences studied. In the linear damage-fluence region the $^4He^+$ dechanneling cross section is a factor of 1.5 higher than for protons suggesting a $\sqrt{Z}$ dependence. This suggests that the nature and the strength of the charged particle collision dominates over critical angle considerations in cluster dechanneling.

REFERENCES

(1) G. Delsarte, J. C. Jousset, J. Mory and Y. Quéré, Atomic Collision Phenomena in Solids, edited by D. W. Palmer, M. W. Thompson, and P. D. Townsend (North-Holland, Amsterdam, 1970).
(2) K. L. Merkle, P. P. Pronko, D. S. Gemmell, R. C. Mikkelson, and J. R. Wrobel, Phys. Rev., B 8, 1002 (1973).
(3) Y. Quéré, Ann. Phys. (N.Y.) 5, 105 (1970).
(4) K. L. Merkle, Phys. Status. Solidi 18, 173 (1966).
(5) K. L. Merkle, to be published.
(6) J. Mory and Y. Quéré, Rad. Effects 13, 57 (1973).
(7) S. T. Picraux, J. Appl. Phys. 44, 587 (1973).
(8) N. Matsunami and N. Itoh, Phys. Letters 43 A, 435 (1973).
(9) K. B. Winterbon, P. Sigmund, and J. B. Sanders, K. Dan. Vidensk. Selsk. Mat. Fys. Medd. 37, 1 (1970).

DISCUSSION

Q: (G. Linker) Do you think that there could be a possibility to determine the nature of damage just by dechanneling cross section measurements?

A: Dechanneling cross section measurements are a first step in this direction. Our objective at this point is to confirm the validity of the technique; however it can be concluded from our measurements that the dechanneling centers are defect clusters rather than displaced atoms. Further information on the defect clusters could be obtained by measuring dechanneling cross sections along a variety of channeling directions to determine their three dimensional character. Comparisons between single and double alignment dechanneling may also help in evaluating the nature of the damage centers.

Q: (Wilkens)   How did you derive the value of 50 keV per cluster?

A: Our TEM observations indicate that the average number of clusters per incident ion is roughly proportional to the damage energy, $N_d = E_D/E_0$, in the energy range 50 to 300 keV, with $E_0 \approx 50$ keV.

HEAVY ION DAMAGE IN THIN METAL FILMS

W. Kesternich and K. L. Merkle

Argonne National Laboratory, Argonne, Illinois   60439

INTRODUCTION

A wide range of recoil energies is usually encountered in a material that is under irradiation with energetic particles. It is of considerable interest to know the damage as a function of recoil or cascade energy. Since isolated Frenkel pairs as well as cascades are usually produced together, it has been difficult to get specific information about the defect production in cascades, for example from studies of the energy dependence in charged paricle irradiations(1,2,3). Especially in the case of electrical resistivity measurements, where only one integral quantity is measured, energy dependence measurements have not been very successful in showing the differences between cascades of various energies. The purpose of the present work was to see whether monoenergetic displacement cascades could be studied by means of electrical resistivity measurements. We can then hope to get specific information about the average displacement cascade of a well defined energy.

It is well known that monoenergetic cascades can be produced by self ion bombardment. The cascade is in this case not initiated by a recoil atom, but by an external beam of self ions. The cascade starts directly below the surface and the cascade energy is given by the energy of the incoming self ion. Self ion bombardments have been very useful in studies of displacement cascades by transmission electron microscopy (TEM)(4) and field ion microscopy (FIM)(5). TEM and FIM have shown the vacancy clusters associated with depleted zones(6). Individual vacancies and interstitials have been identified by FIM. TEM work(7) has indicated that the cascade splits into subcascades as the cascade energy is increased.

Electrical resistivity measurements have in the past been very useful in giving a measure of the total amount of damage that is produced ("No. of Frenkel pairs"). Resistivity measurements have also served as a tool to study defect reactions that involve mutual recombination and annealing at sinks. For monoenergetic cascades we would expect to get information on the total amount of damage in a cascade, on the number of close pairs and on the number of free interstitials and vacancies.

For electrical resistivity measurements on self ion cascades, rather thin specimens are required. In order to make sure that the cascades did not penetrate significantly beyond the back surface of the films we irradiated under random incidence and we used films whose thickness was greater than twice the average damage depth. In this way we inject the damage into the bulk of the specimen and at the same time minimize the defect loss through the surfaces.

There are considerable difficulties in relating the measured resistance changes to defect concentrations or number of defects per cascade. The most important of these are:

1. The measured resistivity changes deviate considerably from the bulk values due to the size effect in the electrical resistivity.

2. The concentration of defects varies as a function of depth below the surface.

3. The electrical resistivity change due to defect clusters is smaller than that due to the sum of the point defects contained in the cluster.

A quantitative evaluation of the absolute number of defects present will be only accurate to the extent to which these various influences are known and can be corrected for. In the present work we have made corrections for the electrical size effect by means of the Sondheimer theory. However, there are considerable improvements possible, regarding a better determination of the size effect parameters as well as in determining the influence of damage inhomogenities. Our present data are, therefore, very rough and rather preliminary. Nevertheless, we shall see that one already can come to valuable conclusions regarding the defect production in cascades, the cascade volume, defect density and number of close pairs in a cascade.

EXPERIMENTAL

Thin Cu, Ag and Au films of (001) orientation were grown by vacuum evaporation on rocksalt cleavage surfaces. The films were transferred to a tantalum holder and the rocksalt was dissolved in water. The surface of the tantalum holder had been oxidized in order to provide an insulating substrate for the specimen. The shape of the films, as determined by a mask during evaporation, allowed four point dc resistance measurement on several sections of each specimen. The thickness of the films was between 1500 and 3500 Å. The average damage depth was between 0.2 and 0.5 of the specimen thickness. The films were irradiated with self ions (Cu^{++}, Ag^{++} and Au^{++}) near 500 keV at approximately 10°K. The <001> direction was tilted 10 to 15° off the incident beam direction. Ion beams in the range $\leq 10^{-10}$ Amps were used. The ion dose was measured by means of the beam intercepted by a Faraday cage. The Faraday cage had a hole through which part of the beam could reach the specimen. An x-y sweep was used in order to insure beam homogenuity over the Faraday cage as well as the specimen. During irradiation the pressure in the target chamber was of the order of 10^{-8} Torr. Resistance increases were measured at dose increments of 3×10^9 ions/cm^2 initially. The dose increments were increased as the irradiation progressed. Figure 1 shows a plot of the resistivity change in Ag as a function of dose. The raw data as well as the data corrected for the size effect are indicated. The positive curvature which is readily noticable, even at the low doses, indicates a strong tendency towards saturation. However, there is no simple exponential saturation behavior. This can be seen in Fig. 2 where we have plotted the damage rate $d\Delta\rho/d\phi$ versus the resistivity $\Delta\rho$ that has been induced by the irradiation. We see an initially strong damage rate decrease which is followed by a much more gradual one at higher dose and resistance values. For a simple exponential saturation we would expect a single straight line with a negative slope in Fig. 2. The fact that we see two regions with approximately straight lines is an indication that at least two processes are responsible for the saturation behavior.

ANALYSIS

1. We shall first consider the case where the mean free path of the conduction electrons is larger than the film thickness. In this case the surface scattering has to be taken into account and we correct our data by means of the Fuchs-Sondheimer theory(8,9). We shall assume that the size effect corrected resistivity changes are equal to the resistivity changes that one would obtain if the defects were smeared out homogeneously across the thickness of our film. This assumption seems justified as long as the mean free path of the conduction electrons is large compared to the film thickness and as long as the damage penetrates into the bulk of the film.

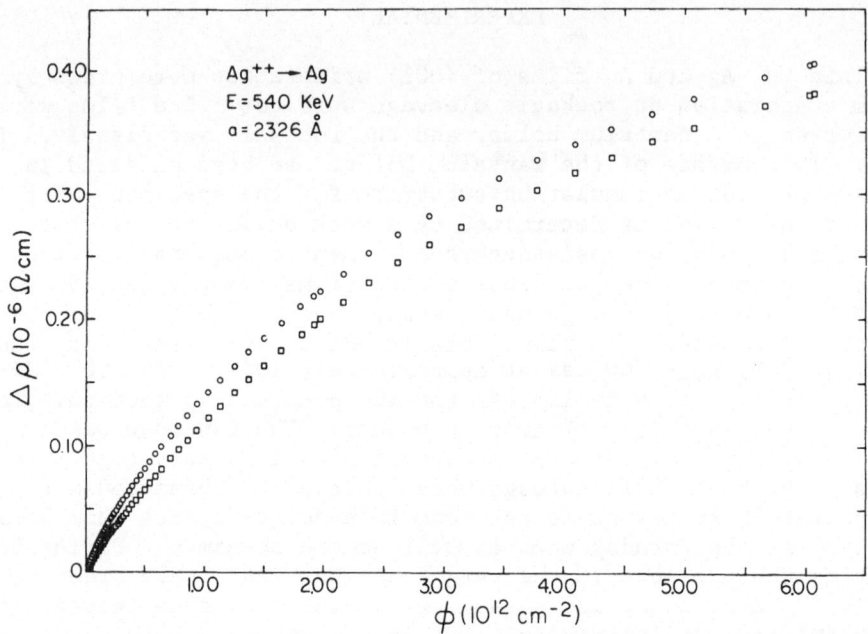

Fig. 1
Resistivity change versus fluence in a silver film irradiated with
540 keV Ag^{++} self ions. The uncorrected data (circles) as well as
the data corrected for the size effect (squares) are indicated.

If all of the defects were present in the form of Frenkel pairs
(FP), we would have the following connection between the resistivity change $\Delta\rho$ and the Frenkel pair concentration C_F

$$\Delta\rho = \rho_F \, C_F \quad ,$$

where ρ_F is the resistivity change per unit concentration of
Frenkel pairs.

Now, we know that in cascades a large fraction of the defects
can be present in the form of point defect clusters. The resistivity change per FP will be reduced due to this effect(10,11).
To take this into account we introduce an efficiency factor ζ. We
now have

$$\Delta\rho = \zeta\rho_F \, C_F \qquad \zeta \leq 1.$$

The resistivity change per incident ion/cm^2 is given by:

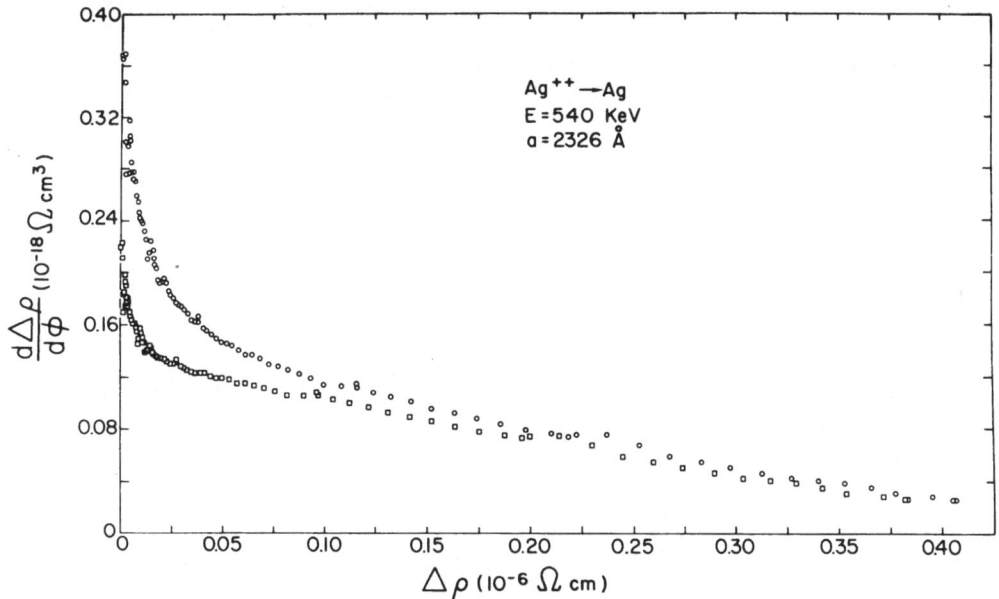

Fig. 2
Damage rate as a function of the induced resistivity for the
same irradiation as in Figure 1.

$$\frac{d\Delta\rho}{d\phi} = \zeta\rho_F \frac{N_F^C}{N\,a} \quad, \tag{1}$$

where ϕ is the ion dose (cm^{-2}), N the number of atoms per cm^3,
N_F^C the average number of FP per cascade, and film thickness a.

Equation (1) is valid as long as the overlap between cascades
can be neglected. Where cascades overlap, the resistivity increment per cascade is reduced. We can assume that on the average no
additional defects are produced in the overlap volume. In this
case the damage is proportional to the undamaged fraction of the
total volume $(1-V_d/V)$. We also have the relation $V_d \cdot C_F^C = V \cdot C_F$
between the damage volume V_d, and the FP concentration in a cascade C_F^C and the total volume V and the average FP concentration in
the specimen C_F. Therefore, we get for the damage rate

$$\frac{dC_F}{d\phi} = \frac{N_F^C}{N\cdot a} \,(1 - C_F/C_F^C) \quad. \tag{2}$$

In terms of resistivity changes we have with $C_F = \dfrac{\Delta\rho}{\zeta \cdot \rho_F}$ and $C_F^C = \dfrac{N_F^C}{N \cdot V_o}$ where V_o is the cascade volume:

$$\frac{d\Delta\rho}{d\phi} = \frac{\zeta\rho_F N_F^C}{N \cdot a} \left(1 - \frac{\Delta\rho V_o N}{\zeta\rho_F N_F^C}\right) \quad . \tag{3}$$

At saturation we get

$$\Delta\rho_{sat} = \frac{\zeta\rho_F N_F^C}{V_o N} \quad . \tag{4}$$

According to equation (3) we can get from a plot of $\dfrac{d\Delta\rho}{d\phi}$ vs. $\Delta\rho$ the number of FP per cascade

$$N_F^C = \frac{N \cdot a}{\zeta\rho_F} \left(\frac{d\Delta\rho}{d\phi}\right)_o \tag{5}$$

and also the cascade volume

$$V_o = \frac{\zeta\rho_F}{\Delta\rho_{sat}} \frac{N_F^C}{N} = \frac{a}{\Delta\rho_{sat}} \left(\frac{d\Delta\rho}{d\phi}\right)_o \quad , \tag{6}$$

where $\left(\dfrac{d\Delta\rho}{d\phi}\right)_o$ is the initial damage rate.

2.   Let us now consider the case where the mean free path of the conduction electrons is small compared to the film thickness. This case is only reached after a rather high concentration of damage has been introduced. For example, at the highest FP densities attainable (~.5%), the mean free path still is of the order of 400 Å. In sufficiently thick films that also exhibit a large variation in damage density as a function of depth we can, however, expect that the measured $\Delta\rho$ is not proportional to the average damage density across the film. For the particularly simple case of no size effect and a layer of thickness $t < a$ being damaged we have the relation:

$$\Delta\sigma_{av} \cdot a = \Delta\sigma \cdot t, \tag{7}$$

where $\Delta\sigma_{av}$ and $\Delta\sigma$ are the average change in conductivity across the whole film and the change in conductivity in the damage layer respectively. If ρ_o is the initial resistivity of the film, the maximum possible resistivity change of the film is

$$\Delta\rho_{av}^{max} = \frac{t}{a-t} \rho_o .$$

Since the induced resistivity changes in our experiments are usually at least a factor of 10 higher than ρ_o, we certainly do

not have this case. However, we might nevertheless expect some
small influence of damage inhomogenuities on the $\Delta\rho$ values measured.

RESULTS

The damage rate vs. induced resistivity curves for Cu, Ag and
Au show an initially strong decrease in the damage rate. The curves
give a more gradual decline at higher $\Delta\rho$ values. Figure 2 shows
an illustration of this. According to eqn. (3) we should get just
one straight line with a negative slope: the fact that we get two
approximately linear regions in Fig. 2 indicates that at least two
mechanisms are responsible for the observed saturation behavior.
We found that the part with the smaller slope is dependent on the
specimen thickness. This is what we expect if we have inhomogeneous
defect distributions and if the layers at smaller depth become
saturated before the layers at the larger depths. We assume that
the initial decrease in the damage rate is due to overlapping cascades. We use the intersects with the two axes of a straight line
fitted to this section to determine the quantities of eqns.
(5) and (6). We have collected these values in Table I, using
$\zeta = 1$ and $\rho_F = 2.5 \times 10^{-4}$ Ω cm. Because there is some uncertainty
regarding the parameters that should be used to make the size
effect correction, we have determined the quantities in Table I
for several values of the size effect parameters. The uncer-
tanties indicated in the table are due to fluctuations in the
results depending on what size-effect parameter was chosen within
the limit of expected values.

DISCUSSION

Let us first look at the <u>damage rates.</u> The number of Frenkel
pairs given in Table I does not vary much from material to material.
We shall now compare these numbers with the predictions of the
modified Kinchin and Pease model. Robinson has indicated that
the number of FP per cascade is roughly proportional to the damage
energy E_D. The damage energy is that fraction of the cascade
energy that goes into nuclear collisions. We calculated this
energy from the Lindhard theory using the expression given by
Robinson(12). If we compare the experimental values with the number of theoretical Frenkel pairs N_F^t we can derive a cascade efficiency factor $\xi = N_F^C/N_F^t$ with $N_F^t = E_D/2E_d$. It turns out these
ξ values range between .2 and .3 (see Table I). It is well known
that the total FP production efficiencies in a number of metals
under light ion bombardment are approximately .3. Our result
means in this context that the average defect production efficiency in energetic cascades as measured by electrical resistivity changes is not much different than it is in charged

W. KESTERNICH AND K. L. MERKLE

TABLE I

Specimen	Film Thickness a (Å)	Cascade Energy E (keV)	Frenkel pairs per cascade N_F^C	Cascade Efficiency ξ	Cascade Volume V_o (cm³)	FP Concentration c_F^C	Stage I Annealing %
Au 395A	3280	500	1340	0.28	$(2.5\pm1)\cdot10^{-16}$	$(1.2\pm1)\times10^{-4}$	3.5
Au 396	1770	500	1070	0.23	$(6.6\pm1)\cdot10^{-17}$	$(2.6\pm.3)\times10^{-4}$	
Ag 44A	2330	540	1210	0.22	$(2.7\pm.2)\cdot10^{-16}$	$(7.5\pm2)\times10^{-5}$	2.3
Cu 45A	3170	560	1220	0.23	$(4.1\pm1)\cdot10^{-16}$	$(5\pm.5)\times10^{-5}$	30.

particle irradiations where the average recoil energy lies in the 100 eV region. From this result it becomes clear why $\Delta\rho$ investigations of the energy dependence in charged particle irradiations have not shown any measurable deviations in the damage rates that might be attributed to energetic cascades. On the other hand, we know from TEM observations in Au, that the efficiencies for vacancy clusters formed at 300°K are of the order of .3. This suggests that our real efficiency values might be significantly higher than indicated in Table I due to the ζ factor which was assumed to be 1 in Table I. A reduction by a factor of 2 in the resistivity due to clustering effects could be quite possible in our case. The clustering effects are expected to be smaller in Cu compared to Ag and Au. At present the ambiguities which arise from the size effect corrections are too large to warrant a more detailed discussion of the absolute damage rates.

In contrast to the damage rates, there is a strong increase in the cascade volume in going from Au to Ag and Cu. We find that the average cascade radius increases from 250 to 460 Å. It should be mentioned that the large discrepancy in V_o between the two Au specimens is probably due to the fact that our analysis breaks down if the specimen thickness is too large. Determination of V_o is much more sensitive to damage inhomogenity than the damage rate. If only part of the specimen is being damaged V_o will be over estimated. We assume that the lower value of V_o^o is closer to the true cascade volume. Our lower value is about a factor of two larger than what is obtained from dechanneling data at 300°K(13). Comparison of the cascade radii with the second moment of the damage distributions of the random cascade theory(14) gives fair agreement in both cases. Our TEM observations(15), however, indicate larger cascades. Also as we go to Ag and Cu the cascade radii deduced from the measured cascade volumes are smaller by about a factor of 2 compared to the ones predicted by the random theory. This can all be understood in terms of subcascade formation. In this case damage free areas can remain between the subcascades. The $\Delta\rho$ measurements see only the damaged volume, while TEM observations as well as the random cascade model relate to the total volume within the boundaries of which heavy damage has taken place. The volumes determined by the $\Delta\rho$ measurements correspond to an average deposited energy per atom of .08, .02 and .007 eV for Au, Ag and Cu respectively.

This energy density which decreases by about a factor of 10 from Au to Cu will reflect itself in the amount of rearrangement that can take place in the defect population while the cascade equilibrates. The most likely candidates to be affected by this "heating effect" are the close pairs. From Table I we see that there is almost no stage I annealing in Au and Ag, while 30% of the damage in copper anneals upon heating above Stage I. We conclude that a significant amount of the damage in Cu is present in the form of close pairs and Stage I interstitials. It is clear that the concept of a "thermal spike" would be completely inappropirate in the case of copper. The energy densities mentioned above are in any case fairly low; however, in the case of Au and Ag thermal heating effects might play a role, as can be seen if we express the average deposited energy per atom $E_a = E_D/NV_o$ in terms of a temperature $T = E_a/k$. We then get 930, 230 and 87°K for Au, Ag and Cu respectively.

CONCLUSIONS

1. Damage production efficiencies in displacement cascades in Au, Ag and Cu are on the order of .25 or greater.

2. The cascade volume increases in going from Au to Ag and Cu.

3. The damage volume is in general smaller than the total cascade volume.

4. The average FP concentrations are on the order of 10^{-4}.

5. The average energy per atom is rather small (<.1 eV). Thermal spike effects could play a role in Au and Ag, but not in Cu.

REFERENCES

1) H. H. Andersen and H. Sørensen, Rad. Eff. 14, 49 (1972).
2) H. E. Schiøtt and P. V. Thomsen, Rad. Eff. 14, 39 (1972).
3) K. L. Merkle, Phys. Stat. Sol. 18, 173 (1966).
4) See for example M. Wilkens in Vacancies and Interstitials in Metals, North Holland, Amsterdam, 1969 p.485; or B. L. Eyre J. Phys. F 3, 422 (1973).
5) D. N. Seidman, J. Phys. F 3, 393 (1973).
6) A. Seeger, Proc. Sec. Intern. Conf. Peaceful Uses of Atomic Energy, Vol. 6, Geneva 1958 p.250.
7) K. L. Merkle, in Radiation Damage in Reactor Materials (International Atomic Energy Agency, Vienna 1969), Vol. 1, p. 159.
8) E. H. Sondheimer, Advan. Phys. 1, 1 (1952).
9) K. L. Merkle and L. R. Singer, Appl. Phys. Let. 11, 35 (1967).
10) L. L. R.Alfred, Phys. Rev. 152, 693 (1966).
11) J. W. Martin, J. Phys. F 2, 842 (1972).
12) M. T. Robinson, in Radiation Induced Voids in Metals, Eds. J. W. Corbett and L. C. Ianniello USAEC (1972), p.397.
13) P. P. Pronko and K. L. Merkle, these conference proceedings.

14) K. B. Winterbon, P. Sigmund and J. B. Sanders, Mat. Fys.
     Medd. Dan. Vid. Selsk. 37, No. 14 (1970).
15) K. L. Merkle, to be published.

DISCUSSION

Q: (M. L. Swanson) Your saturation damage curves are very
similar to those observed during neutron irradiation at ~ 4°K,
which were interpreted in terms of a recombination volume for
spontaneous Frenkel pair annihilation. Have you analyzed your
results in this way?

A: The initial drop in damage rate can not be interpreted in
this way. In our earlier work (9) involving irradiation in which
a high number of Frenkel pairs were produced, we obtained a spontaneous recombination volume that was in good agreement with the
neutron data. I would like to note however that the saturation
value in $\Delta\rho$ that is associated with the spontaneous recombination
of interstitials and vacancies lies at much higher values than the
saturation value that we ascribe to overlapping cascades. Our
present results are only a first attempt in measuring the various
cascade parameters that we discussed. Further experiments will be
required in order to show definitely whether our present assignment of the initial drop in damage rate is correct.

FORMATION OF INTERSTITIAL AGGLOMERATES AND GAS BUBBLES

IN CUBIC METALS IRRADIATED WITH 5 keV ARGON IONS

B. Hertel, J. Diehl, R. Gotthardt and H. Sultze

Max-Planck-Institut für Metallforschung

Stuttgart, Germany

1. Introduction

Former transmission electron microscope (TEM) studies /1,2,3/ on Cu and Au foils, which were bombarded with 1 to 5 keV Ar ions, showed that interstitial clusters (in the configuration of Frank dislocation loops) are formed below the bombarded surface in a depth remarkably larger than the calculated random range of Ar ions. This was interpreted by the propagation of focussing replacement collision sequences (r.c.s.) originating near the end of the heavily damaged layer within the random range of the incoming Ar ions. One of the strongest arguments in support of this interpretation was the dependence of the depth distributions of the interstitial agglomerates on the crystallographic orientation of the foil surface. The distributions revealed only one maximum in {100}-foils, but two distinct peaks in {110}-foils, as one would expect, if the agglomerates are formed at the end of the range of the r.c.s., if the ranges of r.c.s. scatter more or less randomly around an average value and if this average range is independent on the angle between the surface and the ⟨110⟩-directions along which the r.c.s. propagate. Based on these results conclusions on the magnitude of the average range of a r.c.s. were drawn implying that the derived range values are the average r.c.s. ranges in an otherwise undisturbed lattice.

Although this work found considerable attention (see e.g. /4,5,6/) some of the implications and conclusions involved in it were questioned later on, e.g. in /4,5,7,

507

8,9/. Questions arose as regards the effects of the Ar
atoms implanted in the foils and of other impurities, the
influence of not having used magnetic separation of the
Ar ions, and finally the statistical significance of the
orientation dependency of the peaks in the depth distribution histograms. Admittedly in this "first generation"
of experiments the impurity content of the samples was
not controlled very well, since epitaxially grown films
evaporated in conventional high vacuum were used and in
addition the relatively simple equipment applied was not
free of hydrocarbons, so that during bombardment a carbon
layer could be formed on the sample surface. Furthermore,
the arguments on the penetration of the Ar projectiles
were based solely on theoretical calculations.

    For all these reasons it seemed to be feasible to re-
peat some of the original experiments and to extend them
under improved conditions. For the experimental results
reported here a new ion bombardment equipment was built,
providing e/m-separation and avoiding carbon contamination
of the samples. Furthermore, electropolished foils from
high purity materials were used as targets instead of
evaporated films. In the case of copper depth distributions of more than two surface orientations were studied.
It didn't seem to be necessary to repeat also the contrast analysis for determining type and configuration of
the agglomerates. But in addition to imaging by diffraction contrast (revealing defects causing lattice distortions) defocus contrast micrographs (sensitive to distor-
tion free defects with electron densities different from
the bulk material) were taken in order to search for
small Ar bubbles, which could indicate where the incoming
ions came to rest in the sample.

    Supposing that the depth distribution of the inter-
stitial agglomerates are indeed indicative for the existence and the ranges of r.c.s., it seemed to be interesting
also to study a b.c.c. metal with the same technique,
since so far very little is known about r.c.s. in b.c.c.
metals. Therefore, Nb was studied in addition and some of
the results are included in this report. Analysis of diffraction contrast of point defect agglomerates condensed
in dislocation loops is more complex in b.c.c. than in
f.c.c. metals. Therefore, we shall introduce here only
the depth distributions and, as regards the nature and
the configurations of the defects observed in Nb, confine
ourselves to the statement that they were identified to
be dislocation loops (in various configurations) of interstitial type /10/.

2. Experimental

2.1. Ion bombardment apparatus. The ion source consists essentially of a commercial quadrupole gas analyser. The ionizer is operated with an Ar gas pressure of 10^{-4} torr. The ion detector is replaced by an electrostatic lens system, by which the ions, leaving the quadrupole, are focussed onto the samples and accelerated to the desired energy. Behind the lens system the ions pass through a flight tube, 500 mm in length, with apertures of 3 mm and 2 mm diam. at the entrance and the end, respectively. For the suppression of secondary electrons the samples are mounted inside of a cage (at -90 V) with a 3 mm entrance hole. The system is evacuated with a turbo-molecular pump in the lens section (background pressure $2 \cdot 10^{-8}$ torr, operating pressure $5.5 \cdot 10^{-6}$ torr) and in the sample section with a sputter ion pump ($5 \cdot 10^{-10}$ torr background, $3 \cdot 10^{-8}$ torr at operation). By this means the system is kept free of hydrocarbons to avoid surface contamination during bombardment. Typical ion current densities are 2 to $4 \cdot 10^{-8}$ Amp/cm^2.

2.2. Sample materials and irradiations. Informations on the starting materials (foils), the pretreatments of the samples and the irradiation conditions are presented in Tab. 1. Discs, 2.3 mm in diameter, were cut from the foils and electrochemically polished for TEM. After mounting of the polished foils the bombardment system was evacuated and beaked out at 200°C for several hours to obtain clean vacuum conditions.

2.3. Microscopy and evaluation. Cu and Au were examined in a JEM 200A electron microscope at 200 kV, Nb in a JEM 150 at 150 kV. In each case a diffraction pattern was taken before tilting to determine the surface orientation. For the depth distribution curves the depths of 150 (in some cases 300) agglomerates were measured from a stereo micrograph pair, using a Hilger and Watts mirror stereoscope. To obtain the histograms the depth scale was divided into units of 10 Å. Then 1/3 of the number of agglomerates falling into the 10 Å unit x_i was assigned to the unit x_{i-1}, the unit x_i and the unit x_{i+1}, each. In this way the uncertainty in the individual depth measurements was taken into account. In the histograms shown in the following the so determined number $n(x_i)$ is plotted normalized by $\Sigma n(x_i) = N$, the total number of measurements.

Very small gas bubbles are normally not visible in TEM. A clearly detectable contrast is only observable if

Tab. 1. Data on samples and irradiations

Metal and Surface Orientation	Material	Preparation before Irradiation	Ion Current Amp/cm^2	Doses, Irradiation Temp.
Cu {100} {110} {112} {120}	MRC marz grad 5 N	annealed in H$_2$ at 1000°C and in UHV 10^{-9} torr	2–4·10^{-8}	1–2·10^{15} cm^{-2} room temp.
Au {100}	Degussa 5 N	annealed in UHV 10^{-9} torr	3.2·10^{-8}	1–2·10^{15} cm^{-2} room temp.
Nb {100} {110} {135}	MRC marz grad 5 N	annealed and degassed in UHV 10^{-9} torr at 2000°C	1 · 10^{-8}	1–10·10^{13} cm^{-2} room temp., liqu. N$_2$

the micrographs are taken out of focus with a defocus distance of the order of several 1000 Å /11/. Therefore, it was necessary to make "through the focus" series. Several micrographs from the same area of the foil were taken with different positions of the imaging plane relative to the foil surface. If bubbles are present, one expects their contrast to change from a dark spot surrounded by a weak brith ring (overfocus) to a bright spot surrounded by a dark ring (underfocus).

3. Results

3.1. Depth distributions of defect agglomerates in copper. After bombardment under the conditions summarized in Tab. 1 the total number of visible defects per unit surface area was found to be constant (3·10^{11} cm^{-2} for {100}-foils) for ion doses between 1.1 and 2·10^{15} cm^{-2}. Only the size of the defects increased, so that at higher doses the contrasts of the agglomerates begin to overlap.

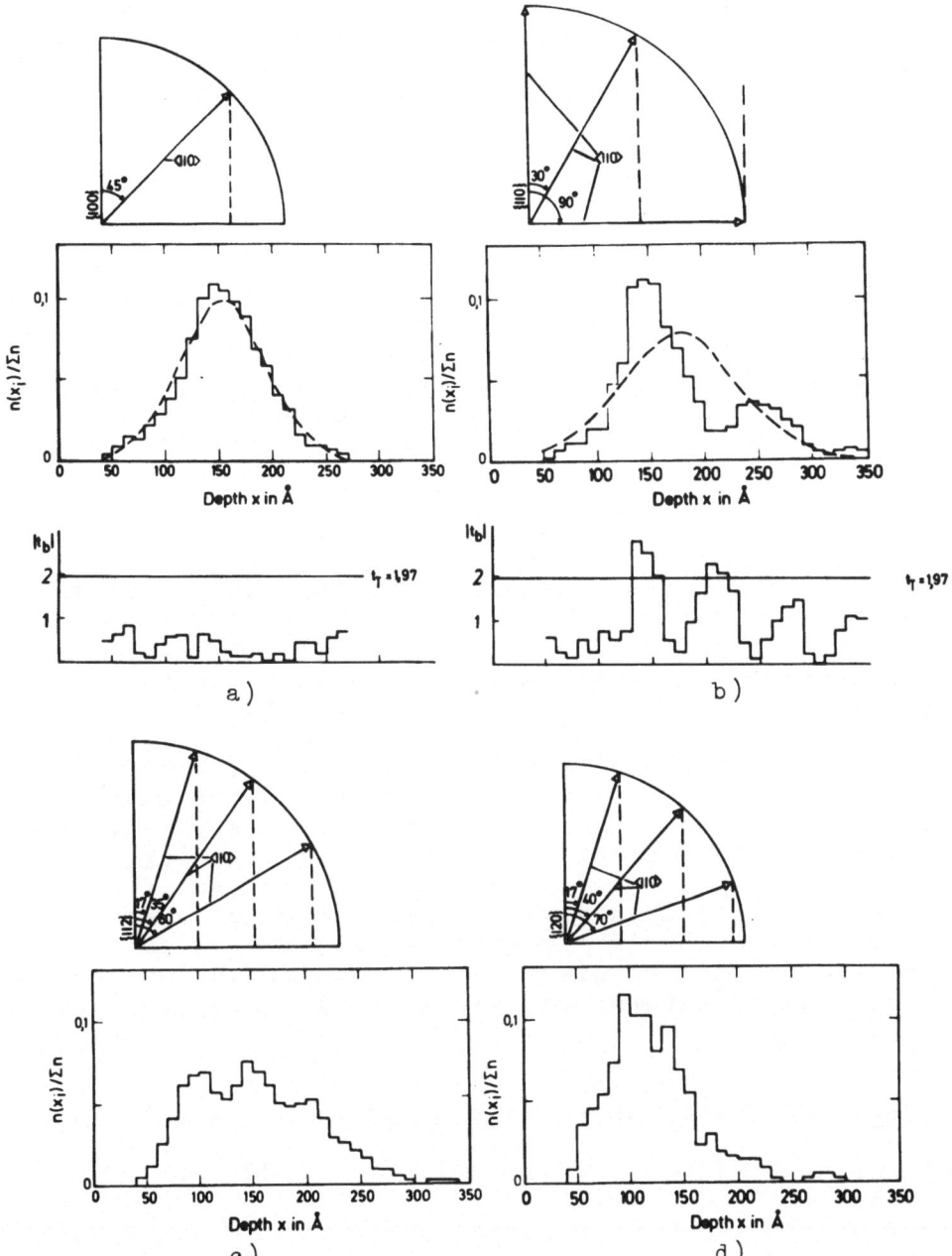

Fig. 1. Depth distributions of interstitial agglomerates in Cu, $1.3 \cdot 10^{15}$ Ar$^+$/cm^2, together with angles between the surface and all {110} directions. Surface orientations: a) {100}, b) {110}, c) {112}, d) {120}. For lower parts of a) and b) see sect. 3.2.

Thus the depth distributions were measured in the mentioned dose range. Foils with the four surface orientations indicated in Tab. 1 were investigated. The results
are shown in Fig. 1.

For the surface orientation {100} the agglomerates
were found in the range from 40 to 260 Å below the bombarded surface. The distribution shows one maximum at 150 Å
with a half width of ~80 Å. In foils with surface orientation {110} agglomerates between 50 and 350 Å were observed. The distribution shows two clearly separated maxima at 140 Å and at 240 Å. The greater one contains about
75 % of all agglomerates and has a half width of ~60 Å.
For surface orientation {112} agglomerates could be seen
between 40 and 330 Å. Three peaks appear at 100 Å, 150 Å
and 200 Å, but they are not as clearly separated from each
other as the two in the {110}-foil. In the foil with surface orientation {120} the distribution of the defect
depths, ranging from 40 to 290 Å, does not reveal a com-
parably well developed peak structure. One maximum can be
seen at 90 Å, there are indications for two others at
~140 Å and ~200 Å.

The upper parts of Figs. 1a - 1d show the attempt to
correlate the maxima in the depth distributions with the
penetration of r.c.s. in ⟨110⟩-directions having different
angles to the surface. (These angles for all ⟨110⟩-direc-
tions are plotted). The simple interpretation adopted in
/2,3/ and outlined in the introduction is applied. Fixed,
orientation independent ranges (arrows) starting at the
end of the Ar penetration (~40 Å) are assumed. The vertical dashed lines then give the projections of these ranges
on the surface normal. For each foil orientation the
range is adjusted to give the best fit to the observed
maxima. The so determined average ranges are presented in
Tab. 2.

Tab. 2. R.c.s. ranges determined from Figs. 1a-1d

Foil orien-tation	{100}	{110}	{112}	{120}	
No. of expec-ted maxima	1	2	3	3	
Range (Å)	170	200	190	170	average: 180 ± 15

3.2. Statistical significance of the depth distributions in Cu.

In order to check as to which degree the maxima in the depth distributions could be caused simply by statistical deviations from a random distribution due to the limited number of agglomerates measured the following test procedure was performed for all 4 histograms: Assuming that the "real" distribution is a Gaussian (normal distribution), approximate values for the mean depth of the distribution and of its standard deviation (variance) were calculated from the individual depth measurements. From plots of the corresponding Gaussians (dashed curves in Figs. 1a and 1b) the deviations of the values of the histograms from these calculated curves were determined in each case. For the {100} and the {110} foils they are shown in the lower parts of Figs. 1a and 1b. In these plots t_b is this deviation normalized by the variance s of each point of the approximated Gaussian $q(x)$ with $s = \sqrt{(q/N)(1-q)}$ and N number of measurements. These t_b-values can then be compared with corresponding tabulated values[+] t_T above which according to mathematical statistics deviations can be declared to be significant with an allowed error probability α (or statistical regularity $1-\alpha$). For $t_b > t_T(\alpha)$ the deviations are of physical significance for $t_b < t_T$ they are not.

For the {100} foil the deviations are rather small, only about half of the allowed ones for $\alpha = 0.05$ ($t_T=1.97$). Even if a probability $\alpha = 0.4$ is permitted ($t_T=0.842$), the total distribution curve follows within statistical limits a Gaussian distribution. Almost the same is true if the first maximum in the depth distribution of the {110} foil is tested separately. However, for the total distribution curve of this foil (Fig. 1b) the first maximum and the minimum between the two maxima are outside the limits allowed, even for $\alpha = 0.05$ and, therefore, the two maxima are clearly of physical significance and can not be caused by statistical deviations.

Not equally clear-cut results are obtained for the {112} and the {120} foils, obviously due to the strong overlap of the different peaks. If we adopt the high value of $\alpha = 0.4$ (which is compatible with the {100} foil), in both cases the three maxima are outside the limits of statistical deviations, whereas this is not the case for $\alpha = 0.05$. Nevertheless, it seems to be reasonable to attribute some physical significance also to the maxima in the histograms of these foils.

+) For this comparison the so-called Student distribution was used.

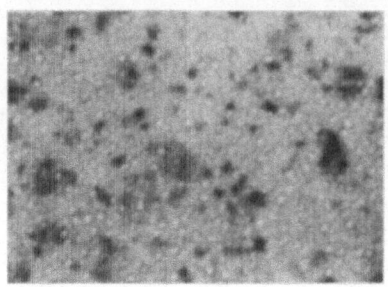 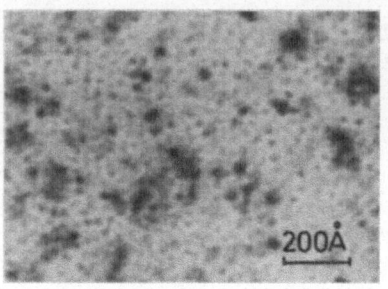

            underfocus   overfocus

Fig. 2. Defocus contrast of Ar bubbles in Au, room temp.,
$1.5 \cdot 10^{15}$ Ar$^+$/cm^2, difference in focus ~2.5 μ.

   3.3. Formation of Argon bubbles. If micrographs of the
agglomerates are taken under in-focus conditions no evidence for bubble formation was found. This was proved for
doses up to $4 \cdot 10^{15}$ Ar$^+$/cm^2. With "through the focus" series bubbles could be made visible at doses above $1.5 \cdot 10^{15}$
Ar$^+$/cm^2 in Cu and also in Au. Since in Cu the background
intensity is relatively high due to the inelastic scattering and the masses of Cu and Ar are similar, the contrast of the bubbles is rather weak. But in Au the bubbles
are clearly visible. An example is given in Fig. 2 for
the under- and the over-focus case. The measured mean diameter of the bubbles is ~20 Å at $1.5 \cdot 10^{15}$ Ar$^+$/cm^2. Visual
observations of stereo pairs give evidence that the bubb-
les are close to the bombarded surface. They are clearly
separated in depth from the interstitial agglomerates.
Because it is relatively difficult to obtain stereo pairs
in exactly the same defocus condition and to make the surface and the bubbles visible in the same micrographs, so
far we did not succeed in evaluating defocus stereo pairs
quantitatively.

   3.4. Depth distributions of defect agglomerates in
niobium. After bombardment under the conditions given in
Tab. 1. the contrast spots observed in Nb could also be
identified to originate from small dislocation loops formed by interstitial atoms /10/. The depth distributions
of the loops were measured under various conditions. The
results from foils with three different surface orientations (see Tab. 1), bombarded at room temperature, are
shown in Fig. 3.

   Although a few agglomerates are found close to the
surface, the majority of them lie - as in the f.c.c. me-

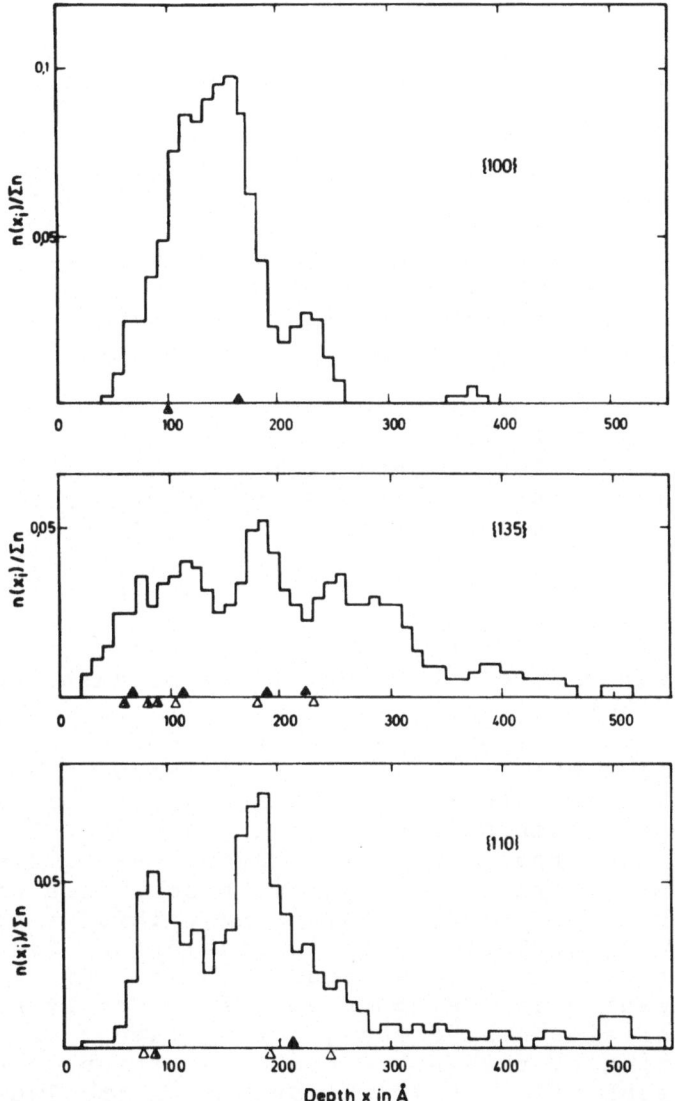

Fig. 3. Depth distributions of interstitial agglomerates in Nb, $1 \cdot 10^{14}$ Ar^+/cm^2, room temperature.

tals - much deeper in the foil than the Argon atoms can
penetrate by random collisions ($\sim$50 Å). Again the depth
dependencies are rather different for different orientations. For the surface orientation {100} one predominant
maximum is observed at $\sim$160 Å, whereas the {110} foil
shows two peaks. The {135} foil reveals a more complicated depth distribution with indications of at least three
peaks. Using the same arguments as in the case of copper
(/2,3/ and sect. 3.2. of present paper) it is thought that
the interstitial atoms forming the agglomerates penetrate
into the lattice by means of r.c.s. also in a b.c.c lattice. If it is assumed again, that peaks in the depth
distributions indicate the average range of r.c.s. in
closed packed directions ($\langle 111 \rangle$ in b.c.c. crystals) a
quantitative interpretation of all experimental results
is not possible in a straightforward manner.

If we assume the average range of r.c.s. (starting
at a depth of 50 Å) to be 200 Å in all possible $\langle 111 \rangle$-
directions (adjusted to data in Fig. 3), peaks in the
curves of Fig. 3 are expected at the depths marked by ▲.
There is a reasonable agreement for the {100} and the
{135} foils, but for the {110} foil only one maximum instead of two observed ones is expected. There are two possible interpretations for the first peak in the {110}
foil. If one takes into account that the surface deviates
by a few degrees from the exact {110}-orientation the $\langle 111 \rangle$
directions in the {110} plane penetrate under a small
angle into the foil and could therefore cause the first
maximum. The symbols Δ in Fig. 3 are based on such a deviation. The first peak could be accounted for in this way
but the second peak should split up into two. In the alternative interpretation it is assumed that r.c.s. can
propagate not only along $\langle 111 \rangle$ but also with a smaller
range along {100}(assisted focussing). This is not unlikely if the work of Nelson /12/ is considered. Assuming a
r.c.s. range in {100} directions of 50 Å, additional maxima are expected at the depth values marked by ▲ in Fig. 3.
No clear decision can be made, but we favour the second
interpretation at present, since in addition to explaining
the first peak of the {110} distribution it also helps to
understand the unresolved peak structure of the {135}
foil at a depth <130 Å.

4. Discussion

The observations on copper show clearly that the ear-

lier results /2,3/ obtained under less clean conditions
were basically correct and not strongly influenced by the
unknown, but certainly higher impurity content in the bulk
material or the carbon deposited on the surface during
bombardment. The facts, that the peak structures in the
depth distributions of the interstitial agglomerates could
be reproduced for the {100} and {110} foils and that they
could be proved by statistics to be of physical relevance,
strengthens the view based originally on these two orientations that the constitutents of the agglomerates are
self-interstitials which penetrate into the foil by means
of r.c.s. and that their average range is indicated by
the maxima in the agglomerate distributions. These conclusions are supported strongly by their consistency with
the peak structures observed in the depth distributions
of foils having a lower symmetry ({112}, {120}). The
possible suspicion that the observed agglomerates were
formed by Ar, became very unlikely by the observations
that a larger portion of the Ar atoms form bubbles much
closer to the surface (in the range of penetration by random collisions) and that Ar bubbles and defect agglomerates show completely different types of contrast.

Even if it is accepted that the interstitial agglomerates are formed at the end of the range of r.c.s.
along $\langle 110 \rangle$ directions, there are still two possibilities
for an effect of impurities on our observations, namely
that the observed ranges are limited by defocussing of the
r.c.s. at impurity atoms and/or that the agglomeration of
the interstitials is nucleated at impurities. Both effects
can not be excluded completely. However, it is unlikely
that they are of importance for the following reasons:
(i) The average r.c.s. range from Tab. 2 (180 $\overset{\circ}{A}$) is somewhat larger than the corresponding value reported in /2/
($\sim$130 $\overset{\circ}{A}$). This difference might originate from impurity
stopping of r.c.s. (part of it could also be due to syste-
matic deviations since for /2,3/ a different stereoscope
was used), but if so, this impurity effect is apparently
not very strong. (ii) The density of observed agglomerates
for room temperature bombardment with $1-2 \cdot 10^{15}$ ions/cm^2
was larger in /3/ than the value reported here. This could
be caused by nucleation at impurities and different impurity contents in both cases. But it can equally well be
explained by homogeneous nucleation, if one takes into
account that interstitials are very mobile at room temperature and that the ion current density was higher in
/3/ by a factor of $\sim$25. Consequently the production rate

and the local density of mobile interstitials must have
been higher in the earlier experiments than in the present ones and thus also the probability for homogeneous
nucleation.

Our conclusions on r.c.s. ranges are in contradiction
to theoretical calculations /13/ in two respects: Our
ranges are much larger than the calculated ones, which
might originate from the choice of the interatomic potentials used in the calculations. Qualitatively there is a
discrepancy as well. The theory predicts a range distribution of interstitial atoms which decreases continuously, having its highest value at zero penetration, whereas
our results indicate a more or less random distribution
around a finite mean value. Our conclusions could be erroneous if strong losses of interstitials or dislocation
loops to the heavily damaged surface zone or the surface
occurred. Such losses by thermal migration of interstitials
are certainly possible. They are not very likely to influence our agglomerate distributions strongly, since e.g.
in the {110} foil the first maximum should be cut off and
shifted more strongly than the second one by such a denudation, which does not seem to be the case. If losses
through the surface due to glissile dislocation loops would
occur, all of these ought to have disappeared, as only
Frank loops were observed. Our conclusions can be understood theoretically if it is assumed that the r.c.s. have
finite ranges already for the minimum transfered energy
necessary to produce replacements in focussing collision
sequences.

The observed formation of Ar bubbles at room temperature can not simply be explained by thermal diffusion,
since according to /14/ such thermal agglomeration should
take place only above 180°C. Therefore, we have to assume
that radiation induced diffusion of Ar due to the kinetic energy dissipation in the damaged surface layer takes
place and causes the bubble formation.

Our result that r.c.s. can be created also in b.c.c.
metals and that they possibly occur not only in the close
packed $\langle 111 \rangle$ direction but also in $\langle 100 \rangle$ is in accordance
with earlier results of Nelson /12/. He concluded from
sputtering experiments that r.c.s. in $\langle 100 \rangle$ and $\langle 110 \rangle$
should be possible, but only at higher energies and with
smaller ranges than in $\langle 111 \rangle$. It should be mentioned that
recently evidence for the existence of r.c.s. was also

obtained from internal friction /15/ and resistivity /16/ measurements on neutron irradiated Fe and Nb. Although these results are less direct, they are consistent with the main conclusions drawn here from the Nb observations.

References

/ 1/ H. Diepers and J. Diehl, phys. stat. sol. <u>16</u>, K 109 (1966).

/ 2/ J. Diehl, H. Diepers and B. Hertel, Can. J. Phys. <u>46</u>, 647 (1968).

/ 3/ H. Diepers, phys. stat. sol. <u>24</u>, 235 (1967).

/ 4/ J.A. Venables and G.J. Thomas, in: Vacancies and Interstitials in Metals, edt. by A. Seeger et al., North-Holland, Amsterdam, 1970, p. 531.

/ 5/ J.A. Venables, in: Atomic Collision Phenomena in Solids, edt. by D.W. Palmer et al., North-Holland, Amsterdam, 1970, p. 132.

/ 6/ B.L. Eyre, J. Phys. F <u>3</u>, 422 (1973).

/ 7/ L.E. Thomas and R.W. Balluffi, Appl. Phys. Lett. <u>9</u>, 171 (1966).

/ 8/ L.E. Thomas, T. Schober and R.W. Balluffi, Rad. Effects <u>1</u>, 257 (1969).

/ 9/ W. Frank and A. Seeger, Rad. Effects <u>1</u>, 117 (1969).

/10/ R. Gotthardt, unpublished.

/11/ M. Rühle and M. Wilkens, Proc. 5th Europ. Congr. on Electron Microscopy, The Institute of Physics, London and Bristol, 1972, p. 416.

/12/ R.S. Nelson, Phil. Mag. <u>8</u>, 693 (1963).

/13/ G. Düsing and G. Leibfried, phys. stat. sol. <u>9</u>, 463 (1965).

/14/ B. Jouffrey, Bull. Soc. Franc. Minér. Crist. <u>87</u>, 557 (1964).

/15/ M. Weller and J. Diehl, Proc. 5th Intern. Conf. Internal Friction and Ultrason. Attenuation, in press.

/16/ D. Keil, Dr.rer.nat.-Thesis, Univ. Stuttgart 1973.

DISCUSSION

Q: (H. J. Smith) If you are sure that you are looking at interstitial clusters, what happens to the vacancies, and how does this picture fit in with stage III annealing in Cu?

A: Vacancies are created in a near surface layer and the surface is a deep sink for the vacancies.

Q: (T. S. Noggle) What is the experimental range of focusing collisions in Au?

A: In the "first generation" of experiments, we found a peak in a (100) foil at ~ 220 Å. The deduced range for f.c.s. then is ~ 200 Å.

Q: (K. L. Merkle) Did you look at the dose dependence of the interstitial cluster formation?

A: The visibility of interstitial clusters begins at approximately 5×10^{14} Ar/cm^2 and we find a saturation at approximately 4×10^{15} Ar/cm^2 with approximately 3×10^{11} clusters/cm^2.

Q: (W. L. Brown) How many interstitials are involved in the clusters formed by a focusing collision segment in Cu?

A: To produce a cluster visible in TEM in Cu, we need 10^4 Ar$^+$ ions for 5 keV implants. For each cluster ~ 100 interstitials are stored.

OBSERVATION OF ION BOMBARDMENT DAMAGE IN A Ni(100) CRYSTAL BY HELIUM ION INJECTION

E. V. Kornelsen and D.E. Edwards, Jr.

Radio and Electrical Engineering Division

National Research Council, Ottawa, Canada

INTRODUCTION

Earlier papers (1) (2) have demonstrated that the entrapment of helium in tungsten is strongly dependent on the presence of lattice damage. It was concluded (2) that the entrapment proceeds via the interstitial diffusion of injected helium at room temperature, and that any helium atoms not encountering a defect during their diffusion escape through the surface. Considerations of helium ion penetration depths suggested that a trap concentration as low as $\sim 10^{-9}$ should have been detectable when the bombardment area was about 0.1 cm^2.

In a subsequent paper (3) an attempt was made to identify the particular lattice defects which led to the bound states observed in tungsten. The present paper is an attempt to make comparable identifications for the case of nickel bombarded with both helium and heavier inert gas ions.

EXPERIMENTAL

The apparatus and technique used were described earlier (2). Briefly, a nickel crystal was prepared in an ultra-high vacuum system which contained a small ion gun and a sensitive mass spectrometer. After surface preparation and annealing, the crystal was bombarded with measured doses of either helium ions alone or with heavy (damaging) ions followed by helium (probing) ions. The crystal was then heated at about 20K/sec and the resulting rate of helium evolution monitored by the mass spectrometer. The sensitivity of the apparatus was such that 1×10^9

desorbing helium atoms gave a clearly detectable peak.

The nickel crystal used was a disc 1 cm in diam. 1 mm thick.
It had a (100) exposed surface prepared by standard metallographic
grinding followed by electropolishing. It was examined by x-ray
back reflection and high energy electron diffraction. The ultra-
high vacuum preparation consisted of heating in oxygen to remove
carbon impurities followed by several cycles of sputter cleaning
and annealing. Temperature was measured by a Pt/Pt 10% Rh thermocouple spot welded to the crystal edge. The bombarded area
(0.125 cm^2) was a disc 4 mm in diameter at the centre of the crystal
face.

The penetration depths of the damaging and the probing ions
were discussed in the earlier papers (2) (3) and are roughly
applicable in the present case also: 5 keV Ne^+ or Ar^+ ions, which
were used for producing damage, are thought to penetrate 50 to
100 Å into the crystal and to create atomic disorder over most of
their range. Helium ions of a few hundred eV energy probably have
a mean penetration of ∿10 to 20 Å, while at 5 keV they may
penetrate a few hundred Å. Any information obtained by the present
technique will thus pertain to a layer of the crystal lying within
a few hundred Å of the bombarded surface.

The atomic displacement threshold energy in nickel measured by
electron impact is 24 eV (4). To transfer this amount of energy,
a He_4^+ ion must have at least 100 eV of kinetic energy, and a He_3^+
ion at least 130 eV.

RESULTS

Helium trapping at damage created by 5 keV Ne^+ ions is
illustrated in Fig. 1. The lower spectrum should be considered a
"background" case. It was obtained by bombarding the undamaged
crystal with 2×10^{12} He_3^+ ions at 200 eV. We believe the main peaks
at 700K and 850K are due to damage created by the helium ions.
It proved impractical to inject helium into nickel without creating
some lattice damage. At ion energies sufficiently low to avoid
atomic displacements in Ni (130 eV for He_3^+) the fraction of the
ions entering the crystal and their depth of penetration have
become small enough that the probing sensitivity is very poor. As
a compromise between sensitivity and lattice damage we chose
200 eV He_3^+ as the probing particles. The binding energies of He_3
and He_4 were found to be the same within our experimental error
(±0.05 eV) for all the bound states observed.

Following the indicated neon bombardment (1×10^{11}, 5 keV) the
same helium probe yielded the upper spectrum of Fig. 1. The
two original peaks (labelled B2 and B3) are seen to be strongly

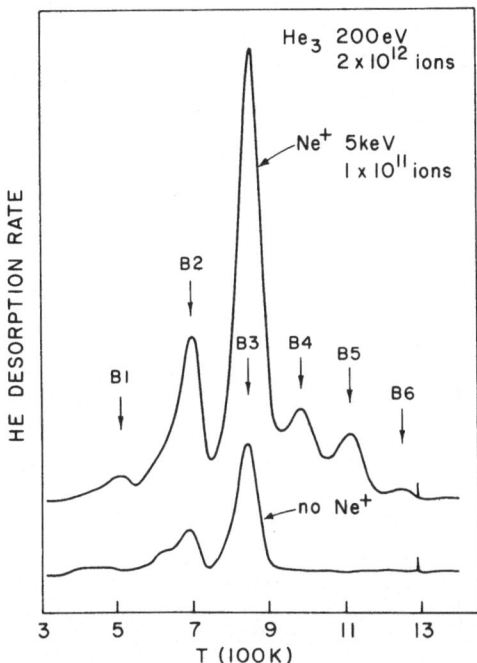

Fig. 1   Helium 3 desorption spectra showing the effect of bombarding
         the surface with 1×10^{11} 5 keV Ne^+ ions before helium injec-
         tion (2×10^{12} 200 eV He_3^+).   Labelled arrows B1 to B6 identify
         peaks discussed in the text.

enhanced and four additional peaks have appeared.   The fraction of
the incident helium trapped has increased from 7×10^{-3} to 6.5×10^{-2}.

    Bombardments with He_4^+ alone at low ion energies yielded the
spectra shown in Fig. 2.   All four peaks prominent here (S1 to S4)
are in fact quite distinct from those of Fig. 1.   Peak S3 can just
be detected on the low temperature shoulder of B2, and S4, while
close to the B3 temperature, is definitely lower by 35 ± 10K, well
outside the experimental uncertainty.   The ion energies were chosen
to show that three different threshold energies exist: one for S1
and S2 between 5 eV and 10 eV, one for S3 between 10 eV and 20 eV
and a third for S4 between 20 eV and 50 eV.   These very low threshold
energies suggest that the helium trapping is taking place in pre-
existing sites within a few atomic layers of the crystal surface.

    Spectra observed for 6×10^{12} incident helium ions of slightly
higher energies are presented in Fig. 3.   At 75 eV only the surface
related states can be seen.   Beginning at 100 eV and increasing
rapidly with energy, peaks B2 and B3 appear.   The surface peaks,
on the contrary, begin to decrease in amplitude.   The coincidence

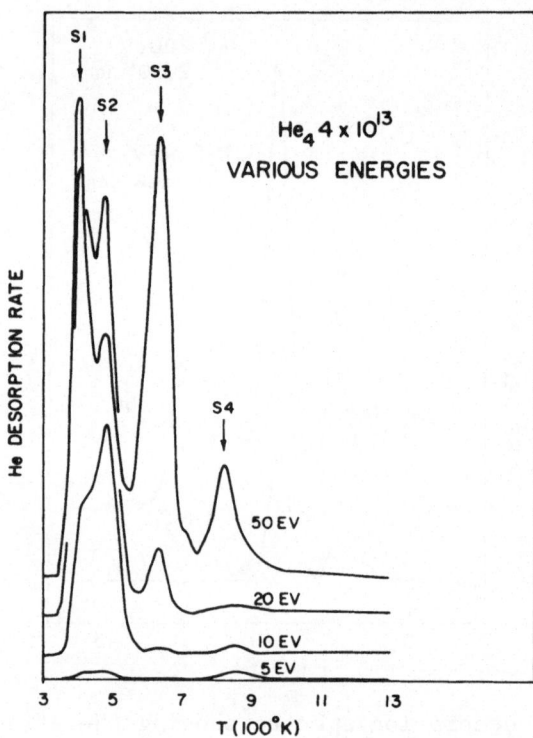

Fig. 2   Desorption spectra following low energy He$_4^+$ ion injection
         with no prior damage. Labelled arrows S1 to S4 indicate
         the positions of peaks not visible in Fig. 1

of the thresholds for B2 and B3 with that for atomic displacement
in nickel (see Sec.2) suggest that they involve trapping in bulk
point defects.

        Additional information concerning the nature of the traps was
obtained by varying the dose of probing ions (He$_3^+$, 200 eV)
following a standard damaging bombardment (Ne$^+$, 5 keV, 1×10^{11}).
The upper spectrum of Fig. 1 is one of the resulting series of
spectra. The populations of the individual peaks were obtained by
deconvolution of the spectra (see ref. 3) and plotted against the
ion dose. Two distinctly different types of behaviour were found,
and are shown in Fig. 4 and Fig. 5. The former demonstrates a
linear relationship between population and dose for peaks B3 and
B4, suggesting single occupation of two different kinds of traps.
The latter shows population varying quadratically with dose for
peaks B2 and B5 corresponding to the occupation of two kinds of
traps by two helium atoms. Peak B6 appeared to vary even more
rapidly with dose than B2 and B5. The data are, however,
inadequate to allow a definite order to be assigned.

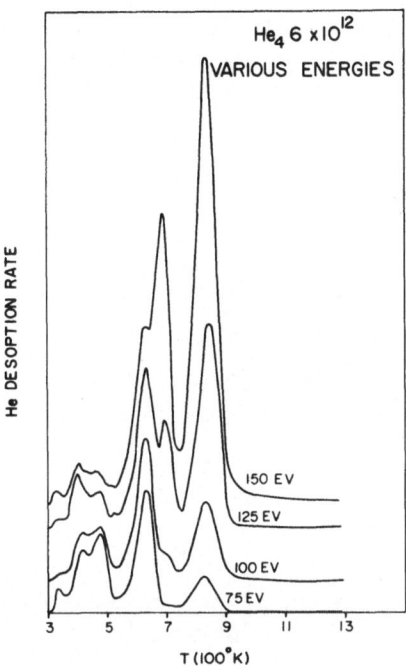

Fig. 3   **Desorption** spectra for 6×10^{12} He$_4^+$ ions of the energies indicated.

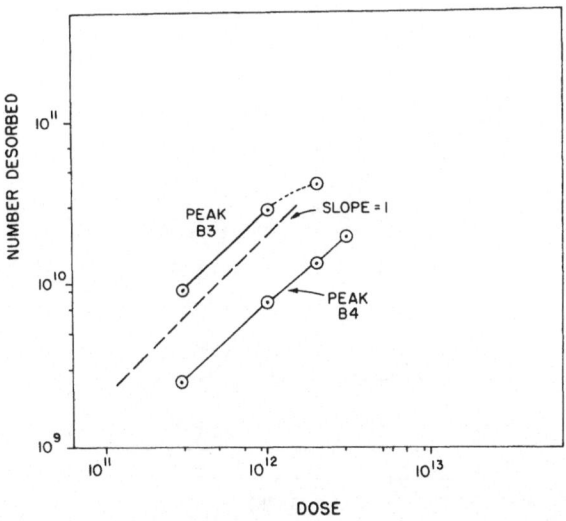

Fig. 4   Population of peaks B3 and B4 as a function of 200 eV He$_3^+$ dose on a surface damaged by 1×10^{11} 5 keV Ne$^+$.

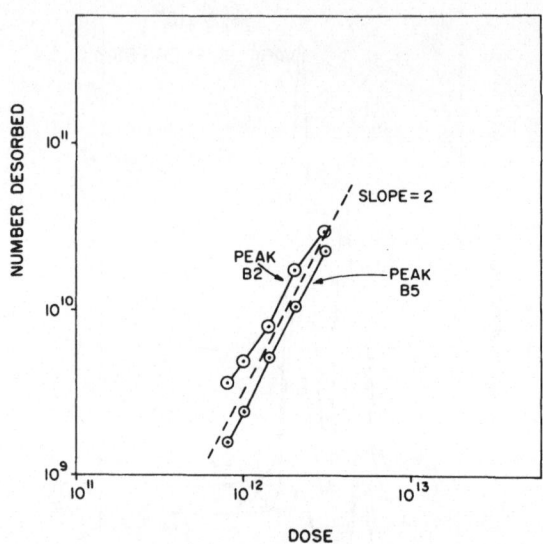

Fig. 5   Population of peaks B2 and B5 as a function of dose for
         the same bombardments as in Fig. 4.

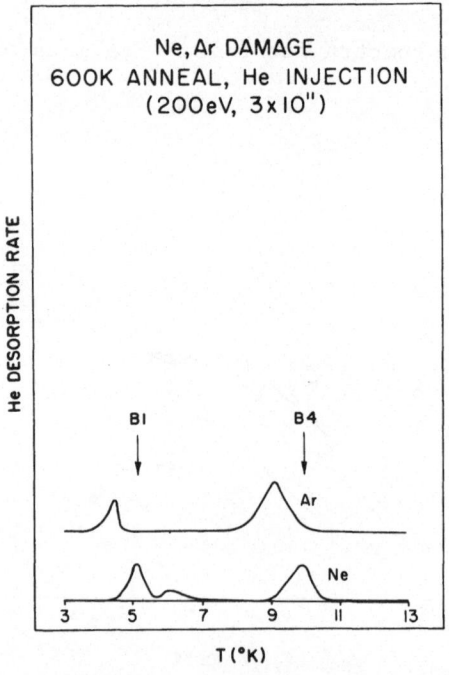

Fig. 6   Desorption spectra following 600K anneal of damage created
         by 1×10^{11} 5 keV Ne^+ and Ar^+. The He_3^+ injection was 3×10^{11}
         at 200 eV in each case and the resulting small B3 peak has
         been subtracted.

Lastly, the effect of 600K annealing of the crystal between damaging and probing bombardments is shown in Fig. 6. For the lower spectrum the damage was caused by 1×10^{11} 5 kev Ne^+ ions and the probe was 3×10^{11} 200 eV He_3^+. The probe bombardment alone gave only a B3 peak whose amplitude was about the same as the B4 peak shown. This peak has been subtracted to show the remaining peaks more clearly. The upper spectrum was obtained from an experiment identical to the above except that Ar^+ ions were used instead of Ne^+ in the damaging bombardment. The two major peaks are seen to be shifted to lower temperature for the heavier damaging ion, allowing the conclusion that the relevant traps are both associated with the inert gas impurity atoms in the crystal.

DISCUSSION

Interstitial lattice defects involve regions of reduced inter-atomic spacing in the crystal. Helium atoms, which undergo only repulsive interactions with the lattice atoms, would find such a region of higher potential energy and be repelled. It does not seem likely, therefore, that any interstitial would act as a helium trap. Since peaks B2 and B3 appeared as soon as point defect production became possible (see Fig. 3), we propose that the relevant traps are single lattice vacancies in both cases. From Figs. 4 and 5, it seems reasonable that peak B3 corresponds to the release of the helium atom from a singly-occupied vacancy, and B2 to the release of the first of two helium atoms from a doubly occupied vacancy. Consistent with Fig. 6 we propose that the traps leading to peaks B1 and B4 are vacancies bound at the nearest and second nearest neighbor distances respectively to a substitutional neon impurity atom. In each case occupation by a single helium atom is involved. From the dose dependence (Fig. 5) and the high binding energy (Fig. 1) we suggest that peak B5 might involve the double occupation of a lattice divacancy, while on the same arguments, B6 might involve multiple occupation of a trivacancy or other vacancy cluster.

Binding energies and rate constants were measured for peaks B2 and B3 by varying the rate of temperature increase during desorption. The rate constant was found to be $3 \times 10^{12 \pm 1}$ in both cases, and the binding energies 1.70 eV for B2 and 2.10 eV for B3. Assuming the same rate constant to apply, peaks B4 and B5 would have binding energies 2.35 eV and 2.70 eV respectively. More careful measurements will be necessary before other binding energy assignments can be made.

CONCLUSIONS

As was found to be the case in tungsten (3), helium atoms migrate freely as interstitials in nickel at room temperature but become trapped if they encounter defects containing one or more lattice vacancies.

On the basis of desorption spectra obtained while varying helium ion energy and dose and damage annealing temperature, it has been possible to assign some of the observed desorption peaks tentatively to specific trap configurations.

Regardless of the accuracy of the assignments made the results clearly demonstrate, as in the case of tungsten, that helium trapping in nickel is very sensitive to the presence of ion-induced lattice damage and that the helium binding energy is highly specific to the detailed atomic configuration of the trap.

REFERENCES

1)   K. Erents, G. Farrell and G. Carter. Proc. 4th Int. Vac. Congress (Institute of Physics and Physical Society, London) p.145 (1968).

2)   E.V. Kornelsen, Can. J. Phys. 48, 2812 (1970).

3)   E.V. Kornelsen, Radiation Effects 13, 227 (1972).

4)   P.G. Lucasson and R.M. Walker. Phys. Rev. 127, 485 (1962).

DISCUSSION

Q: (G. L. Kulcinski) Have you considered the possibility that once a He atom is released from a trap it may recombine with another trap on the way to the surface? This would presumably alter the desorption at high (dT/dt) rates.

A: The shape of the desorption peaks at low dose suggest that no significant retrapping effects are occurring. At doses $> 10^{14}/$ cm^2 the peaks broaden rapidly, presumably because of such effects.

Q: (L. C. Feldman) What is the He bombarding energy dependence for the relative amount of He trapped in the (He+vacancy) site as compared to the (2 He + vacancy) site?

A: Up to keV helium ion energy we find that the multiple occupation states become more prominent with increasing energy at constant fluence (about $10^{13}/cm^2$).

CHAPTER VII

ION IMPLANTED GAS BUILDUP

HELIUM IMPLANTATION EFFECTS IN VANADIUM AND NIOBIUM

G. J. Thomas and W. Bauer

Sandia Laboratories, Livermore, CA   94550

I.   INTRODUCTION

There are a number of materials problems associated with the controlled thermonuclear reactor (CTR). One of these involves energetic particle (hydrogen isotopes and helium) fluxes which are high enough to attain considerable fluence levels in relatively short times (days or weeks of operation). The effects of this particle bombardment on first wall, collector and divertor assemblies must be considered, and consequently there have recently been a number of studies[1-3] in the area of high dose hydrogen isotope and helium irradiation of metals. These studies show the manifestation of the gas accumulation - such as bubbles, blisters, exfoliation - to have a considerable influence on the material behavior and on gas re-emission characteristics. Both surface morphology and gas re-emission are important variables to understand in order to achieve viable CTR component designs.

We have examined the effects of high fluence 300 keV He implantation in Nb and V at implantation temperatures ranging from 400°C to 1200°C. These materials (and their alloys) are considered as possible candidates for CTR applications and this temperature region spans the estimated CTR operating temperature range. Implantation to doses up to 4×10^{18} He atoms/cm^2 were made, corresponding to $\sim$ 40 dpa and 0.1-0.5 He atom fraction in the implanted region. Gas re-emission measurements were made during implantation and samples were examined by scanning (SEM) and transmission (TEM) electron microscopy. Considerable similiarity in the behavior of these two materials was found, as well as with 316 SS[2] and Mo[3]. A strong temperature dependence was observed in the re-emission characteristics, surface features, and gas bubble morphology within the implanted layer. As might be expected the

533

low initial He re-emission rate found at the lower implantation
temperatures results in gas accumulation and subsequent surface
deformation. The essential features of this deformation are
affected by the way in which the gas agglomerates at lower concentration. The temperature dependence of the re-emission and surface
deformation then is primarily dependent on the mobility of the
He and radiation produced vacancies and vacancy complexes.

II. EXPERIMENTAL

Samples prepared from MARZ grade material were polished and
mounted on a resistively heated Ta foil - annealing of the sample
prior to implantation was performed in-situ. The He re-emission
was measured during implantation by means of a differentially
pumped system which effectively isolated the sample chamber and He
mass spectrometer vacuum from the accelerator beam tube. Several
different methods were used to obtain an absolute calibration of
this system. These are discussed in more detail in a previous
publication.[2]

III. RESULTS

The results of He re-emission measurements on V and Nb for
a variety of implantation temperatures are shown in Figures 1 and 2.

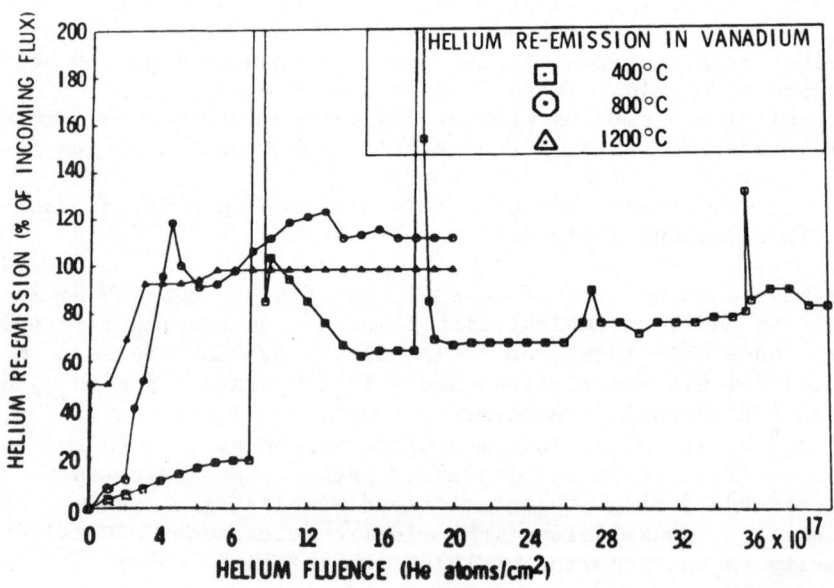

FIGURE 1

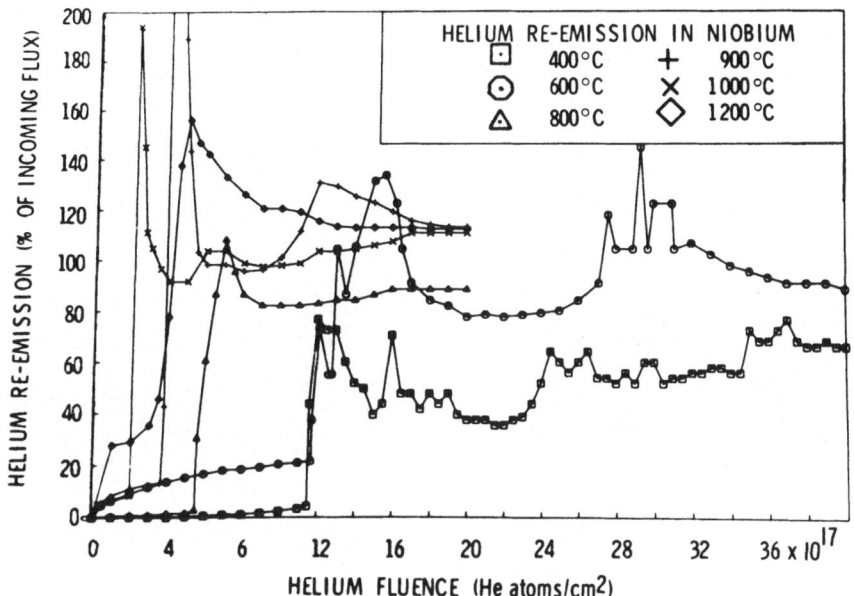

FIGURE 2

Several common features are evident from these plots. First of all,
it is seen that, in general, the re-emission is low until a certain
critical dose is reached, at which time the re-emission abruptly
rises to a higher value. The critical dose decreases with increasing
implantation temperature, and the re-emission after the critical dose
is generally greater at the higher implantation temperatures.
The critical dose behavior is associated with the abrupt onset of
surface deformation. In Figures 3 and 4 scanning electron micrographs are shown of V and Nb samples after implantation to doses of
$2-4 \times 10^{18}$ He atoms/cm^2 (well beyond the critical dose). A strong
temperature dependence is readily apparent. At 400°C, large scale
flaking of the surface is found. The release of large amounts of
untrapped gas when this occurs accounts for the gas bursts in the
re-emission observed at this temperature. At 800°C, only a very
low density of small blisters and holes were found on the surface
of V. No 800°C micrograph in the case of Nb is included because of
the extremely low density of surface features. However, a swelling
of the entire implanted region was observed, and this feature will
be discussed later. The primary surface deformation at 1000°C and
1200°C consists of faceted holes a few tenths of a micron in
diameter. It is clear in this case that the critical dose is
associated with the formation of these holes and that the high,

HELIUM IMPLANTATION OF VANADIUM

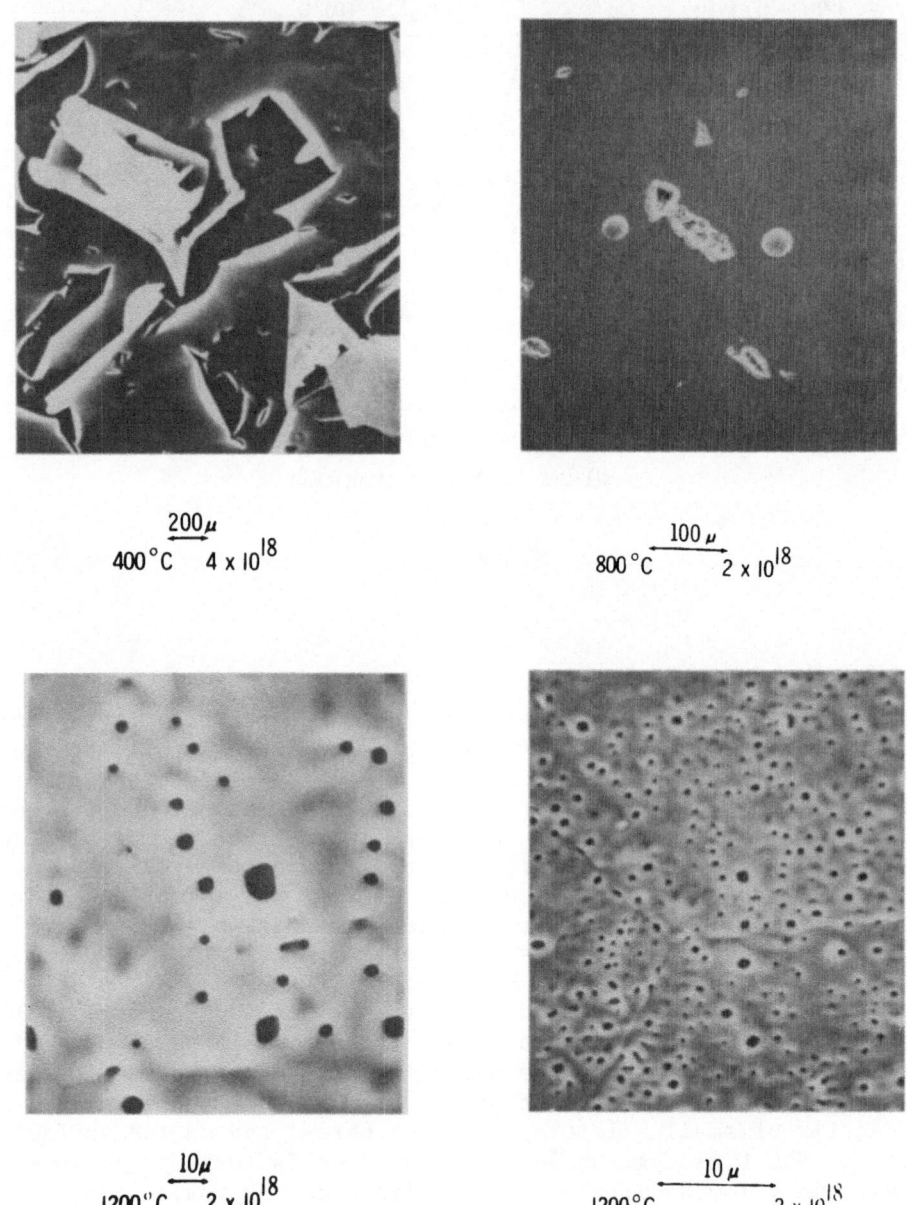

FIGURE 3

HELIUM IMPLANTATION OF NIOBIUM

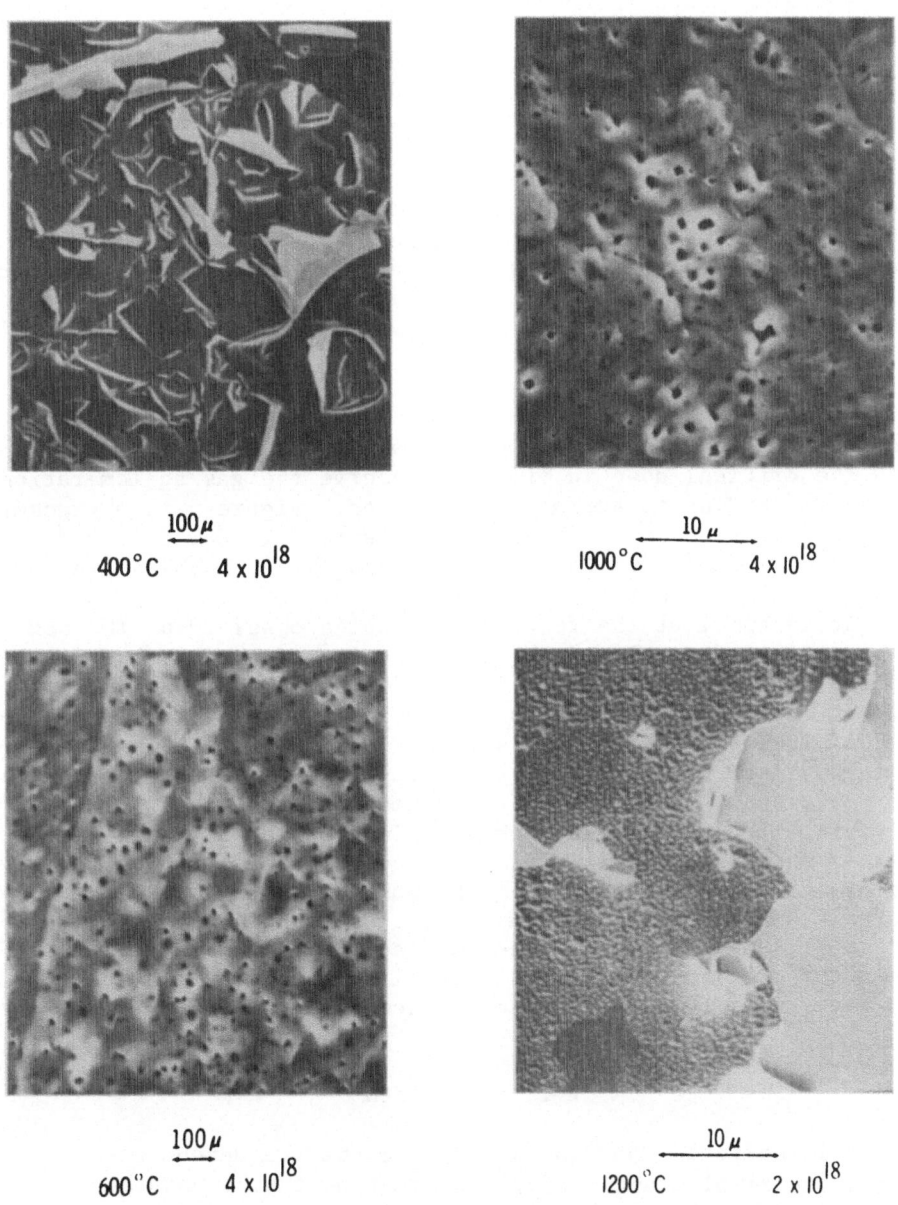

FIGURE 4

fairly constant gas re-emission beyond the critical dose is due to this porous structure allowing diffusion of the gas to the surface.

It is apparent in Figures 1 and 2 that in some cases the measured re-emission is sometimes greater than 100% of the incoming beam. A portion of this effect may be due to the absolute calibration uncertainty. However, it is believed that some of this release is due to radiation enhanced diffusion of He trapped during the earlier portion of implantation via a vacancy assisted mechanism. After a dose of 2×10^{18} He atoms/cm^2 at 1200°C, the He beam was replaced by a 150 keV proton beam. A proton energy of 150 keV resulted in proton vacancy production near the initial implanted He. With a proton flux of 2×10^{15}/cm^2 sec, a He release of approximately 4% on the same scale as in Figures 1 and 2 was found. This direct test tends to confirm that at least a portion of the release at the higher temperatures is due to radiation enhanced diffusion of initially trapped He.

To investigate bulk features, observations were also made by transmission electron microscopy on samples implanted to just below the critical dose in order to observe the gas agglomeration mechanisms leading to surface deformation. Figure 5 is a transmission electron micrograph of a Nb sample implanted at 1200°C to a dose of 4×10^{17} He atoms/cm^2. Large faceted gas filled bubbles are found ranging in size from about 250 Å to 2500 Å. It would appear from this micrograph that the formation of holes observed by the SEM can be explained simply by the intersection of these large bubbles with the sample surface. In Figure 6 is shown a transmission electron micrograph of a Nb sample implanted at 800°C to a dose of 5×10^{17} He atoms/cm^2. In this case, faceted gas bubbles approximately 300 Å in diameter with a density of 5×10^{14} cm^{-3} are found. This corresponds to a fractional volume increase, $\Delta V/V$, of about 1%, giving rise to the observed swelling previously discussed. Figure 7 shows a micrograph of a sample implanted at 600°C to a dose of 1×10^{18} He atoms/cm^2. The bubbles in this case are considerably smaller (~ 35 Å diameter), of much higher density ($\sim 5 \times 10^{16}$ cm^{-3}), and show no evidence of faceting. Presumably, with higher doses an interconnected gas filled region is formed which then gives rise to the observed surface flaking.

IV. DISCUSSION

It is believed that the strong temperature dependence of the re-emission results and surface deformation can be understood in terms of the bubble growth rate. At the lowest temperatures studied, 400°C - 600°C, vacancies are expected to have little mobility. He migration occurs primarily by an interstitial mechanism, but strong He trapping would be expected to occur in implantation produced

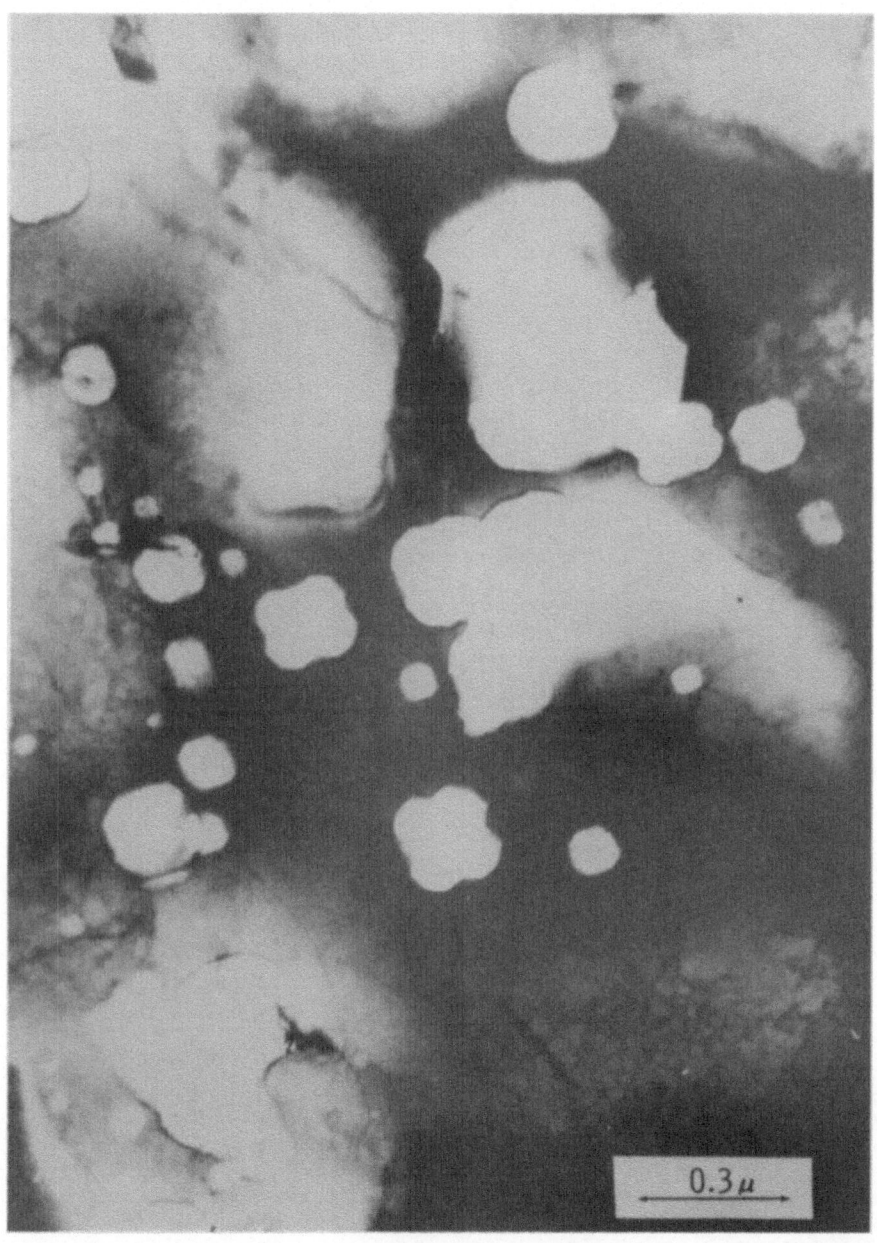

FIGURE 5

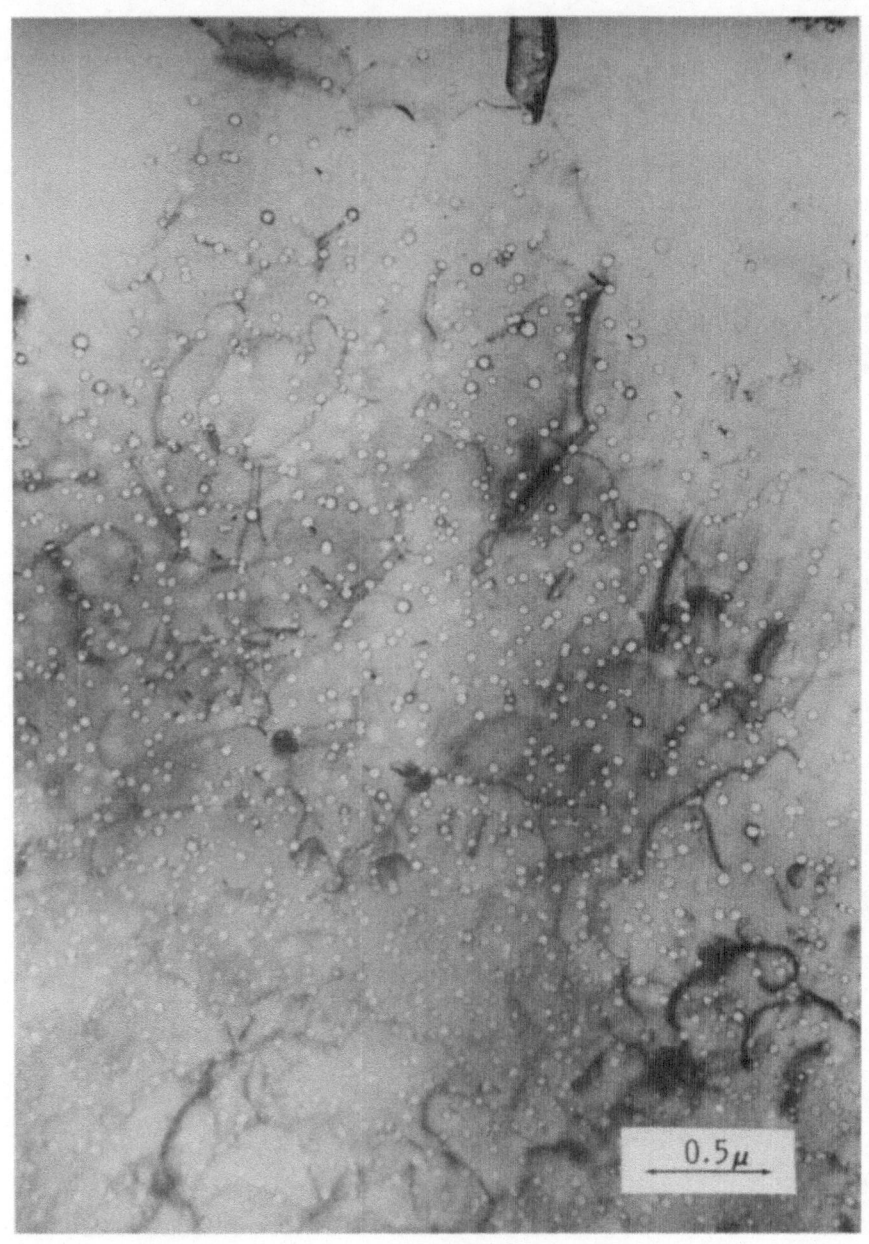

FIGURE 6

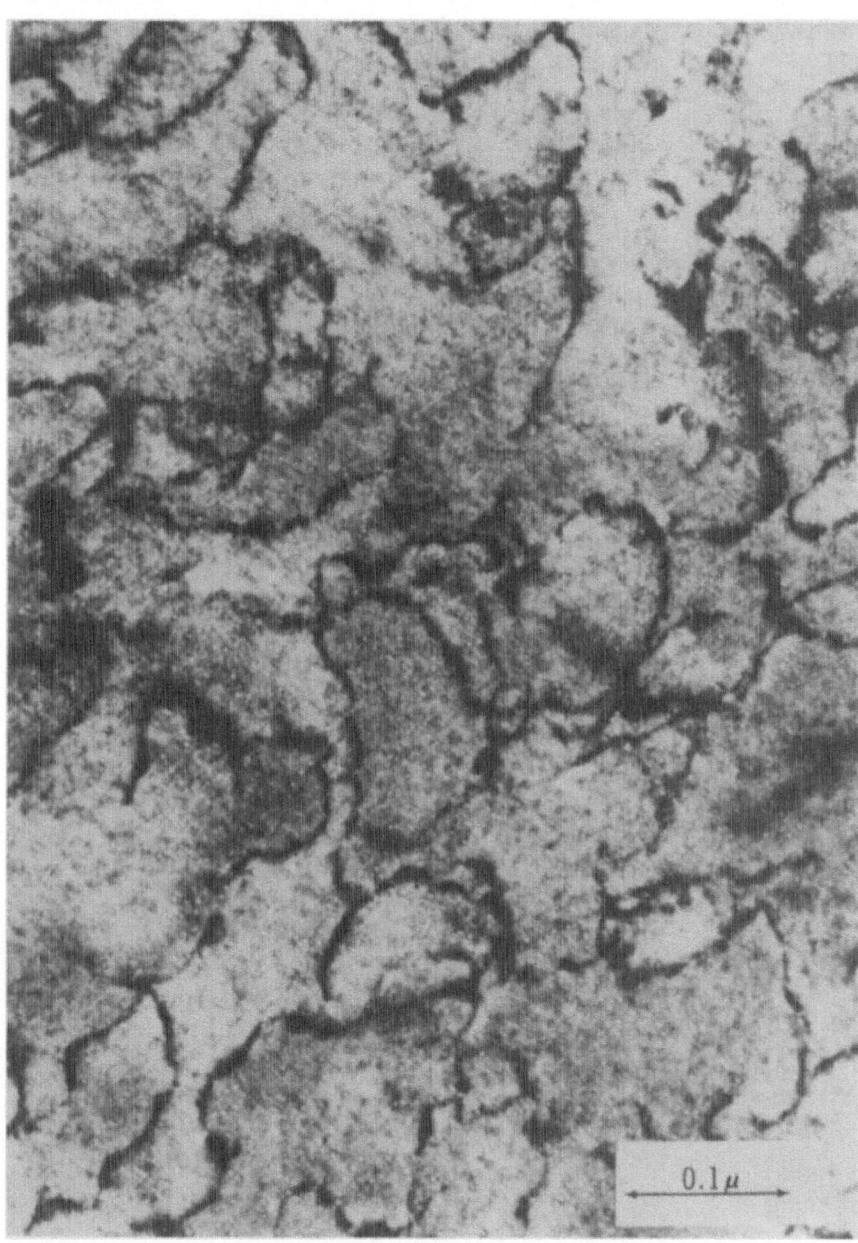

FIGURE 7

vacancies. Thus, one observes (Figure 7) slow bubble growth and a
high bubble density (nucleation probability). The strong trapping
prevents release of the implanted gas, however, and an interconnected
gas filled region is ultimately formed. It is not known whether
this occurs by actual bubble overlap during growth or if there is
some critical spacing at which individual bubbles interact. At the
intermediate temperature region of 800°C-900°C, vacancies are highly
mobile allowing the formation of much larger bubbles which are in
equilibrium with the lattice, and therefore, generating considerably
less lattice strain. The large volume increase due to the bubbles
produces a macroscopic swelling of the implanted region. At the
highest temperatures studied (1000-1200°C) the high mobility of gas,
vacancies, and presumably small bubbles produces very large bubbles
which continue to grow until they eventually intersect the surface.
The resultant porous structure allows the rapid release of implanted
He so that the re-emission value is high and further bubble growth
is inhibited.

REFERENCES

1. See the preceding three papers in this conference.

2. W. Bauer and G. J. Thomas, J. Nucl. Mat. $\underline{47}$, 241 (1973).

3. W. Bauer and G. J. Thomas, Proceedings of the 1973 International
    Conference on Defects and Defect Clusters in b.c.c. Metals and
    Their Alloys, (National Bureau of Standards, Gaithersburg,
    Maryland, 1973).

DISCUSSION

Q: (M. Kaminsky) In some of your studies you found that the
helium remission rate exceeds the helium implantation rate. Did
you have a chance to do these studies at different dose rates?

A: No, we have not examined the dose rate dependence of these
effects at this time.

EFFECT OF He$^+$ AND D$^+$ ION BEAM FLUX ON BLISTER FORMATION IN NIOBIUM AND VANADIUM[†]

S.K. Das and M. Kaminsky

Argonne National Laboratory

Argonne, Illinois 60439

ABSTRACT

The effect of incident ion beam flux on the blister formation in annealed polycrystalline niobium and vanadium has been investigated for 0.5-MeV He$^+$ and 0.25-MeV D$^+$ projectiles. For the He$^+$ ions the flux was varied from 1 x 10^{13} ions/(cm^2-sec) to 1 x 10^{15} ions/(cm^2-sec), the targets were held at 900°C, and the total dose was varied from 0.1 to 1.0 C/cm^2. For the D$^+$ ion irradiation the flux was varied from 1 x 10^{14} to 1 x 10^{15} ions/(cm^2-sec), the niobium targets were held at 700°C, and the total dose was 2.0 C/cm^2. For both the helium implanted vanadium and niobium the blister density shows a stronger flux dependence than the average blister diameter. For deuteron implanted niobium the blister size increased as the flux was increased from 1 x 10^{14} to 1 x 10^{15} ions/(cm^2-sec).

INTRODUCTION

Irradiation of metal surfaces such as niobium and vanadium with He$^+$ and D$^+$ projectiles are known [1-9] to form blisters on surfaces. Earlier blistering studies in other metals can be found in Refs. 1-9 and the references cited therein. The results of these studies [1-9] reveal that the shape, size, density and the degree of blister formation (defined here as the fraction of total irradiated area that is occupied by blisters) depend on such parameters as projectile energy [3,6], target temperature [2-8], channeling condition of the projectile [2,3], orientation of the irradiated surface planes [2,3,6,9-12], initial defect concentration in the target [1,4], and total dose [1,3,4,6]. Another important irradiation parameter is the projectile flux. To our knowledge

the effect of flux on blister formation has received little
attention. At a particular irradiation temperature the rate of gas
build-up near the implant depth and the subsequent blister formation will depend on the incident ion flux and the rate of gas
release from the surface. Depending on the balance of gas trapping
and gas release, there may or may not be an effect of the
projectile flux on blister formation. In the present paper this
effect of projectile flux on blister formation in annealed
polycrystalline niobium and vanadium was investigated for two
gases - namely helium and deuterium; the former having extremely
low permeability and the latter having extremely high permeability.

Such studies of blistering in niobium and vanadium by implantation with helium ions with energies ranging from 0.1 to
3.5-MeV are of interest to the controlled thermonuclear fusion
reactor program. For example, blistering processes can play an
important role in the operation of D-T fusion reactors [12-14] in
that they lead to severe erosion of the wall surfaces exposed to
impact by energetic projectiles such as helium.

EXPERIMENTAL TECHNIQUES

The polycrystalline niobium and vanadium foils were obtained
from Materials Research Corporation (Marz grade). The foils were
first given a fine metallographic polish and then annealed at
1200°C for 2 hours in a vacuum of 3-5 x 10^{-7} Torr and finally electropolished by techniques described earlier [1,4]. The polished
areas had an average surface roughness of less than 0.02 μm as
determined by a scanning electron microscope with a resolution
limit of about 0.02 μm. A mass analyzed beam of $^4He^+$ ions with
energy of 0.5-MeV, or of D^+ ions with energy of 0.25-MeV from a
2-MeV Van de Graaff accelerator was highly collimated (half angle
of divergence = 0.01°) and was incident parallel (within $\sim$0.1°) to
the surface normal of the target. For high temperature irradiations
thin targets ($\sim$25 μm thick) were used and were heated by ohmic
heating. The target temperature was measured with an optical
pyrometer; a correction was applied for absorption in the window
of the target chamber. For He^+ ion irradiations the ion flux was
varied from 1 x 10^{13} ions/(cm^2-sec) to 1 x 10^{15} ions/(cm^2-sec) and
for D^+ ion irradiation the ion flux was varied from 1 x 10^{14} to
1 x 10^{15} ions/(cm^2-sec). In each case the irradiated area on each
target was a circular spot of 2 mm in diameter. During the irradiation the vacuum in the target chamber was maintained at about
1-2 x 10^{-8} Torr by ion pumping. The irradiated target surfaces were
examined with a scanning electron microscope, Cambridge Stereoscan
model Mark IIA. The size distribution of the blisters were measured with a Zeiss particle-size analyzer from enlarged micrographs.
For blisters with irregular shapes the average diameter of the
blister was defined as the diameter of a circle having the same

area as the blister. The mean values of the average blister
diameters presented in this paper are the weighted average values
determined from the size distribution of the blisters. The results
on blister size, density and fraction of total irradiated area
occupied by blisters given in this paper represent average values
from at least two irradiation runs. In cases where considerable
scatter was observed the runs were repeated three or four times.

RESULTS

The blister formation in annealed polycrystalline niobium and
vanadium samples will be described separately for the two types of
projectiles used, $^4He^+$ and D^+ ions.

$^4He^+$-Ion Irradiation of Niobium

The scanning electron micrographs in Figs. 1(a), 1(b) and 1(c)
show the blisters formed after annealed polycrystalline niobium
surfaces at 900°C were irradiated with 0.5-MeV $^4He^+$ ions to a
total dose of 0.1 C/cm^2 for fluxes of 1 x 10^{13}, 1 x 10^{14} and
1 x 10^{15} ions/(cm^2-sec), respectively. In each micrograph two
types of blisters can be distinguished. The smaller size (type I)
blisters have diameters ranging from approximately 0.3 to 1.0 μm
(see insets in Figs. 1a-c) and are found to be nearly independent
of the fluxes used. The other type (type II) of blisters have
larger diameters, ranging from 3 to 15 μm. The value for the
average blister density of the smaller size blisters is found to
be approximately (5 ± 3) x 10^6 blisters/cm^2, and appears to be
independent of the three dose rates studied within the quoted
experimental error. For the larger blisters the blister density
increases from (3 ± 2) x 10^4 blisters/cm^2 to (6 ± 3) x 10^4
blisters/cm^2 as the flux is increased from 1 x 10^{13} to 1 x 10^{14}
ions/(cm^2-sec), but a further increase in flux to 1 x 10^{15}
ions/(cm^2-sec) does not seem to change this value within the
quoted error limits. The fraction of total irradiated area that
is occupied by the blisters of both types shows a slight increase
with increasing flux. The values increase from approximately
(1.5 ± 0.5)% to (3.5 ± 1.0)% as the flux is increased from
1 x 10^{13} to 1 x 10^{15} ions/(cm^2-sec).

The scanning electron micrographs in Figs. 2(a) and 2(b) show
blisters formed on polycrystalline niobium surfaces after irradiation under similar conditions as described above, except for a
higher total dose of 1.0 C/cm^2 and for fluxes of 1 x 10^{14} ions/
(cm^2-sec) and 1 x 10^{15} ions/(cm^2-sec), respectively. Again two
types of blisters can be distinguished. For the flux range studied
the average diameters of the smaller size blisters range from 0.3 to
1.8 μm while those for the large size blisters range from 3-15 μm.
A comparison of the blister diameters for a given flux but for the
two total doses of 0.1 and 1.0 C/cm^2 shows that the range of blister

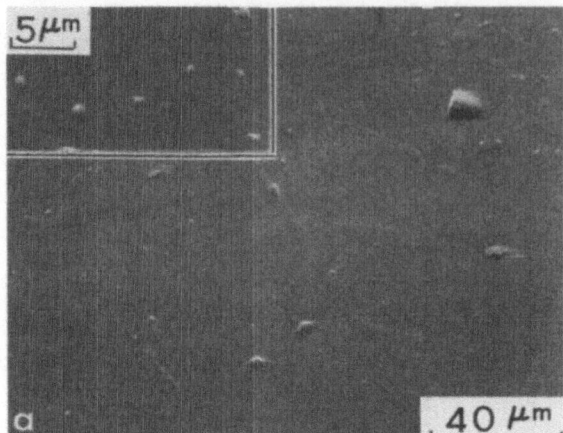

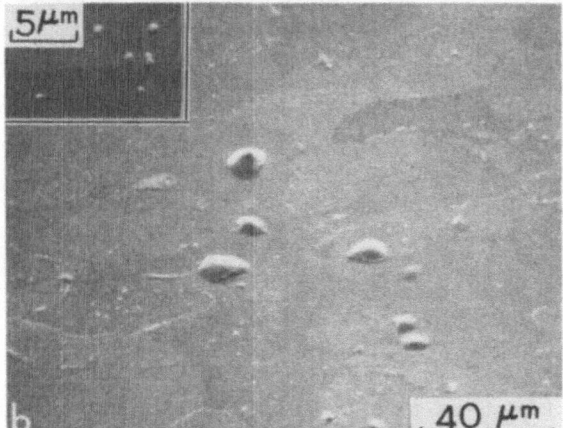

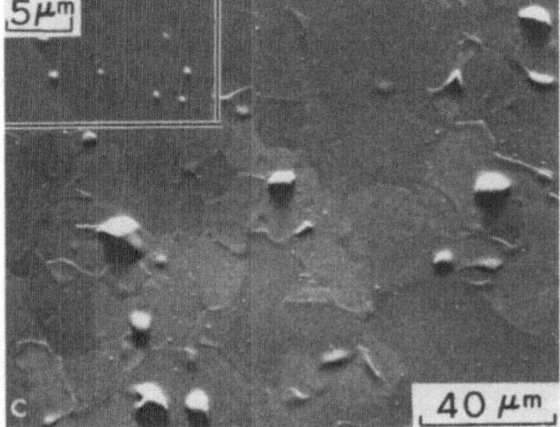

Fig. 1. Scanning electron micrographs (SEM) of surfaces of annealed polycrystalline Nb after irradiation at $900^{\circ}C$ with 0.5-MeV $^{4}He^{+}$ for a total dose of 0.1 C/cm^{2} at fluxes of:

    (a) 1 x 10^{13} ions/ (cm^{2}-sec)

    (b) 1 x 10^{14} ions/ (cm^{2}-sec)

    (c) 1 x 10^{15} ions/ (cm^{2}-sec)

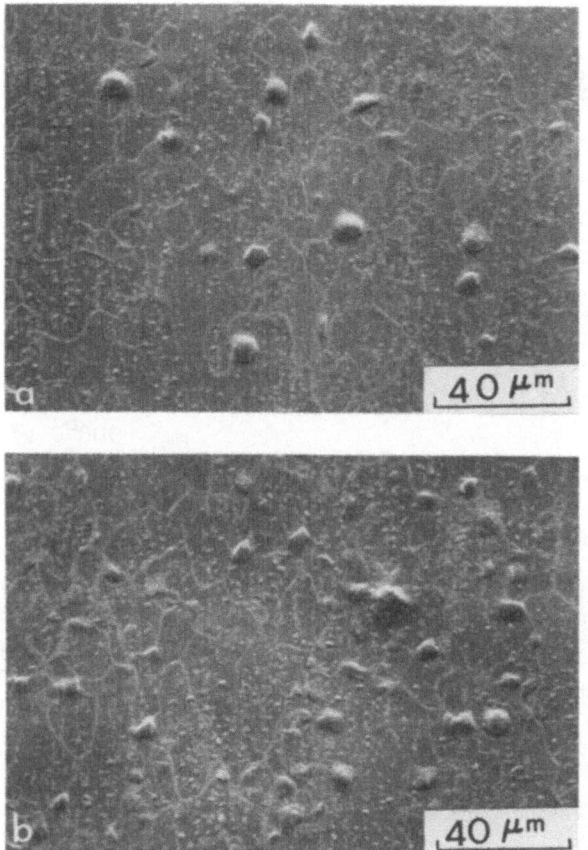

Fig. 2. SEM's of surfaces of polycrystalline Nb after
irradiation at 900°C with 0.5-MeV ^{4}He$^+$ for a total dose
of 1.0 C/cm^2 at fluxes of (a) 1 x 10^{14} ions/(cm^2-sec)
and (b) 1 x 10^{15} ions/(cm^2-sec).

diameters has practically the same value for the two doses studied.
The values for the blister densities for both the smaller and the
larger size blisters are approximately (8 ± 2) x 10^6 blisters/cm^2
and (1.5 ± 0.5) x 10^5 blisters/cm^2, respectively, and show (within
the quoted error limits) no significant flux dependence. The
average value for the fraction of total irradiated area that is
occupied by blisters of both size classes is (9 ± 2)% and is,
within the quoted error limits, the same for the flux range
studied.

^{4}He$^+$-Ion Irradiation of Vanadium

The micrographs in Figs. 3(a), 3(b) and 3(c) show typical
blisters formed after irradiating annealed polycrystalline

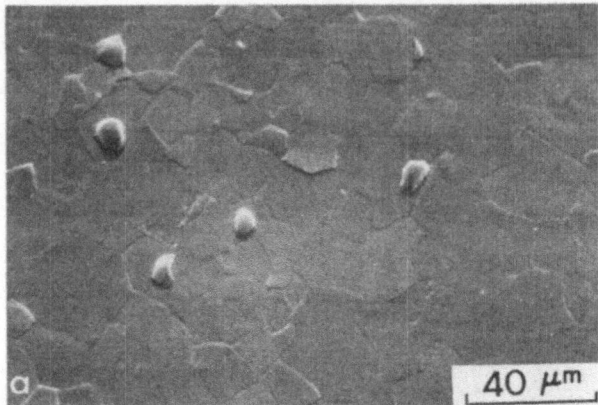

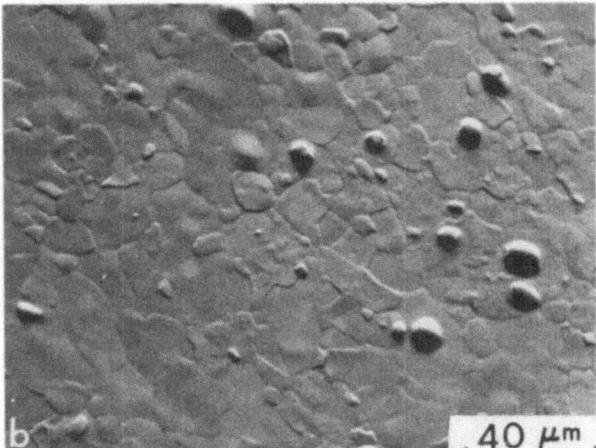

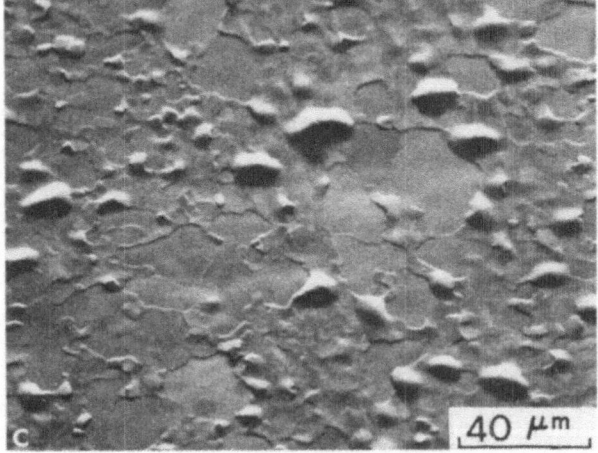

Fig. 3.   SEM's of surfaces of annealed polycrystalline V after irradiation at 900°C with 0.5-MeV ^{4}He$^+$ for a total dose of 0.1 C/cm^2 at fluxes of:

   (a) 1×10^{13} ions/(cm^2-sec)

   (b) 1×10^{14} ions/(cm^2-sec)

   (c) 1×10^{15} ions/(cm^2-sec)

vanadium surfaces held at 900°C with 0.5 MeV $^4He^+$ ions to a total dose of 0.1 C/cm^2 for fluxes of 1 x 10^{13}, 1 x 10^{14}, 1 x 10^{15} ions/(cm^2-sec), respectively. The average diameters of blisters of both types do not seem to depend significantly on beam flux. For the small size blisters the value for the average diameter ranges from approximately 0.1 to 0.3 µm and for the large size blisters the average diameter ranges from about 3 to 15 µm for the range of fluxes studied. It is observed that the value for the blister density for the small size blisters is (5 ± 2) x 10^6 blisters/cm^2 and that within the error limits quoted this value is the same for the three fluxes. For the larger size blisters, however, a dependence of the blister density on the flux is observed. The values increase from (1.0 ± 0.5) x 10^4, to (3 ± 1) x 10^4 and to (7 ± 2) x 10^4 blisters/cm^2 as the flux increases from 1 x 10^{13}, to 1 x 10^{14} and to 1 x 10^{15} ions/(cm^2-sec), respectively.

Figure 4 shows blisters formed after irradiating annealed polycrystalline vanadium surfaces under the same irradiation condition as those described above (Fig. 3), but for a higher total dose of 1.0 C/cm^2. The micrographs in Fig. 4(a) and 4(b) show typical blisters observed for fluxes of 1 x 10^{14} and 1 x 10^{15} ions/(cm^2-sec), respectively. It is found that the blister diameters for both types of blisters do not show an observable flux dependence. For the smaller size blisters the value for the diameter ranges from 0.5 to 1.0 µm and for the larger size blisters it ranges from 3 to 15 µm for the two flux values studied. The blister density for the small size blisters is about (8 ± 3) x 10^6 blisters/cm^2 for the two fluxes studied. The value for the large size blisters, however, increases from (2 ± 1) x 10^4 to (7 ± 2) x 10^4 blisters/cm^2 as the flux is increased from 1 x 10^{14} to 1 x 10^{15} ions/(cm^2-sec). As the dose is increased from 0.1 to 1.0 C/cm^2 the range of the values of the average blister diameters does not show an appreciable change but the mean value of the average blister diameters shows a slight increase from 6.2 to 7.4 µm. It is of interest to note that the values for the average blister diameters for both types of blisters are similar to those observed for niobium surfaces (Fig. 2) within the ranges quoted. The fraction of total irradiated area that is occupied by both types of blisters is found to increase from approximately 5% to 15% as the flux is increased from 1 x 10^{14} to 1 x 10^{15} ions/(cm^2-sec).

D^+-Ion Irradiation of Niobium

The micrographs in Figs. 5(a), 5(b) and 5(c) show blisters formed on the surfaces of annealed polycrystalline niobium at 700°C after irradiation with 250-keV deuterons to a total dose of 2.0 C/cm^2 for the fluxes of 1 x 10^{14}, 4 x 10^{14} and 1 x 10^{15} ions/(cm^2-sec), respectively. It has been observed earlier [6,9] that the blister shape strongly depends on the orientation of the grains.

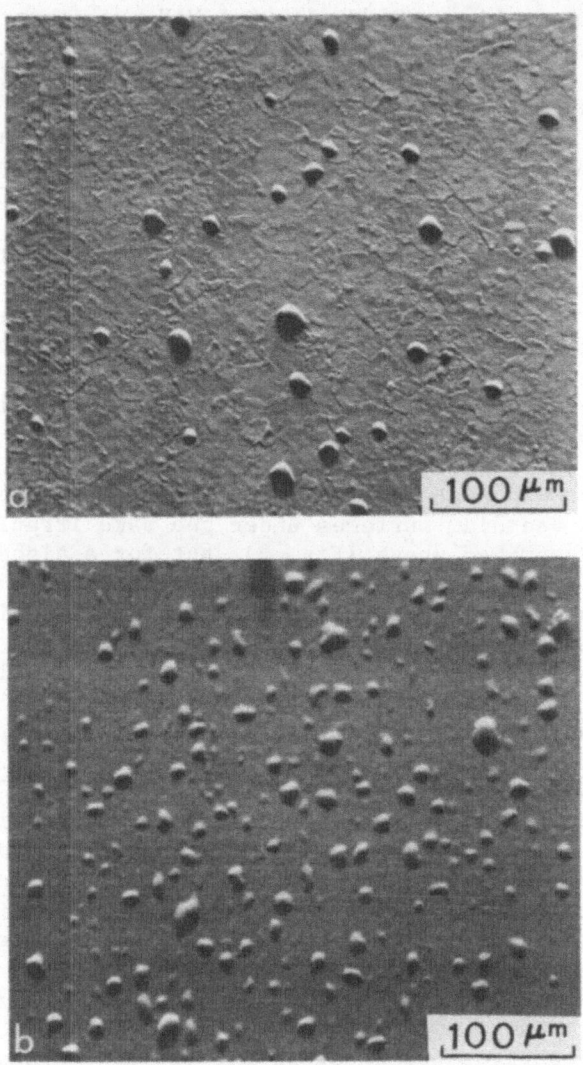

Fig. 4. SEM's of annealed polycrystalline V surfaces
after irradiation at 900°C with 0.5-MeV $^4He^+$ for a
total dose of 1.0 C/cm^2 at fluxes of (a) 1 x 10^{14}
ions/(cm^2-sec) and (b) 1 x 10^{15} ions/(cm^2-sec).

In Fig. 5(a) most of the blisters in one of the grains exhibit
three fold symmetry resembling the "crow-foot" shaped blisters
described earlier [2,3,6]. Since the orientation of this particular
grain is unknown, one can only conjecture that the prongs of the
"crow-foot" shaped blisters are aligned along the [$\bar{1}2\bar{1}$], [$\bar{1}\bar{1}2$] and
[$2\bar{1}\bar{1}$] directions of the niobium lattice was observed previously

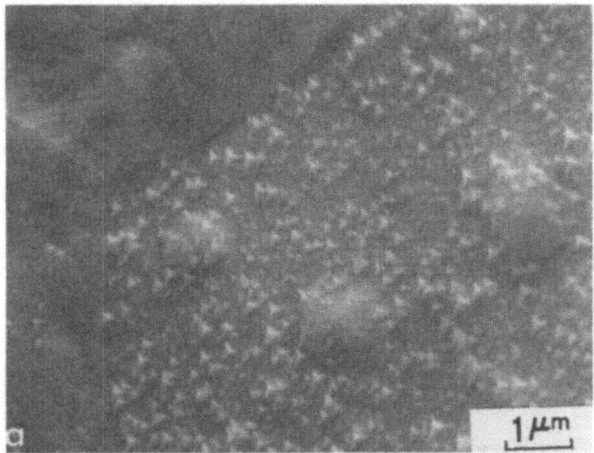

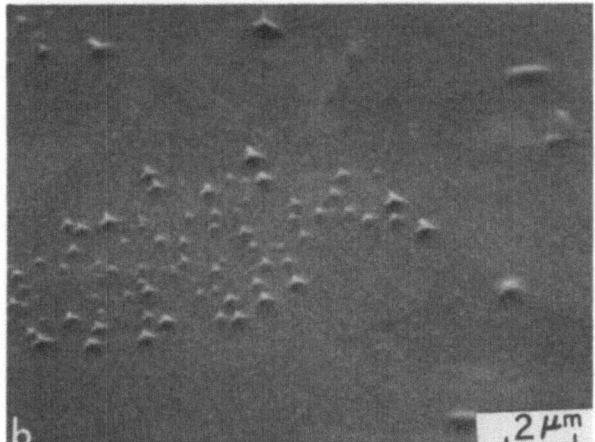

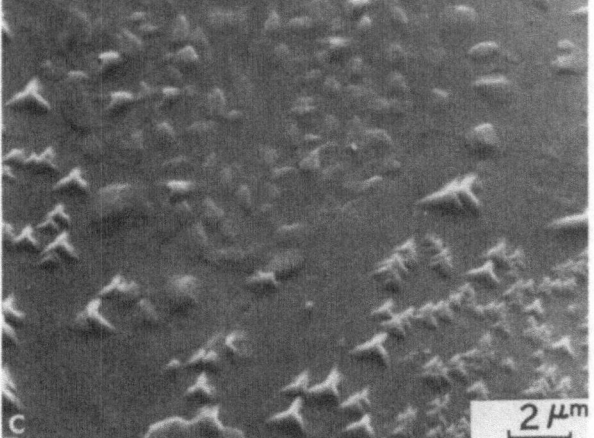

Fig. 5. SEM's of annealed polycrystalline Nb surfaces after irradiation at 700°C with 250-keV D^+ ions to a total dose of 2.0 C/cm^2 at fluxes of:
- (a) 1×10^{14} ions/$(cm^2$-sec)
- (b) 4×10^{14} ions/$(cm^2$-sec)
- (c) 1×10^{15} ions/$(cm^2$-sec)

[2,3,6] for a (111) surface plane of a monocrystalline niobium target irradiated at 900°C with ^{4}He$^+$ ions.

If one compares the size of such crow-foot shaped blisters for the different fluxes, it is seen that the length of each of the three prongs of a particular blister increases by approximately a factor of three as the flux is increased from 1 x 10^{14} to 1 x 10^{15} ions/(cm^2-sec) as can be seen in Figs. 5(a) and 5(c). The density of the "crow-foot" shaped blisters decreases from approximately (3 ± 1) x 10^8 to (1.0 ± 0.5) x 10^8 blisters/cm^2, as the flux increases from 1 x 10^{14} to 1 x 10^{15} ions/(cm^2-sec). It may be noted that the blister density for grains of other orientations can be different by many orders of magntidue [see also Fig. 3 in Ref. 6].

DISCUSSION

The results presented here show that the blister density for the large size blisters (type II) is flux dependent for the case of niobium and vanadium irradiated at 900°C with helium ions. Such a dependence can in part be related to the increase in displacement rate of lattice atoms with increasing flux. For the flux range studied the displacement rate can vary approximately from 10^{-4} to 10^{-2} dpa/sec. As the displacement rate increases the vacancy supersaturation increases and this in turn can cause a sharp increase in the void nucleation rate for the ranges of temperature and displacement rates considered here (see discussion by Wiedersich and Katz [15] for nickel at 600°C). The effect of higher displacement rates on the nucleation rates of helium bubbles may be similar. However, at this time it is not quite clear why for the helium ion irradiated niobium samples the effect of flux is less pronounced. It is possible that the differences in the interstitial impurity concentrations in niobium and vanadium may be responsible in part for the observed differences.

For helium ion irradiation at 900°C the blister size for both types of blisters does not change with flux significantly for vanadium, but for niobium the blister size changes slightly with flux. The blister size for 900°C irradiation is considerably smaller than for room temperature irradiation [16], which is in part due to the enhanced helium release [5,18] at 900°C. The simultaneous processes of gas implantation, diffusion and surface erosion have been discussed by Biersack [19]. For our irradiation conditions the loss of blister skin due to sputtering amounts to about 0.1-0.2 percent of the total skin thickness. Therefore, this sputtering erosion process plays an insignificant role in the blister formation and rupture in the present experiments. It appears that the blister size is much more strongly dependent on the irradiation temperature than on the incident beam flux (for the flux range

studied), even though a higher beam flux can cause an increased
helium release via an increased vacancy production.

The blisters formed in annealed polycrystalline niobium after
deuteron irradiation show an increase in size as the flux is
increased (Fig. 5(a)-5(c)). As has been pointed out earlier [6]
the smaller amount of blistering by deuteron irradiation than by
helium-ion irradiation even for doses which are much higher for
deuteron than for helium, can be related to the fact that the
solubility and diffusivity in niobium is many orders of magnitude
larger for deuterium than for helium. The diffusion coefficient
of deuterium in niobium [20] is $D_D = 1.3 \times 10^{-4}$ cm^2/sec at 800°C
while that for helium in niobium [21] it can vary from 10^{-19} to
10^{-14} cm^2/sec as the temperatures are changed from 600°C to 1200°C.

Since the deuterium trapping rate in the lattice has to be
larger than the deuterium release rate, in order to build up large
enough bubbles for blister formation, it can be expected that the
bubble nucleation and growth rate will be flux dependent. This is
in accordance with our observation that the deuteron blister size
increased as the flux was increased (for the flux range studied).
In addition, the increase in blister size with increasing flux can
in part be caused by the lowering of the threshold dose for blister
formation with increasing flux. For example, Verbeek and Eckstein
[22] observed for the case of 15-keV proton irradiation of molybdenum, that the threshold dose decreased with increasing flux.
Qualitatively, a similar dependence can be expected to occur for
our case of deuteron irradiated niobium.

The observation that the fraction of the irradiated area
occupied by both types of helium blisters in vanadium formed at
900°C increases with increasing flux is related to the flux
dependence of blister density and blister size, as described above.

ACKNOWLEDGEMENTS

The authors would like to thank Mr. P. Dusza, Mr. T. Dettweiler
and Mr. W. Aykens for their help during the experiments.

REFERENCES

†Work performed under the auspices of U. S. Atomic Energy
Commission.

[1] S. K. Das and M. Kaminsky, J. Appl. Phys., 44, 25 (1973).
[2] M. Kaminsky and S. K. Das, Appl. Phys. Lett. 21, 443 (1972).
[3] S. K. Das and M. Kaminsky, J. Appl. Phys., 44, 2520 (1973).

[4] S. K. Das and M. Kaminsky, in Defects and Defect Clusters in B.C.C. Metals and their Alloys, Nuclear Metallurgy Vol. 18, edited by R. J. Arsenault, National Bureau of Standards, Gaithersburg, Maryland, p. 240 (1973).

[5] W. Bauer and G. J. Thomas, ibid, p. 255.

[6] M. Kaminsky and S. K. Das, Rad. Effects, 18, 245 (1973).

[7] J. M. Donhowe and G. L. Kulcinski, in Fusion Reactor First Wall Materials, edited by L. C. Ianiello, U. S. Atomic Energy Commission Report No. WASH-1206, p. 75 (1972).

[8] J. M. Donhowe, D. L. Klarstrom, M. L. Sundquist and W. J. Weber, Nucl. Tech., 18, 63 (1973).

[9] M. Kaminsky and S. K. Das, Appl. Phys. Lett., 23, 293 (1973).

[10] M. Kaminsky, Adv. Mass Spectrometry, 3, 69 (1964).

[11] L. H. Milacek and R. D. Daniels, J. Appl. Phys., 39, 5714 (1968).

[12] M. Kaminsky, IEEE Trans. Nucl. Sci., 18, 208 (1971).

[13] M. Kaminsky, in Proceedings of the International Working Sessions on Fusion Reactor Technology, Oak Ridge National Laboratory, Oak Ridge, Tennessee, 1971, U. S. Atomic Energy Commission Report No. CONF-719624 p. 86 (1971).

[14] M. Kaminsky, in Proceedings of the 7th Symposium on Fusion Technology, 24-27 October 1972, Grenoble, France, (Commission of the European Communities, Luxembourg, 1972), p. 41.

[15] H. Wiedersich and J. L. Katz in Ref. 4, p. 530.

[16] S. K. Das and M. Kaminsky, in Proceedings of the Texas Symposium on the Technology of Controlled Thermonuclear Fusion Experiments and the Engineering Aspects of Fusion Reactors, 20-22 November 1972, Austin, Texas (USAEC, National Technical Information Service, Springfield, Virginia 22151, to be published).

[17] M. Kaminsky and S. K. Das, Paper presented to Am. Nucl. Soc. 1973 Winter Meeting, San Francisco, Nov. 11-16, 1973, to appear in Trans. Am. Nucl. Soc. (1973).

[18] W. Bauer and D. Morse, J. Nucl. Mat. 44, 337 (1972).

[19] J. P. Biersack, Rad. Effs, 18, (1973).

[20] G. Schaumann, J. Völkl and G. Alefeld, Phys. Stat. Sol 42, 401 (1970).

[21] S. Blow, J. Brit Nucl. Energy Socl, 11, 371 (1972).

[22] H. Verbeek and W. Eckstein (these proceedings).

DISCUSSION

Q: (R. S. Blewer) Can you state the "as implanted" volume concentration of helium at the implant depth? Do these values exhibit a good correlation with bubble nucleation and subsequent blister formation for implants at different energies?

A: The "as implanted" concentrations of helium at the implant depth is about 50 at.% of helium for a total dose of 0.1 C/cm^2 at room temperature, if one assumes 20% straggling and no helium release. If one considers the entire irradiated volume, the helium concentration is approximately 10 at.% for the conditions quoted above. We observe that (see Refs. 3 and 6) for a dose of 1.0 C/cm^2 the blister size increases with increasing projectile energy even though the helium concentrations will decrease at the implant depth.

Q: (G. L. Kulcinski) Since you have such high dpa rates in the surface (10^{-4} to 10^{-2} dpa-sec^{-1}) would you expect to see any effect of a 10^{-6} dpa-sec^{-1} damage rate from neutron irradiation (aside from different spike size)?

A: Due to the lack of data on the synergistic effects caused by the simultaneous irradiation with 14-MeV neutrons and energetic helium projectiles on blister formation, it is difficult to predict if any significant effect can be expected for the damage rates you quote.

Q: (E. V. Kornelson) When blisters form at higher temperature, e.g., 600°C in Nb, does the increased gas kinetic pressure influence the onset of blisters?

A: The total dose values used in the studies presented here were too high to study the "on-set" of blistering and the effect of temperature on it. The "on-set" of blistering has been found to be strongly temperature dependent (see Ref. 6). However, our observations of the size and density of blisters, and of the skin exfoliation for different target temperatures (R.T. to 900°C) indicate that no simple relationship exists between these quantities, the target temperature and the kinetic pressure of the gas in a blister. For example, for annealed polycrystalline vanadium irradiated with 0.5-MeV He$^+$ ions to a total dose of 1.0 C/cm^2, it is found that the blister size increases as the target temperature increases from room temperature to 600°C but decreases by several orders of magnitude as the target temperature is increased further to 900°C (Ref. 4). One reason for the observed decrease at 900°C is that the helium release has increased significantly. The large skin exfoliation at 600°C is in part due to the significant decrease of yield strength of annealed vanadium at this temperature as compared to the room temperature value. A rise of gas pressure in the blister as the temperature is raised from room temperature to 600°C may aid the exfoliation process.

DEPTH DISTRIBUTION AND MIGRATION OF IMPLANTED

HELIUM IN METAL FOILS USING PROTON BACKSCATTERING[*]

Robert S. Blewer

Sandia Laboratories, Albuquerque, New Mexico 87115

ABSTRACT

Proton backscattering at 2.5 MeV has been used to measure the mean depth and profile of implanted helium distributions as a function of implant energy, implant fluence, and post-implant anneal temperature in copper foils of varying thickness. Distributions implanted at 54 keV, 104 keV, and 158 keV agree (to within 100 Å) with calculated projected ranges for helium in copper at each energy. At room temperature the shape of the distributions is approximately Gaussian with no evidence of a supertail or of the peaks being skewed either toward the surface or the interior of the foils. Implantation of some foils was performed at two energies (highest energy first) using both equal and unequal doses. Resultant profiles were those expected from overlapping Gaussians centered at the predicted depths. Implanted helium fluences ranging from 5×10^{16} He$^+$/cm^2 to 3×10^{17} He$^+$/cm^2 result in backscattering peaks for helium which increase in magnitude in the proper proportion to increasing fluence. Detection sensitivity of 1 at. % He in Cu has been demonstrated. In addition, profiling of other low Z elements (e.g. oxygen, carbon and deuterium) in the foils or on their surfaces is also described.

The effect of <u>in situ</u> isochronal and isothermal annealing on the disposition of the implanted helium has also been observed. Above temperatures of 200°C, the peak of the helium distribution decreases in magnitude, but no lateral spreading of the profile (as expected in Fick's Law diffusion) is observed. Moreover, the

[*] This work was supported by the U. S. Atomic Energy Commission.

helium peak height decreases steadily for each of the isochronal temperature plateaus between 200°C and 450°C. Isothermal annealing at 225°C and 400°C produces almost no additional change in the magnitude of the helium peak at the given temperature over three anneal periods of increasing duration. Throughout annealing, the symmetric form of the Gaussian distribution is retained. There is, as yet, no evidence of preferential diffusion of the implanted helium either into the undamaged depths of the foils or through the residual ion-implantation-induced damage between the helium implanted layer and the foil surface. These observations could be explained if the helium were trapped at or near its end-of-range location at room temperature and then released in proportionate fractions at progressively higher temperatures by formation and subsequent rupture of bubbles developing in the implanted layer. Evidence has been obtained by scanning electron microscopy which supports this hypothesis.

INTRODUCTION

The behavior of helium in solids has become a subject of increasing technological importance, notably with respect to materials problems of fission and fusion reactors. Experimental determination of the initial depth distribution of implanted helium and its subsequent migration behavior has been of particular concern to those working in reactor materials research and related activities. In earlier studies[1,2] we observed the influence of implanted helium dose, implant energy and post anneal temperature on dimensional expansion and on the formation of helium filled surface bubbles in erbium metal. Similar (surface bubble) studies have also been conducted on other materials by several investigators.[3] By fracturing our implanted thin film samples[1] we could determine the depth of the subsurface bubble layer by transverse section viewing of the sample in the SEM. This technique provided a detailed view of the bubble layer and allowed an accurate measure of its depth but it required samples implanted to fluences or annealed to temperatures sufficient to precipitate the helium atoms from the lattice into bubbles,[2] which is a step removed from profiling the as-implanted dispersed atom distribution. Though the depth of the center of the bubble layer, so determined, was close to theoretical values of projected range, this procedure was not capable of yielding quantitative information on the implanted helium depth profiles. For this reason other avenues of approach were sought to extend our investigations.

Helium depth profiling using the ion microprobe was attempted but this instrument was found too insensitive to helium to be of value in measuring helium distributions.[4] Alternatively the use

of ion backscattering seemed inappropriate because of the small
value of the elastic scattering cross sections for low Z elements
in relation to that of higher Z host materials;[5] it was expected
that the backscattering yield from implanted light atoms could not
be discerned above the thick target background of metal hosts.
However, using a thin foil, Mervine et. al. (unpublished)[6] did
succeed in detecting implanted helium in Pd by proton backscattering. We have developed the helium depth profiling potential
of proton backscattering and recently reported its attributes for
the first time in the open literature.[7] Proton backscattering
from thin foils has been applied in this article to obtain data
on the depth distribution and migration (upon annealing) of helium
in copper. Helium has also been detected in metals by other
methods such as He $\rightarrow$ He transmission coincidence scattering at
48.5 MeV by Pieper and Theus (unpublished)[8] and more recently by
using the ^{3}He (d, p) ^{4}He nuclear reaction by Pronko[9] and by
Picraux and Vook[10].

EXPERIMENTAL

Polycrystalline copper foils from 0.5 to 2.5 μm thick were
mounted in circular metal frames. The foils were implanted with
^{4}He at energies between ~ 50 keV and ~ 150 keV using sufficiently
low beam current densities (< 5 μA/cm^2) to insure that the foils
remained near room temperature (< 40°C as measured by thermocouple)
during implantation. A frame bearing one of the implanted foils
was then attached to the holder and heater assembly shown at the
top of Figure 1 for backscattering analysis. This assembly has
been so designed that protons backscattered in the foil at 164°
(lab) are counted by a surface barrier detector, but those which
pass through the foil are trapped inside the assembly and cannot
be backscattered (by single scattering events) into the detector.
The holder and heater assembly were attached to a precision
vacuum manipulator to assure ease of sample and holder alignment
with the incident proton beam and to permit probing of different
parts of the foil. The detector was collimated and exhibited a
resolution (FWHM) of less than 10 keV. The holder and heater
assembly was surrounded by a cylindrical Faraday cup through a
solid angle of 2π steradians (except for small holes for the
incident beam and the detector). Pumping holes of increasing
diameter were positioned along the length of the holder in order
to avoid destructive pressure differentials across the foil surface
during evacuation of the target chamber. A conventional 2.5 MeV
Van de Graaff accelerator was used to supply a well-collimated
mass-analyzed monoenergetic beam of protons[11] of 0.5 mm^2 cross
sectional area to the foil. Low beam currents (< 5 nA) were used
to avoid beam heating the foils and to circumvent pulse pile-up
problems. Care was also taken to limit the total proton charge

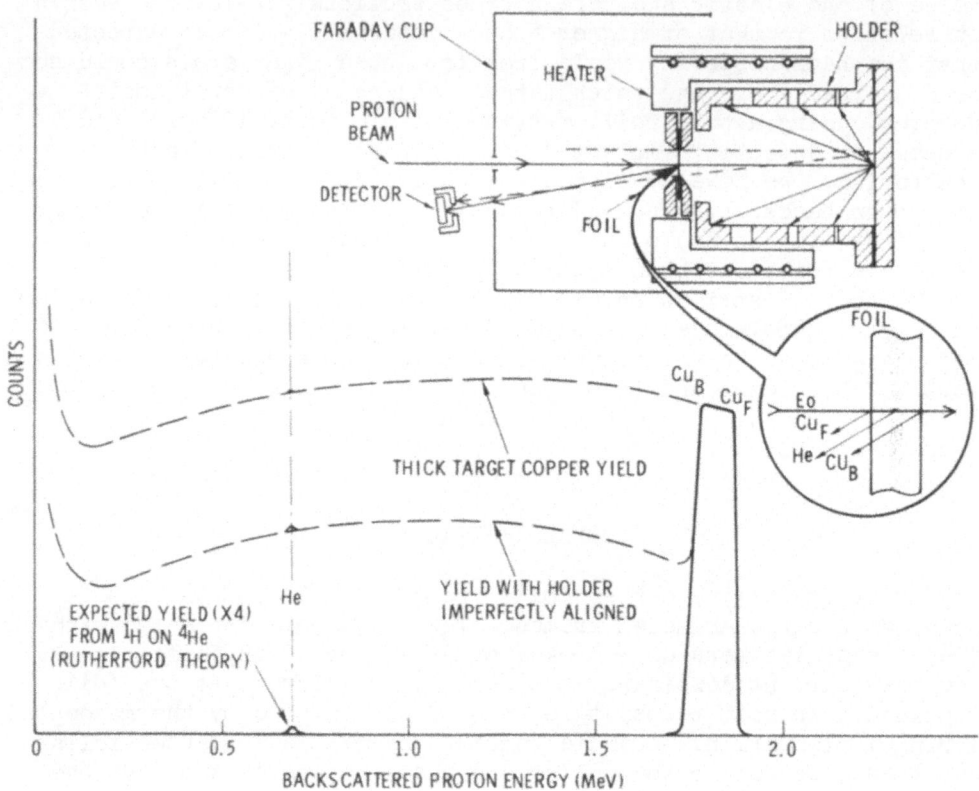

Figure 1:   Schematic of an implanted foil installed in the holder
            and heater assembly (upper right); energy spectrum
            expected for 2 MeV protons backscattered from the
            implanted foil (lower left).

incident on the implanted foils to < 10 μC in order to minimize
redistribution of the implanted helium atoms. The leading edge
of the copper peak was used each run to check the energy calibration of the multichannel analyzer data. The analyzer itself
was checked with a precision pulser on several occasions to assure
·continuing linearity.

                          RESULTS AND DISCUSSION

    In Figure 1 (lower part), the expected backscattering
spectrum of 2 MeV protons incident normally on a helium implanted

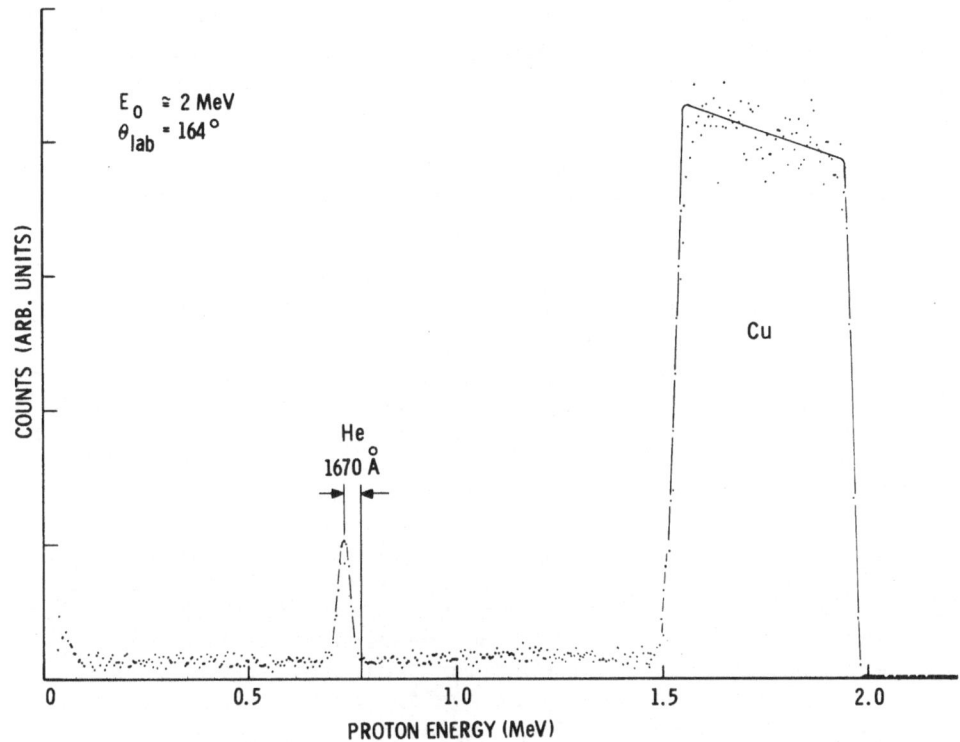

Figure 2: Multichannel analyzer output exhibiting the number of
           protons backscattered from a copper foil (implanted
           with He to a fluence of 3 x 10^{17} He+/cm2 at 54 keV) as
           a function of energy.

copper foil is illustrated. As expected from the Rutherford*
scattering cross sections the proton backscattering yield from
the copper atoms in the foil should be clearly evident, but the
yield from protons backscattered from helium atoms within the
copper foil would be barely detectable (i.e. the Rutherford proton
yield from helium atoms should be 210 times smaller than the
Rutherford proton yield from copper atoms in equal concentration).
Such a small yield (especially when reduced by the implanted helium
dilution factor) would be lost in the background counts from a
thick copper target (shown by the upper dashed line). However, by
utilizing a properly aligned holder as described above, together

---

*It is known that yields somewhat in excess of those predicted by
 Rutherford theory occur for proton backscattering from low-Z
 elements. Nevertheless, it is useful to employ the easily calculable Rutherford yields in this illustration for comparative
 purposes.

with a thin implanted foil, the thick target background can be reduced to near zero thereby fulfilling at least a necessary condition for observing implanted helium.

In Figure 2 is shown an actual data plot which exhibits the needed reduction of background over the thick target case, and which also reveals a helium yield orders of magnitude in excess of that predicted by the Rutherford cross sections. This enhanced yield renders the implanted helium profile plainly visible and is due to resonance scattering similar to that found in neutron-helium scattering by Staub and Stephens.[12] The vertical line to the right of the helium peak indicates the energy at which protons would appear if they were backscattered from helium atoms lying on the foil surface. The energy offset between this line and the observed peak of the helium distribution can be converted into an implant depth through a knowledge of the stopping cross section of protons in copper in this energy range.[13]* It can be deduced in this manner that the peak of the helium distribution lies 1670 Å ± 100 Å beneath the surface of the copper foil. The probable error in the applicable proton stopping cross sections[13] is ± 2.5% which corresponds to an uncertainty of no more than ± 40 Å in the measured projected range. The ± 1σ width of the distribution can likewise be deduced to be 590 Å ± 100 Å. The measured projected range R_p compares favorably with the theoretical value of 1700 Å but the experimentally determined range straggling ΔR_p is below the calculated value of 725 Å for helium implanted in copper at 54 keV.[15] It is noteworthy that the shape of the helium peak is the approximate Gaussian form expected with single energy implants in polycrystalline materials. There is no evidence of enhanced room temperature diffusion to a level nearer the surface (where the maximum in the instantaneous implantation damage occurs), at least at concentrations above 1 at. % He. Neither is there an indication that the implanted helium diffuses at room temperature deeper into the foil than its end of range position, even though Kornelsen[16] observed that at lower doses He injected at low energy into tungsten single crystals diffuses at room temperature to depths of 1 μm in the undamaged lattice.

Some of the foils were implanted to multiple doses at two energies (highest energy first). The results of the backscattering analysis of one such sample are exhibited in Figure 3. As shown

---

*Correction factors which influence this conversion have been included in the calculations. These include the added stopping produced by helium atoms in the copper[14] and also the energy dependence (change with penetration depth) of the effective stopping cross section. Corrections of these types were found to be almost negligible in this experiment but were nonetheless taken into account.

significantly greater precision than would be expected from the use of the detector resolution (FWHM) as an indicator of the best depth sensitivity achievable. This favorable situation arises because both the detector response and the helium distribution are symmetric so that the mean of the profile can be calculated using standard statistical methods. Under these circumstances, the energy corresponding to the peak of the helium distribution can be specified with a precision equal to that with which the peak of the detector response can be specified and the uncertainty in energy of the detector response maximum is a small fraction of the FWHM width of the detector function. The degree of agreement between experiment and theory is therefore not surprising.

Measured range straggling is consistently below theoretical values at each energy, the difference increasing with increasing energy. As stated previously, the total proton charge was kept low with the result that counting statistics limit the accuracy with which the peak height (and thus the width at 60.5% of full maximum (i.e., ± 1 σ)) can be measured. For this reason, the accuracy of the determination of ΔR_p is less than that for R_p. Increased accuracy requires simply that the total proton charge applied to the foil be increased.

One of the most attractive advantages of this proton back-scattering technique is the ability to observe changes in the helium profile as a function of temperature excursions or aging tests in an essentially non-destructive manner. In Figure 5 the helium concentration (± 1 σ) retained in the foil is plotted as a function of isochronal anneal temperature. In the inset are exhibited the changes in the depth distribution of the room temperature helium implant as a result of 35 minute isochronal anneals at 200°C, 300°C, 400°C and 450°C. Several noteworthy features are observed: (1) retained helium decreases linearly with anneal temperature beginning at ∼ 200°C, (2) the distribution does not lose its "as implanted" symmetry upon annealing but instead simply shrinks in magnitude, (3) the profile peak remains at essentially the initial implant depth beneath the foil surface and (4) the width of the distribution also remains almost constant.

The absence of lateral spreading of the distribution is contrary to what would be expected if Fickian diffusion were occurring. Although there is no indication of diffusion of the helium toward the surface, the total amount of implanted helium retained in the foil nonetheless decreases throughout the duration of the isochronal annealing schedule (i.e. the helium peak shrinks in place), having already started with the first anneal temperature and continuing in a linear fashion with each new increasing temperature plateau. (Conversely, isothermal annealing at 225°C and 400°C produces almost no additional change in the magnitude of the

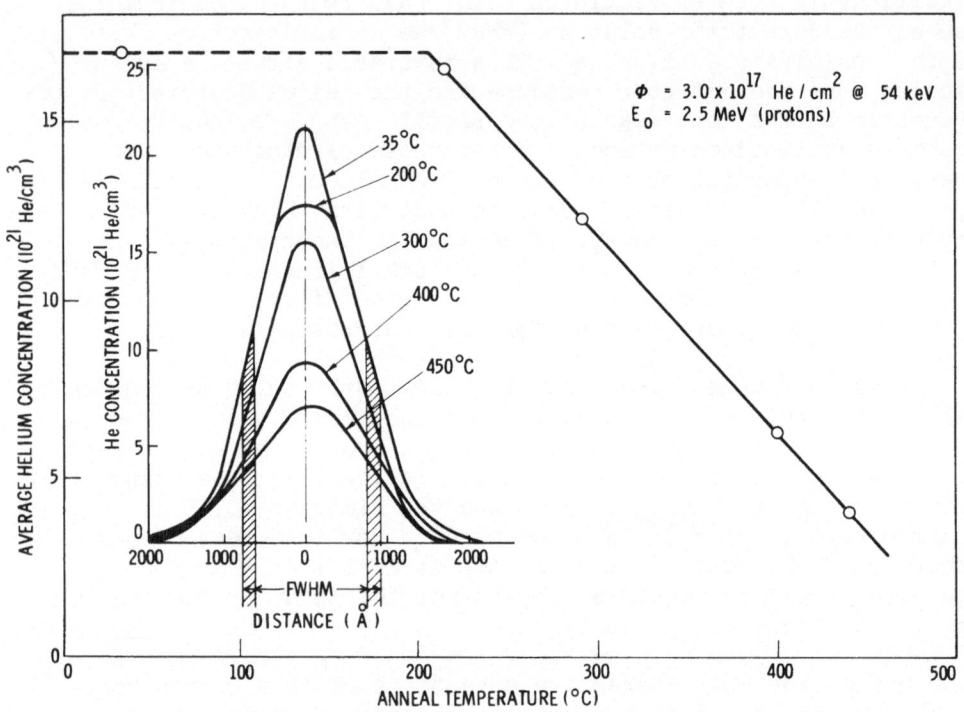

Figure 5:   Plot of average volume concentration ($\pm$ 1 σ) of implanted
            helium retained in the foil versus anneal temperature.
            Inset displays depth profile of implanted helium at
            progressively higher anneal temperatures.

retained helium at the given temperature over three anneal periods
of increasing duration). The shape of the Gaussian is distorted
somewhat at 200°C; at this temperature it appears that helium
escapes only from the depth corresponding to the peak of the distribution resulting in a blunted peak. However the original
Gaussian form is observed at each of the other temperatures but it
may be slightly offset toward the surface at 450°C.

    The type of behavior observed in our experiments could be
explained if the helium atoms were trapped at or near their end-
of-range locations at room temperature and then released in proportionate fractions at progressively higher temperatures by
formation and subsequent rupture of bubbles developing in the
implanted layer. Such behavior would be consistent with data
obtained in our previous high dose experiments in Er[1,2,19,20] and
those of others in a variety of other materials.[16,17,21,22] Evidence has been obtained by scanning electron microscopy which

 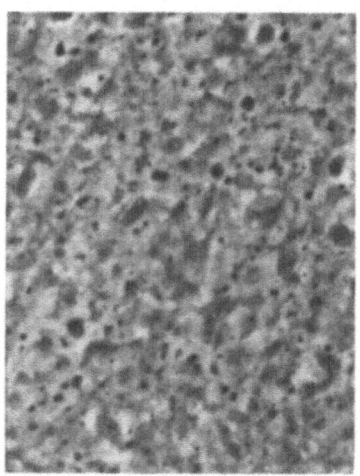

UNIMPLANTED   IMPLANTED

3 μm

Figure 6:   Scanning electron micrographs of masked vs implanted
            areas of the foil used in obtaining the results of
            Figure 5 (after the 450°C anneal).

supports this hypothesis, as shown in Figure 6. Exhibited are
adjacent sections of the foil from the anneal run illustrated in
Figure 5 after the 450°C anneal was completed. Ruptured surface
bubbles from 1000 Å to 8000 Å diameter can be seen at high magnification. Subsequent annealing of the foil at 900°C in a gas
mass spectrometer confirmed that the amount of helium remaining in
the film when added to that released during the in situ anneals
equaled the amount originally implanted.

If the helium were released from the foil by rupture of subsurface bubbles, the rate-determining step would be the failure of
the metal over each bubble. Apparently when the temperature reaches
a value at which the pressure in the bubble exceeds the ultimate
yield strength of the metal comprising the bubble lid, the bubble
ruptures and releases the helium contained within it. Such a
hypothesis suggests that helium release would be less sensitive to
annealing at increasing temperatures than thermally activated
diffusive release and would likewise be insensitive to the time
duration of isothermal anneals, which is what is observed.

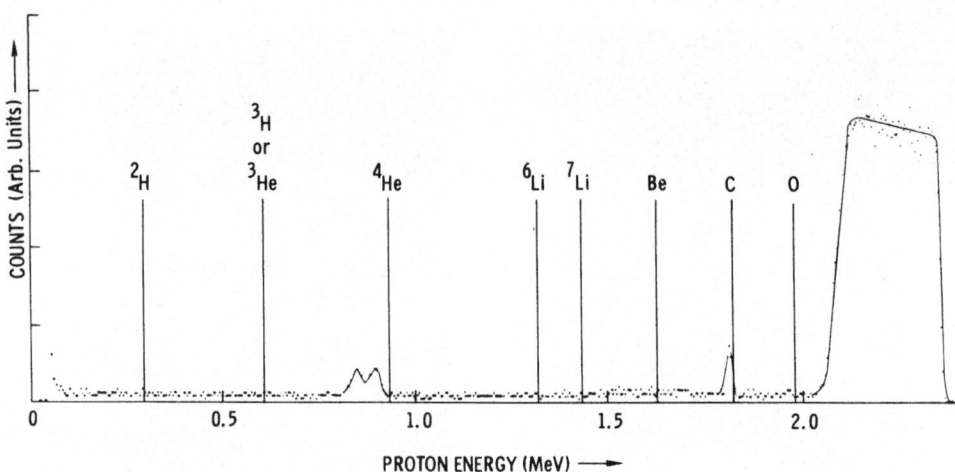

Figure 7: Typical proton backscatter yield from helium implanted
           copper foil as a function of energy. The energies at
           which protons backscattered from other low-Z elements
           would appear (assuming adequate yield) are indicated
           by the labeled vertical lines.

        From these proton backscattering experiments it is known that
for relatively high dose implants the helium remains essentially as
initially distributed (centered about the projected range R_p).
Thus the bubbles which form apparently release the helium contained
within them not by migration of the bubbles to the foil surface,
but instead by rupture at the implant depth. This greatly clarifies
the situation with respect to the mechanism by which implanted
helium, once it has precipitated into bubbles, escapes from the
solid. Long range interstitial or substitutional diffusion either
does not occur in these experiments or is sufficiently small in
magnitude to be unobservable.

        An important ancillary advantage in the use of this technique
is the simultaneous detection and profiling of other low-Z elements
in or on the foil which may influence the helium or hydrogen isotope distribution or migration measurements. These other elements
are observable because of the low number of background counts over
the energy range from near zero up to the energy of the metal foil
peak. This capability is important in implantation studies where
radiation enhanced oxidation or collection of carbon deposits on
the foil surface may occur. Also, the importance of thermally
induced surface oxide layers on the migration characteristics of
implanted helium can be monitored together with the changes in the
rare gas depth profile. In Figure 7 is exhibited a typical

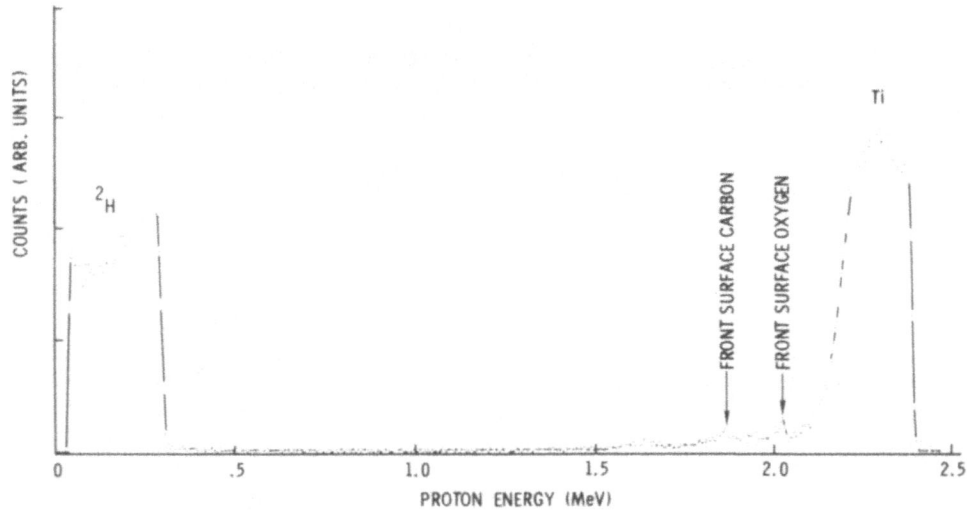

Figure 8:   Proton backscattering spectrum of titanium deuteride
            using 2.5 MeV incident protons.

backscattering spectrum from a copper foil with helium implanted
at two energies. Surface carbon from the implantation is also
clearly detectable. Surface oxygen has likewise been noted on
other foils which contained polymer layers on the front and rear
surfaces.[7] The position of each of several low-Z elements and
isotopes which would appear if present is also indicated. Because
of the low background, observation of these other elements is
favored even if the yield is not enhanced to the same extent as
for 2.5 MeV protons on helium. Certainly, other techniques have
been reported which are capable of detecting one or more of these
low-Z elements, but the author knows of no other which is capable
of resolving them all while simultaneously profiling a helium
distribution.

       Finally, though this paper is devoted principally to helium
distribution profiling, it perhaps will be of interest to those
conducting CTR materials research that depth profiling of hydrogen
isotopes also lies within the scope of this technique. For example,
in Figure 8 is exhibited the proton backscattering spectrum from a
titanium deuteride foil. Note that, in addition to surface oxygen
and carbon and a degree of impurity content extending through the
foil, the deuterium peak is also clearly resolved. Its atomic
number is lower than helium, but its yield with respect to 2.5 MeV
protons is nearly as great as that of the titanium. Deuterium
distributions have also been observed in metals and metal hydrides
by other techniques.[23-28]

CONCLUSION

By utilizing the appropriate experimental design and sample geometry, the depth distribution of helium implanted in copper (and its changes on annealing) have been successfully investigated. Projected range calculations agree with experimentally measured values for implant energies between $\sim$ 50 keV and $\sim$ 150 keV though measured values of range straggling are consistently below the calculated values. Upon annealing, the implanted helium is observed to escape from the foil without apparent diffusion of the helium to the surface. It can be deduced by the non-Fickian nature of the helium migration (release) and by scanning electron micrographs of the samples after annealing that the implanted helium is escaping from bubbles which rupture at the original implant depth. The presence and behavior of carbon, oxygen, deuterium and (in principle) any other low-Z element except ^{1}H in or on the samples can be detected and profiled concurrently with tests which are performed to study the behavior of helium distributions. Likewise the proton backscattering technique can be used to study the distribution of other low-Z elements introduced intentionally into hosts of any mass which can be obtained as a thin foil (e.g., nitrogen in Ta, carbon in stainless steel, or deuterium and/or tritium in metal hydrides).

The non destructive and in situ characteristics of this method are particularly well suited for sequential physical-event investigations such as diffusion. Studies of the behavior of deuterium, [3]He (or tritium), and [4]He in heavier host materials are open to the power and utility of the backscattering technique and should serve to mutually complement recently reported [3]He and deuterium lattice location studies.[10] These tools should be fully exploited to resolve the many remaining questions about the behavior of helium and hydrogen isotopes in metals.

ACKNOWLEDGEMENT

It is a pleasure to acknowledge Drs. E. P. EerNisse, S. T. Picraux, R. A. Langley and S. M. Meyers for the use of the Sandia 300 keV Positive Ion Accelerator and the 2.5 MeV Van de Graaff Accelerator facilities and also for helpful discussions. Technical assistance by J. M. McDonald, C. M. Fuller and B. G. Self is also gratefully acknowledged. The scanning electron micrograph was made by J. K. Maurin.

REFERENCES

1. R. S. Blewer and J. K. Maurin, J. Nucl, Mat. 44, 260 (1972).
2. R. S. Blewer, Proceedings of the International Conference on

Ion-Surface Interaction, Sputtering and Related Phenomena, Garching, West Germany, Sept. 25-28, 1972, Rad. Eff. 18-21 (1973).

3. For example, see the papers contained within the "Ion Implanted Gas Buildup Session" of this Conference Proceedings.
4. J. W. Guthrie and R. S. Blewer, unpublished data, May 1971.
5. J. W. Mayer, L. Eriksson and J. A. Davies, Ion Implantation in Semiconductors, Academic Press, New York (1970) p. 16.
6. L. R. Mervine, R. C. Der, R. J. Fortner, T. M. Kavanagh and J. M. Khan, Lawrence Livermore Laboratories, Report UCRL-73087, Feb. 23, 1971.
7. R. S. Blewer, Appl. Phys. Lett., Dec. 1, 1973.
8. A. G. Pieper and R. B. Theus, NRL Memo Report 2394, Naval Research Laboratory, Washington D.C., Feb. 1972.
9. P. P. Pronko and J. G. Pronko, Proc. Int'l. Confr. Ion Beam Surf. Layer Anal., June 18-20, 1973, Yorktown Hts., N.Y. (to be published) Th. Sol. Films 16 (1973).
10. S. T. Picraux and F. L. Vook, these Conference Proceedings.
11. For further details on the accelerator and energy calibration see R. A. Langley and R. S. Blewer, Proc. Int'l Confr. Ion Beam Surf. Layer Anal., June 18-20, 1973, Yorktown Hts., N.Y. (to be published) Th. Sol. Films 16 (1973).
12. H. Staub and W. E. Stephens, Phys. Rev. 55, 131 (1939). N. P. Heydenburg and N. F. Ramsey, Phys. Rev. 60, 42 (1941).
13. Ward Whaling, Handbuch der Physik XXXIV, Springer Verlag, Berlin (1958), Sect. 2., p. 193.
14. S. Furukawa, H. Matsumura and H. Ishiwara, Proc. Int'l Confr. on Ion Beam Surf. Layer Anal., Yorktown Hts., June 18-20, 1973 (to be published) Th. Sol. Films 16 (1973).
15. D. K. Brice, Rad. Eff. 11, 227 (1971).
16. E. V. Kornelsen, Can. J. Phys. 48, 2812 (1970).
17. S. K. Erents and G. M. McCrackan, Rad. Eff. 18, 191 (1973).
18. W. White and R. M. Mueller, Phys. Rev. 187, 499 (1969).
19. R. S. Blewer and W. Beezhold, Rad. Eff. 19, 49 (1973).
20. R. S. Blewer and J. K. Maurin, "30th Ann. Proc. Electron Microscopy Soc. Amer.," Los Angeles, Calif., 1972, C. J. Arceneaux, ed. p. 444.
21. W. Bauer and G. J. Thomas, J. Nucl. Mat. 42, 96 (1972).
22. S. K. Das and M. Kaminski, J. Appl. Phys. 44, 25 (1973).
23. R. S. Langley, S. T. Picraux and F. L. Vook, 5th Ann. Confr. on Surf. Studies (unpublished) Sept. 5-7, 1973, Rocky Flats, Colo.
24. E. A. Wolicki, NRL Report 7477, Naval Research Laboratory, Washington, D.C., December 1972.
25. G. M. Padawer, D. J. Larson and P. N. Adler, Met. Trans. 2, 2287 (1971).
26. D. A. Leich and T. A. Tombrello, Nucl. Instru. Meth. 108, 67 (1973).
27. B. L. Cohen, C. L. Fink and J. H. Degnan, J. Appl. Phys. 43, 19 (1972).
28. C. M. Bartle, N. G. Chapman and P. B. Johnson, Nucl. Instr. Meth., 95, 221 (1971).

DISCUSSION

Q: (J. A. Davies) The excellent agreement between your measured R_p values and those calculated by Brice provides strong support for the accuracy of the predicted path-length-to-projected-range conversion. On the other hand, this conversion is the dominant factor in determining the <u>projected</u> range straggling. Hence, the large discrepancy between your measured and calculated straggling values would suggest that the error lies in the experimentally derived straggling. Would you agree?

A: I think this technique provides a good means to measure ΔR_p experimentally and to check the validity of assumptions made in projected range and projected range straggling calculations, as you suggest. It should be remembered that the low total counts used in our initial experiments limit the accuracy with which ΔR_p has been determined in contrast to the more accurate determination of R_p, which is not as sensitive to counting statistics. For this reason, more accurate data is needed before one draws firm conclusions as to this question.

Q: (F. Morehead) What was the standard deviation of your resolution function in Å and did you take it into account in calculating ΔR_p of He?

A: The detector response (FWHM) corresponded to ~ 350 Å; the $\pm 1\sigma$ value of the helium distribution for 54 keV He^+ was ~ 1180 Å. The effect of the detector resolution function on the measured value of ΔR_p was therefore small ($< 5\%$) but was taken into account in the data reduction. The effect of the detector response on the precision with which R_p could be determined was significantly less than for ΔR_p, as indicated in the text.

BLISTERING OF NIOBIUM DUE TO LOW ENERGY HELIUM ION BOMBARDMENT INVESTIGATED BY RUTHERFORD BACKSCATTERING.

J. Roth, R. Behrisch, B. M. U. Scherzer

Max-Planck-Institut für Plasmaphysik, EURATOM
Association, D-8046 Garching, Germany

ABSTRACT

Radiation damage and blister formation in a Nb single crystal due to room temperature bombardment with 2-4 keV He^+ ions has been investigated by Rutherford backscattering in double aligned technique ($\langle 100 \rangle / \langle 111 \rangle$ and $\langle 111 \rangle / \langle 111 \rangle$) of 150 keV protons. By this technique it is possible to observe <u>in situ</u> the development of blisters. The critical ion dose for blister formation is about 10^{17} He^+/cm^2 increasing with energy. It is higher for He^+ bombardment in aligned than in random directions. The mean thickness of the cover of the blisters increases from 234Å at 2 keV to 420Å at 4 keV He^+ bombardment in random direction. The average blister size as measured in the SEM was 600Å at 2 keV and 1400Å at 4 keV random bombardment. In $\langle 100 \rangle$ aligned bombardment it increases by roughly a factor of 2. The results of our backscattering measurements before the appearance of blisters may be interpreted by the assumption that small Helium bubbles are formed at the penetration depth of the He ions. The measured penetration depth is considerably larger than predicted by theory (Schiøtt).

INTRODUCTION

Blistering of metals due to bombardment with energetic H, D, T, and He ions may lead to serious wall erosion in controlled thermonuclear reactors /1/. It has been estimated /2/ that during He bombardment the exfoliation of a metal surface due to blistering may exceed the erosion due to sputtering by orders of

573

magnitude. The phenomenon of blistering due to ion bombardment
has been predicted as early as 1912 by J. Stark and G. Wendt /3/.
In the last years a great number of investigations has been performed on blistering of metals by light ion bombardment covering
an energy range from 7 keV to several MeV /4-11/. In most of
these experiments blister formation is observed either optically
or with a scanning electron microscope. Kaminsky /5/ first
observed gas bursts during bombardment of Cu and Ag with D^+
and related this phenomenon to the gas release from rupturing
blisters. Recently Erents and McCracken /9/ as well as Bauer
and Thomas /7/ measured the Helium gas reemission from the
target during bombardment as a function of ion dose parallel to
the observation of blisters. According to their measurements
a strong increase of He reemission takes place at the moment
of blister appearance, that is at a critical bombarding dose. It
is of interest, to supplement the rather time consuming technique
of investigating the surface with optical or scanning electron
microscope after bombardment by other techniques which allow
to follow the process of blister formation during bombardment.
This is especially valuable in measurements with very low energy
of the primary ions (~ 1 keV) where blister size becomes so small
that they are hardly detectable by optical or even scanning electron
microscopy.

We have applied Rutherford backscattering of 150 keV protons
from a Nb single crystal using double alignment technique to investigate the development of radiation damage and the onset of
blister formation due to the bombardment with 2-4 keV ions.

EXPERIMENTAL

The experiment was performed on PHARAO, a 50-150 keV
accelerator for H^+ and He^+ which has been described elsewhere /12/. The experimental set up is shown schematically in
Fig. 1. Two magnetically analysed ion beams enter the target
chamber at right angles. The production of damage and blisters
is performed by bombarding the target with a 1-10 keV He^+ beam
collimated by two diaphragms giving a circular beam cross section
of 3 mm diameter at the target. The current density is about
10^{-6} A/cm^2. The beam current is monitored periodically by
moving a Faraday cup into the beam right behind the second dia-
phragm.

A 150 keV proton beam was used to analyse the damage by
Rutherford backscattering combined with channeling and blocking.

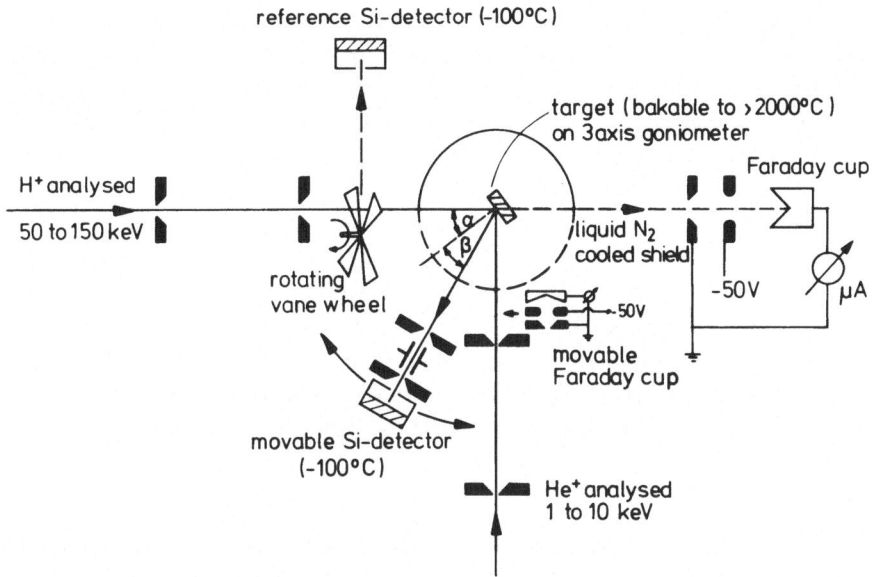

<u>Fig. 1</u>   Schematic of experimental set up

The proton beam is collimated by two diaphragms to a 0.3 mm
diameter which is 10% of the He beam diameter. The primary
ion dose used for backscattering spectrum is determined by
scattering a certain fraction of the primary beam into a reference
detector by means of a rotating vane wheel /13/. This device is
calibrated with a Faraday cup which collects the primary beam
if the target is removed from the beam line. The hydrogen particles backscattered from the target are detected and energy analysed by a cooled Si surface barrier detector with an energy resolution of 3 keV /14/. Its angle of acceptance is $4.0 \cdot 10^{-5}$ sr.
The detector may be moved around the target on a vertical and
horizontal axis respectively.

The target is a Nb crystal with a (100) surface provided by
courtesy of Dr. R. E. Reed, Oak Ridge National Laboratory. It
was chemically polished for 1-3 minutes in 70 vol. % HNO_3 (65%),
30 vol. % HF (75%). The crystal is mounted on a 3 axis UHV
goniometer /15/. It is annealed in situ at 2200°C for several
minutes prior to bombardment. The crystal was bombarded with
2-4 keV He^+ at room temperature in random and $\langle 100 \rangle$ direction
and backscattered energy distribution taken periodically in double
alignment ($\langle 100 \rangle$ / $\langle 111 \rangle$ and $\langle 111 \rangle$ / $\langle 111 \rangle$). The minimum yield
$\chi(x)$ in $\langle 111 \rangle$ / $\langle 111 \rangle$ double alignment as a function of depth x
of the annealed unbombarded crystal is only less than a factor

of two higher than predicted by theory (Behrisch et al. /19/
eq. 23). The agreement is much better than in our previous work
due to the better crystal used in the present work. Radiation
damage introduced into the crystal by the 150 keV H$^+$ beam for
Rutherford backscattering analysis is negligible. The energy
distribution of the backscattered protons can be transformed into
a depth distribution /16/. The stopping power for protons in Nb
/17,18/ is nearly constant in the energy range of interest (60-
150 keV). The energy resolution of 3 keV of the detector results
in a depth resolution of ~50Å for the backscattering analysis. The
measurements were performed in a vacuum of ~10^{-7} torr, main
residual gas being mercury. The two second important residual
gas species (H$_2$0 and C0) were well below 10^{-8} torr.

RESULTS

A set of backscattered energy distributions taken in ⟨100⟩/⟨111⟩
double alignment with 150 keV protons after damaging the crystal
by different amounts of 4 keV He$^+$ bombardment in a random
direction is shown in Fig.2. For comparison a random spectrum
has been added. The development of the ion induced radiation
damage takes place in two steps: At low doses up to 1.4·10^{17}
He$^+$/cm^2 the influence of radiation damage on the backscattered
energy spectra is manifested as a broad shallow hump extending from the region beneath the surface peak to a depth of ~700 Å.
Further, strongly enhanced dechanneling occurs as shown by the
increase of backscattered intensity at larger depth. After a dose
of 1.4·10^{17} He$^+$/cm^2 the shape of the backscattering spectrum is
remarkably changed. Backscattered intensity from a surface
layer of some 400-500 Å has strongly increased forming a broad
peak. With further bombardment the backscattering spectrum
reaches a steady state at about 4.0·10^{17} He$^+$/ cm^2 where the
damage peak has approximately reached random intensity.

The random spectrum which was taken on the highly damaged
crystal after 6·10^{17} He$^+$/cm^2 random bombardment shows a step
of approximately equal width as the damage peak.

In order to investigate the dependence of damage production due
to helium ion bombardment on ion energy and direction of incidence
additional measurements were performed with 2 keV He$^+$ at ran
dom incidence and with 4 keV He$^+$ at ⟨100⟩ aligned incidence. The
results are shown in perspective graphs, Fig. 3 - 5. A third axis
has been added to demonstrate the development of the double
aligned Rutherford backscattering spectra with He ion dose.

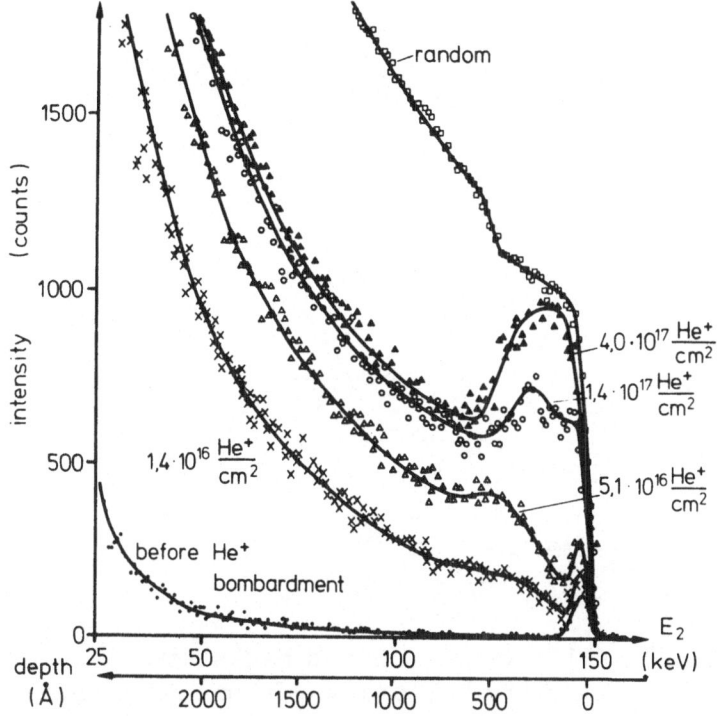

Fig. 2   ⟨100⟩/⟨111⟩   double aligned backscattered energy
distributions of 150 keV protons from a Nb single crystal after
damage production by different amounts of 4 keV He$^+$ bombardment in a random direction (incidence 12° to surface normal) at
room temperature.

Since the damage introduced by 2 keV He$^+$ bombardment was
analysed in ⟨111⟩/⟨111⟩ alignment whereas the 4 keV damage
was analysed in ⟨100⟩/⟨111⟩ alignment the depth scales have been
fitted in order to facilitate comparison.

Optical and scanning electron micrographs were taken of the
bombarded surface after different bombarding He doses. While
no visible change of the surface could be observed after bombardment with less than 10^{17} He$^+$/cm^2, blisters were always found
after bombardment with several 10^{17} He$^+$/cm^2 when the broad
surface peak in the Rutherford backscattering spectrum had developed. The blisters are of circular shape and nearly uniform
size. Their average size increases with energy from ∼600 Å at
2 keV to ∼2000 Å at 4 keV He bombardment. The influence of

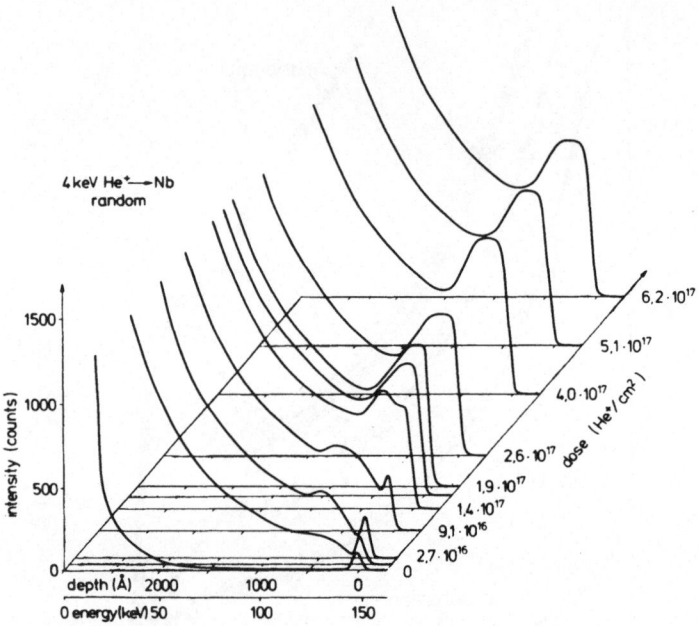

Fig. 3   ⟨100⟩ / ⟨111⟩ double aligned backscattered energy and
depth distributions of 150 keV protons from a Nb single crystal
after damage production by 4 keV He⁺ at random incidence.

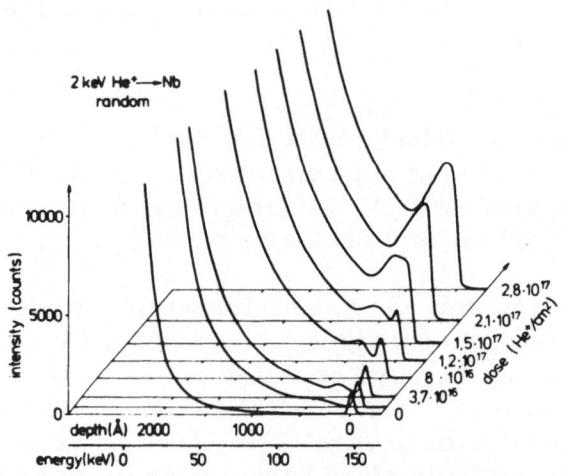

Fig. 4   ⟨111⟩ / ⟨111⟩ double aligned backscattered energy and
depth distributions of 150 keV protons from a Nb single crystal
after damage production by 2 keV He⁺ at random incidence.

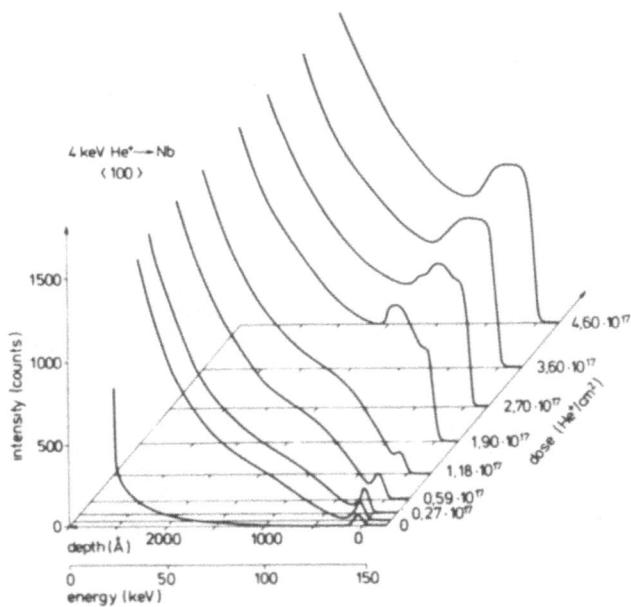

Fig. 5   ⟨100⟩ / ⟨111⟩ double aligned backscattered energy and
depth distributions of 150 keV protons from a Nb single crystal
after damage production by 4 keV He$^+$ at ⟨100⟩ aligned incidence.

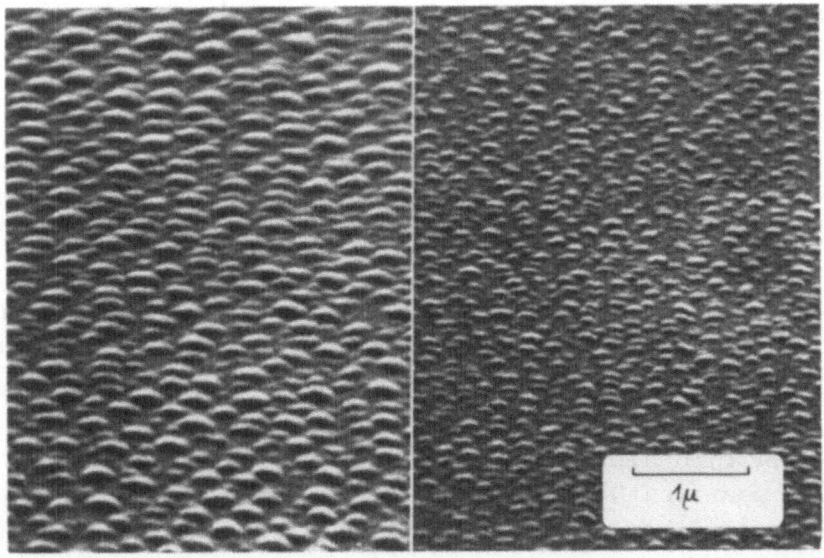

Fig. 6   Blisters after bombardment of Nb with 4 keV He$^+$ in
⟨100⟩ and random directions.

direction of incidence is demonstrated in Fig. 6. Aligned He bombardment produces considerably larger blisters than random bombardment.

DISCUSSION

The high dose peak may be explained by enhanced backscattering near the surface due to the deformation of a certain fraction of the surface by blisters. The bulging of the surface layer, forming the cover of the blisters, leads to misalignment with respect to the incident beam, while the underlying lattice remains unaffected.

The width of the blister peak is a measure of the thickness of the blistered layer. It depends on primary energy and to a somewhat smaller extent on the direction of incidence. In Tab. 1 critical doses and thickness values are given for the different energies and directions of He bombardment:

E(keV)	direction of incid.	critical dose (He^+/cm^2)	width of blister peak	theor. penetration depth (Schiøtt)
2	random	$1.26 \cdot 10^{17}$	235Å	85Å
4	"	$1.4 \cdot 10^{17}$	420Å	140Å
4	100	$1.9 \cdot 10^{17}$	470Å	—

Tab. 1   Thickness of blister peak due to He^+ bombardment of Nb.

The measured peak widths are considerably larger than theoretical penetration depths /20/. The reasons for this discrepancy are not yet well understood. It may partly be due to the fact that:

1) In a well ordered lattice a large fraction of the incident ions is channeled along close packed lattice directions or planes even in the case which we denominate as "random incidence". At these low energies critical angles for channeling may be quite large.

2) The electronic stopping power of low energy helium is overestimated by Lindhard's theory so that theoretical penetration depths are somewhat too small.

3) The stopping power of H^+ in He implanted Niobium may be higher than in the pure metal. But this effect cannot account for the observed descrepancy of a factor of 3 in penetration depth. Even a 100 atomic percent He concentration in Niobium increases

the stopping power of 150 keV protons by only $\sim 14\%$ if Bragg's rule of additivity of stopping powers is assumed using stopping power data by Northcliffe and Schilling /17/.

In a previous paper /18/ a method has been described to calculate stopping power values from backscattered energy distributions. Applying this method to the random spectra taken after He bombardment (Fig. 2) we find a 17% higher stopping power in the helium implanted region, defined by a step in the random spectrum, than at larger depth. This does, however, not necessarily mean that we have a more than 100 atomic percent helium concentration in the implanted region because the decrease in backscattered intensity of the random spectra may also be due to partial alignment of the covers of the blisters with the incident ion beam.

The fraction of the surface covered by blisters can be assessed from the intensity of the "blister peak". The fact that it is close to random intensity (Fig. 2) means that coverage attains nearly 100% in accordance with the surface micrographs (Fig. 6).

The explanation of the backscattering spectra after bombardment with $< 10^{17}$ He^{+}/cm^{2} is less straightforward. Three mechanisms may contribute to the increase in backscattered proton intensity:

1) Protons may be directly backscattered from Nb atoms located interstitially between lattice rows due to radiation damage.

3) Small gas bubbles will lead to enhanced direct backscattering due to additional surfaces in the bulk formed by these bubbles.

3) Lattice strain due to radiation damage and gas take up causes dechanneling rather than direct backscattering from the aligned beam.

While the first two mechanisms should produce a peak in the backscattered energy distribution at the depth of maximum radiation damage the third mechanism only leads to enhanced dechanneling increasing the slope of the spectrum in the region of maxiumum lattice strain.

Comparing the thickness of our blister peaks (Fig. 2-5) with the low dose spectra we find that it roughly coincides with the maximum of the hump which appears after low dose bombardment. On the other hand the maximum slope of the distribution appears at much smaller depth. This indicates that a large part of the hump is due to direct backscattering. Pronko et al. /21/ have

obtained similar spectra by low temperature (25°K) bombardment
of Nb single crystals with 300 keV Ar. They show that all interstitials are annealed at room temperature. From these results
we conclude that direct backscattering in our measurements is due
to gas bubbles which are formed by He agglomeration at the penetration depth of the He ions. The depth at which these bubbles
occur increases with primary He ion energy, their depth distribution is larger in aligned than in random incidence.

The separation of the low dose phase and the onset of blister
formation is not very sharp. Thus, the critical doses as given in
Tab. 1 only define a certain dose range in which blisters start to
develop.

CONCLUSION

We have shown that Rutherford backscattering in double
alignment on metal single crystals is a sensitive technique to
observe the early stages of blister formation due to ion bombardment. Critical dose and blister thickness can be determined in
situ. The technique is probably also applicable to observation
of void formation since double aligned yield is sensitive to additional surfaces in the bulk of the material. Some details of the
observed spectra are not yet well understood, e. g. the fine structure of the blister peak in its starting phase and the apparent
discrepancy between thickness and theoretical penetration depth.
For clarification of these points the work is continued.

Acknowledgements: We thank Dr. H. Vernickel for his continuous
interest in this work and Dr. H. Verbeek for valuable discussions.
H. Wacker was responsible for the operation of the accelerator
and the target preparation, H. Schmidl was in charge of the electronic equipment, Mrs. C. Drewes made the diagrams, and Miss
I. Wunderlich typed the manuscript. Their contributions are
gratefully acknowledged.

REFERENCES

/1/   R. Behrisch, Nuclear Fusion 12, 695 (1972).

/2/   S. K. Das, Private communication.

/3/   J. Stark, G. Wendt, Ann. d. Physik 38, 921 (1912)

/4/   W. Primak, J. Appl. Phys. 34, 3630 (1963)

/5/   M. Kaminsky, Adv. Mass Spectrometry 3, 69 (1964)

/6/ M. Kaminsky, S. K. Das, J. Appl. Phys. <u>44</u>, 25 (1973)

/7/ W. Bauer, G. J. Thomas, J. Nucl. Mat. <u>47</u>, 241 (1973)

/8/ R. D. Daniels, J. Appl. Phys. <u>42</u>, 417 (1971)

/9/ S. K. Erents, G. M. McCracken, Rad. Effects <u>18</u>, 191 (1973)

/10/ R. C. Mikkelson, J. W. Miller, R. E. Holland, D. S. Gemmell, J. Appl. Phys. <u>44</u>, 935 (1973)

/11/ W. Eckstein, H. Verbeek a) Verh. DPG (VI) <u>8</u>, 439 (1973)
                              b) these conference proceedings.

/12/ R. Behrisch, Vak. Technik <u>10</u>, 250 (1967)

/13/ B. M. U. Scherzer, Thesis, Technische Universität, München (1969)

/14/ H. Schmidl, IPP-Report 9/3 (1971)

/15/ R. Behrisch, G. Mühlbauer, B. M. U. Scherzer, J. Phys. E 2, <u>2</u>, 381 (1969)

/16/ E. Bøgh, Can. J. Phys. <u>46</u>, 653 (1968)

/17/ L. C. Northcliffe, R. F. Schilling, Nucl. Data Tables A7, 233 (1970)

/18/ R. Behrisch, B. M. U. Scherzer, Thin Solid Films (in press)

/19/ R. Behrisch, B. M. U. Scherzer, H. Schulze, Rad. Effects <u>13</u>, 33 (1972)

/20/ H. Schiøtt, Mat. Fys. Medd. Dan. Vid. Selsk. <u>35</u>, no. 9 (1966)

/21/ P. Pronko, J. Bottiger, J. A. Davies, J. B. Mitchell, Rad. Effects (in press)

/22/ W. Whaling, Handbuch der Physik <u>34</u>, 193, Springer Verlag, Berlin (1958)

DISCUSSION

Q: (K. L. Merkle) How do you account for the fact that on the one hand the bending of the surface layer due to blisters occurs only over a fraction of the total surface area while on the other hand the backscattering spectra almost come up the random yield?

A: The backscattering yield from the layer corresponding to the

blister covers is about 10% lower than the random yield. Only 60
to 80% of this layer has been bent during the blister formation,
however the whole layer is heavily damaged which also contributes
to the high yield.

Q: (W. K. Chu) Two variables are used in your dechanneling and
disorder measurement, namely energy (2 keV and 4 keV) and dose
(10^{16} - 10^{18}/cm^2). Can you combine the two variables into one,
namely concentration (He/cm^3)? This might give a clearer picture.

A: In principle one should be able to derive a critical number
for blister formation, which is a combination of the ion energy
and the ion dose. However at these low energies, an appreciable
amount of 10% to 30% of the incident ions are backscattered in the
collision cascade. This is energy dependent. It has to be subtracted to obtain the number of gas ions collected in the target.
Further, the mechanical strength of the target material will play
a role. This will be modified by the actual range distribution
of the incident ions in the target.

RADIATION DAMAGE AND GAS DIFFUSION IN

MOLYBDENUM UNDER DEUTERON BOMBARDMENT

G M McCRACKEN and S K ERENTS

UKAEA Culham Laboratory

Abingdon, Berkshire, England

1. Introduction

The release of gas from metals after implantation has been studied for many years[1]. A case of interest both from a fundamental point of view and from the application to controlled thermonuclear experiments is the bombardment of metals with hydrogen ions. This case is unique because of the high diffusion coefficient and high solubility of hydrogen in metals. In previous papers we have examined in detail the trapping and release of deuterium in nickel.[2,3] It was shown that the release rate is much less rapid than expected on the basis of thermal diffusion and there is considerable evidence that radiation damage by the incident ions produced trapping sites which inhibited subsequent diffusion by the implanted gas atoms. Similar evidence has been obtained for the helium tungsten system by Kornelsen[4] who was also able to identify the types of defect responsible for sites of different binding energies. In the investigation of nickel, estimates were made of the binding energies of different sites and their population as a function of incident ion dose. A similar study has now been made in molybdenum in which it is shown that the release rate is critically dependent on the damage and annealing history of the sample, and that under certain conditions release rates close to those predicted by thermal diffusion can be obtained.

2. Experiment

The experimental equipment is similar to that described previously (fig 1).[2,5] An ion beam in the energy range 5 to 35 keV is mass analysed and focussed on a target in the target

chamber at a pressure $\sim 10^{-9}$ torr. The chamber is continuously
pumped by an aperture of known pumping speed and the gas release
is measured by a quadrupole mass filter. The target chamber is
surrounded by an outer chamber which has a pumping speed for
hydrogen of $\sim 10^4$ ℓ/sec, obtained by using a large titanium sub-
limation pump.

    Measurements consist basically of observing the instantaneous
release rate of deuterium from the surface as a function of time
after the beam is turned on, for different target conditions. This
is frequently followed by recording a post bombardment thermal
desorption spectrum as the target temperature is increased linearly
with time. The vacuum time constant of the system is .03 secs.
The target consists of a strip of molybdenum (99.9%) 1 mm thick and
3 mm wide which is prepared by electropolishing and is heated
directly by passing ac through it. It has been shown previously[6]
that the target readily saturates so that there is an equilibrium
between the rate of arrival of ions and the rate of re-emission of
gas from the surface. The ion beam current is in the range 1-10 μA
and the beam diameter is typically 3 mm. The area of the beam has
an uncertainty of 20% leading to a systematic error in the estimates
of ion dose per cm^2.

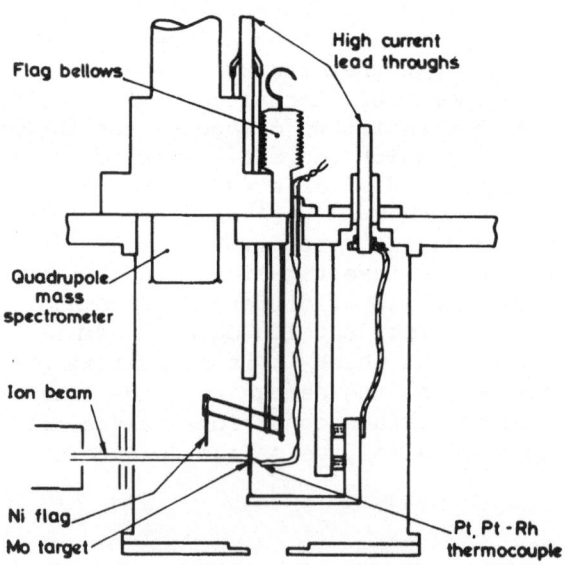

Fig 1   Schematic of target chamber.

3.   Results

(i) Dose dependence

  The release rate of the implanted gas during bombardment was found to be a function of the previous history of the sample - in particular on the amount of damage or annealing which the target had been subjected to. Results for the sample in different conditions are shown in fig 2. The curve (a) is for a sample in which the damage is saturated by prolonged ion bombardment so that further damage causes no further change in the release behaviour. Gas implanted during the damaging stage is thermally released as described in (iii). Curve (b) is fully annealed as described later (ii). It is seen that the release rate is changed by more than two orders of magnitude from the damaged to the fully annealed states. Both of these states are completely reproducible in the same sample.

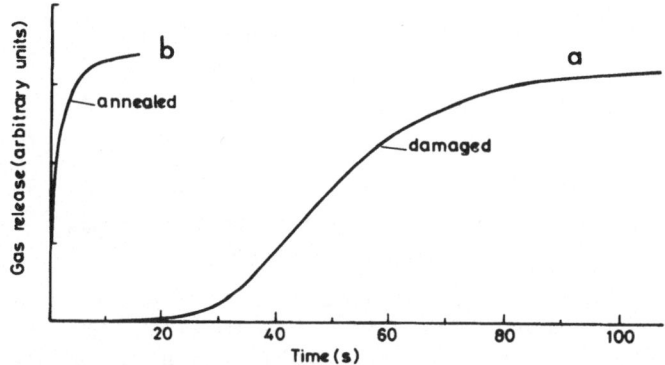

Fig 2 Gas release during 20 keV D^+ implantation of damaged and
    annealed Mo.

(ii) Damage annealing

  The effect of temperature on the annealing of the sample is illustrated in fig 3. A target is damaged to saturation by a long period of bombardment with deuterons - typically 800 µA- min per cm^2 at 400K. The sample is then annealed for 10 minutes at successively higher temperatures and in between each anneal the target is bombarded for a short time at 2.5 µA in order to produce a characteristic re-emission curve as in fig 2. (The anneal removes the gas implanted during the damaging stage.) The number of ions trapped in the target during this test bombardment was measured by integrating the thermal release curve and this was used as a measure of the number of damage sites remaining in the sample. As seen in fig 3 there was a marked annealing effect at $\sim$ 1500K and the target was fully annealed at 1700K. The time and temperature agrees well with the recrystallization time for cold worked molybdenum specimens.[7]

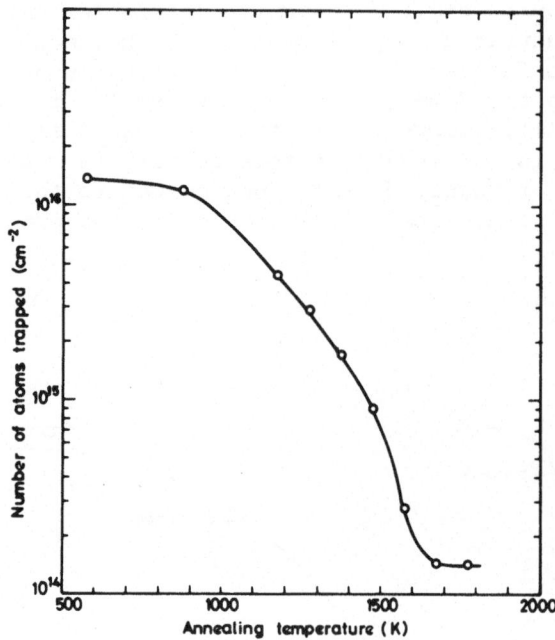

Fig 3   Annealing curve for molybdenum after damage induced by
        20 keV D^+ ions. Annealing time 10 mins. Ordinate is the
        gas desorbed after a test implantation to gas saturation
        at 523K.

(iii) Post implantation thermal desorption

     For the annealing tests to be meaningful it is clear that the gas injected during the short test implantations between anneals must be completely removed during the annealing process. Post implantation thermal release spectra using a linear temperature ramp were therefore measured for initially annealed specimens after a variety of implantations, as shown in fig 4. It is clear that all gas is released from the target at temperatures above 750K and hence the annealing curve shown in fig 3 is meaningful above that temperature.

     It is clear from the thermal desorption spectra that there is an increasing population of atoms in damaged sites with increasing

dose and it is assumed that this is due to the increasing number
of sites.   It is also clear that there is some thermal desorption
at the implantation temperature as release is observed immediately
after implantation stops for some seconds.   In fact in an earlier
investigation[8] implantation was carried out at 77K and a continuous thermal desorption spectrum was observed from 100K to 700K.
The major peaks were then observed between 150K and 250K.   An
attempt was made to analyse the present spectra in terms of individual peaks in the same way as was previously done for deuterium
implanted in nickel.[3]   However it was found necessary to introduce at least 6 peaks to obtain good agreement between theoretical
and experimental results and since they were not at all well
resolved it was virtually impossible to do this in an unique way.
Thus the attempt to obtain individual site populations was
abandoned.

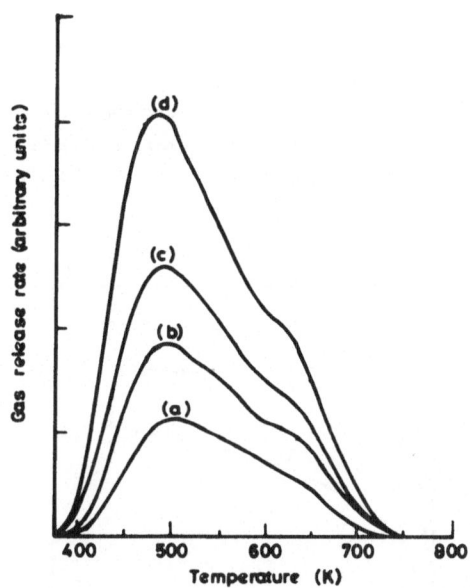

Fig 4   Post bombardment thermal release of gas from Mo after
        implantation of 20 keV D^+ ions at 373K.   Incident dose
        (a) 1.9×10^{16} ions cm^{-2} (c) 9.5×10^{16} ions cm^{-2}
        (b) 4.7×10^{16} ions cm^{-2} (d) 1.9×10^{17} ions cm^{-2}

(iv) Damage production rate

     The fact that gas can be released at a much lower temperature
than that at which the damage is annealed suggests a technique for
measuring damage rates.   An annealed target is damaged by ion
bombardment for a fixed time at a temperature below the annealing
temperature.   Then by heating the damaged target just above the

gas release temperature (750K in this case) the gas implanted
during damage production is removed. The target is then bombarded
with a short test dose at a well defined temperature and the re-
emission rate during implantation is measured. Finally the target
is heated and the thermal desorption spectrum is obtained. From
integration of the re-emission curve during implantation and the
thermal desorption curve an estimate is made of the number of atoms
trapped in the lattice. Since the number trapped in an annealed
target is normally small (cf fig 2) all these trapped atoms can
reasonably be assumed to be trapped at damage sites. Furthermore
if the test implantation is continued until the target is
saturated then it is probable that all damage sites which can be
occupied at that temperature are filled. Thus the number of atoms
desorbed, which can be quite accurately measured absolutely, is a
direct measure of the number of damage sites produced in the
lattice.

    Results for such an experiment are shown in fig 5 for a molyb-
denum target bombarded by deuterons at three different energies.
Initially the damage increases linearly with dose and a measurable
amount of damage is produced in a few seconds at the current
density used of $\sim$ 25 μA/cm^2. As the dose increases the rate of
damage decreases and the target reaches a saturation level. This

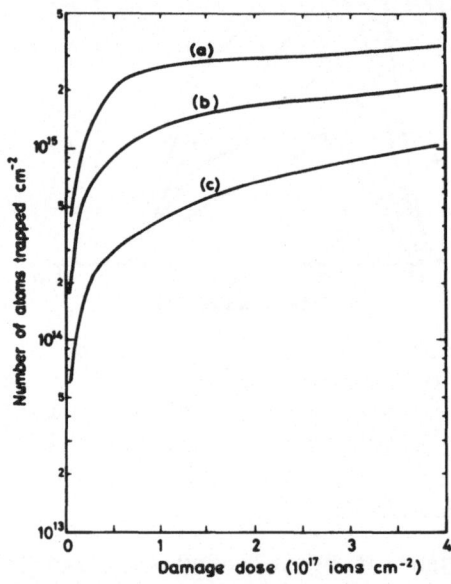

Fig 5   Trapping sites produced by D$^+$ ions in molybdenum. The
        ordinate is the gas desorbed after a test implantation
        at 623K.   (a) 35 keV   (b) 20 keV   (c) 7 keV

saturation level is apparently higher for larger incident ion
energies. The initial linear increase is a measure of the number
of trapping sites which have been produced by a given dose, ie, a
given number of incident ions. Thus the slope of the curve can be
directly converted to damage rate in terms of defects/incident ion.
In this way the damage rate has been plotted as a function of
energy as shown in fig 6. In addition to the three energies
directly measured, the threshold for defect production has also
been plotted assuming a displacement energy of 40 eV. It is seen
that the damage rate increases roughly linearly with energy, and
that the damage rate is remarkably low. Even at 30 keV the rate
is less than 0.5 defects/ion. This low defect production rate is
consistent with the fact that atoms do diffuse out of an annealed
sample - for if the defect production rate was greater than unity
then one would expect most incident ions to be trapped. No
attempt has been made to calculate the defect production rate
directly but it is generally accepted that with light ions the
primary energy loss mechanism is electronic and that the
proportion of the total energy lost by nuclear collisions is
small.

    The temperature at which the measurements of damage pro-
duction rate are made are obviously critical. These were chosen
for practical reasons to be sufficiently high that diffusion out
of the sample was rapid so that there were no atoms which were
simply diffusion trapped. However since there are a range of

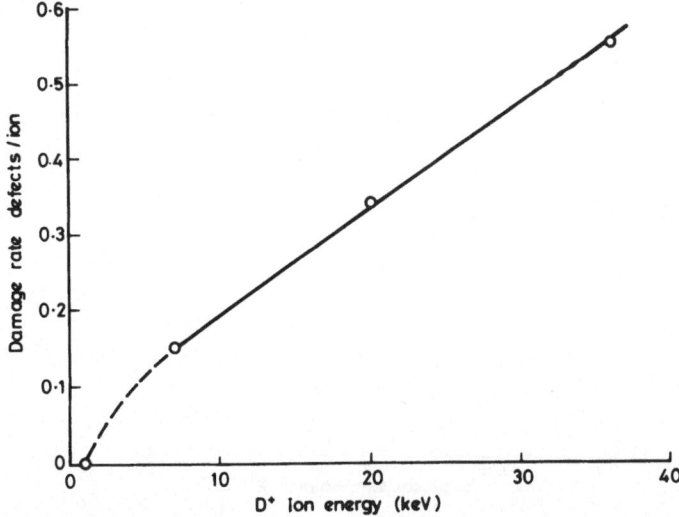

Fig 6   Rate of production of defects in Mo under bombardment by
         D$^+$ ions. Defects measured at 623K by implantation and
         subsequent thermal desorption of a test desorption of a
         test dose of D$^+$ ions.

defect binding energies only those sites which have a binding
energy above a given level can be occupied at any given temperature. Thus the measurements outlined above measure only the
number of sites with binding energies for deuterium above a
certain value. There will be many more sites with energies below
this value which are not detected. Obviously it would be useful
to repeat the measurements over a range of temperatures.

(v) Temperature Effects

     As discussed above temperature has an important effect on
release rate both through the temperature dependence of diffusion
rate and due to the range of binding energies of the deuterium
atoms at different types of defect. Measurements of the number
of atoms trapped in the surface have therefore been made both for
an initially fully annealed sample and for a sample which had been
damaged to saturation. The results are presented in fig 7.

     In the case of the damaged sample, since the test implantation
continued until the target saturated, these figures represent
the maximum number of trapping sites which can be occupied at
any particular temperature. The number of these sites increases
as the temperature decreases down to the lowest temperature

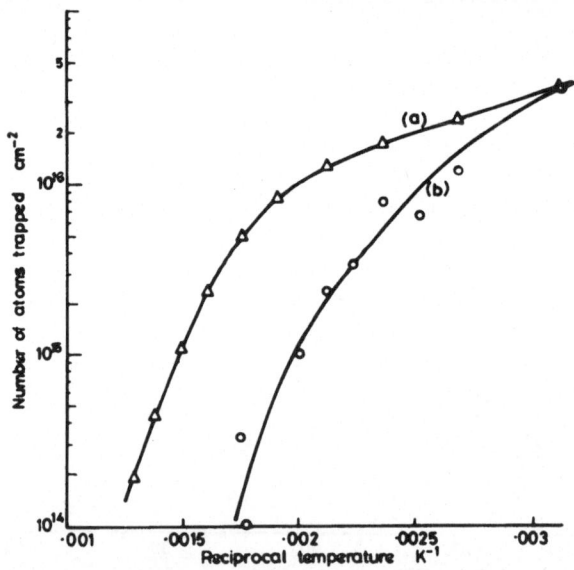

Fig 7   Number of atoms trapped in Mo during bombardment by
        20 keV D^+ ions at various temperatures
        (a) Mo initially damaged by 3×10^{17} ions cm^{-2}
        (b) Target annealed at 1770K for 10 mins between runs.

investigated of 293K. The concentration of damage sites is then
3.5×10^{16} cm^{-2}, or assuming a damage range equal to the range
($\sim$ 1000 A$^{\circ}$) the concentration is 3.5×10^{21} cm^{-3}. This is
of course extremely large and suggests that there may be many
deuterium atoms trapped per defect site.

The result for the annealed sample does not necessarily
represent atoms trapped at residual defects but, at least at
high temperatures, may be simply the number of atoms which are
present in interstitial sites and have not diffused out within
the time taken to make the measurement. However as temperature
decreases the time to reach saturation increases as the diffusion
coefficient decreases. Thus at lower temperatures there is a
significant amount of damage introduced in the sample in the time
required to make the measurement (the sample is annealed between
measurements at different temperatures). At the lowest temperatures the number of atoms trapped was virtually identical to that
in the damaged sample thus indicating that the target had been
saturated and the number of atoms trapped was determined by the
trapping sites only.

Discussion

The results obtained considerably clarify the situation
regarding release of hydrogen from metals during surface bombardment. It is now clear for example why the diffusion model discussed previously[2] does not fit the results in practice even
qualitatively. In the case of an already damaged target incident
ions may go into the lattice and stay there permanently without
contributing to the diffusive flow and there will be a finite dose
required to saturate the trapping sites before any diffusive flow
will be observed. For a fully annealed target where the diffusion

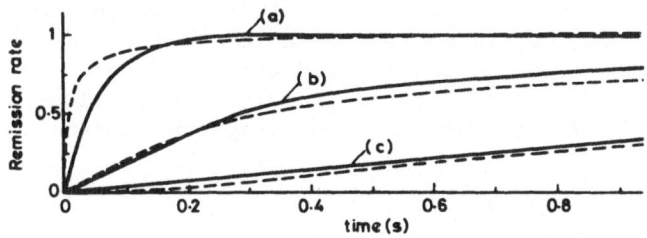

Fig 8   Deuterium re-emission rate from annealed Mo during
        bombardment by 20 keV D^{+} ions

experimental	calculated
(a) 600K	$D = 1 \times 10^{-8}$ cm^2 sec^{-1}
(b) 523K	$D = 2.8 \times 10^{-10}$ cm^2 sec^{-1}
(c) 473K	$D = 4.4 \times 10^{-11}$ cm^2 sec^{-1}

coefficient is high, diffusive release may occur in a time short
compared with that required to produce a significant number of
damage sites and in this case the release should follow the
diffusion model. The agreement though not completely satisfactory
is not unreasonable, fig 8, and the diffusion coefficient obtained
is a factor of 5 lower than that obtained in conventional
thermal diffusion experiments[9] at 600K though a factor of 200
lower at 473K where trapping may be significant.

The presence of trapping sites has been deduced in direct
measurements of thermal diffusion coefficients particularly in
the case of iron[10,11,12] and also in the case of tungsten.[13]
In the case of iron the effect is particularly important below
500K with activation energies varying between 2 and 16 kcal/mole
in different samples. The trapping sites have been attributed to
microcracks, pores and occluded impurities and have been shown to
be strongly dependent on the mechanical and thermal history of
the iron. In tungsten it was found that there could be two kinds
of site, one interstitial and another of unknown nature. The concentration of interstitial hydrogen was $\sim 10^{-3}$ of the total
hydrogen concentration. An analysis of diffusion in the two types
of site was given which provided a qualitative explanation of the
diffusion behaviour observed.

The filling of damage sites with hydrogen atoms and their
subsequent thermal desorption appears to be a potentially powerful
technique for detection of lattice defects in metals. Provided
that the test hydrogen implant is small so that it produces little
damage itself or else it is implanted at an energy below that
required to cause a lattice displacement (~ 1000 eV for H^+ on Mo)
then the sensitivity of the technique might be as high as 1 defect
in 10^6 lattice atoms. A similar technique to examine defects has
been suggested by Kornelsen[4] with helium ion implantation. The
advantage of hydrogen over helium is that the hydrogen can be
desorbed at a temperature well below that at which the damage
anneals, whereas this is not the case for helium. On the other
hand the disadvantage of hydrogen is that surface adsorption is
a competing process for hydrogen trapping and may lead to large
backgrounds if UHV techniques are not employed during implantation.

Conclusions

The diffusion of hydrogen in metals is critically dependent
on the amount of damage in the lattice. The effective solubility
of hydrogen is similarly affected. In molybdenum the effects are
particularly of importance at temperatures below 750K.

Measurements have been made of the rate of defect production
by incident ions of a type of trapping site with a binding energy

above a certain critical value. This value is determined by the temperature at which the experiment is carried out and examination of a wide range of temperatures should be investigated in the future.

These experiments show that hydrogen implantation is a potentially valuable tool for measuring defect concentrations although the interpretation of the types of defect detected has not yet been made. The technique should be equally applicable to damage produced by radiation other than ions, eg., to electrons and neutrons.

REFERENCES

1. CARTER, G and COLLIGON, J S.   1968 Ion bombardment of Solids Ch 8, Heinemann London.

2. ERENTS, S K and McCRACKEN, G M.   1969 Brit J Appl Phys $\underline{2}$, 1397-1405.

3. ERENTS, S K and McCRACKEN, G M.   1970 Radiation Effects $\underline{3}$, 123-129.

4. KORNELSEN, E.   1972 Radiation Effects $\underline{13}$, 227-236.

5. McCRACKEN, G M, MAPLE J H C and WATSON H H H.   1966 Rev Sci Instrum $\underline{37}$, 860-6.

6. McCRACKEN, G M and MAPLE, J H C.   1967 Brit J Appl Phys $\underline{18}$, 919-30.

7. KOHL, W H.   1960 Materials and Techniques for electron tubes pg 319, Reinhold New York.

8. ERENTS, S K.   1972 unpublished.

9. JONES, P M S, GIBSON R and EVANS, J A.   1966 UKAEA Report No AWRE O-16/66.

10. HEUMANN, Th and DOMKE, E.   1972 Proceedings of the International Conference on Hydrogen in Metals, Jülich, Vol 2, pg 492.

11. McNABB, A and FOSTER, P K.   1963 Trans Met Soc AIME $\underline{227}$ 618.

12. ORIANI, R A.   1970 Acta Met $\underline{18}$, 147.

13. FRAUENFELDER, R.   1969 J of Vac Sci and Tech $\underline{6}$, 388-397.

DISCUSSION

Q: (G. L. Kulcinski) I think your results may be very important
in interpreting void formation by proton bombardment. It may be
possible that in the process of diffusing out of the irradiated
sample, the proton could get trapped, even temporarily, and thereby act as nucleation sites for void formation. Would you care to
comment on this possibility in refractory metals or austenitic
steels?

A: It is possible that there may be sufficient hydrogen occupation
of damage sites to play a role in void nucleation in some cases.
However, the evidence from the present thermal desorption experiments indicates that in molybdenum the hydrogen will be released
at a temperature below the main void formation temperature range.

RADIATION BLISTERING AFTER H^+, D^+, and He^+ ION IMPLANTATION INTO SURFACES OF STAINLESS STEEL, Mo, AND Be

H. Verbeek and W. Eckstein [+]

Max-Planck-Institut für Plasmaphysik, EURATOM

Association, D-8046 Garching

ABSTRACT

H^+, D^+, and He^+ ions with energies of 15 and 150 keV were implanted into surfaces of stainless steel, Mo, and Be. The occurrence of blisters has been observed by scanning electron microscopy and by optical interference microscopy.

In all investigated metals severe blistering was observed after He irradiation to fluences above 3×10^{17} ions/cm^2.

On H^+ and D^+ bombarded stainless steel irregular shaped blisters were observed mainly at special sites such as grain boundaries and precipitations. The critical fluence for blister formation by 15 keV D^+ ions was in the order of 5×10^{18}/cm^2.

For polycrystalline Mo the shape and the average size of hydrogen blisters depends strongly on the pretreatment of the targets. For annealed targets the blister size is much smaller than for unannealed material. The shape and the critical fluence depend strongly on the orientation of individual grains. The critical fluence, defined as the value at which blisters are first observed in some grains, decreases with increasing current density of the bombarding particles.

On D^+ bombarded Be, blisters occurred at fluences below $3 \cdot 10^{17}$ ions/cm^2.

+) Presently on leave of absence at IBM Research Laboratories, San Jose, Ca., USA.

INTRODUCTION

It is well known that implantation of light ions into metals to high fluences can result in the formation of blisters, which release bursts of the implanted gas. This has been observed by mass spectrometry /1, 2, 3, 6/. The bombarded surfaces have also been investigated by interferometric methods /4/ and the blisters have been observed directly by scanning electron microscopy /5, 6, 7/.

As was pointed out earlier /8, 5b, 6/, the formation of blisters may severely impair the lifetime of the first wall of a future fusion reactor.

The injected gas ions come to rest in a certain depth of the solid. When the gas is insoluable in the metal (as is the case especially for the noble gases) it is collected in gas bubbles with diameters below $\approx 50 \text{\AA}$. These have been observed in transmission electron microscopes /9, 10/. If their density exceeds a certain value, these bubbles coalesce to form blisters. For low energies the blistering phenomenon is restricted to light ions such as H^+, D^+, and He^+ since for the heavier ions the surface layer is carried away owing to sputtering before a sufficient amount of gas is collected beneath the surface.

As yet little is known about the basic mechanism of blister formation. The critical fluence for the occurrence of blisters and their sizes and shapes depend on several parameters: 1) the range of the ions, 2) the diffusion constant and 3) the solubility of the injected gas in the metal, 4) the yield strength of the metal, 5) the temperature during bombardment (which influences parameters 2) to 4), and 6) the dose rate of the bombarding ions. In most cases no reliable data about the solubility and the diffusion constants are available. The latter are influenced, especially in bcc-metals, to a great extent by radiation damage /11, 12/, which is necessarily connected with the ion implantation.

Thus, up to now no theoretical predictions about critical fluences, blister shapes and sizes for any gas-metal combination could be made. A method which gives some insight in the mechanism for blister formation is to observe the dechanneling in Rutherford backscattering as was recently done by Roth et al. /13/.

The d - t fusion reactor will be operated at temperatures which correspond to mean particle energies in the 10 to 20 keV region. Until now, however, most of the blistering work has been done with ions of much higher energies. This is why we started our work with H^+, D^+, and He^+ ions of 15 keV.

EXPERIMENTAL PROCEDURE

The 15 keV ion implantation was performed in the BOMBARDON apparatus described elsewhere /14/. The targets were bombarded with a normally incident mass analysed ion beam with current densities up to $4 \times 10^{-4} A/cm^2$. At a distance of 3mm from the target was an aperture 0.4mm in diameter which defined the bombarded area. Between this aperture and the target a second aperture 0.8mm in diameter was mounted. With a potential of -30V this served to suppress the secondary electrons from the target. Thus, by digitizing and counting the incoming charge the total bombardment dose could be measured. The target was mounted on a linear motion feedthrough, while the aperture remained fixed. Up to six bombarded spots could thus be produced on one specimen.

The ion beam was focused on the first aperture with an electrostatic einzel lens. But even for the highest current densities used the lens was slightly defocused to achieve a rectangular current distribution on the targets.

During irradiation the pressure was below 5×10^{-8} torr. All bombardments, with the exceptions indicated, were done at room temperature.

In addition, some implantations were performed with the PHARAO accelerator /15/ at 150 keV. In this case the doses were obtained from a calibrated reference counter.

All targets were mechanically polished, with a final 0.1μ grain size alumina finish.

The bombarded targets were investigated by optical microscopy with differential interference contrast (INCO insert on a Zeiss Universal microscope) and by a scanning electron microscope (SEM) (Cambridge Stereoscan). In several cases the optical microscope was advantageous compared with the SEM. Especially the very shallow hydrogen blisters gave a much higher contrast in the optical microscope, whose resolution, however, is much less than the SEM's.

RESULTS AND DISCUSSION

He and D Irradiated Stainless Steel

The stainless steel targets were cut from ordinary 4301 material containing 18% Cr, 9% Ni, and < 2% Mn. Fig. 1 shows

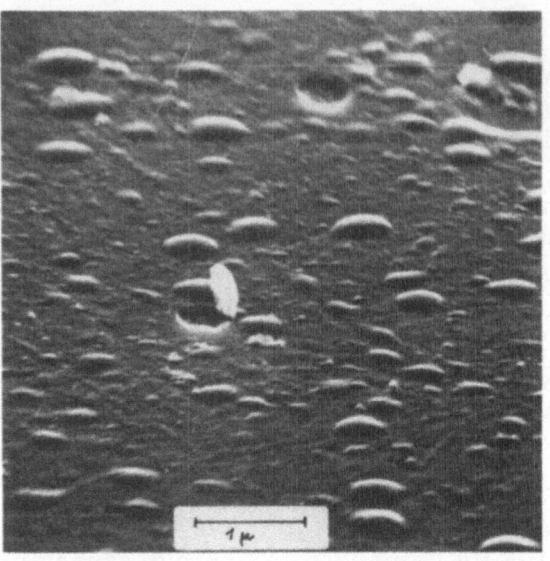

Fig. 1   SEM micrograph of stainless steel irradiated at 300 K by 15 keV He ions to 3×10^{18} ions/cm^2. $j = 1.2\times10^{-4}$A/cm^2, $\alpha = 70^0$.

a scanning electron micrograph [+)] of a stainless steel surface bombarded with 15 keV He ions to a total fluence of 3×10^{18} ions/cm^2. Over the whole bombarded spot dome shaped blisters with an average size of 0.3μ are distributed. Some of them ruptured and lost their covers. The critical fluence for the occurrence of the blisters is between 1.2 and 4×10^{17} ions/cm^2. 15 keV D^+ bombarded stainless steel surfaces are shown in Fig. 2. With the same current density (1.4×10^{-4} A/cm^2) as for the He^+ bombardment the critical fluence is $\approx 5\times10^{18}$ ions/cm^2, i.e. an order of magnitude larger than for He bombardment. The irregularly shaped blisters, some of which are apparently ruptured, occurred only at special sites, while larger areas of the bombarded spot remained unaffected. Often the blisters are arranged on straight lines, which may be grain boundaries. In several cases blisters surrounded precipitations, which could be identified by x-ray analysis in the SEM to be MnS.

[+)] For all micrographs in this paper no correction for the angle α between the investigated surface and the probing electron beam is provided. Thus, the indicated scale is valid only in the horizontal direction.

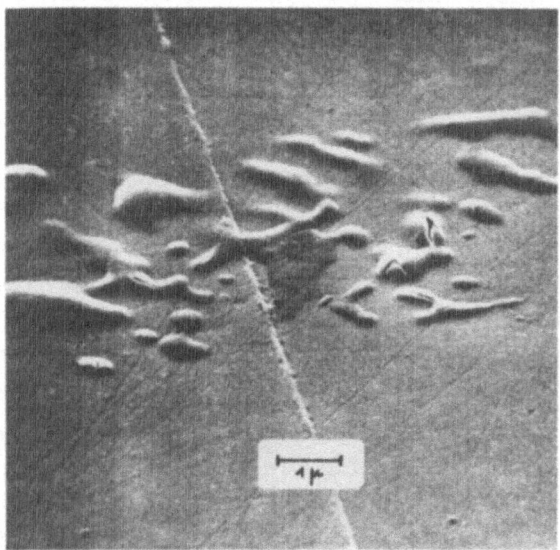

Fig. 2   SEM micrographs of stainless steel
irradiated at 300 K by 15 keV D ions to 5×10^{18} ions $/cm^2$.
$j = 1.4 \times 10^{-4}$ A/cm^2, $\alpha = 60^{\circ}$.

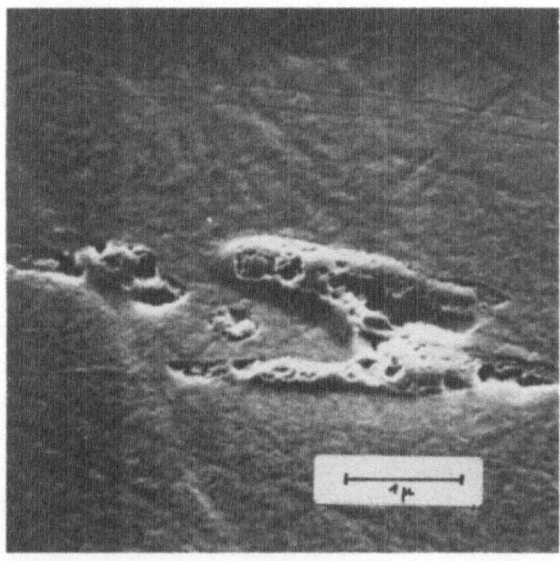

Fig. 3   SEM micrograph of stainless steel irradiated at 620K by
15 keV D ions to 6×10^{18} ions$/cm^2$. $j = 1.4 \times 10^{-4}$ A/cm^2, $\alpha = 60^{\circ}$.

When the bombardment was performed at 350°C the resulting blisters were of more rugged appearance than at room temperature as is shown in Fig. 3.

Molybdenum

For the molybdenum targets 99.95% purity material (MRC VP grade) was used, except for the bombardments with 150 keV. After mounting in the vacuum of the bombarding apparatus the targets were generally annealed at ≈1600°C by electron bombardment. This caused recrystallization resulting in grain sizes in the range of 0.01 to 0.1mm. Apart from the grain boundaries, no structure could be recognized on the annealed target surfaces.

He irradiated Mo. With 15 keV He bombarded polycrystalline Mo we can essentially confirm the results of Erents and McCracken /6/. Fig. 4 shows a) optical and b) scanning electron micrographs. The blisters are regularly dome shaped; some of them have lost their covers. In the optical micrograph Fig. 4a. The edge of the irradiated area is seen and the unbombarded area is shown for comparison at the bottom of the picture. The grain boundaries due to recrystallization during the annealing before irradiation are clearly visible. It is also seen that the blister size and density depend strongly upon the orientation of the individual grains, a feature not reported by Erents et al. /6/.

As we had no means of determining the orientation of the grains, we made some additional investigations on a Mo single crystal with a (100)-surface. In this case the He ion energy was 150 keV. The crystal could be oriented by Rutherford backscattering /15/. We observed no differences in the appearance of the blisters for bombardment aligned along ⟨100⟩ and random directions. The average blister diameter, however, was 6.0μ for the aligned bombardment, while it was 4.8μ for random implantation. Moreover, from SEM micrographs in large magnification the thickness of the cover of ruptured blisters could be estimated at ≈ 4000 Å for the aligned and ≈ 3000 Å for the random bombardment. This was to be expected because of the enhanced range of the ions incident along the ⟨100⟩ direction. The thickness of the blisters due to random bombardment agrees fairly well with the range given by Schiøtt /16/.

H$^+$ and D$^+$ irradiated Mo. The blistering of hydrogen irradiated Mo depends strongly on the pretreatment of the specimens. On unannealed targets taken from rolled sheeting as supplied by the manufacturer and polished large blisters with average dia-

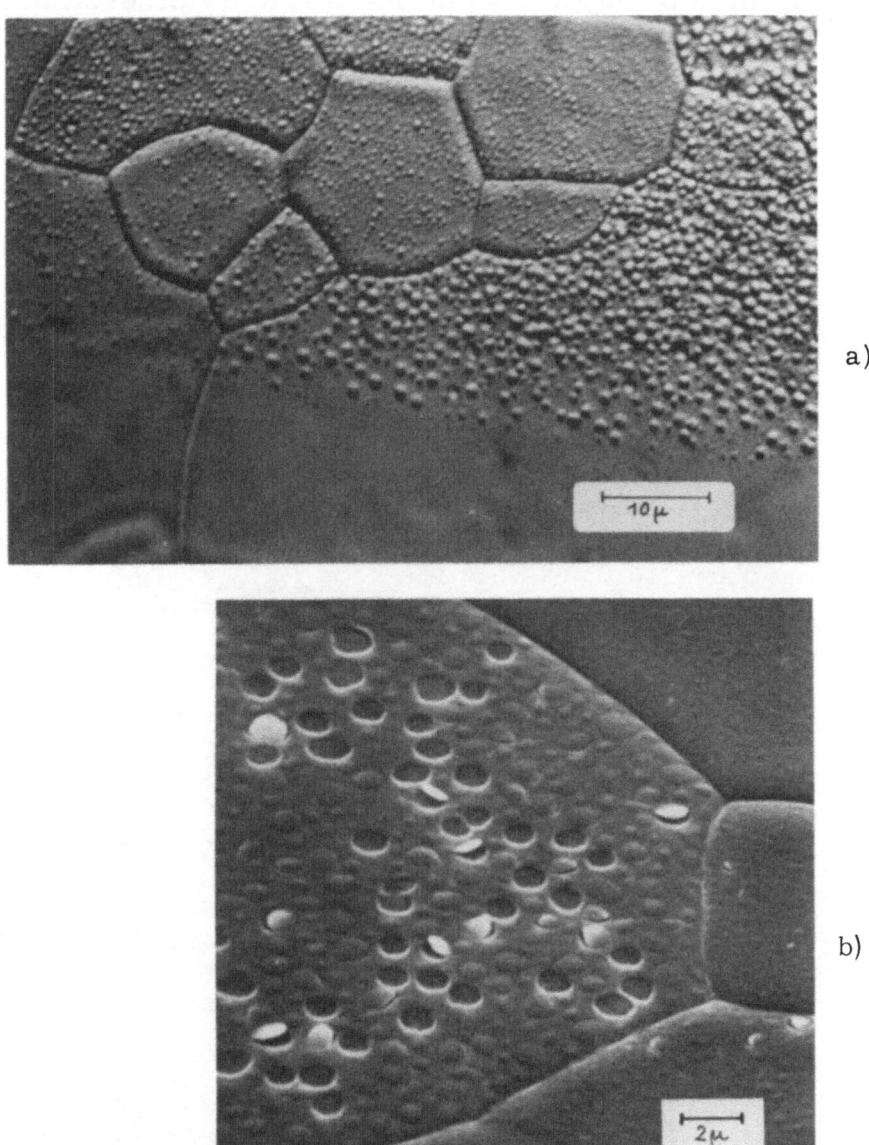

Fig. 4   a) Optical and b) SEM micrograph with $\alpha = 60°$ of annealed Mo irradiated by 15 keV He ions $2.5\text{x}10^{18}$ ions/cm^2. $j = 1.2\text{x}10^{-4}$ A/cm^2.

meters of 2μ could be observed after 15 keV D^+ implantation, as shown in Fig. 5. Blisters observed on annealed material had mean diameters of only 0.3μ, in contrast to the observations on He irradiated cold-worked material /5/ where smaller blisters than on annealed material were found.

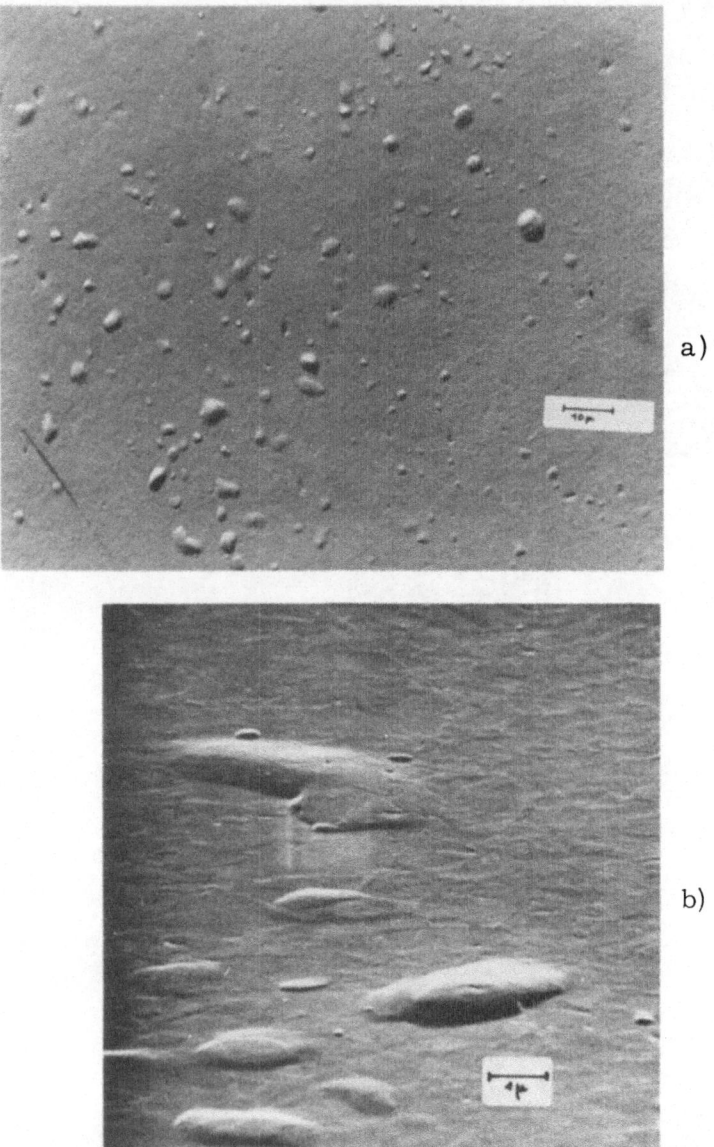

Fig. 5   a) Optical and b) SEM micrograph with $\alpha = 70^o$ of unannealed Mo irradiated by 15 keV D ions to 8×10^{18} ions/cm^2. $j = 12 \times 10^{-4}$ A/cm^2.

In investigating hydrogen implanted annealed polycrystalline
molybdenum we found no significant differences between proton
and deuterium irradiated specimens. The shape and size of the
blisters and the critical fluence for their occurrence depend
strongly on the grain orientation. This effect is even more pronounced than for He implantation. Some examples of the various
shapes and densities of blisters due to 15 keV D^+ irradiation are
shown in Fig. 6a, b and Fig. 7. In Fig. 6a a rather severely blistered

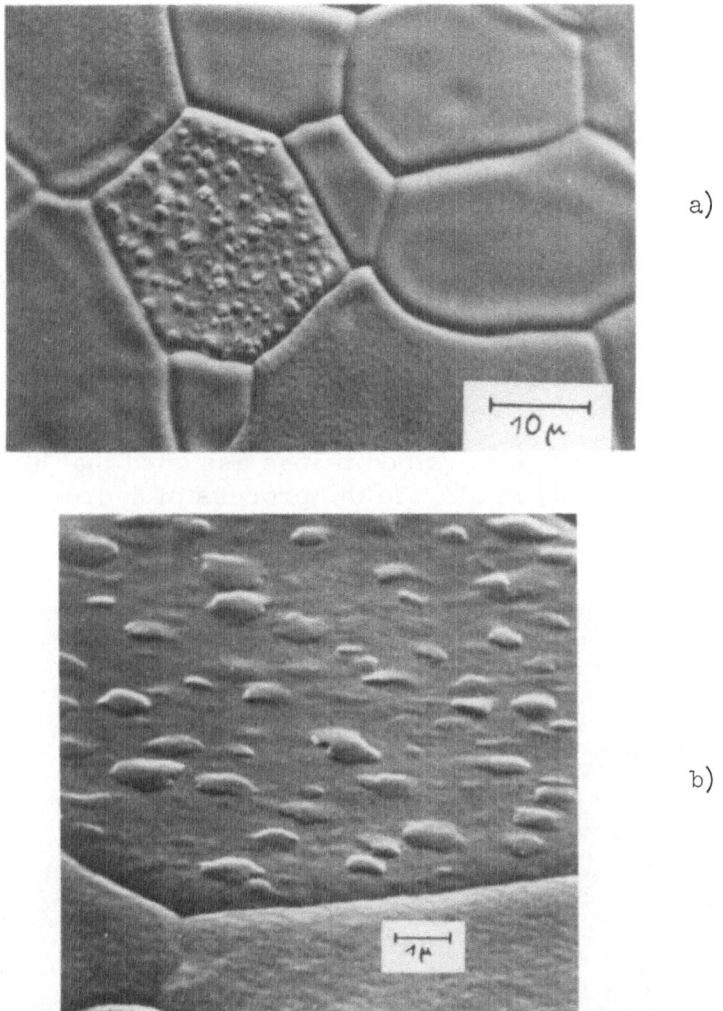

a)

b)

Fig. 6   a) Optical and b) SEM micrograph with α = 70° of annealed
Mo irradiated by 15 keV D ions to 2×10^{18} ions/cm^2.
j = 3.3×10^{-4} A/cm^2.

grain is shown surrounded by grains which are nearly unaffected. Fig. 7 shows crowfoot-shaped blisters with the prongs of the blisters oriented in a manner characteristic of each grain. For the relatively low energy of 15 keV the critical angles for channeling are of the order of several degrees. Thus it is very likely that the incoming ions are channelled even in polycrystalline material. This may cause the large differences in the appearance of the individual grains.

We also bombarded a Mo single crystal with 150 keV protons. But as for He bombardment we did not find a significant difference in the appearance of the blisters for implantations along ⟨100⟩ and random directions.

The critical fluence F_c for the occurrence of blisters, defined as the value at which blisters can first be observed in any grain, depends on the flux j of the bombarding particles. Fig. 8 shows this dependence for 15 keV H^+ irradiation. F_c was determined by irradiating several spots on one target by increasing the fluence stepwise at constant flux. As this is a rather time consuming procedure, rather large steps had to be used and the error bars are considerable. But it is clearly seen that F_c decreases with increasing j.

This is readily understood if it is assumed that hydrogen diffuses considerably in Mo. In the process of hydrogen build-up the implantation competes with the loss by diffusion. Thus, with

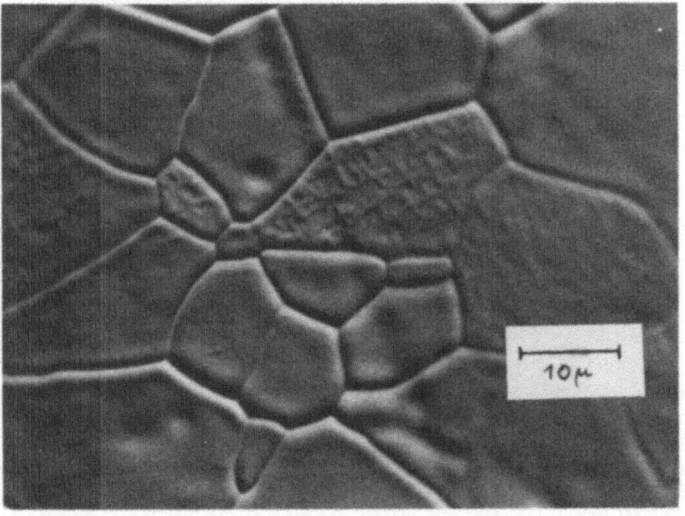

Fig. 7 Optical micrograph of a different area of the same specimen as in Fig. 6.

a high incoming current the critical concentration is reached earlier than with low currents.   When the bombarding energy is increased the range straggling is also increased.   This causes the implanted hydrogen to be distributed over a larger depth range and thus the critical concentration is reached at higher fluences as observed for 150 keV (Fig. 8).

Beryllium

Be targets were cut from sintered 99. 8% purity material supplied by Berylco. Besides polishing no further treatment was performed before irradiation.   Both D^+ and He^+ ion implantations with 15 keV cause blisters at fluences lower than 3×10^{17} ions/cm^2. An example of D^+ irradiated Be is shown in Fig. 9. Most of the blisters show cracks.   The He blisters are of similar shape, but are slightly smaller.

The blistering of Be is of considerable interest since Be as a low Z material is often used as substrate for Rutherford backscattering measurements /17/.Blistering of the substrate may seriously damage the surface layer to be investigated.

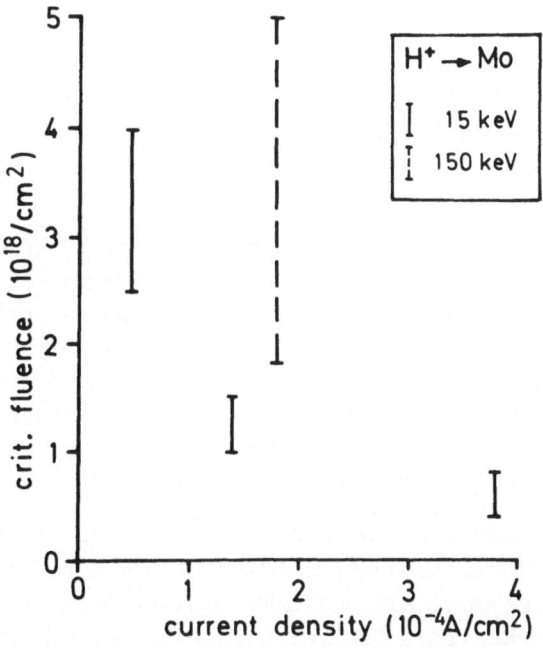

Fig. 8 Dependence of the critical fluence for the formation of blisters on the ion dose rate.

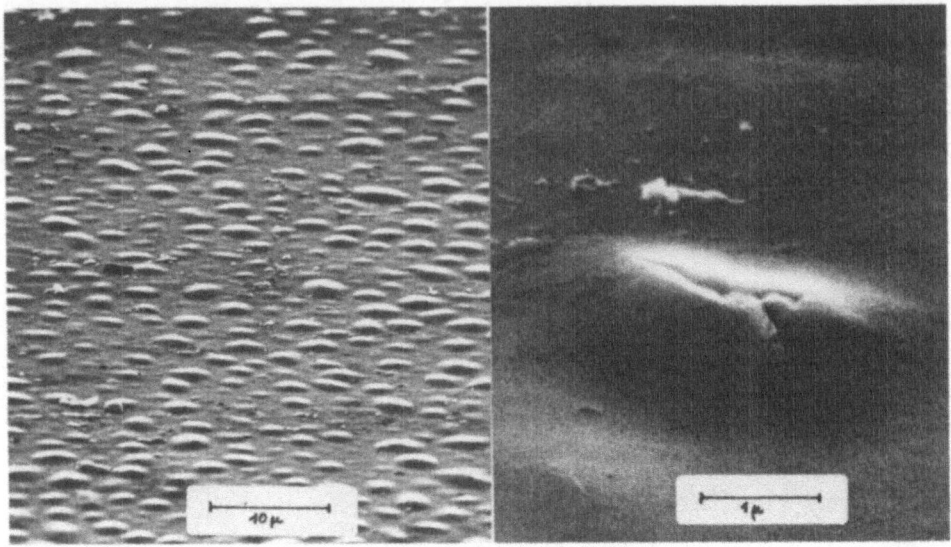

Fig. 9   SEM micrographs of Be irradiated by 15 keV D^+ ions to 10^{18} ions/cm^2. j = 1.2x10^{-4} A/cm, α = 70°.

CONCLUSIONS

For 15 keV He irradiation in the investigated materials stainless steel, Mo, and Be severe blistering was observed at fluences above $3x10^{17}$ ions/cm^2. Many of the blisters ruptured and lost their cover. Thus, a substantial erosion of the material by the blistering effect can be anticipated.

For 15 keV the phenomenon is similar for H^+ and D^+ bombardment. It is different and more complicated than with He bombardment. For stainless steel the hydrogen blisters occur mainly at grain boundaries and precipitations. In Mo the critical fluence decreases with increasing dose rate. Various shapes and sizes of the blisters depending on the crystallite orientation were observed. The blisters often show cracks but in no case a substantial loss of material could be observed.

Thus, it is doubtful, whether hydrogen blistering contributes considerably to the erosion of the bombarded surfaces. It may be, that if the blisters once are formed their covers may protect the underlying material from further implantation and blistering. To evaluate the degree to which hydrogen blistering contributes to the wall erosion in a fusion reactor, it is necessary, however, to irradiate more materials at higher temperatures to much higher fluences, since the particle flux on to the wall of a controlled thermonuclear reactor is estimated to be up to $10^{16} cm^{-2} s^{-1}$ /8b/. To this avail further experiments in our laboratory are prepared.

ACKNOWLEDGEMENTS

We are grateful to R. Behrisch for his continuous interest in this work. To him and to B. M. U. Scherzer, who performed the 150 keV implantations, thanks are due for numerous discussions. The targets were investigated with the scanning electron microscope in the laboratory of Dr. H. Klingele, München, Adelgundenstr. 8. We are indebted to him and B. Böhlken for the SEM micrographs. We thank also Mrs. G. Knoebel for preparing the targets, and R. Hippele and S. Schrapel for their skilful technical assistance.

REFERENCES

/1/   M. Kaminsky, Adv. Mass Spectr. 3, 69 (1963).
/2/   R. D. Daniels, J. Appl. Phys. 42, 417 (1971).
/3/   W. Bauer, G. I. Thomas, J. Nucl. Mat. 47, 241 (1973).
/4/   W. Primak, J. Luthra, J. Appl. Phys. 37, 2287 (1966).
/5/a   S. K. Das, M. Kaminsky, J. Appl. Phys. 44, 25 (1973).
   b   M. Kaminsky, S. K. Das, Rad. Eff. 8, 245 (1973).
   c   S. K. Das, M. Kaminsky, J. Appl. Phys. 44, 2520 (1973).
/6/   S. K. Erents, G. M. McCracken, Rad. Eff. 18, 191 (1973).
/7/   R. S. Blewer, paper presented at the Int. Conf. on Ion Surf. Interaction, Garching Sept. 72, Rad. Eff. in press.
/8/a   R. Behrisch, W. Heiland, Proc. 6th Symp. on Fusion Techn., Aachen 1970, EUR 4593e, p. 461.
   b   R. Behrisch, Nucl. Fusion 12, 691 (1972).
/9/   R. S. Barnes, D. J. Macey, Proc. Roy. Soc. 275, 47 (1963).
/10/   R. S. Nelson, Phil. Mag. 9, 343 (1964).
/11/   H. K. Birnbaum, C. A. West, Ber. Bunsen Gesellsch. 76, 806 (1972).
/12/   E. V. Kornelsen, Rad. Eff. 13, 227 (1972).
/13/   J. Roth, R. Behrisch, B. M. U. Scherzer (this Conference).
/14/a   W. Eckstein, H. Verbeek, Report IPP 9/7, June 1972 and Vacuum (in press).
   b   J. Vac. Sci. Techn. 9, 612 (1972).
/15/a   R. Behrisch, Vak. Technik 10, 250 (1967)
   b   B. M. U. Scherzer, Thesis TU München 1969, IPP Report 2 /80.
   c   R. Behrisch, B. M. U. Scherzer, H. Schulze, Rad. Eff. 13, 33 (1972).

/16/a M. E. Schiøtt, Mat. Fys. Medd. Dan. Vid. Selsk. 35, No. 9
       (1966)
    b M. E. Schiøtt, Rad. Eff. 6, 107 (1970).
/17/   W. Eckstein, B. M. U. Scherzer, H. Verbeek, Rad. Eff. 18,
       135 (1973).

DISCUSSION

Q: (M. Kaminsky) Do you know what the initial defect concentration was in your cold worked and annealed metal samples?

A: No.

Q: (T. D. Ryan) Can you explain the grain boundary visibility in your optical micrographs and the apparent rumpling adjacent to the grain boundary?

A: This is not due to the bombardment, it is also seen in the unbombarded area. The rumpling is coming from the differential interference contrast used with the optical microscope.

Q: (J. P. Biersack) From your graph depicting the critical fluence vs. current density, a critical gas concentration might be extracted. Did you analyze your data in this sense?

A: This is not possible because of the unknown diffusion coefficient for hydrogen in damaged molybdenum. It is well known that the diffusion coefficient, especially in bcc metals, is strongly influenced by radiation damage.

CHAPTER VIII

VOIDS AND IMPLANTATION SIMULATION OF NEUTRON DAMAGE

A REVIEW OF ION SIMULATION OF HIGH TEMPERATURE NEUTRON DAMAGE AND VOID FORMATION

G. L. Kulcinski

Nuclear Engineering Department, The University of

Wisconsin, Madison, Wisconsin 53706

Introduction

The use of beams of charged particles to simulate high temperature neutron damage is only four years old. This technique was introduced in 1969 by Nelson and Mazey (1) and it was clearly aimed at obtaining information on the behavior of cladding material for fast reactors. Since that time, over 60 papers have been published in the open literature and at conferences (2-6) in addition to countless internal company reports. The nature of the studies have ranged from purely theoretical to quite empirical with some countries actually using the experimental programs to guide the design of multi-million dollar fast reactors.

It is difficult to gain the proper perspective of a field of research which is so young. Part of this difficulty stems from the fact that some of the important information is not being released because of commercial considerations, and part of it comes from the fact that many of the programs are still identifying experimental problems, let alone solving them. Nevertheless, tremendous progress has been made in a relatively short time and the future of this technique is indeed quite bright.

The object of this paper will be to sketch a brief historical description of how the field evolved and to summarize the various types of studies that have been performed. The bulk of the paper will be devoted to highlighting recent successes of neutron damage simulation. In addition to listing the advantages of this technique, we will attempt to be quite candid about its many limitations. Finally, a few of the more promising future studies will be explored.

The Necessity and Validity of Such Studies

It was not long after the 1966 discovery that high temperature neutron irradiation of metals could cause significant
swelling (7), that designers of fast breeder reactors realized the
tremendous economic impact of that phenomena. The early expecta-
tion of the swelling in 316 stainless steel, the primary cladding
material for the U.S. LMFBR program, is shown in Figure 1. This
figure shows that swelling values of 2-15% were expected in the
FFTF reactor and 7-60% may be possible in commercial LMFBR's.
These values were put into persepctive by Huebotter and Bump (8)
who stated that if 5% swelling must be accomodated in all the fast
reactors built in the period 1970-2020, the present worth of a
solution was $864,000,000. It was also stated that if the swelling
amounted to 15%, then a 1970 present worth value was $5,600,000,000.
Obviously, there was a sufficient economic incentive to solve the
problem.

Unfortunately it appeared that it would take 5-10 years or
more to obtain confirming data for Figure 1 if fast test reactors
such as EBR-II and DFR were used. It was necessary to know the
answer sooner than that.

Not very much was known in the 1966-1969 period about how voids
formed.   However, scientists have now shown that even though
vacancies and interstitials are produced in equal numbers during
irradiation, more interstitials than vacancies are absorbed by the
dislocations in the solid. This imbalance leaves an excess of
vacancies in the lattice which builds up to a critical supersaturation such that the vacancies finally precipitate into voids.

Many authors, too numerous to list here but referred to in
(2-6) have developed theories to explain the swelling in reactor
components. We will use the ideas developed by Bullough and co-
workers (9-12). The main features of their theory are summarized
in the equation 1 which applies after a threshold damage of Kt_0
has been reached where K is the displacement rate in dpa sec^{-1} and
t_0 is the time required to reach the threshold damage state. The
per cent swelling due to the production of voids is then,

$$\frac{\Delta V}{V} (\%) = (Kt)SF \qquad\qquad (1)$$

where

S is a function of the material properties such as sink
density and size includes the dislocation density

and

F is mainly a function of the irradiation temperature,
the migration energy of the vacancies and has a weak,
complex dependence on K.

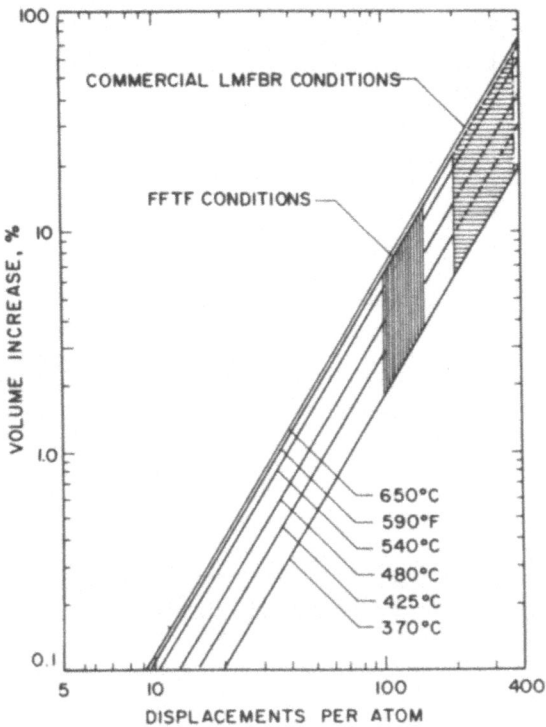

Figure 1.   Swelling Estimates of 20% Cold Worked 316 Stainless Steel
            FFTF Interim-Design Basis (August 1969) Reference 8.

Although this theory was only fully developed in the last few years,
Nelson and Mazey (1) recognized one of its key features   in 1969.
It is noted that once nucleation has been accomplished, the swelling
is proportional to the product of K and t and does not, to a first
approximation, depend on these two quantities separately.   This
means that if one wants to shorten the irradiation time, then it
is only necessary to raise K.   The early experiments used high
intensity beams of heavy ions whose displacement effectiveness per
particle is approximately 10^2-10^6 times that of equivalent neutron
fluxes   (Figure 2). When this displacement effectiveness was
multiplied by currently available beams of charged particles,
scientists were able to raise the displacement rates by factors of
approximately 10^3-10^4.   By using such high displacement rates, the
irradiation time required to reach a given damage state was lowered by the same amount and designers could then have reasonable
swelling information in a matter of days compared to the many years
required for neutron irradiation.

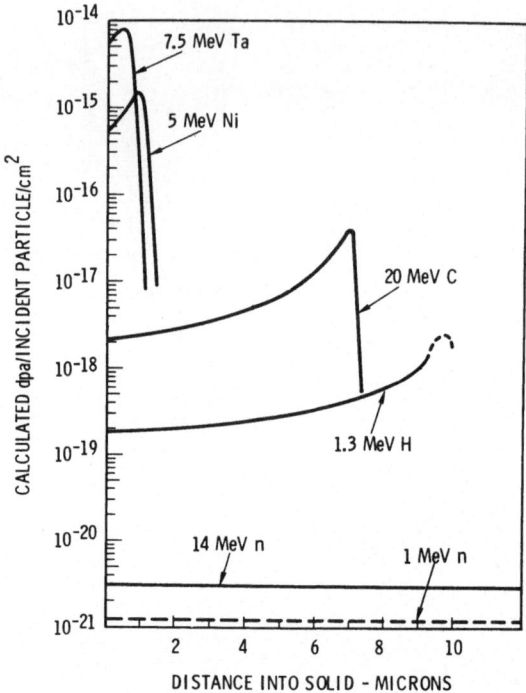

Figure 2.   Displacement Effectiveness for Various Energetic
             Particles in Nickel.

There is one correction that must be made because of the high
displacement rates and that is the effect of K on F in equation 1.
It was recognized by Bullough and Perrin (10) that the important
quantity that must stay constant in ion bombardment experiments is

$$\frac{K_1}{D_{v1}} = \frac{K_2}{D_{v2}} \tag{2}$$

where
             when K_1, K_2 are the defect production rates in two
             different irradiation environments.
             D_{v1}, D_{v2} are the diffusivities of vacancies during
             the experiments and proportional to exp − $(Q_m^v/kT_{1,2})$

This means that slight adjustments in the irradiation temperature
would have to be made to maintain the equality in equation 2.   The
calculated peak temperature shift, while perhaps too simplified
in terms of more recent theories but nevertheless still useful, is
shown in equation 3.

$$\frac{T_1 - T_2}{T_2} = \frac{\left[\dfrac{kT_2}{Q_m^v} \ln\left(\dfrac{K_1}{K_2}\right)\right]}{\left[1 - \dfrac{kT_2}{Q_m^v} \ln\left(\dfrac{K_1}{K_2}\right)\right]} \tag{3}$$

The significance of equation 3 is that in order to simulate damage at some displacement rate K_2 and temperature T_2 by irradiation at a higher displacement rate of K_1, one must irradiate at a slightly higher temperature T_1. We will see an example of this shift later in the paper but Figure 3 from Brailsford and Bullough (11) shows the effect of K on F for a factor of 10^3 difference in displacement rate.

Types of Simulation Studies Which Have Been Performed

An attempt has been made to summarize in Table 1 all of the simulation studies reported to date. Chronologically, the earliest studies were performed with relatively low energy (~100 keV) beams

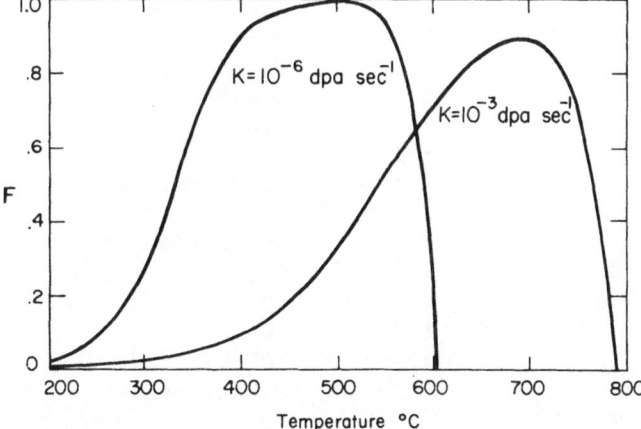

Figure 3. Effect of Displacement Rate on the Temperature Dependence of Void Swelling in Metals-Brailsford and Bullough (11)

of H, C, O, and Fe at Harwell (1). These were quickly replaced
with 20 MeV carbon ion studies (23-29) which were carried on for
1-2 years before switching to a 45-50 MeV Ni ion beam which is
used almost exclusively now. Investigators at Atomics International started soon after the initial Harwell studies with beams of
0.75 to 1.4 MeV protons (30-35). Their work has been almost
entirely directed toward the austenitic stainless steels. Scientists
at Battelle Northwest Laboratories were the first to use high energy
(5-11 MeV) heavy ion beams of Cu, and Ni on the austenitic stainless steels (35-39) but their work is now directed toward pure metals,

Table I
Summary of Continuing Ion Bombardment Studies
Directed at Void Formation

Laboratory	Ions	Ion Energy-MeV	Materials Studied	Reference
Harwell	H,C,O,Fe	0.1	316SS	1
	C	0.1	Co	14
	C,Cu,Ni	0.1	Ni,Cu,Ni-Cu	15
	Al	0.4	Al	16
	N	2	Mo,TZM	17-22
	C	20	316SS,Ni,PE16	23-29
	Ni	46	Ni,Ni alloys	29
AI	H	0.75-1.4	304SS,316SS,321SS	30-34
	"	"	Ni	35
	"	"	Ta	34
BNW	Cu	5	316SS	35-37
	Ni	5	316SS	38,39
	Se	6-11	Ni	40-42
	Ni	5-6	Ni,Nb,Mo,TZM	39,42-46
	Ta	7.5	Mo,Nb,V,TZM, Nb-1Zr,Ta	42-48
Saclay	Ni	0.1-0.5	Ni	49-51
	Cu	0.5	Cu	49,52
ANL	H	1.25	304SS	53
	Ni	3.5-4	304SS,316SS,321SS	53-55
	Ni	3.25-4	V,V alloys	56-57
	Ni	3	Nb	58
G. E.	D	12.3	304SS,321SS	59
	Ni	5	304SS,316SS,321SS	60-65
NRL	Ni	2.8	Ni,Fe,Fe alloys	66-67
Wisconsin	Al	2	Al	68
	Cu,Nb	15	Al,V	69

i.e. Ni, Nb, and Mo   (39-46). A 7.5 MeV beam of Ta ions was also
used to study refractory metals and alloys and currently the Ni and
Ta beams are used almost exclusively   (43-48). A group at Saclay,
headed by Adda, reported using 0.1-0.5 MeV beams of Ni and 0.5 MeV
beams of Cu on Ni and Cu respectively and this work is still
continuing   (49-52). Another group at ANL has reported the use of
1.25 MeV hydrogen (53) and 3.5-4 MeV beams of Ni on austenitic
stainless steels   (53-55). More recently the 3-4 MeV beams of Ni
ions has been used to study void production in V alloys (56,57) and
Nb (58).   Initial work at G. E. utilized a 12.3 MeV beam of deuterium
ions to bombard austenitic stainless steels (59) but a large amount
of work has recently been done with a 5 MeV beam of Ni ions from
the same accelerator that the BNW group used at High Voltage
Engineering Corporation.   (60-65) More recently a group at Naval
Research Laboratory has used a 2.8 MeV beam of nickel to study Ni,
Fe and Fe alloys (66,67). Work at the University of Wisconsin
has been started using a 2 MeV beam of Al (68) and a 15 MeV beam
of Cu and Nb ions to study pure Al and V (69).

Irradiation, Analysis and Calculation Techniques

The use of high energy charged particles from accelerators
to produce high damage states in materials has been treated previously (23,30,39). One drawback (or advantage depending on the
type of study) to ion simulation work is the necessity to supply
gases to the solid which are normally generated neutronically. Preinjection with helium to 1-100 ppm has been described by several
authors and the reader is referred to these references (59,70,71).
While it is common practice to preinject the samples with helium
before irradiation, such a practice must be closely studied,
especially at high helium content because it may influence the
nucleation at an early stage of irradiation. A much better technique
would be to inject simultaneously with helium from an auxilliary
accelerator.

Because of the limited range of the charged particles, the
analysis of the microstructure by transmission electron microscopy
(TEM) is often a time consuming task. Methods for precise removal
of damaged layers and subsequent analysis of the size and density
of voids have also been described in the literature (23,30,39,41,
72). An interesting technique which can measure overall swelling
without resorting to precise TEM analysis has been described by
Johnston et al., (60-62). Their technique relies on measuring the
difference in the step height of a surface which has been irradiated
as opposed to a surface which was shielded during irradiation. The
bombarded area tends to swell and protrude from the surface in relation to the shielded region (Figure 4).

Finally, the correlation between neutron irradiation work and
ion bombardment studies is made through the common unit of damage

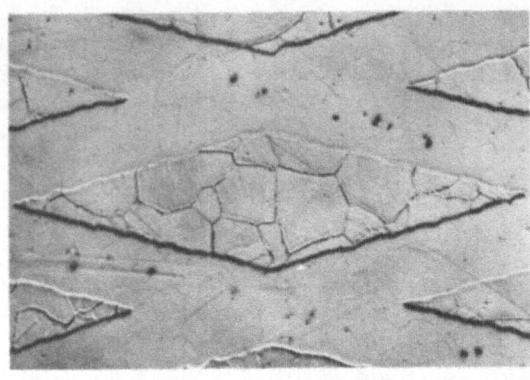

100μ

Figure 4.   Surface of a 304SS Specimen that was Covered with a
            Mask Containing Diamond Shaped Holes During 5 MeV
            nickel bombardment at 660°C. The bombarded regions have
            been elevated as a result of the swelling (Reference 61).

called the dpa, for displacements per atom. Methods for calculating
dpa values differ from group to group and standardization in this
field is sorely needed. Nevertheless the reader can obtain some
idea of how the damage is calculated in various ion bombarded samples
by reading references 19,23,54,62,73-75.

Some of the Noteable Successes of the Ion Bombardment Technique

       One of the earliest conclusions of the simulation studies had
to do with the role of gases in nucleating voids in metals. It was
known that helium atoms were generated in reactor components and it
was normally assumed, prior to 1969, that these gas atoms were the
nuclei for producing voids in metals. The simulation studies provided the first unambiguous test of this theory because the experimenter could choose to dope with or exclude helium from the sample
during the experiment. Early studies revealed that helium gas was
not required for void nucleation in steels, Ni, Mo, V, and Nb
(1,17,23,36,37,43). On the other hand if helium was present, it
could have a very dramatic effect on the nucleation as shown in
Figure 5. More recent studies by Johnston et al. (60,61,63-65)
have shown that in stainless steels, samples containing helium
gas swelled more than those that did not. In a somewhat unrelated
study, Loomis et al. (58) found that in the absence of helium,
void nucleation was inhibited by increasing amounts of oxygen
impurities in Nb.

       Another early success of this technique was the discovery of
the void superlattice in several pure materials; Al (16),Ni
(41,42,46),Nb (42,44-47),Mo (17-22, 46,48). It was found that
if pure metals are irradiated to a high damage state (>>1 dpa)

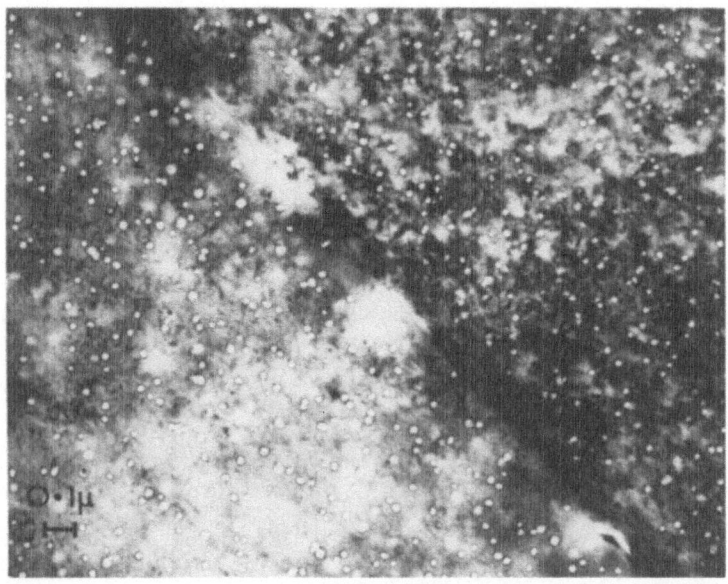

0.1 ppm He

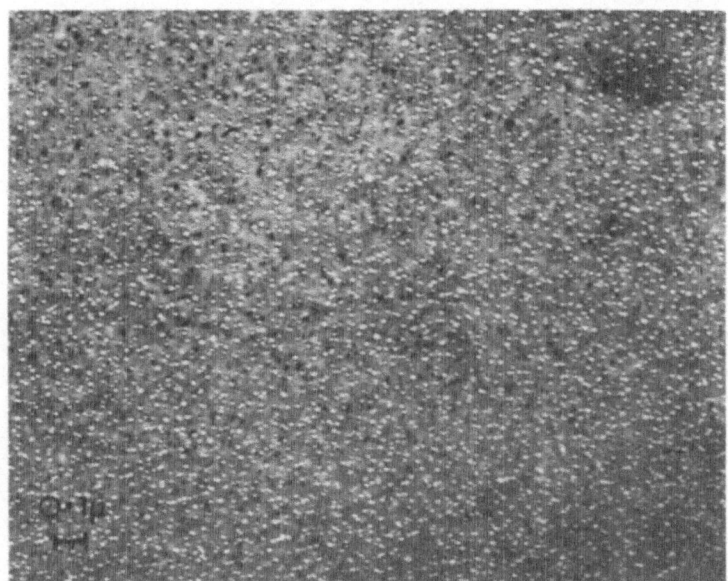

100 ppm He

Figure 5. Effect of helium on void nucleation in 316 SS at 550°C during 20 MeV carbon ion bombardment. Damage level is 15 dpa (23).

near the temperature threshold for void formation (~$0.3T_m$) that
the voids would form arrays consistent with the crystal structure
of the parent lattice. For example, voids in N^+, Ni^{++} or Ta^{++}
bombarded Nb and Mo formed a bcc array (Figures 6 and 7) and those
in Al and Ni formed fcc superlattices. This unique structure was
discovered at about the same time an equivalent structure was found
in neutron irradiated metals by Wiffen (76) and Erye (77). Since
none of the ion bombarded samples contained preinjected helium it
was clear that helium was not responsible for the observed effect.
It is also evident that the ordered void structure is not peculiar
to high damage rate processes because those formed during neutron
irradiation did so at a damage rate of 1/1000 to 1/10,000 that of

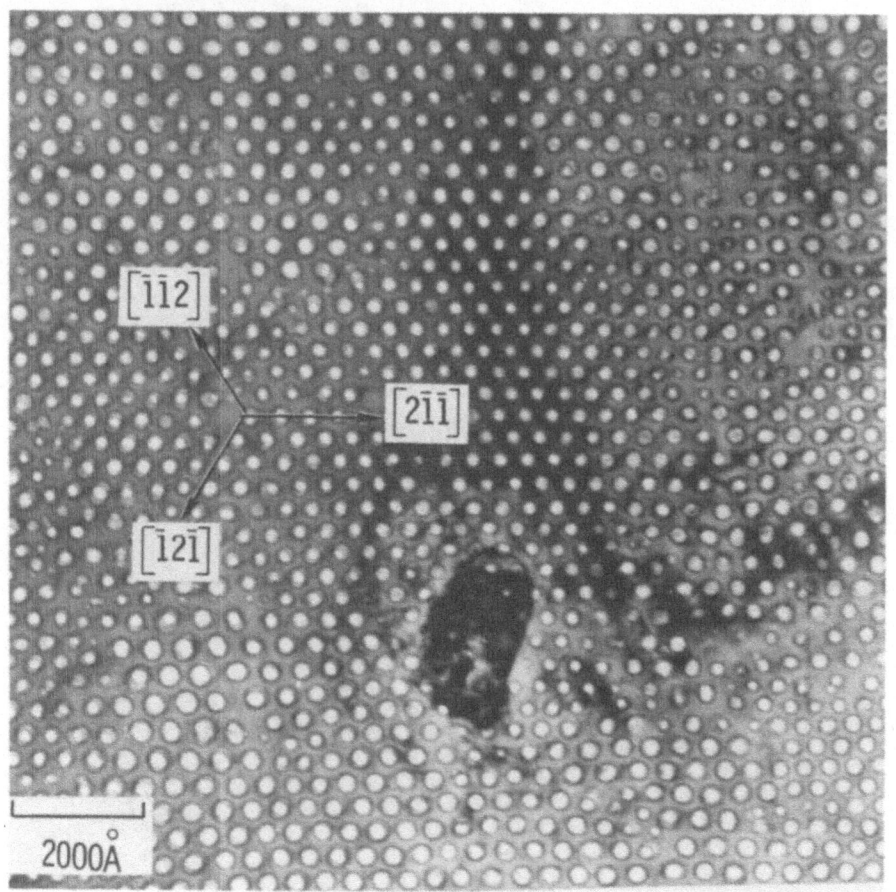

Figure 6.   Superlattice of voids in Nb bombarded at 800°C to 300 dpa
             with 7.5 MeV Ta ions. The superlattice is being viewed
             along the <111> direction and is bcc in nature (46).

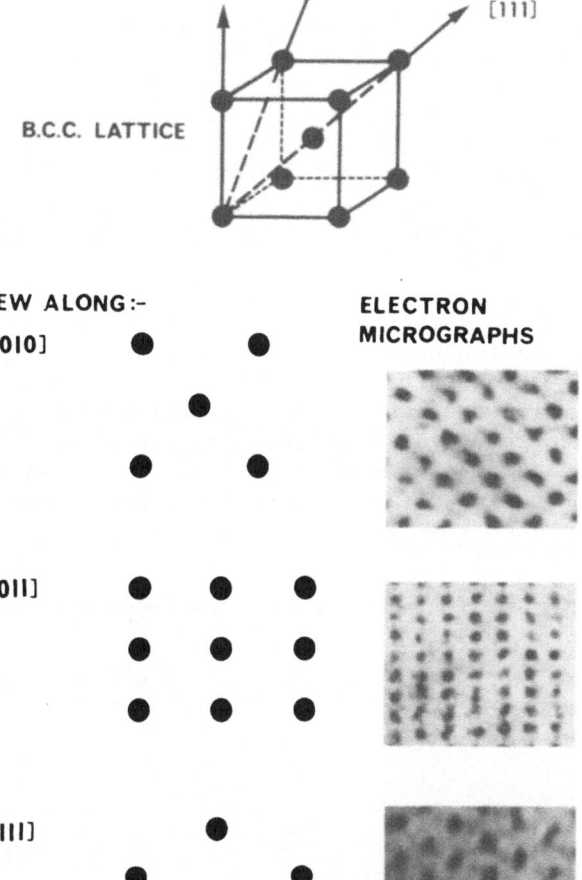

Figure 7. Superlattice of voids in Mo bombarded at 870 C to 100 dpa with 2 MeV nitrogen ions. The various projections of the bcc superlattice are shown (20).

the ion bombardment studies. There are several current theories
to explain the superlattice formation (78-81) but the exact situation is not clear at this time. Updated information from a recent
review article (46) on ordered structures is given in Table II.

Another area which ion bombardment studies have proved to be
extremely important is that of swelling saturation. Because of the
extremely high damage rate in ion bombardment experiments it was
quite easy to surpass the damage levels of the present and get a
preview of what might happen at extremely high damage states.
Early work by Harwell and BNW (25,41,42) showed a saturation in the
swelling behavior of Ni (Figure 8) and the Harwell work even
showed a saturation in swelling behavior of 316 stainless steel (25)
(Figure 9). Later work has shown a similar saturation in Nb(47)
(Figure 10).

The cause of the saturation is not clear at this time, but
the onset of saturation corresponds to the formation of a superlattice in the case of pure Ni and Nb. It is speculated that when
the density of voids reaches such a high level, the dislocations
are absorbed into the voids thus removing the bias for void formation. The close spacing of the voids also makes them the predominant sinks for <u>both</u> interstitials and vacancies thus prohibiting
void growth.

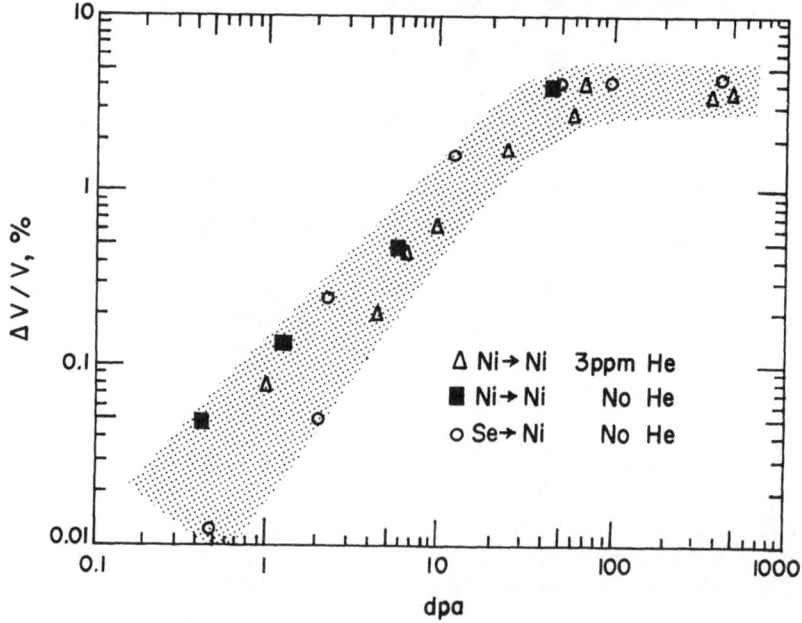

Figure 8. Saturation in swelling of Ni after heavy ion bombardment
           (42).

Table II
Summary of Data on Ordered Void Lattices

Material	Bombarding Species	Irradiation Temp. °C	Damage Level dpa	Super-lattice Structure	a_o Void Lattice Constant Å	$2r_v$ Ave. Void Size Å	$\frac{a_o}{r_v}$	Reference
Ni	5 MeV Ni	525	360	fcc	620	250	5.0	41,42
Ni	6 MeV Se	525	400	fcc	660	180	7.3	41
Al	400 KeV Al	50-75	40-80	fcc	600-800	100-140	8.5-12	16
Mo	neutrons	585	36	bcc	265	64	8.3	76
Mo	neutrons	790	36	bcc	328	72	9.1	76
Mo	neutrons	650	30-60	bcc	340	40	17	77
Mo	2 MeV N	870	100	bcc	220	40	11	20
Mo	7.5 MeV Ta	900	130	bcc	310	60	10	46
Mo (a)	7.5 MeV Ta	900	150	bcc	460	140	6.6	48
Mo-0.5 Ti	neutrons	585	36	bcc	215	69	6.1	76
Mo-0.5 Ti	neutrons	790	36	bcc	315	72	8.8	76
TZM	2 MeV N	870	400	bcc	220	60	7.3	20
Nb	7.5 MeV Ta	800	140	bcc	340	125	5.4	46
Nb	7.5 MeV Ta	800	290	bcc	380	110	6.9	46
Nb	7.5 MeV Ta	900	300	bcc	750	250	6.0	46
Nb	5 MeV Ni	800	5	bcc	350	45	15.6	42
Nb	neutrons	790	34	bcc	665	186	7.2	76
Ta	neutrons	585	20	bcc	205	61	6.7	76
W	neutrons	550	15	bcc	195	30	13	84
W	neutrons	800	15	bcc	250	40	12.5	84

a) single crystal

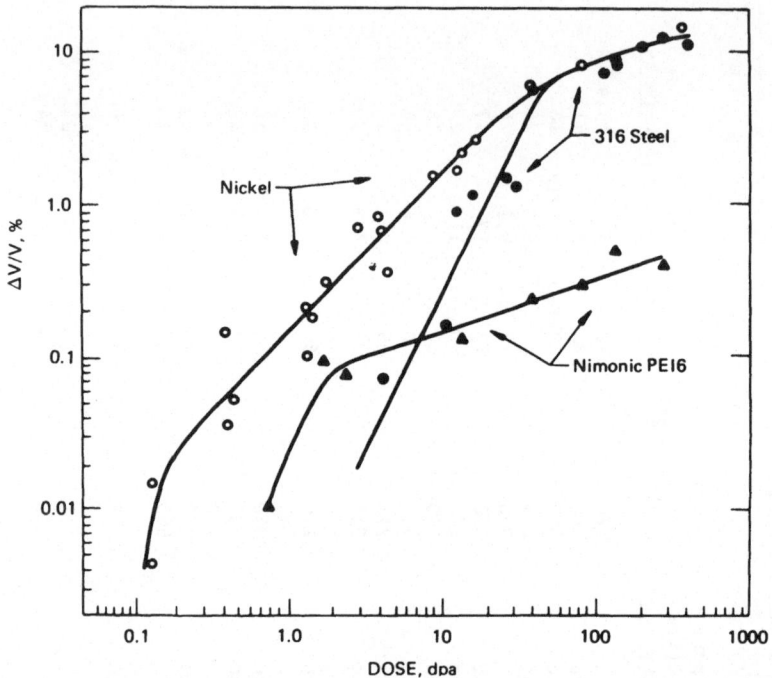

Figure 9. Swelling behavior of Ni, 316 SS and PE-16 during 525 C
            bombardment with 20 MeV carbon atoms. All materials
            contain 10 ppm He. (25)

This reasoning does not hold true for the saturation in 316
stainless steel because no void superlattice has been found in this
material. There has been a recent suggestion that the carbon atoms
which are producing the damage may build up to such a level as to
suppress the void formation. Recent work by Steigler et al. (82)
on very low carbon containing 316 stainless steel would tend to
support the carbon suppression theory. The discrepancy could be
solved by irradiating to very high fluences with self ions. Such
experiments have been performed by Johnston et al., (63-65) and the
ANL group (53-55). Neither of these latter groups finds such a
saturation   (Figure 11). The lack of a saturation in steel is also
supported by proton work at AI   (34). However, the possibility exists
that in the proton experiments hydrogen may be promoting swelling
before it diffuses from the sample. Finally, recent studies of 316
stainless steel by high voltage electron microscopy (83) (HVEM)
also show a continuing swelling with fluence. Considerable work is
needed here before one will be able to confirm or deny the saturation
phenomena in steel.

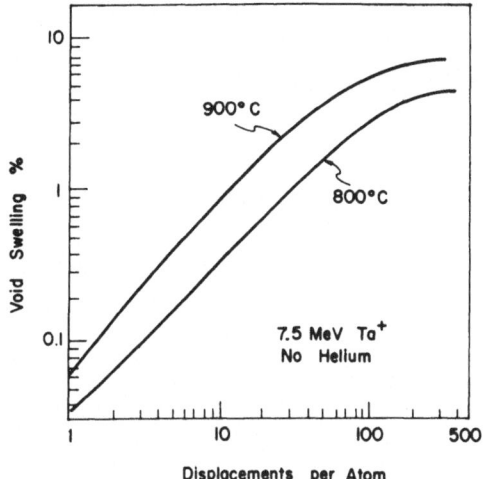

Figure 10. Saturation of swelling in Nb during 7.5 MeV Ta ion bombardment (47).

Probably the most important success of the simulation techni-que from an economic standpoint is the extension of present swelling information on fast reactor cladding. The four most active groups in this area are Harwell, G. E., AI and ANL and some of their results are shown in Figures 9,11-12. Harwell and G. E. have also used the simulation technique to screen various alloys with respect to their swelling behavior and some typical results are shown in Figure 12. The effects of thermomechanical treatments such as cold working 316 SS (64) (Figure 12) and formation of coherent precipi-tates in PE16(25) (Figure 9) have been shown to be quite beneficial in reducing the swelling behavior. The fact that one can study several alloys within the period of a few days at the same closely controlled conditions, means that large, costly, and time consuming screening programs with radioactive metals can be avoided.

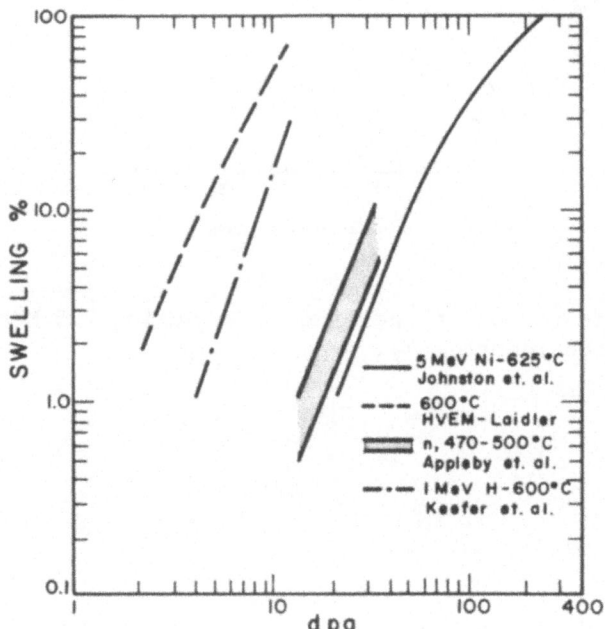

Figure 11.   Comparison of actual neutron induced swelling in
              ST 316 SS to that determined by H+, Ni+, and
              HVEM simulation studies. Note that the ion bombard-
              ment studies were conducted at a high enough
              temperature to overcome the temperature shift.
              The dpa values were corrected to recent recommendations
              by D. G. Doran for protons, electrons, and neutrons
              (47 dpa per 10^{23} n/cm^2, E > 0.1 MeV).

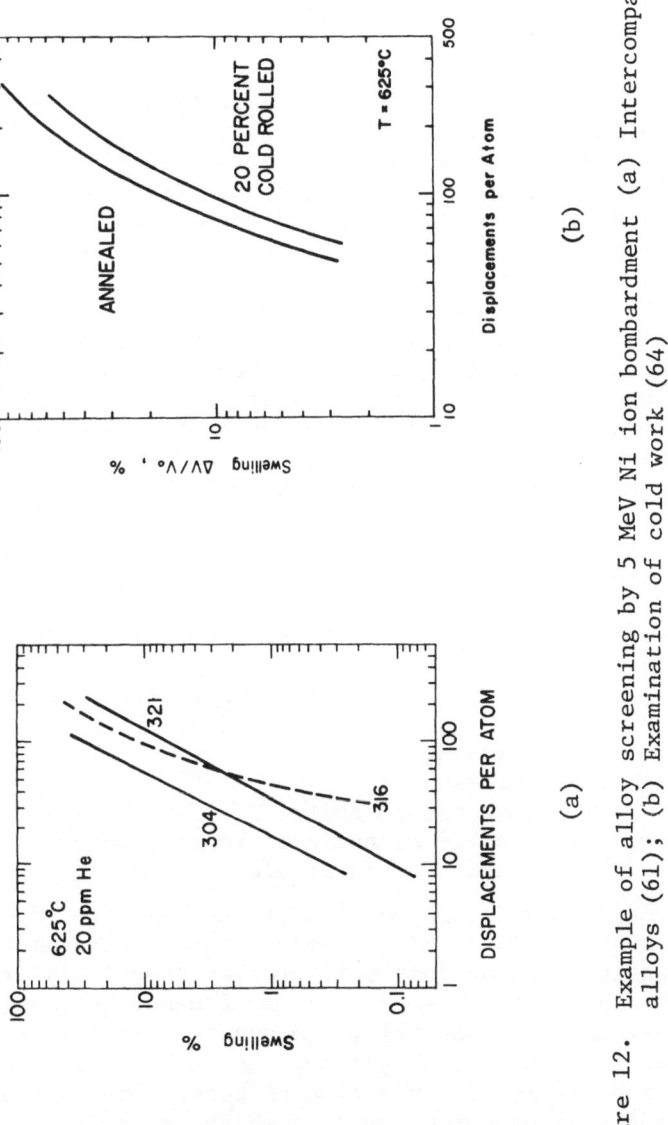

(a)

(b)

Figure 12. Example of alloy screening by 5 MeV Ni ion bombardment (a) Intercomparison of alloys (61); (b) Examination of cold work (64)

Once promising alloys are identified, concentrated study under reactor conditions can be pursued. Herein lies one of the greatest
promises of this simulation technique.

The group at Harwell has used this approach in coming up with
a more swelling resistant alloy, PE16, a nimonic nickel based alloy.
It can be seen from Figure 9 that this alloy swells almost a factor
of 30 less than solution treated 316 stainless steel at a 525°C
irradiation temperature.

There have been other discoveries in this field which deserve
mention here and will undoubtedly be investigated more fully in
the future. These include the early detection of a reversion of
the austenitic phase to ferrite at high damage states, which could
have severe implications if the trend continues (31). Several
groups (BNW, ANL, NRL, and Harwell) are investigating the effects
of alloying on void suppression for non LMFBR materials. These
studies could lead to a better fundamental understanding of the
void formation process in all metals. Finally, a group at NRL
has discovered that rastering the ion beam across the surface of
a sample can give results completely atypical of defocussed beam
conditions (66). The effect can be traced back to the fact that
even though the average dpa rate is the same for both samples,
it can be considerably higher instantaneously and locally in the
rastered sample. Void formations under extremely high dpa rates
can have technical significance in future fusion reactors as we
shall see later.

Limitations of the Ion Bombardment Technique

The major limitations that are known at this time fall into
5 categories:
            1. Alloying
            2. Stress
            3. Temperature shift
            4. Lack of Mechanical Property Data
            5. Excess Interstitials

We have already discussed some of the possible problems with
respect to a change in chemical composition. The experimenter
must constantly guard against this problem especially if he wants
to make any sense out of temperature or fluence dependence of the
void size or density. Normally, TEM analysis is performed at a
position far from the end of path of the incident ion in order to
minimize the possible contamination effects. One must be especially
careful about studies which use ions which are mobile during
irradiation because they could migrate from the point of deposition to areas of TEM analysis.

Another complicating feature of the ion studies is that because of the limited range of the atoms (of the order of a few microns at most) the majority of the sample is undamaged. When considerable swelling is induced, the undamaged part of the sample can impose large constraints on the expanding region. Such effects are quite evident in Figure 4 and the effect of stress on swelling must be understood before the magnitude of this problem can be assessed in ion bombarded samples.

We have already seen that one must raise the ion bombardment temperature with respect to that being simulated because of the rate effect. The higher temperature irradiation could give atypical results if this temperature causes microstructural changes in the material not seen at the temperature to be simulated. A good example of this effect can be seen in 20% cold worked 316 stainless steel. Straalsund and Brager (85) have shown that 100 hours at 650°C is sufficient to cause some recovery of the cold work in 316 stainless steel. Since the high dislocation content tends to suppress void formation, its removal during high temperature ion bombardment studies may give misleading results. A typical temperature shift for a displacement rate ratio of 10^3 in steel is approximately 100–150 C which means that IB studies would probably be limited to simulating temperatures less than 500–550°C during neutron irradiations. Similar reservations must be stated for alloys with complex precipitates that could be dissolved at high temperature.

The limited range of charged particles also means that the samples cannot be used for mechanical property measurements. Even the 45–50 MeV Ni ions used by the Harwell group have only a 5 micron range in steel (29). Perhaps the only chance for such measurements will be the high energy (>20 MeV) proton or deuteron beams such as those used at ANL (86).

One final concern that must be mentioned for self ion irradiation (i.e. Ni→Ni or SS) is the addition of excess interstitials to those already produced by irradiation. It has been speculated that if the ratio of injected atoms to displaced atoms is more than 1:100, the biasing effect of the dislocations could be overcome and the swelling suppressed. However, such an effect has not been seen in studies conducted to date.

Possible Directions For Future Simulation Studies

It is expected that the majority of simulation studies in the immediate future will be aimed at predicting fast reactor cladding behavior. Particular attention will be paid to the potential saturation effects and the screening of alloys for advanced reactor concepts.

Basic studies will probably concentrate on the effect of metallic impurities and gas atoms on the nucleation of voids in metals. These studies should be able to contribute significantly to the development of a complete theory of void swelling in metals.

The study of void superlattices should lead to a consistent theory for their formation. It is hoped that such theories may also be able to predict uses for ordered structures in such areas as superconductivity (i.e. flux pinning sites), high strength metals or for studying the interaction of dislocations with a regular array of defects. One interesting feature of this work is that such ordered arrays can be formed quickly, are quite stable thermally, and could be formed in metals without introducing radioactivity.

The advent of fusion reactor technology will certainly benefit from this simulation technique. Alloy screening studies will help narrow the potentially promising alloys to a few which can then be studied in depth when 14 MeV neutron test facilities become available. The generation of gas in such a hard neutron spectrum is truely a unique problem for fusion reactors as seen in Figure 13 where the ratio of the helium gas to dpa values are given for V in thermal and fast fission reactors as well as for fusion facilities (87). It can be seen that gas to dpa ratio in fusion reactors is 2 to 3 orders of magnitude higher than in fission reactors which means that it will not be possible to simply test potential fusion reactor materials in fission facilities. However, by coupling two

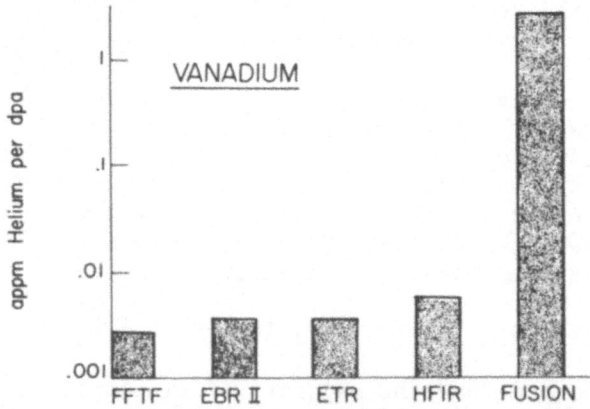

Figure 13. Ratio of the atomic parts per million of helium gas generated per dpa in vanadium for various nuclear facilities (80).

accelerators, one generating helium ions and the other generating heavy ions it will be easy to generate the proper ratios typical of any nuclear facility.

Finally, and perhaps one of the most promising future areas of study will be the simulation of extremely high damage rates typical of pulsed or laser fusion reactors (Figure 14). Steady state magnetic reactors will have damage rates comparable to current fission facilities ($\sim 10^{-6}$ dpa sec^{-1}) (87) but pulsed systems will have instantaneous rates of $\sim 10^{-5}$ to 10^{-4}/dpa sec^{-1} separated by 3 seconds of no damage at all (88). This is even more aggravated in laser systems where the damage rates are ~ 0.1 dpa sec^{-1} for approximately 1 microsecond. Such laser systems may be pulsed on the order of 10 times a second (89). It will be very risky to extrapolate the current radiation damage data over such a wide range of damage rates and much theoretical and experimental work is needed to assess this situation.

Conclusion

It has been shown that in the short period of four years the ion bombardment simulation technique has made notable discoveries

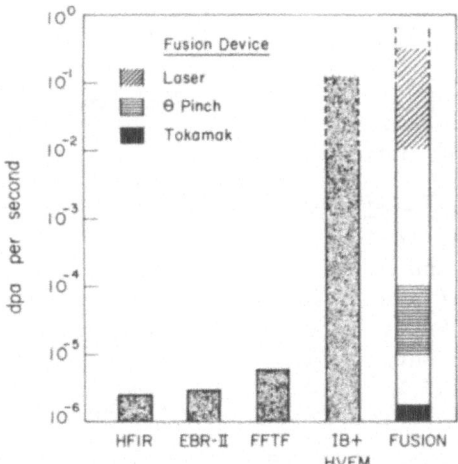

Figure 14.   Representative displacement rates in 316 SS for several nuclear facilities. Note that instantaneous dpa rates for fusion reactors can be quite large even though a time average value is 10^{-6} (87).

about the formation of voids in metals. The simulation studies
are now giving us a preview of what to expect after long term
high temperature neutron bombardment and it is a quick, easy, and
economical tool to screen new alloys. Future work of a basic
nature will undoubtedly contribute to void nucleation and growth
theories. It is possible that the simulation technique will be
even more valuable to the controlled thermonuclear reactor program
than it has been to the LMFBR program.

Acknowledgements

The author wishes to gratefully acknowledge support from the
United States Atomic Energy Commission under Contract AT(11-1)-2206.

REFERENCES

1) R. S. Nelson and D. J. Mazey, P. 157 in Proc. IAEA Symp.
    on Radiation Damage in Reactor Materials, Vol. 2, IAEA,
    Vienna, 1969.
2) Proc. of ASTM Conf. on Irradiation Effects on Structural
    Alloys for Nuclear Reactor Applications, Niagara Falls,
    N.Y., June 1970, ASTM-STP-484, 1971.
3) Proceedings of British Nuclear Energy Society European
    Conference, Voids Formed by Irradiation of Reactor Materials,
    Ed. by S. F. Pugh, M. H. Loretto, and D. I. R. Norris.
    Reading, England, March 24, 1971.
4) Proceedings of Int. Conf. on Radiation Induced Voids in
    Metals, ed. by J. W. Corbett and L. C. Ianniello, Albany,
    N.Y., June 9, 1971, CONF-710601, 1972.
5) Proceedings of ASTM Conf. on Effects of Radiation on
    Substructure and Mechanical Properties of Metals and Alloys,
    Los Angeles, Calif., June 1972, ASTM-STP-529, to be published.
6) Proceedings of the 1973 Int. Conf. on Defects and Defect
    Clusters in B.C.C. Metals and Their Alloys, ed. by R. J.
    Arsenault, National Bureau of Standards, Gaithersburg,
    Maryland, Aug. 14-16, 1973.
7) C. Cawthorne and E. J. Fulton, Nature, 216, 575, 1966.
8) P.R. Huebotter and T. R. Bump, P. 84 in Reference 4.
9) R. Bullough, B. L. Eyre, and R. C. Perrin, J. Nucl. Appl.
    and Tech., 9, 346, 1970.
10) R. Bullough and R. C. Perrin, P. 769 in Reference 4.
11) A. D. Brailsford and R. Bullough, J. Nucl. Mat., 44,
    121, 1972.
12) A. D. Brailsford and R. Bullough, P. 493 in Reference 6.
13) A. D. Brailsford and R. Bullough, A.E.R.E. Report T.P.
    527, 1973.

14) S. A. Manthorpe and S. N. Buckley, P. 239 in Reference 3.
15) D. J. Mazey and F. Menzinger, J. Nucl. Mat., 48, 15, 1973.
16) D. J. Mazey, S. Francis and J. A. Hudson, J. Nucl. Mat., 47, 137, 1973.
17) J. H. Evans, Nature, 229, 403, 1971.
18) J. H. Evans, Radiation Effects, 10, 55, 1971.
20) J. H. Evans, R. Bullough, and A. M. Stoneham, P. 522 in Reference 4.
21) J. H. Evans, A.E.R.E.-R 6733, 1972.
22) B. L.Eyre and J. H. Evans, in Reference 5.
23) R. S. Nelson, D. J. Mazey and J. A. Hudson, J. Nucl. Mat., 37, 1, 1970. See also R. S. Nelson, A.E.R.E.-6151, 1969.
24) D. J. Mazey, J. Nucl. Mat., 35, 55, 1970.
25) R. S. Nelson, J. A. Hudson and D. J. Mazey, P. 430 in Reference 4 and P. 191 in Reference 3.
26) G. P. Walters, P. 223 and 231 in Reference 3.
27) T. M. Williams, P. 205 in Reference 3.
28) J. A. Hudson, D. J. Mazey and R. S. Nelson, P. 213 in Reference 3.
29) J. A. Hudson, S. Francis, D. J. Mazey and R. S. Nelson in Reference 5.
30) D. W. Keefer, A. G. Pard, and D. Kramer, P. 332 in Reference 2.
31) D. W. Keefer, A. G. Pard, C. G. Rhodes and D. Kramer, J. Nucl. Mat., 39, 229, 1971.
32) D. W. Keefer, H. H. Neely, J. C. Robinson, A. G. Pard, and D. Kramer, P. 511 in Reference 4.
33) D. W. Keefer, A. G. Pard, and D. Kramer in Reference 5.
34) D. W. Keefer and A. G. Pard, J. Nucl. Mat., 47, 97, 1973.
35) H. H. Neely and K. Herschbach, Radiation Effects, 7, 187, 1971.
36) G. L. Kulcinski, H. R. Brager, and J. J. Laidler, P. 405 in Reference 2.
37) G. L. Kulcinski, J. J. Laidler and D. G. Doran, Radiation Effects, 7, 195, 1971.
38) H. R. Brager, H. E. Kissinger, and G. L. Kulcinski, Radiation Effects, 5, 281, 1970.
39) G. L. Kulcinski, A. B. Wittkower and G. Ryding, Nucl. Instrum. Methods, 94. 365, 1971.
40) G. L. Kulcinski, J. L. Brimhall and H. E. Kissinger, Trans. Am. Nucl. Soc., 13, 555, 1970.
41) G. L. Kulcinski, J. L. Brimhall, and H. E. Kissinger, J. Nucl. Mat., 40, 166, 1971.
42) G. L. Kulcinski, J. L. Brimhall, and H. E. Kissinger, P. 429 in Reference 4.

43) G. L. Kulcinski and J. L. Brimhall, Trans. Am. Nucl. Soc.,
    14, 604, 1971.
44) G. L. Kulcinski and J. L. Brimhall, P. 291 in Defects in
    Refractory Metals, Ed. by R. deBatist, J. Nihoul, and L.
    Stals, D/1972/0327/1, 1972.
45) G. L. Kulcinski, P. 184 in Proc. of Int. Working Session on
    Fusion Reactor Technology, CONF-710624, 1972.
46) G. L. Kulcinski and J. L. Brimhall, in Reference 5.
47) J. L. Brimhall and G. L. Kulcinski, to be published in
    Radation Effects.
48) J. L. Brimhall and E. P. Simonen, P. 321 in Reference 6.
49) Y. Adda, P. 31 in Reference 4.
50) C. W. Chen, A. Silvent, and G. Sanfort, J. Nucl. Mat.,
    46, 353, 1973.
51) C. W. Chen, A. Mastenbroek, and J. D. Elen, Radiation
    Effects, 6, 127, 1972.
52) L. G. Glowinski, C. Fiche, and M. Lott, J. Nucl. Mat., 47,
    295, 1973.
53) A. Taylor and S. G. McDonald, P. 499 in Reference 4.
54) L. K. Mansur, P. R. Okamoto, A. Taylor, and Che-Yu Li, P.
    509 in Reference 6.
55) S. G. McDonald and A. Taylor, in Reference 5.
56) A. T. Santhanam, A. Taylor, and S. D. Harkness, P. 302 in
    Reference 6.
57) A. T. Santhanam, A. Taylor, and S. D. Harkness, to be
    published.
58) B. A. Loomis, A. Taylor, T. E. Kilppert, and S. B. Gerber,
    P. 332 in Reference 6.
59) J. S. Armijo and T. Lauritzen, P. 479 in Reference 4.
60) W. G. Johnston, J. H. Rosolowski, A. M. Turkalo, and K. D.
    Challenger, Scripta Met. 6, 999, 1972.
61) W. G. Johnston, J. H. Roslowski, and A. M. Turkalo, in
    Reference 5.
62) W. G. Johnston, J. H. Rosolowski, A. M. Turkalo, and T.
    Lauritzen, J. Nucl. Mat. 46, 273, 1973.
63) W. G. Johnston, J. H. Rosolowski, A. M. Turkalo, and T.
    Lauritzen, J. Nucl. Mat., 47, 155, 1973.
64) W. G. Johnston, J. H. Rosolowski, and A. M. Turkalo, J.
    Nucl. Mat., 47, 155, 1973.
65) W. G. Johnston, J. H. Rosolowski, A. M. Turkalo, and T.
    Lauritzen, to be published J. Nucl. Mat.
66) J. A. Sprague, J. E. Westmorland, F. A. Smidt Jr., and P. R.
    Malmberg, Trans. Amer. Nucl. Soc., 16, 71, 1973.
67) F. A. Smidt Jr., J. A. Sprague, J. E. Westmoreland and P. R.
    Malmberg, P. 341 in Reference 6.
68) M. L. Sundquist, to be published.

69) G. L. Kulcinski, R. G. Lott, H. V. Smith Jr., and R. A. Dodd to be published.
70) D. Kramer, H. R. Brager, C. G. Rhodes, and A. G. Pard, J. Nucl. Mat., 25, 121, 1968.
71) J. H. Worth, A.E.R.E. -R-5704, 1968.
72) R. A. Spurling and C. G. Rhodes, J. Nucl. Mat., 44, 341, 1972.
73) D. G. Doran and G. L. Kulcinski, Radiation Effects, 9, 283, 1971.
74) D. K. Brice, Radiation Effects, 11, 51, 1971.
75) I. Manning and G. P. Mueller, to be published Radiation Effects.
76) F. W. Wiffen, P. 386 in Reference 4.
77) B. L. Eyre, P. 323 in Reference 3.
78) A. M. Stoneham, J. of Physics, F, Metal Physics, 1, 778,1971.
79) K. Malén and R. Bullough, P. 109 in Reference 3.
80) A. J. E. Foreman, A.E.R.E.-R7135, 1972.
81) V. K. Tewary and R. Bullough, UK A.E.R.E.-TP-479, 1972.
82) J. O. Steigler, J. M. Leitnaker, and E. E. Bloom, Int. Conf. on Physical Metallurgy of Reactor Fuel Elements, Berkely Nuclear Laboratories, Sept. 2, 1973 to be published.
83) J. J. Laidler, Ibid.
84) J. Moteff in Reference 5.
85) J. L. Straalsund and H. R. Brager, Amer. Nucl. Soc. Trans., 15, 251, 1972.
86) S. D. Harkness, AML-7688,7705,7726 and 7776, 1970.
87) G. L. Kulcinski, D. G. Doran and M. A. Abdou, Univ. of Wisconsin Fusion Design Memo, UWFDM-15, May 1972.
88) K. I. Thomassen and R. A. Krakowski, LA-UR-73-1365, 1973.
89) J. Williams, Engineering System Studies Review Meeting; Germantown, Md., Sept. 19, 1973.

ION RADIATION DAMAGE[*]

O. S. Oen, J. Narayan, and T. S. Noggle

Solid State Division, Oak Ridge National Laboratory

Oak Ridge, Tennessee 37830

ABSTRACT

The depth distribution of damage energy deposited in solids by energetic ions has been calculated and results are compared with the experimental damage found in copper which has been irradiated with 10^{16} 1-MeV protons at ambient temperature. The general form of the experimental damage profile from the transmission electron microscope measurements agrees well with that calculated.

INTRODUCTION

Charged particle accelerators are now being used for the rapid simulation of the radiation damage in metals which occurs in nuclear reactors and leads to swelling at high fluences. The relation of the ion damage to equivalent reactor neutron damage is dependent on theoretical treatments of the slowing down of energetic (MeV) particles in solids. The theory is largely untested by experiment, particularly with respect to the nuclear collision component of the stopping which produces the atomic displacements that generate the damage in metals. The work reported here represents initial results of a combined theoretical and experimental program to obtain information on the relation between the nuclear collisions and damage retained in a solid. In the first part of this work a calculation of the depth distribution of damage will be presented. In the second part some of these calculations will be compared with the damage observed in copper which has been irradiated with 10^{16} 1-MeV

---
[*]Research sponsored by the U. S. Atomic Energy Commission under contract with Union Carbide Corporation.

protons per cm^2 at ambient temperatures. Several papers calculating the depth distribution of damage have appeared recently.[1-3] The
treatment presented here is based largely on the slowing down of
energetic ions developed by Lindhard et al.[4,5]

SLOWING DOWN OF ENERGETIC IONS

Energetic ions slow down in solids through collisions with
lattice atoms and these collisions in general are inelastic in that
electrons are excited simultaneously with the deflection of the projectile. Following the work of Lindhard et al.[4] (LSS) the collisions may be separated into two parts: 1) elastic nuclear scattering of the projectile by the target atoms, and 2) electronic energy
loss due to electron excitation and ionization. Both components
play roles in the slowing down of the projectile, but it is only the
nuclear scattering which leads to displacement damage in metals.

LSS described the differential nuclear scattering cross section in numerical form using a Thomas-Fermi interaction potential.
An analytical approximation of their results found by Winterbon et
al.[6] is

$$d\sigma = \frac{\pi a^2}{2} \lambda \, t^{-4/3} \, [1 + (2 \, \lambda \, t^{2/3})^{2/3}]^{-3/2} \, dt \tag{1}$$

with $\lambda = 1.309$, $t = \varepsilon^2 \sin^2\theta/2$, θ = the center of mass scattering
angle. ε (reduced energy) and a (screening length) are defined as

$$a = 0.468(Z_1^{2/3}+Z_2^{2/3})^{-1/2} (\text{Å}) \quad , \quad \varepsilon = EM_2 a[(M_1+M_2)Z_1 Z_2 e^2]^{-1} = E/E_L .$$

Here Z_1, M_1, Z_2, M_2 are the atomic and mass numbers of the projectile ion and target atom respectively, e is the electronic charge
and E is the energy of the projectile ion. For large values of t
Eq. 1 reduces to Rutherford scattering. The nuclear stopping power,
(average nuclear energy loss per unit path length) may be explicitly
determined for the above scattering law. In reduced units it is

$$\left.\frac{d\varepsilon}{d\rho}\right)_n = \frac{N \, R_L}{E_L} \int_{\gamma^2}^{\varepsilon^2} \frac{T_m t}{\varepsilon^2} \frac{d\sigma}{dt} \, dt = \frac{9}{8\varepsilon}\left\{ \ln\left[\frac{(2\lambda)^{1/3}\varepsilon^{4/9} + \sqrt{1+(2\lambda)^{2/3}\varepsilon^{8/9}}}{(2\lambda)^{1/3}\gamma^{4/9} + \sqrt{1+(2\lambda)^{2/3}\gamma^{8/9}}} \right] \right.$$
$$\left. + \frac{(2\lambda)^{1/3}\gamma^{4/9}}{\sqrt{1+(2\lambda)^{2/3}\gamma^{8/9}}} - \frac{(2\lambda)^{1/3}\varepsilon^{4/9}}{\sqrt{1+(2\lambda)^{2/3}\varepsilon^{8/9}}} \right\} \tag{2}$$

where $\gamma = \varepsilon(T_L/T_m)^{1/2}$, T_m is the maximum energy transferrable to a
lattice atom and T_L is a lower cutoff in the transferred energy which
may be zero. The reduced length, $\rho = RN\pi a^2 4M_1 M_2 (M_1+M_2)^{-2} = R/R_L$.
Here N is the number density of atoms and R is the path length of
the projectile in units consistent with those of N and a.

Nuclear stopping dominates the slowing down process when $\varepsilon \ll 1$. At higher energies electronic stopping becomes more important and eventually at medium and high energies dominates the slowing down. The electronic stopping is not as well understood as the nuclear elastic stopping. For low energies the electronic stopping cross section is an increasing function of the energy of the projectile. The LSS theory predicts the average electronic stopping as

$$\left.\frac{d\varepsilon}{d\rho}\right)_e = k \ \varepsilon^{1/2} \quad , \quad \varepsilon \overset{\sim}{<} 10^2 \tag{3}$$

where k is a constant depending on Z_1, Z_2, M_1 and M_2. Values of k typically range from 0.1 to 0.2 unless $Z_1 \ll Z_2$ when k can be greater than unity. The above theory is in fair agreement with experiment although in some cases the predicted values of k may differ up to 30 to 40% from experimental values.[7] Northcliffe and Schilling[8] have made an extensive compilation of electronic stopping powers using theoretical guidelines to correlate existing data and to predict unmeasured stopping powers. Their results indicate that the LSS theory predicts a value of k which is 20% too low for the case of aluminum slowing down in aluminum, and for low energy protons slowing down in medium and heavy elements it may predict k values up to 50% too low.

ION RANGES

Knowing the electronic and nuclear stopping it is straightforward to calculate the average total path of an incident bombarding ion. It is given by

$$\bar{\rho} = \int_0^{\varepsilon_0} \left[\left.\frac{d\varepsilon}{d\rho}\right)_n + \left.\frac{d\varepsilon}{d\rho}\right)_e \right]^{-1} d\varepsilon \tag{4}$$

The above formula is not quite precise since the average path length per unit energy loss is not precisely the reciprocal of the average energy loss per unit path length,[9] because the slowing down involves statistical fluctuations in the energy loss. The penetration depth of an ion (projection of the total path along the incident direction of motion) is usually of greater experimental interest than the total path. The differences between these range quantities may be considerable, especially when the mass ratio $A = M_2/M_1$ is large and the energy small. The penetration depth is more difficult to calculate than the total path. LSS and Lewis[9] have developed methods to calculate the penetration depth. Following the method of Lewis the average penetration depth to first order is

$$\bar{\rho}_d = \int_0^{\bar{\rho}} <\cos \phi> d\rho \tag{5}$$

where $<\cos \phi>$ is the mean scattering angle given by

$$\langle\cos\phi\rangle = \exp\left\{-\left[\int_{o}^{\rho}\frac{d\rho}{\lambda_{\tau r}}\right]\right\}\text{with }\lambda_{\tau r}^{-1} = R_{L}N\int_{o}^{\epsilon^{2}}(1-\cos\phi)\frac{d\sigma}{dt}\,dt\ . \quad (6)$$

It may be mentioned that whereas the average total path as a function of energy depends on a single parameter k (for an electronic stopping power proportional to velocity) the average penetration depth depends on the mass ratio, A, as well as on k. Figure 1 shows the ratio of penetration depth to total path using the methods of LSS and of Lewis for the case of equal masses. For this case the two methods differ at most about 10% which occurs at the lowest energy. In principal the LSS method is more accurate as it is not restricted to the continuous energy loss approximation as is the method of Lewis and that of Eq. 4. However the method of Lewis will be used here as it lends itself more readily to the calculations which follow and because the results differ little at the relatively high energies of interest here. The single Monte Carlo[10] point at ϵ = 1.0 included only nuclear stopping and is expected therefore to be somewhat too low.

An important quantity in radiation damage studies using ions is the rate of nuclear energy loss as a function of the penetration depth. In terms of previously defined quantities it is given by

$$\frac{d\epsilon}{d\rho_{d}} = \frac{1}{\langle\cos\phi\rangle}\left.\frac{d\epsilon}{d\rho}\right)_{n}\ . \quad (7)$$

Equation 7 is illustrated as a dashed line in Fig. 2 (together with Eq. 2, solid line) for the case where the incident ion has a initial reduced energy of 173.5 and is of the same atomic species as that of the target. It may be noted from comparing the two curves that most of the deflection of the projectile ion occurs near the end of its range.

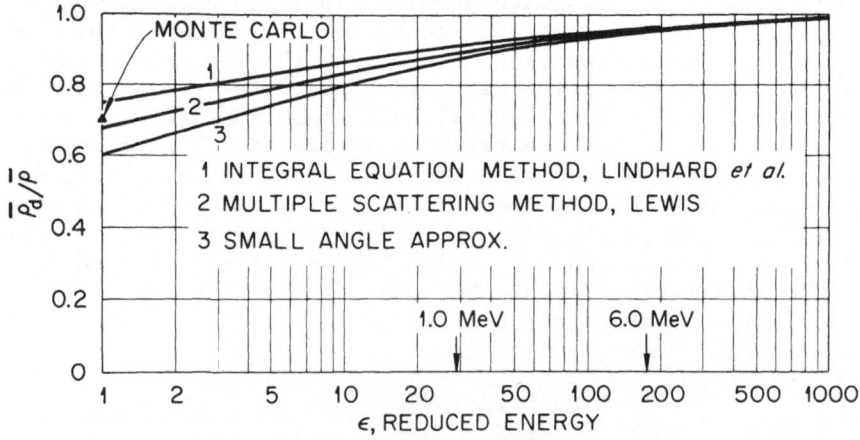

FIG. 1. Ratio of penetration depth to total path versus energy,
         κ = 0.17, A = 1.0. The 1.0 and 6.0 MeV points are for Al.

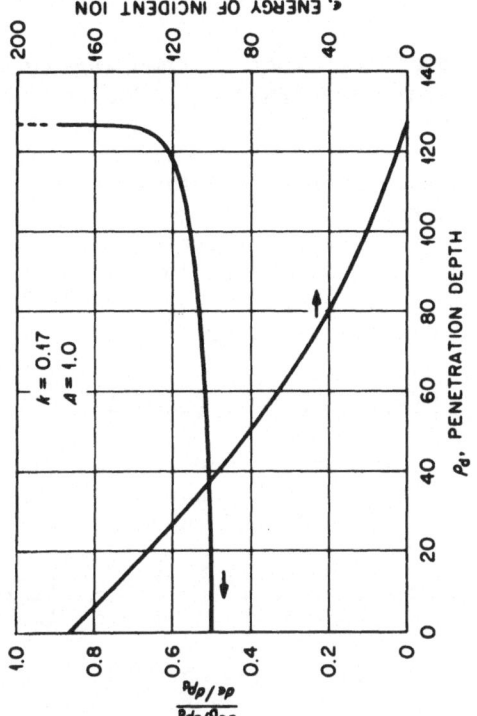

FIG. 3. Ratio of "damage energy stopping" to nuclear stopping vs. penetration depth.

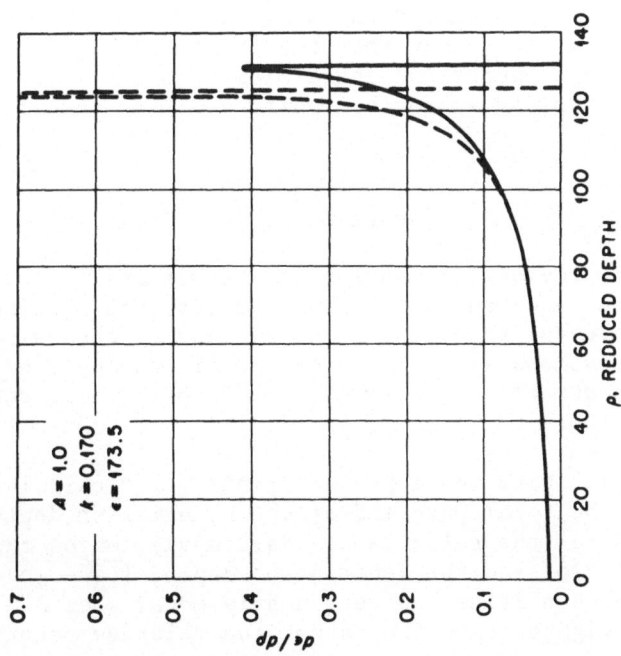

FIG. 2. Nuclear stopping vs. depth.

SPECIFIC DAMAGE ENERGY

Although the projected nuclear stopping given by Eq. 7 is a measure of the spatial dependence of radiation damage, a better measure is the specific damage energy deposited, especially if the primary knock-on energy is much greater than the displacement threshold energy. The damage energy is defined as that portion of the kinetic energy transferred to the target atoms that is available for producing displacements after correcting for the electronic losses of all the displaced atoms. The following expression which is due to Robinson[11] based on the work of Lindhard et al.[5] gives the fraction of energy going into damage from a recoil atom of energy T

$$\frac{T_D}{T} = [1 + k(3.4008 \ \varepsilon^{1/6} + 0.40244 \ \varepsilon^{3/4} + \varepsilon)]^{-1} . \qquad (8)$$

Here T_D is the damage energy and the reduced energy ε corresponds to a lattice atom slowing down in a monatomic lattice. The mean damage energy per unit penetration depth deposited by an incident ion is

$$\frac{d\varepsilon_D}{d\rho_d} = \frac{N}{E_L} \frac{R_L}{<\cos \phi>} \int_o^{\varepsilon^2} \frac{T_D}{T} \frac{T_m t}{\varepsilon^2} \frac{d\sigma}{dt} \ dt . \qquad (9)$$

In the above expression the target material is restricted to monatomic materials, although the bombarding ion may be different than the target. The ratio of the "damage stopping" (Eq. 9) to the projected nuclear stopping (Eq. 7) as a function of ion penetration depth is shown in Fig. 3. In this example the initial energy of the ion is 173.5 reduced units and of the same atomic species as the target. From the figure it is seen that the fraction of nuclear stopping going into damage energy increases slowly at the beginning from a value of 0.5 and rapidly approaches unity near the end of the range.

STRAGGLING

In all of the previous discussion it was assumed that monoenergetic ions have a unique range whereas in reality the slowing down is a statistical process resulting in straggling of the various range quantities. Fluctuations in both the electronic and nuclear energy loss contribute to straggling; however, the former do not seem to be important for $\varepsilon \lesssim 100$.

Lindhard et al.[4] have developed a formalism for calculating the higher moments of the total path and of the penetration depth. Comparison of results for the relative standard deviation of the total path, $[\overline{\rho^2} - (\bar{\rho})^2]^{1/2}/\bar{\rho}$, and the penetration depth, $[\overline{\rho_d^2} - (\bar{\rho}_d)^2]^{1/2}/\bar{\rho}_d$, shows that the two quantities are very nearly equal when $A \lesssim 1$. When the projectile is lighter than the target the relative penetration straggling increases, whereas the relative straggling in the total path decreases. For the latter it has been shown[4] that $\overline{\rho^2} - (\overline{\rho})^2 = \frac{4A}{(1 + A)^2} f(k,\varepsilon)$. No such simple formula exists for the variance in the penetration depth. For high ion energies and where straggling in the electronic stopping may be neglected, Winterbon[12] has shown that the distribution of penetration depths is Gaussian. Using this result the average damage energy deposited per unit penetration depth is approximately

$$\overline{\frac{d\varepsilon_D}{d\rho_d}} (\rho_d) = \int_0^\infty \frac{d\varepsilon_D}{d\rho_d} (\rho_d'-\rho_d) \frac{1}{\sqrt{2\pi}\ \alpha} \exp - \frac{(\rho_d'-\overline{\rho}_d)^2}{2\alpha^2}\ d\rho_d' \quad , \tag{10}$$

where α $\sqrt{\overline{\rho_d^2}-(\overline{\rho}_d)^2}$. Figure 4 shows as an example the specific damage energy (Eq. 10) for the equal mass case. The density of Frenkel pairs is obtained   by multiplying the specific damage energy (Eq. 10) by $\kappa/2E_d$ where E_d is the displacement threshold energy and κ is a constant asymptotically independent[13] of energy with a value of about 0.8.

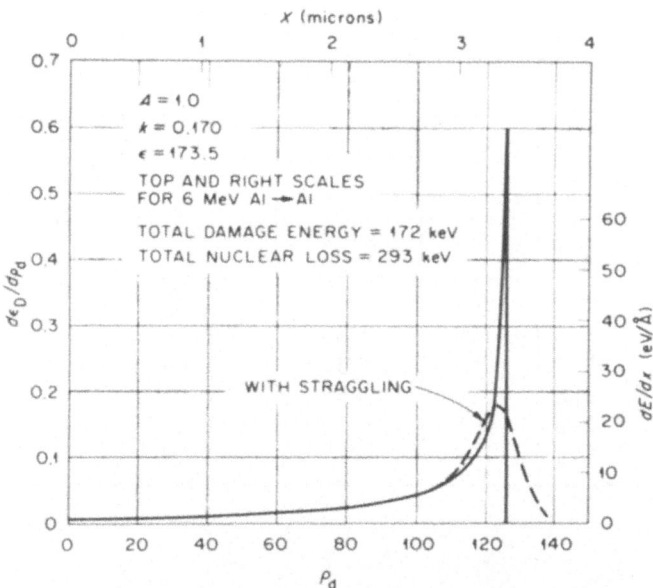

FIG. 4. Mean specific damage energy vs. penetration depth.

In the preceding discussion the depth distribution of damage energy was calculated assuming that energy transport by the recoiling atoms could be neglected. This assumption should be good when the recoil atoms have ranges small compared to that of the irradiating ion. Brice[2] has calculated the spatial distribution of damage using a different approach than the present one, but has also made the above assumption. Sigmund et al.[1] have included the energy transport of the recoiling atoms in their energy depth distribution calculations, but have limited their treatment to the case where the bombarding ion is of the same species as the target. For the equal mass case the latter authors concluded that the energy transport can be neglected for reduced energies, $\varepsilon > 10$.

It has been assumed in the preceding treatment that the number of defects is proportional to the damage energy which is a basic assumption in current displacement damage theory.[13,14] However it should be mentioned that computer simulation studies[15,16] of defect production which allow for close pair recombination find that the number of defects produced increases less than linearly with the damage energy. In view of this finding a note of caution should be exercised when using the damage energy as a measure of defect production.

In the experimental work to be discussed in the next section, copper was bombarded with 1 MeV protons and some calculations related to that system are presented in the following table.

E' (MeV)	R (μm)	X (μm)	$<\cos\phi>$	$N \int_{E_d}^{T_m} T\,d\sigma$ (eV/Å)	$\dfrac{2E_d \int_{E_d}^{2E_d} d\sigma + \int_{2E_d}^{T_m} T\,d\sigma}{\int_{E_d}^{T_m} T\,d\sigma}$	F	$N \int_0^T T\,d\sigma$ (eV/Å)
1	0.0	0.0	1.0	.0055	1.04	.87	.0085
.59	3.3	3.3	.98	.0086	1.04	.87	.0133
.29	5.1	5.1	.96	.0152	1.04	.88	.0231
.15	5.9	5.8	.93	.0258	1.05	.88	.0387
.029	6.5	6.3	.75	.0820	1.06	.89	.122
.0059	6.6	6.4	.27	.177	1.08	.90	.282

Here E' is the residual energy of the proton at a penetration depth, X. The fifth column gives the nuclear energy loss above a displacement threshold energy, $E_d = 25$ eV. The sixth column is the ratio

of the number of displacements using the Kinchin and Pease model to
that computed from column five. The seventh column gives the ratio
of specific damage energy to nuclear stopping, (Eq. 9)/(Eq. 7). The
ratio of the mean penetration depth for 1 MeV protons in copper to
the mean total path calculated by the methods of Lindhard and Lewis
agree to 1% and for the latter is 0.95. The electronic stopping used
in these calculations were taken from Northcliffe and Schilling[8]
(proton energies below 0.1 MeV) and from the compilations of Janni[18]
(above 0.1 MeV). For these relatively energetic protons most of the
range straggling is caused by fluctuations in the electronic energy
losses. The straggling parameter, α, is estimated at 0.22 microns
using the tables of Sternheimer[17] for the electronic contribution
and adding a correction due to nuclear straggling.

EXPERIMENTAL

The experimental part of this program to date has been directed
to the study of copper irradiated with 1 MeV protons. Copper single
crystals (10 mm square by 2 mm thick) after irradiation are electro-
plated with $\sim$ 0.1 μm of brass followed by copper plating to build up
a deposit approximately equal to the original thickness. The crystal
plus deposit is then sectioned perpendicular to the original surface
to give approximately 4 mm square by 0.5 mm thick plates suitable
for thinning into electron microscope specimens. The brass layer
provides a sharply defined interface in the electron microscope such
that the position of the damage can be measured relative to the surface to better than .02 μm.

Figure 5 shows results obtained with this technique. The copper
crystal was irradiated with 10^{16} 1-MeV protons/cm^2 at ambient temperature incident in a random direction. The micrograph shows directly
the large variation in the amount of damage generated along the range
of the ions and the histogram in the figure gives the measured densities of point defects in clusters in 0.5 μm intervals along the range
of the ions. Comparison is made with theoretical calculations as
given by the smooth curve.

The range of the ions as indicated by the cutoff in the defect
clusters and the general form of the experimental damage distribution are in quite good agreement with the theoretical results. Calculations without straggling gives poorer agreement with the experimental results. It should be kept in mind that the apparent quantitative agreement as to the amount of damage reflects the adjust-
ment of the verticle scale lengths and that if the damage energy is
converted to displaced atoms using reasonable values for the displacement threshold energy, one finds that the experiment is detecting only about 1% of the calculated number of displaced atoms.

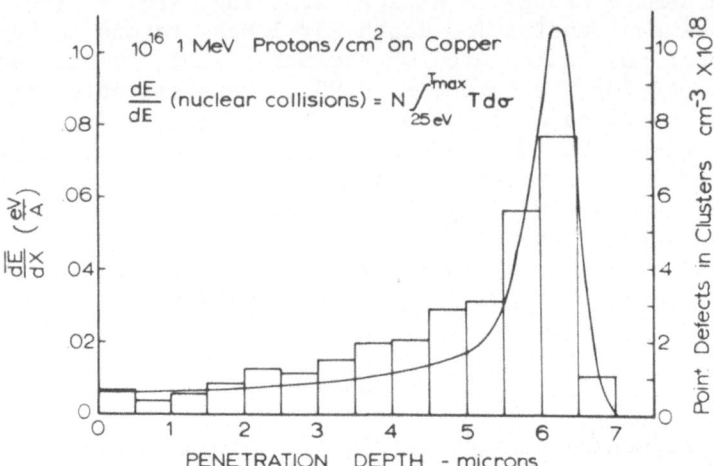

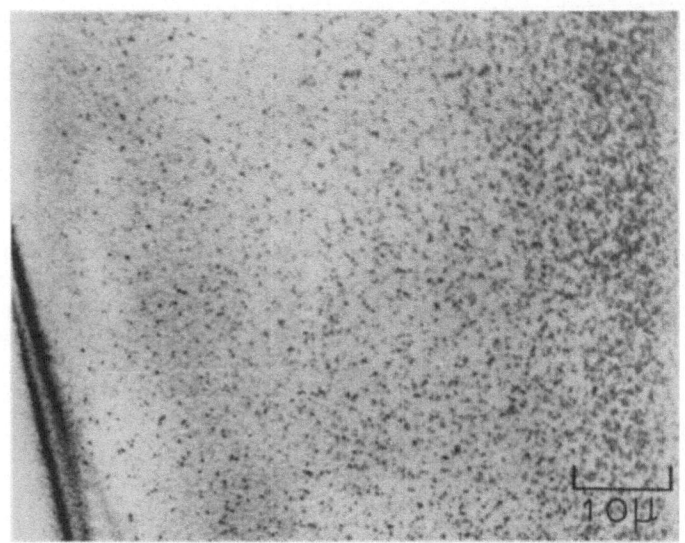

FIG. 5.   Depth distribution of defects in copper irradiated at room
temperature with protons. Micrograph shows the variation
of spot damage with penetration depth and histogram above
shows the density of point defects in clusters to same
penetration depth scale as micrograph. Smooth curve is
calculated nuclear stopping power including straggling.

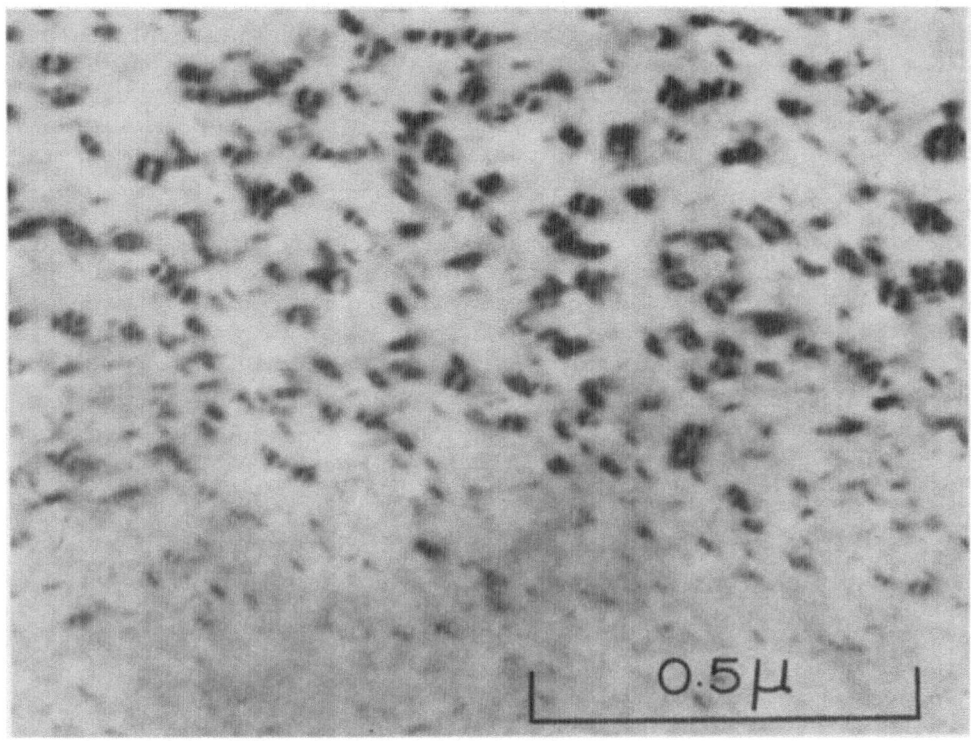

FIG. 6.   Defect clusters near end of range of 1-MeV protons in
          copper. Parenthesis appearance is characteristic of inclined
          dislocation loops. Loops are interstitial in character.

     As can be seen in Fig. 6 the defect structures are dislocation
loops which upon analysis proved to be of interstitial character.
This contrasts strongly with the mixture of vacancy and interstitial
type dislocation loops seen in copper irradiated with fast neutrons
at room temperature. The low retention of defects ($\sim$ 1%) also contrasts strongly with results obtained in fast neutron irradiated
copper in which approximately 10% of the Frenkel pairs produced are
retained for ambient temperature irradiations and are observed in
the electron microscope in the form of defect clusters. The reason
for this difference in the structures and the retention of damage
between the proton and fast neutron irradiations has not yet been
established; however, it seems likely that it is related, in part
at least, to differences in the collision spectra, since the nuclear
scattering of the protons is peaked at small energy transfers. This
in turn leads to the situation that a relatively large fraction of
the damage energy appears as more or less isolated Frenkel pairs
along the path of the ion, with only occasional displacement cascades
due to energetic collisions of the type present in fast neutron irra-

diation. The displacement cascades produced by energetic collisions
are known to reduce the fraction of the damage which anneals compared
to the annealing of isolated Frenkel pairs.[19] The absence of visible
vacancy clusters may also be related to this aspect of the damage
since vacancies produced more or less uniformly in the material which
survive do not have sufficient mobility at the irradiation temperature to move far enough for significant clustering to occur. In the
displacement cascades from fast neutron damage the clustering of
vacancies is enhanced by the local high concentrations which require
only limited mobility in order for clustering to occur. It can be
anticipated that there will be dose and dose rate effects which will
affect the damage structures and studies are in progress to evaluate
these effects. Other variables which will be investigated are irradiation temperature and ion species.

REFERENCES

1. P. Sigmund, M. T. Mathies and D. L. Phillips, Rad. Effects 11, 39 (1971).
2. D. K. Brice, Rad. Effects 11, 227 (1971); ibid., 11, 51 (1971).
3. G. L. Kulcinski, J. J. Laidler and D. G. Doran, Rad. Effects 7, 195 (1971). D. G. Doran and G. L. Kulcinski, ibid., 9, 283 (1971).
4. J. Lindhard, M. Scharff, and H. E. Schiott, Kgl. Danske Videnskab. Selskab, Mat.-fys. Medd. 33, No. 14 (1963).
5. J. Lindhard, V. Nielsen, M. Scharff, and P. V. Thomsen, Kgl. Danske Videnskab. Selskab, Mat.-fys. Medd. 33, No. 10 (1963).
6. K. B. Winterbon, P. Sigmund, and J. B. Sanders Kgl. Danske Videnskab. Selscab, Mat.-fys. Medd. 37, No. 14 (1970).
7. H. E. Schiott and P. V. Thomsen, Rad. Effects 14, 39 (1972).
8. L. C. Northcliffe and R. F. Schilling, Nuc. Data, A7: 233 (1970).
9. H. W. Lewis, Phys. Rev. 85, 20 (1952); ibid., 78, 526 (1950).
10. O. S. Oen and M. T. Robinson, J. Appl. Phys. 35, 2515 (1964).
11. M. T. Robinson, in Nuclear Fusion Reactors, p. 364, British Nuclear Energy Society, London, 1970.
12. K. B. Winterbon, Rad. Effects 15, 73 (1972).
13. M. T. Robinson, in Radiation Induced Voids in Metals, edited by J. W. Corbett and L. C. Ianniello (U.S.A.E.C., CONF-710601, 1972) p. 397.
14. M. J. Norgett, M. T. Robinson, and I. M. Torrens, to be pub- lished.
15. J. R. Beeler, Jr., Phys. Rev. 150, 470 (1966).
16. M. T. Robinson and I. M. Torrens, to be published.
17. R. M. Sternheimer, Phys. Rev. 117, 485 (1960).
18. J. F. Janni, AFWL-TR-65-150 (AD-643837) p. 184 (1966).
19. W. Schilling, G. Burger, K. Isebeck and H. Wenzl, in Vacancies and Interstitials in Metals, edited by A. Seeger, D. Schumacher, W. Schilling, and J. Diehl, North Holland Pub. Co., Amsterdam, p. 255 (1970).

4 MeV IRON ATOM BOMBARDMENT OF IRON

J. R. Beeler, Jr.* and M. F. Beeler**

*Materials Engr. and Nuclear Engineering Depts.

**Engr. Research Dept., North Carolina State University

1. INTRODUCTION

Four MeV iron atom bombardment of fcc iron was studied using the computer experiment approach. Entire 4 MeV collision cascades were simulated by computing, in sequence, each atom-atom collision in a cascade on an individual basis. Firsov's model was used to compute the inelastic energy loss for each collision, and the Erginsoy-Vineyard potential was used as the basis for computing the atom kinetic energy changes during collision.

The two main reasons for making this study were to estimate the amount and extent of damage energy transport in 4 MeV cascades in iron and to compare the results of the Firsov and Lindhard prescriptions for inelastic losses.

2. COMPUTATIONAL METHOD

The 4 MeV collision cascade computer experiments were performed using the COLLIDE Program. COLLIDE simulates atomic collision development on the basis of the branching sequence of binary collisions approximation.[1] In this approximation, it is assumed that an energetic atom (projectile) strikes a stationary target atom in each of its collisions. Whenever a target atom receives an amount of kinetic energy, T, greater than the displacement energy, E_d, it is assumed to be displaced from its normal site in the solid. Each displaced atom subsequently participates in the cascade development as a projectile atom as long as its kinetic energy exceeds E_d. An extensive list of projectile atoms is generated rapidly in

a cascade because, at first, each collision tends to yield two atoms
sufficiently energetic to be classed as projectiles. COLLIDE continues to compute atomic collisions, one at a time, until no more
projectiles remain that are sufficiently energetic to maintain a
collision trajectory. The particular projectile selected each time
for a collision calculation is the one in the projectile list with
the largest velocity.

The kinetic energy transferred from a projectile to its target
is computed according to classical collision mechanics. The collision is determined by an assumed interatomic potential, and the collision angle and collision time integrals are evaluated by Gaussian
quadratures. In the present work, the Erginsoy-Vineyard potential[2]
for iron was used. Inelastic energy loss during a collision is computed on the basis of Firsov's model[3] as modified by Torrens and
Robinson[4]. We will call this particular atom collision model the
Firsov-Erginsoy-Vineyard (FEV) model. We worked closely with Manning and Mueller[5] in comparing the results given by the FEV model
with those given by the Lindhard-Scharff-Schiott (LSS) model[6].
Their E-DEP-1 program computes energy deposition in a cascade on
the basis of LSS theory.

3. RESULTS

The damage energy fraction, $\eta(E)$, is defined as the fraction of
initial kinetic energy possessed by an energetic atom which contributes to displacement production in the collision cascade it initiates. Equivalently, it is the fraction of the initial kinetic energy which is not dissipated in inelastic events during the cascade
evolution.

The damage energy fraction, $\eta_F(E)$, given by the FEV model was
computed by running ten cascades at each of the following energies:
100, 200, 300, 400, 600, 800, 8000, 10^5, 10^6, and 4×10^6 eV. The
average damage energy fraction, $\eta_F(E)$, is listed in Table I and
plotted in Fig. 1. The damage energy fraction, $\eta_L(E)$, given by LSS
theory is also plotted and listed along with the ratio η_L/η_F. The

TABLE I. Selected Damage Energy Fraction Values for Iron.

PKA Energy in keV	$\eta_L(E)$	$\eta_F(E)$	η_L/η_F
0.1	0.871	0.978	0.891
1.0	0.814	0.950	0.857
10.0	0.742	0.900	0.825
100.0	0.618	0.800	0.773
1000.0	0.351	0.533	0.659
4000.0	0.162	0.219	0.738

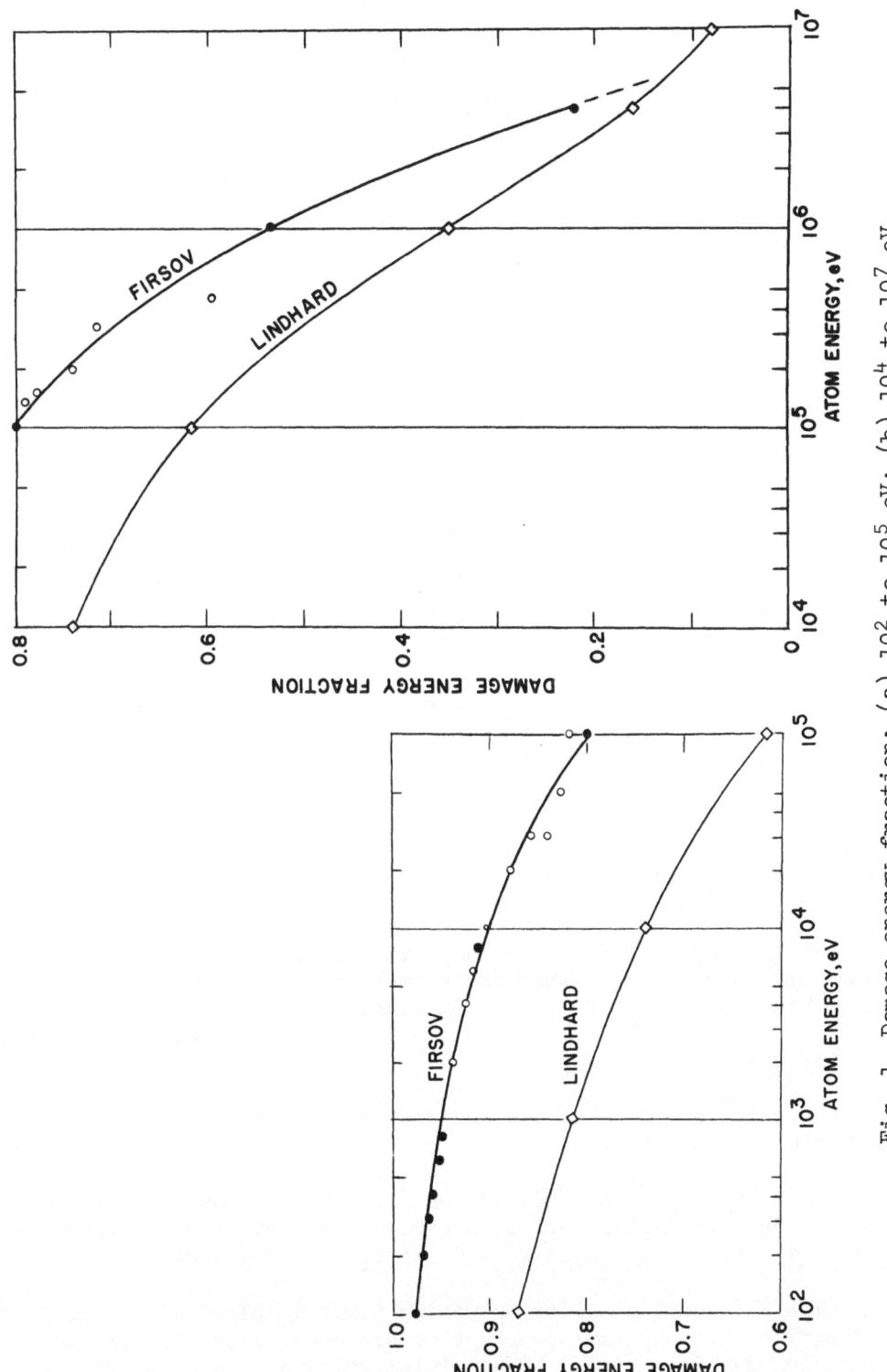

Fig. 1 Damage energy fraction: (a) 10^2 to 10^5 eV; (b) 10^4 to 10^7 eV.

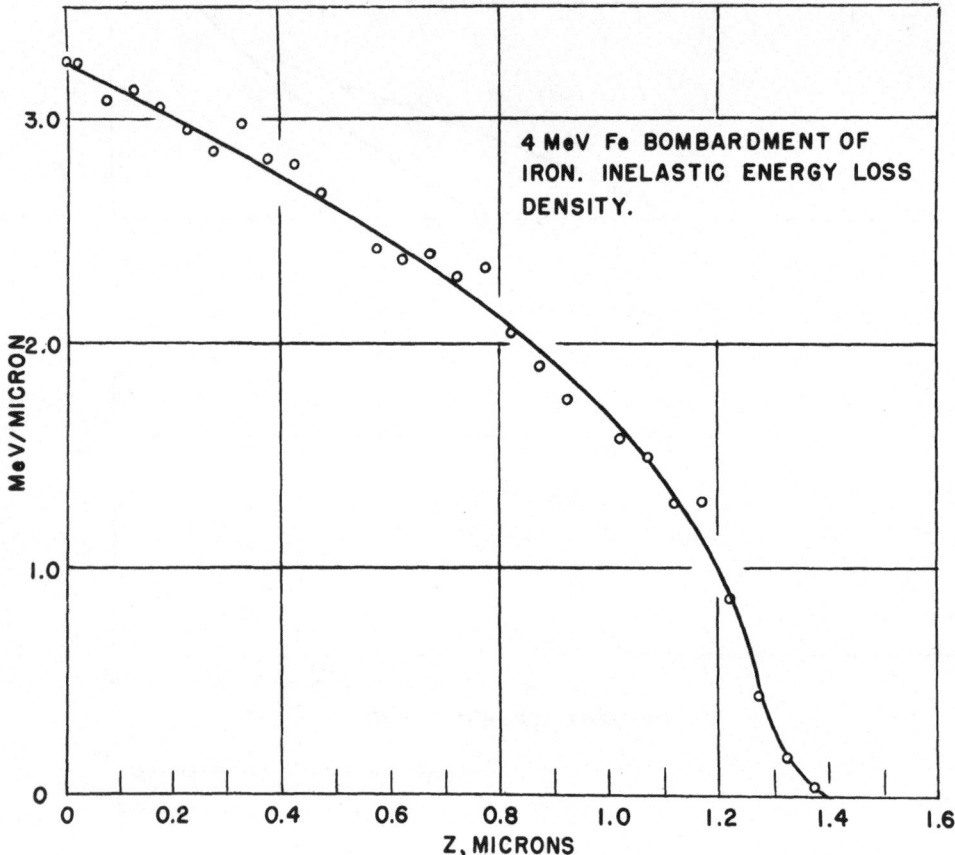

Fig. 2 Inelastic stopping power for an iron atom with initial energy of 4 MeV as a function of normal penetration distance in iron.

smaller inelastic loss per cascade, in the case of the FEV model,
leads to a longer mean range and a smaller amount of range straggling than is predicted by the LSS model. The LSS model gives

$$\eta_L = 1/(1 + kg(\varepsilon)) \tag{1}$$

where $\varepsilon = (E, eV)/(174143 \ eV)$ and $g(\varepsilon) = 32.9258$ for 4 MeV iron
atom bombardment of iron.

Computer experiment simulation gives $\eta_F = 0.219$ for a 4 MeV
cascade in iron. This implies $k = 0.108$ from Eq. (1). However,
a priori, LSS theory gives $k_L = 0.156$ for 4 MeV cascades in iron.

The FEV model inelastic stopping power is plotted in Fig. 2 as
a function of the normal penetration distance, z, of the primary
radiation particle (PRP). The initial value at z = 0 is propor-

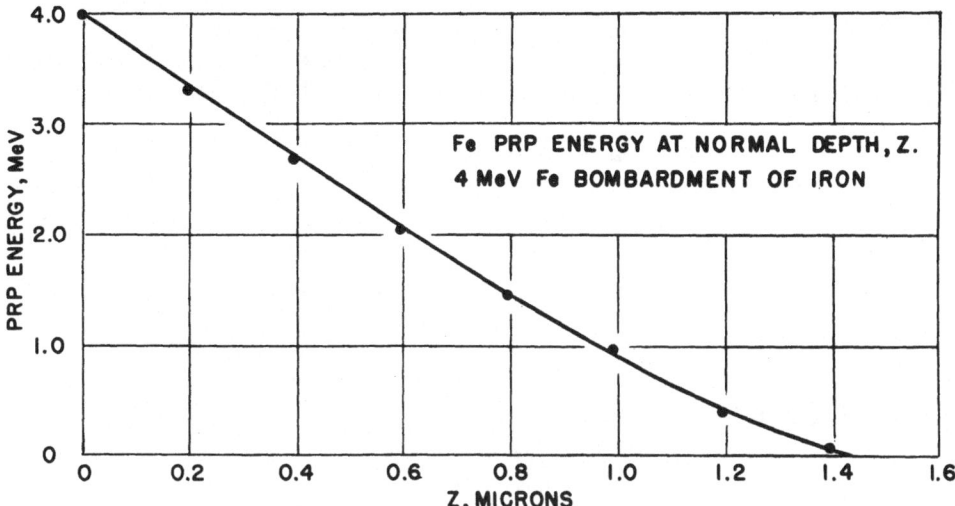

Fig. 3 Energy of a 4 MeV iron atom PRP as a function of the normal
penetration distance, E_z, in iron as given by FEV model computer
experiments.

tional to the square root of the PRP initial energy. The proportionality constant, k, is 0.11 on the basis of the FEV model inelastic stopping power.

The PRP energy, E_z, at normal penetration depth, z, is plotted
in Fig. 3. Manning and Mueller obtain the same E_z curve when they
assign k = 0.108 in their LSS theory E-DEP-1 program. In addition,
they also obtain the same mean PRP range, 1.2 μ, as the FEV model
gives in computer experiment simulation.

The average number of displacement events per cascade is always proportional to the average damage energy per cascade in a
sharp-cut-off displacement energy model simulation. In the simulations concerned here, the sharp-cut-off displacement energy was
E_d = 25 eV. COLLIDE counts a displacement event whenever the energies of both the projectile and the target atom after collision
exceed E_d. It also records the position at which each displacement event occurs. On this basis, COLLIDE constructs a spatial
distribution of the vacancy (displacement) number density. In the
simulation concerned here, the average damage energy per vacancy
was 53 eV independently of the energy of the atom which initiated
the cascade. Because the damage energy per vacancy is a constant,
for all cascades, the true damage energy density distribution,
$S_V(z)$, can be obtained by multiplying the vacancy number density
distribution by 53 eV/vacancy. This true damage energy density
distribution provides a reference for assaying the accuracy of a

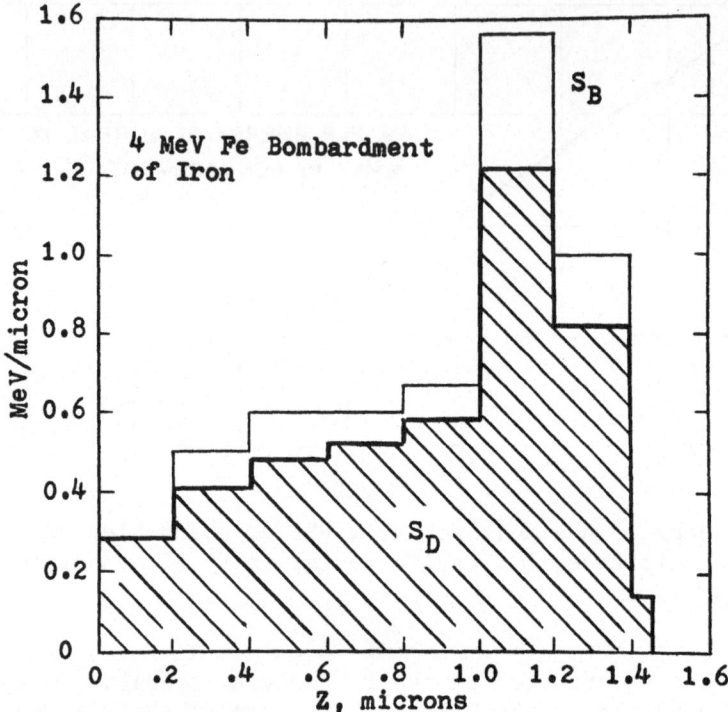

Fig. 4 PKA energy density distribution, S_B, and pre-transport damage energy density distribution, S_D, for 4 MeV iron atom bombardment of iron. FEV model computer experiment results.

given prescription for approximating the true damage energy density
distribution.

    Historically, there have been two stages of approximation to
the true damage energy density. The first was to approximate $S_V(z)$
by the energy transferred per unit length from a PRP to its PKA.
This PKA energy density is denoted by $S_B(z)$. The second was to approximate $S_V(z)$ by the fraction of $S_B(z)$ ultimately associated with
elastic energy transfers, without regard for any possible redistribution of 'damage energy' in collision cascades. This energy
density distribution is called $S_D(z)$. One can regard $S_D(z)$ as being the <u>pre-transport</u> damage energy density distribution and $S_V(z)$
as the <u>post-transport</u> damage energy distribution.

    A comparison of $S_B(z)$ and $S_D(z)$ given by the FEV model appears
in Fig. 4. The integral of $S_B(z)$ is 1.07 MeV and that of $S_D(z)$ is
0.88 MeV. The difference between these integrals is the inelastic
energy loss in the PKA cascades.

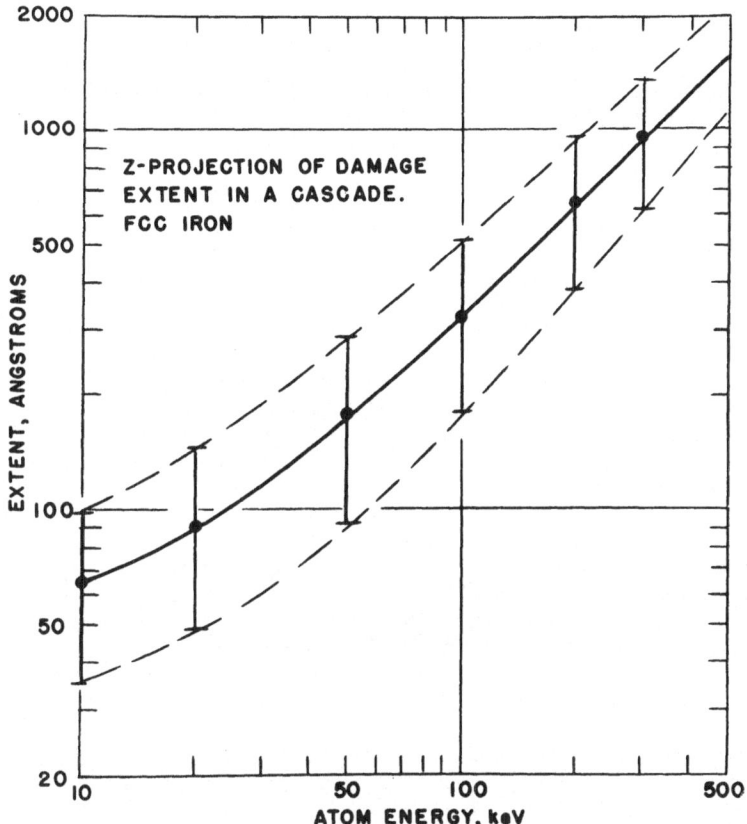

Fig. 5 Projection of the damage extent in an atomic collision cascade in fcc iron on the initial direction of the initiating PKA.
The solid curve describes the average extent. The dashed lines
describe the maximum and minimum extents observed.

   The kinetic energy transferred to a given PKA by the PRP can be
dispersed over a linear extent of hundreds of angstroms within the
cascade it initiates. The projection of the extent of vacancy (displacement) production in a cascade onto the initial direction of the
initiating PKA is described by Fig. 5. At 10 keV, for example, the
average extent (filled circle) is 55 Å, the minimum extent is 35 Å,
and the maximum extent is 100 Å. At 100 keV the mean extent is
about 1000 Å. Using the mean range of a 4 MeV PRP, $\overline{R}$ = 12,000 Å, as
a reference length, one sees that the extent of damage energy transport is significant in the 10^2–10^3 keV range.

   $S_V(z)$ is plotted with $S_B(z)$ in Fig. 6 to show the difference
between the PKA energy density distribution and the true damage energy density distribution. The difference is caused by inelastic
energy losses and damage energy transport.

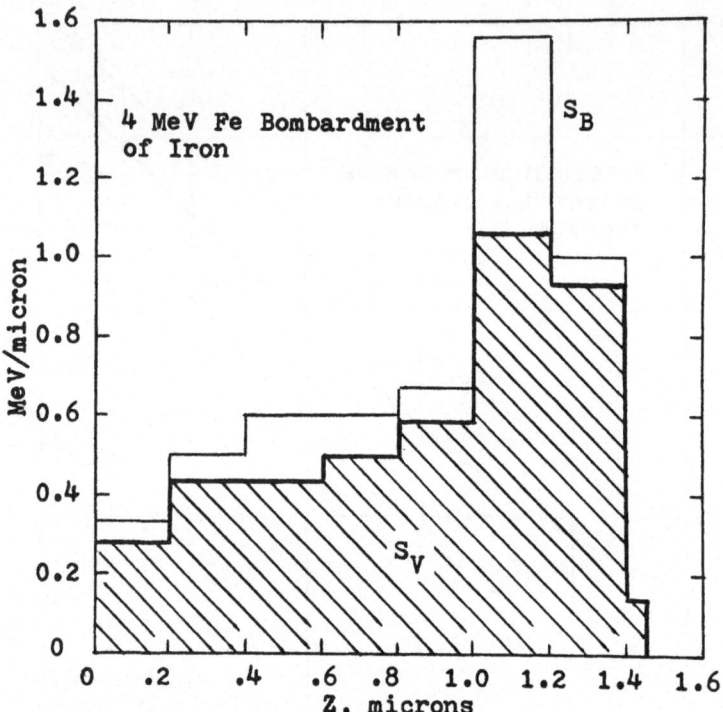

Fig. 6 PKA energy density distribution, S_B, and true damage energy
density distribution, S_V, for 4 MeV iron atom bombardment of iron.
FEV model computer experiment results.

     Damage energy transport in 4 MeV cascades is described by
Fig. 7. Here S_D and S_V given by computer experiment simulation are
compared. The integral of each distribution is 0.88 MeV. Up to
about $2\bar{R}/3$ (8000 Å), the two distributions do not differ significantly. A considerable amount of net damage energy transport occurs in the end-of-range region, however. Damage energy transport
occurs in both the positive and negative z-directions. The damage
energy in the peak region <u>after</u> transport is 16% less than was deposited. In contrast, the damage energy in the regions to the left
and right of the peak region <u>after</u> transport was larger than that
initially deposited.

     A comparison of the true damage energy density contribution by
PKA with energies below 10 keV (low-energy) and energies above 10
keV (high-energy) appears in Fig. 8. Over the 0-8000 Å interval,
high-energy PKA contributed 56% of the true damage energy and low-
energy PKA contributed 44%. Over the 8000-14000 Å interval, high-

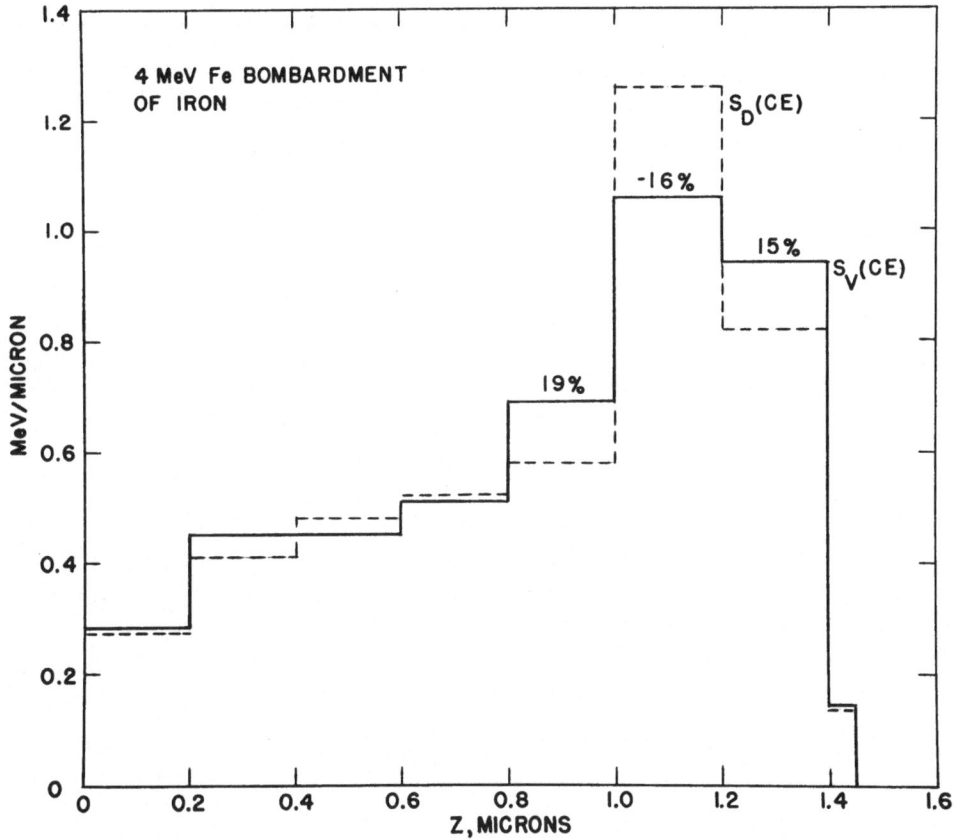

Fig. 7 Pre-transport damage energy density distribution, S_D, and true damage energy density distribution, S_V, for 4 MeV iron atom bombardment of iron. FEV model computer experiment results. Note that the net damage energy transport is negligible for z less than 0.8 µ.

energy PKA contributed 68% of the true damage energy and low-energy PKA contributed 32%. It appears that net transport in the 0–8000 Å region is negligible because in this region the PKA energy spectrum is relatively soft and the PRP trajectory is not much deviated from the z-axis (incident beam direction). In the 8000–14000 Å region, more than 2/3 of the true damage energy density is contributed by high-energy PKA with a hard spectrum. In addition, the PRP trajectory deviates markedly from the z-axis. This allows ejection of PKA with significant trajectory projections along the positive and negative z-directions.

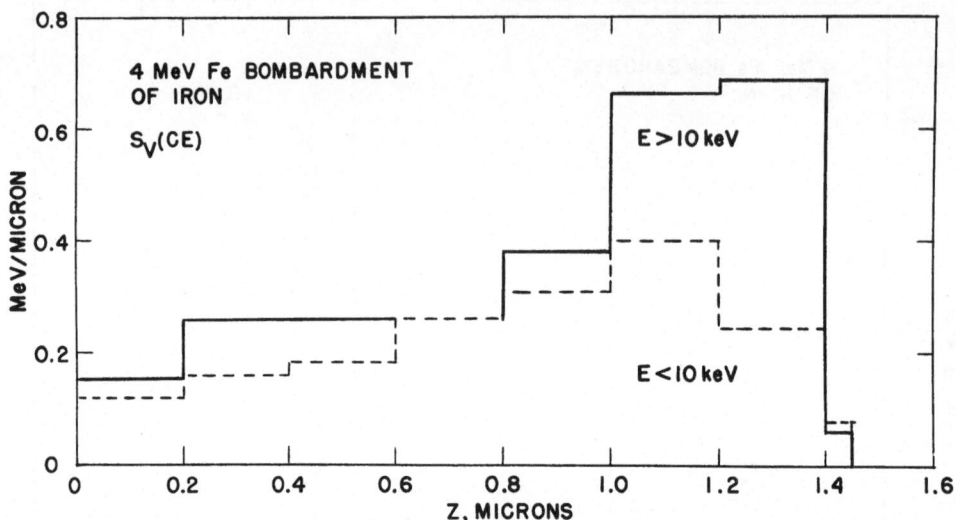

Fig. 8 Contribution to S_V made by PKA with energies below 10 keV
and that made by PKA with energies above 10 keV for 4 MeV iron atom
bombardment of iron. FEV model computer experiment results.

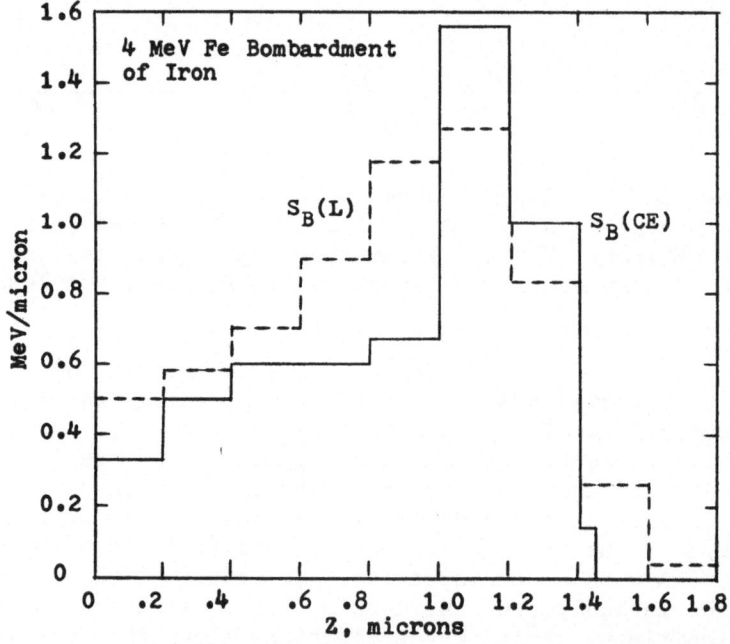

Fig. 9 PKA energy density distribution, $S_B(L)$, computed by Manning
and Mueller, on the basis of LSS theory, for k = 0.108 (dashed
line). PKA energy density distribution, $S_B(CE)$, given by FEV model
computer experiments (solid line).

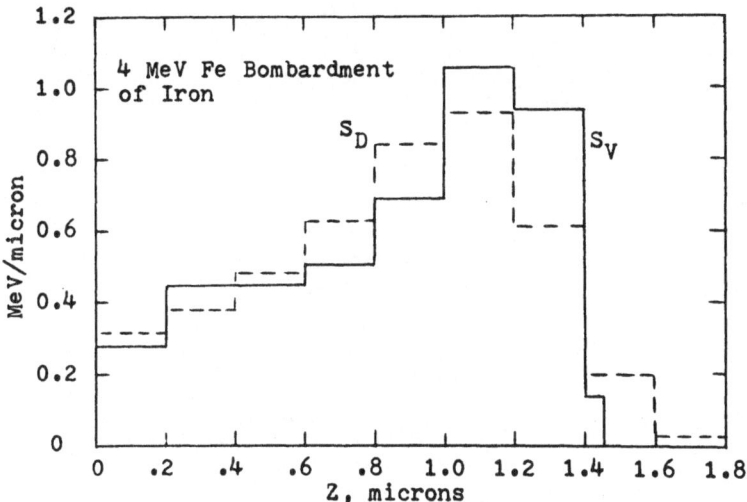

Fig. 10 Pre-transport damage energy density distribution computed
by Manning and Mueller on the basis of LSS theory for k = 0.108
(dashed line). True damage energy density distribution, S_V, given
by FEV model computer experiments. Note that these distributions
are in good agreement for z less than 0.6 μ.

4. COMPARISON WITH LSS THEORY

Computer experiment runs using the FEV atom collision model
give k_F = 0.108 for 4 MeV cascades in iron. LSS theory gives
k_L = 0.156 for the same case.

Manning and Mueller computed $S_D(z)$, $S_B(z)$, the mean range, and
the range straggling using an assigned value of k = 0.108 in their
E-DEP-1 program. They obtained a mean range of 1.2 μ. Their PKA
energy density distribution $S_B(L)$ is compared with the computer experiment distribution, $S_B(CE)$, in Fig. 9. These two distributions
have different integrals. The integral of $S_B(L)$ is 1.26 MeV, while
the integral of $S_B(CE)$ is 1.07 MeV. Their $S_D(z)$ distribution (see
Fig. 10), however, was similar to the true damage energy density
distribution $S_V(z)$ given by the computer experiments. The two distributions had the same integral (0.88 MeV) and agreed well on the
initial interval 0-6000 Å. The agreement occurs over the region in
which the PKA energy spectrum is relatively soft.

5. ANNEALING

Annealing simulation indicated that, at a finite temperature, the fraction of defects surviving short-term annealing (about a microsecond), and their clustering mode, are functions of penetration distance. In MeV-energy atom bombardment, many of the cascades produced by high-energy PKA can interact during short-term annealing. These particular PKA tend to be produced in the final 1/3 of the primary range. This circumstance gives rise to a type of short-term annealing environment not found in other types of primary particle irradiation. One consequence of this is the formation of interstitial dislocation loops at an earlier time than would be the case were there no interaction among cascades during annealing.

ACKNOWLEDGMENTS

The authors are deeply indebted to I. Manning and G. P. Mueller for their continuing suggestions and help during the course of this study; their participation was invaluable. We are also most grateful to M. T. Robinson, N. D. Dudey, and M. J. Fluss for their helpful suggestions. All of the work was supported by U. S. Atomic Energy Commission Contract AT-(40-1)-3912. The work was importantly motivated by a request from T. C. Reuther, Jr., of the USAEC.

REFERENCES

1. J. R. Beeler, Jr., Phys. Rev. 150, 470 (1966).

2. C. Erginsoy, G. H. Vineyard and A. Englert, Phys. Rev. 133, A595 (1964).

3. O. B. Firsov, Soviet Physics JETP 36, 1076 (1959).

4. I. M. Torrens and M. T. Robinson, in Interatomic Potentials and Simulation of Lattice Defects, edited by P. C. Gehlen, J. R. Beeler, Jr. and R. I. Jaffee, Plenum Press (1972), pp. 423-438.

5. I. Manning and G. P. Mueller, Naval Research Laboratory, private communication; NRL Memorandum Report 2555, edited by L. E. Steele and E. A. Wolicki, February, 1973.

6. J. Lindhard, M. Scharff and H. E. Schiott, Kgl. Danske Videnskab Selskab 33, No. 14 (1963).

FLUX (DOSE RATE) EFFECTS FOR 2.8 MeV ^{58}Ni IRRADIATIONS OF PURE Ni

J. E. Westmoreland, J. A. Sprague, F. A. Smidt and
P. R. Malmberg
Naval Research Laboratory, Washington, D. C. 20375

INTRODUCTION

Five years ago a phenomenon now known as swelling was found[1] to occur in reactor materials under prolonged intense neutron irradiation. This swelling, which may represent a 10% or greater volume change, results from condensation of vacancies to form voids. The void formation presents severe limits on possibilities for advanced nuclear reactors. Subsequently it was found that this swelling could be simulated, at least qualitatively, by bombardment with heavy ions in a small fraction of the time required with neutron irradiation. Two conferences[2,3] have summarized early work of this type. However, the higher (10^3 - 10^5 times) production rate of the initial damage which makes simulation feasible may also obscure subtle microstructural changes. Thus, the dose rate effects need to be understood. Qualitative theoretical predictions of these effects have been made.[4,5] The purpose of the present experiment was to obtain two quantitative curves of swelling versus irradiation temperature in pure nickel where flux (dose rate) is the only variable. That is, the flux was varied by a factor of 100 while the bombarding projectile, its energy, and the total fluence of incident ions was maintained the same. Comparisons are made between experimental results and the theoretically predicted effects of flux on swelling.

EXPERIMENTAL

Pure nickel foils (0.1 mm thick, Materials Research Corporation, 99.995% pure) have been irradiated with 2.8 MeV ^{58}Ni$^+$ ions from the NRL 5-MV Van de Graaff accelerator. Fluences of 5.68 x

10^{15} ions/cm^2 have been used, except where otherwise noted. The two fluxes employed were (5.5 - 6.0) μA/cm^2 (high dose rate, "hdr") and (55 - 60)nA/cm^2 (low dose rate, "ldr"). The magnetically analyzed ^{58}Ni$^+$ beam was electrostatically defocused using one stage of a standard electrostatic quadropole strong focusing lens doublet. The uniformity of the beam across the sample was monitored on line by a Danfysik beam profile monitor of the vane type. This method insured uniformity of flux and fluence across the sample surface to $\gtrsim$ 15%. The entire irradiation chamber (heater, thermocouple, cooling system) was electrically isolated, and the ion beam current was integrated directly from this deep Faraday cage. A defining aperture, an aperture to stop stray beam, a secondary electron suppressor, and a ground plane were positioned in that order approaching the profile monitor beginning 15 cm away. The profile monitor was itself then 15 cm upstream from the turret, which can hold up to ten samples for sequential irradiations without breaking vacuum. After defocusing, the beam was electrostatically deflected to reduce the effect of any neutral or multiply charged nickel ions that might have been present in the beam. The vacuum at the sample during irradiation was < 5 x 10^{-6} mm Hg.

An estimate of the rise in sample temperature during irradiation was made by spot welding a thermocouple directly to a test sample surface. A $\sim 25^{\circ}$C rise was noted for the high dose rate beam and a negligible rise was assumed for the low dose rate beam.

Prior to irradiation the nickel foils were vacuum annealed for two hours at 1000°C. After irradiation (4500 ± 500) Å of the front surface of each foil was electropolished away. The thickness measurement, employing the electrolytic cell as one arm of an interferometer, will be described elsewhere.[6] The samples were then jet polished from the rear until perforation, and the sample was analyzed by transmission electron microscopy. The thickness of the area examined was determined by stereomicroscopy. Values were typically (1000 - 3000) Å.

DAMAGE CALCULATION

The damage level and damage dose rate were calculated employing the computer program E-DEP-1,[7] which calculates the deposition of initial damage energy.* The result for ^{58}Ni → pure Ni (8.98 x 10^{22} atoms/cm^3) is shown in Figure 1. The maximum value occurs at a depth of $\sim$ 5000 Å with a FWHM of $\sim$ 2000 Å. This peak value of 1.264 MeV/μ is converted to displacements per atom (dpa) with the

---

*This is an approximation to the amount of energy available to create initial displacements at a given depth in the sample. All inelastic effects are allowed for but transport of damage energy by recoiling secondary atoms is neglected.

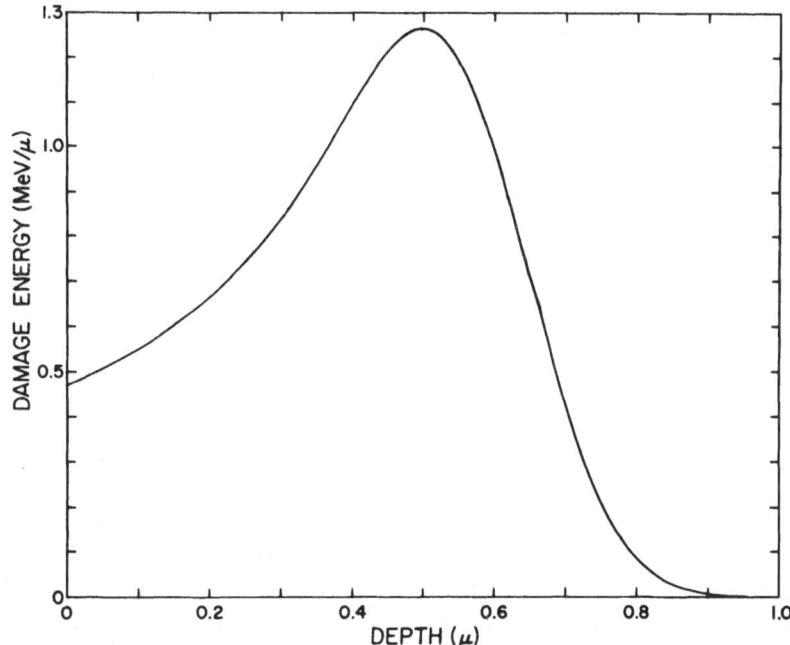

Fig. 1. Initial damage energy as calculated by E-DEP-1 (Ref. 7) for 2.8 MeV ^{58}Ni incident on pure Ni (atomic number density 8.98 x 10^{22} atoms/cm^3).

Kinchin and Pease formula[8] modified by Torrens and Robinson[9] as follows: $n_d = \kappa \nu(E)/(2E_d)$ where n_d is the number of displacements, $\kappa = 0.8$, and a value of 25 eV is employed for E_d. An atomic number density of 8.98 x 10^{22} atoms/cm^3 is used for Ni to convert n_d to dpa. A fluence of 5.68 x 10^{15} ions/cm^2 is thus equivalent to a maximum calculated damage density of 13 dpa; the low dose rate (1dr), to ~ 7 x 10^{-4} dpa/sec; the high dose rate (hdr), to ~ 7 x 10^{-2} dpa/sec.

RESULTS

Values have been obtained for void size distributions and calculated void number densities and swelling for high dose rate foils irradiated at 325 - 725°C and low dose rate foils irradiated at 300 - 600°C. An approximate correction for not counting those voids $\gtrsim$ one-half of a void diameter out of either of the sample surfaces is employed. This is the use of a corrected number of voids for each size class $\hat{n}_i = n_i t/(t-d_i)$ where n_i is the actual number of voids counted of diameter d_i, and t is the thickness of the sample. The swelling is given as a percentage of the final

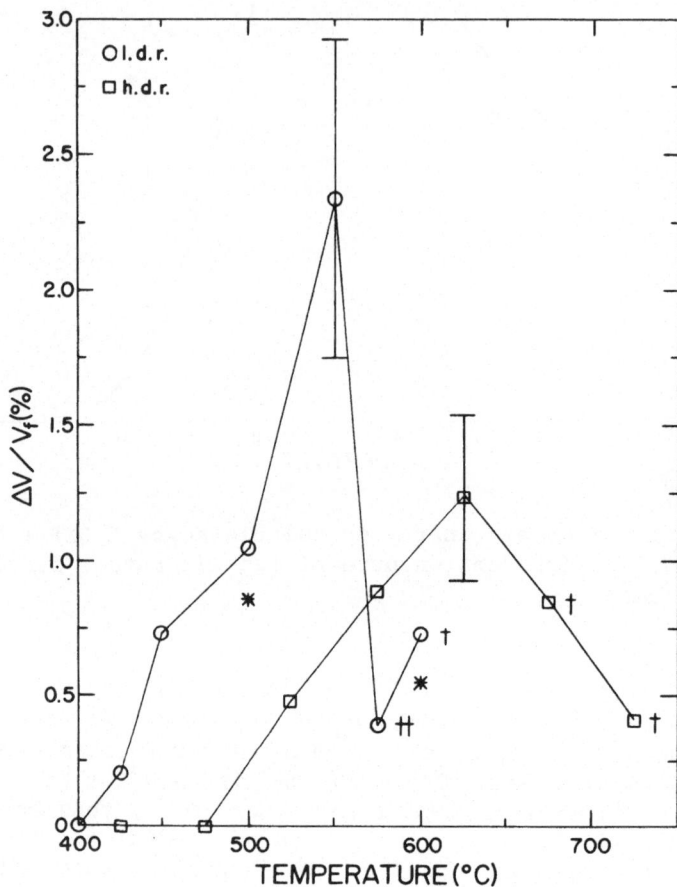

Fig. 2. Swelling: percentage void volume of final sample volume
versus irradiation temperature. Pure Ni irradiated to a fluence
of 5.68×10^{15} 2.8 MeV ^{58}Ni ions/cm^2 (13 dpa). l.d.r. = 7.10^{-4}
d.p.a./sec—h.d.r. = 7×10^{-2} d.p.a./sec. (†) Poor statistics as
few voids, see Fig. 3. (††) Perhaps too deep to be 13 dpa.

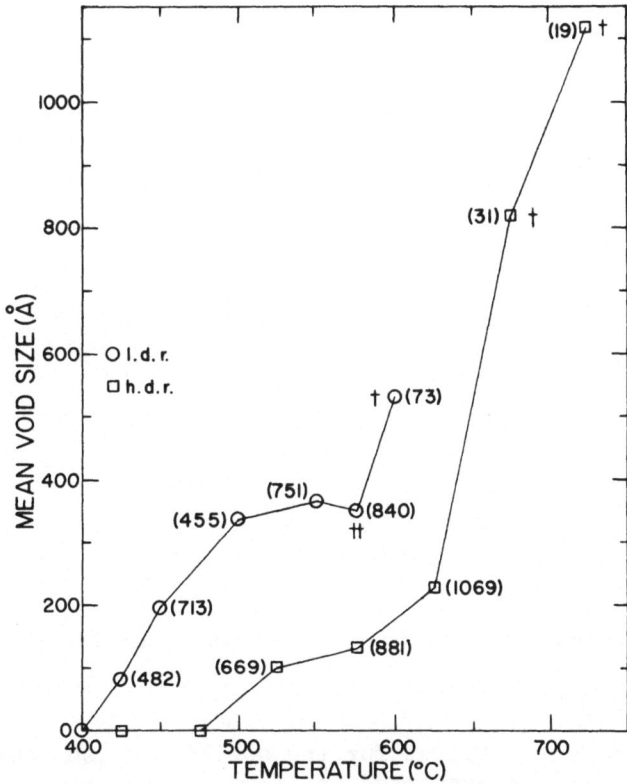

Fig. 3. Mean void size (diameter) for same samples as Fig. 2.
Numbers in ( ) are number of voids actually counted in sample.

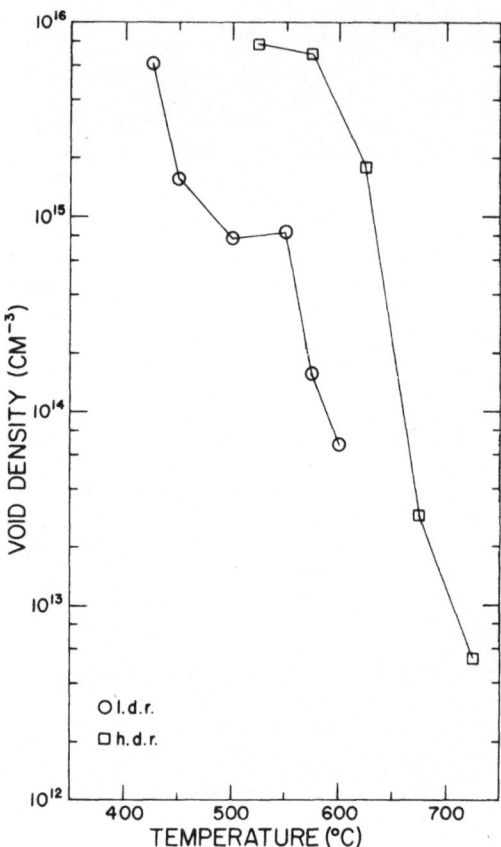

Fig. 4.   Void number density (ρ_v) for same samples as Fig. 2.

volume of the sample. Values for the swelling versus temperature
for the two fluxes are presented in Figure 2, the mean void
diameter, in Figure 3, and the void number density, in Figure 4.
An estimated error of ± 25% is indicated in Figure 2 for reference,
although this choice is somewhat arbitrary. In hdr foils, no
swelling is observed for foils irradiated at T ≤ 475°C; voids are
observed for T ≥ 525°C. For ldr foils, no swelling is observed
for T ≤ 400°C; voids are observed for T ≥ 425°C. An estimate of
the temperature shift of the swelling curve of (100 ± 25)°C has
been obtained from the portions of the two curves of Figure 2
where the swelling is still increasing.

The doses in ions/cm^2 for the samples yielding the ldr data
points for 500°C and 600°C were respectively 4.59 x 10^{15} (80.8% of
goal dose) and 4.28 x 10^{15} (75.4% of goal dose) due to ion source

filament burnout. The actual calculated swellings are indicated in Figure 2 with an asterisk (*). These points have then been scaled linearly[10] with dose to yield the points on the curve. The actual calculated mean void diameter has been scaled as $(dose)^{1/3}$ and only the scaled points are shown on Figure 3. The sample yielding the $575^\circ C$ point on the ldr curve is suspected to have been front thinned too deep into the sample to accurately represent damage at the calculated peak of the damage distribution due to difficulties with the thinning process on the day that sample was electropolished. However, no means is presently available to determine the actual depth, so the point is shown as obtained. Also, the relatively small number of voids counted in the highest temperature ldr sample and the two higher temperature hdr samples imply additional uncertainty in these points.

The dislocation density was obtained for some of the samples of this experiment using the line intersection method. The effective sink strength expression for the voids[5] $4\pi r_v \rho_v$ was evaluated using the corrected void number density ρ_v versus void radius r_v distribution. These results are presented in Table 1.

Table 1

$\varphi(dpa/sec)$	$T(^\circ C)$	$4\pi r_v \rho_v (cm^{-2})$	$\rho_d (cm^{-2})$
7×10^{-2}	525	4.9×10^{10}	9.0×10^9
7×10^{-2}	575	5.7×10^{10}	8.5×10^9
7×10^{-2}	675	1.5×10^9	5.0×10^8
7×10^{-4}	500	1.4×10^{10}	4.8×10^9
7×10^{-4}	575	3.4×10^9	4.8×10^9

Figure 5 shows transmission electron micrographs typical of the voids and dislocations observed in the samples of this experiment. Micrographs of the ldr ($550^\circ C$) and hdr ($575^\circ C$) samples are shown.

Grain boundary denuding (absence of voids on either side of a grain boundary) has been observed, for example, in neutron irradiated nickel.[11] Figure 6 shows a transmission electron micrograph of a grain boundary observed in the ldr $500^\circ C$ sample. This boundary exhibits denuding up to ~ 1800 Å away. Similar denuding has been observed in the ldr $600^\circ C$ (~ 3600 Å), the ldr $550^\circ C$ (~ 2500 Å), the ldr $450^\circ C$ (< 1000 Å), the hdr $675^\circ C$ (~ 5000 Å), and the hdr $525^\circ C$ (~ 350 Å).

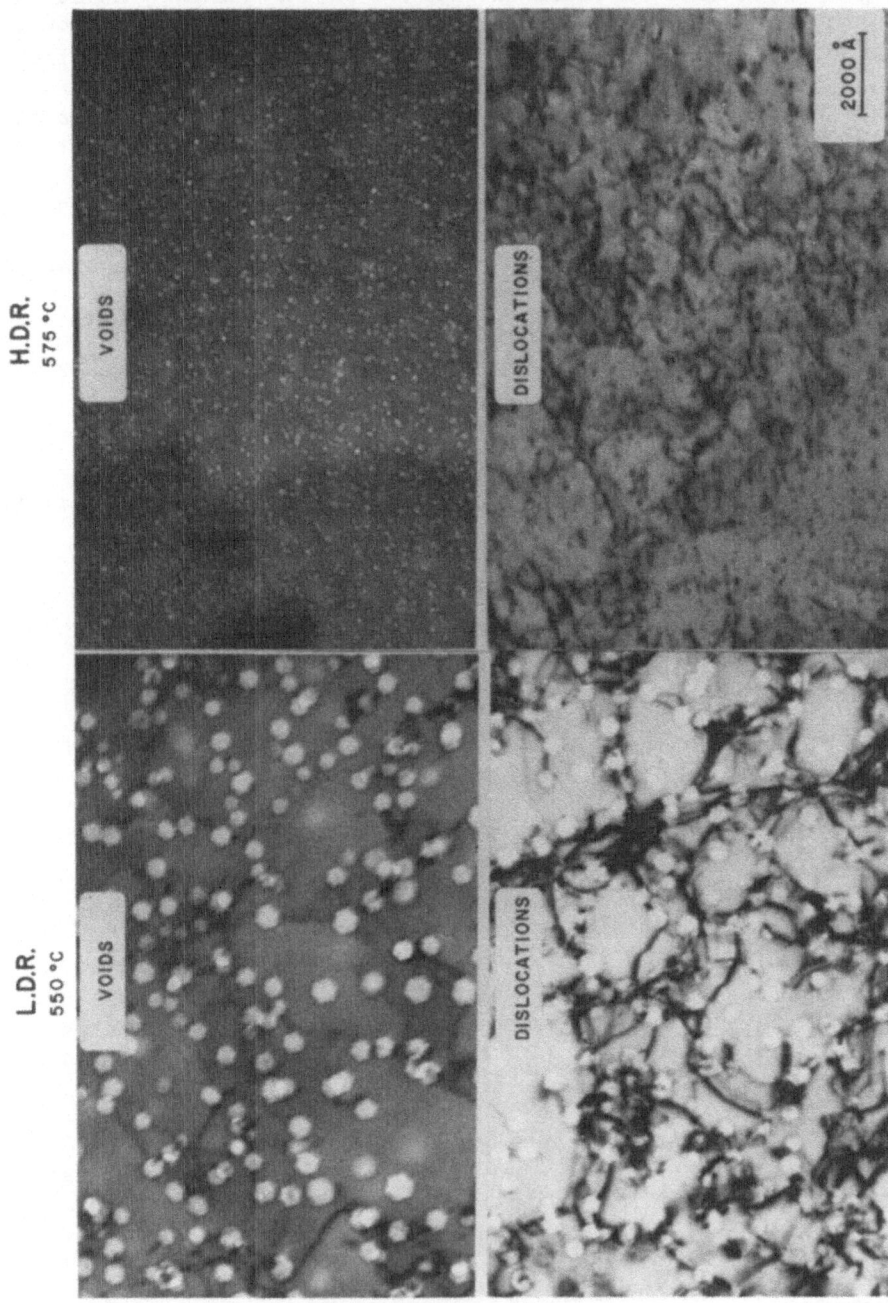

Fig. 5. Transmission electron micrographs showing voids and dislocations as observed in l.d.r. (550°C) and h.d.r. (575°C) points of Fig. 2.

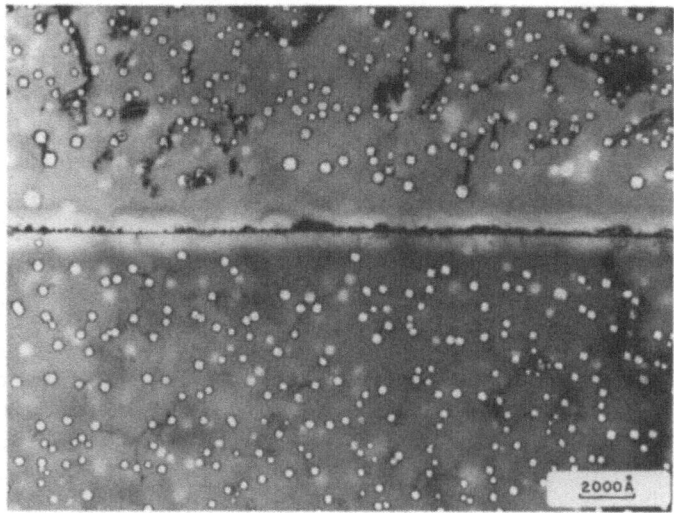

Fig. 6. Transmission electron micrograph of a grain boundary in
l.d.r. (500°C) illustrating grain boundary denuding.

DISCUSSION

Briefly,[12] the concepts of void growth are as follows: The
concentrations of vacancies C_v and interstitials C_i are assumed to
be distributed throughout the system with concentration gradients
existing only around the various defect sinks. The form of a
concentration gradient is obtained by solving the steady rate
diffusion equation with assumed boundary conditions. The steady
state is achieved through a balance between the point defect
generation rate K, the recombination rate, and the rates of migration to sinks.

The explanation for the observed shift of the swelling curve
to higher temperatures with increasing flux (dose rate) is as
follows:[4] At low temperatures recombination, proportional to
$C_v C_i$, is the dominant mechanism that determines the concentration
of vacancies and interstitials in the material. With increasing
displacement rate (flux of a given bombarding particle, different
particle, etc.) both C_v and C_i increase, resulting in increased
recombination, hence reduced void growth (swelling). At high
temperatures in bulk material, the extent to which C_v exceeds its
thermal equilibrium value is the dominant mechanism, and increasing
the displacement rate increases C_v, allowing swelling to take place
at higher temperatures.

Several formulas have been used in an attempt to scale to an
equivalent neutron fluence and spectrum radiation damage produced
by various projectiles other than neutrons. The simplest method

calculates a damage production rate for a given bombarding particle, then employs the following equation to relate this damage rate to some desired neutron rate:

$$1/T_2 = 1/T_1 + (k/Q)\ln(K_1/K_2) \tag{1}$$

where k is Boltzmann's constant, and for the displacement rates K_i, the T_i are the temperatures for equivalent swelling. The quantity Q depends on the dominant mechanism controlling the swelling. Thus, Q can be shown to be the vacancy migration energy in the recombination dominated ("low temperature") regime under the assumption that equivalent swelling is obtained for temperatures and fluxes such that the ratio of the sink loss rate to the recombination rate is a constant. It has been shown[13] that for metals under certain restrictions on the void diameter and dislocation density, Eq. (1) holds for scaling the high temperature limit at which $\Delta V/V = 0$ with Q the self-diffusion energy for the material. A more detailed expression relating swelling and dose rate has also been obtained.[5] Values of these expressions will be compared with the present experimental results.

Using for Ni a value of Q of 1.4 eV for the vacancy migration energy[5] and the ratio of displacement rates of 100 employed in this experiment in Eq. (1) yields for $T(ldr) = 400°C$, $\Delta T = 160°C$, for $T(ldr) = 450$, $\Delta T = 185°C$, and for $T(ldr) = 500°C$, $\Delta T = 215°C$. Although the experimental value of ΔT does appear to increase slightly with increasing temperature in Figure 2 as do these calculated values, the calculated values are considerably larger than the measured value.

The Brailsford and Bullough expression $F(\eta)$ relating equivalent swelling temperature and flux[14] has been evaluated for the two sets of values of ρ_d, $4\pi r_{vo}v$, and K of this experiment and in each case an assumed ratio of dislocation loop density ρ_d^l to total dislocation density ρ_d of 10^{-2}. All other parameters are as in Ref. 5. The two resulting curves are shown in Figure 7. These curves are shifted at least 200°C higher than the experimental data.

One can calculate in an approximate way the effect of the surface as a sink from Foreman's expression[15] for a thin foil. This yields for $T \lesssim 600°C$ a width $\lesssim 500$ Å for the region adjacent to a surface where concentrations are changed from bulk values by $\sim 5\%$. Alternatively, one can regard the width of grain boundary denuding as comparable to that of the surface denuding. This yields an effective depth of denuding (ldr) for 500°C of ~ 1800 Å. Either result suggests that for the lower temperature region ($T \lesssim 600°C$) the present experimental data are representative of bulk material. The implication of the discrepancy between the experi-

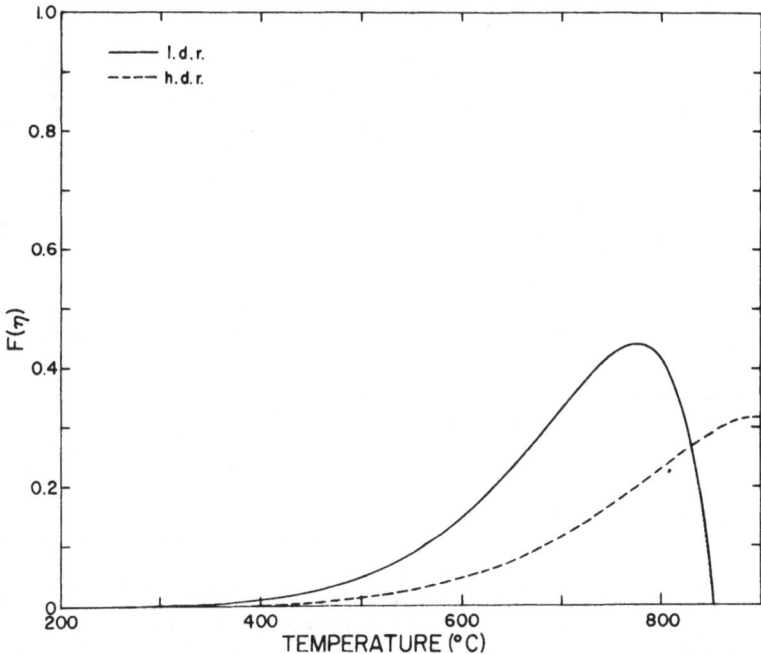

Fig. 7. Calculated values of Eq. (51), Ref. 5 (in notation of Ref. 5), with no precipitates, $\rho_d^L/\rho_d = 10^{-2}$, $\bar{C}_{vL} = 0$ (no vacancy emission by loops), $\bar{C}_{vs} = C_v^e = \exp(-1.6\ eV/kT)$ (vacancy concentration at void-matrix interface in thermal equilibrium), $Z_i = 1.01$, $Z_v = 1.00$, $\alpha/D_i = 10^{17}/cm^2$, $D_v = 0.6 \exp(-1.4\ eV/kT) cm^2/sec$. Solid line: l.d.r. $K = 7 \times 10^{-4}$ d.p.a./sec, parameters from 500°C sample as follows: $\rho_d = 4.8 \times 10^9/cm^2$, $4\pi r_v \rho_v = 1.40 \times 10^{10}/cm^2$. Dashed line: h.d.r. $K = 7 \times 10^{-2}$ d.p.a./sec, parameters from 525°C sample as follows: $\rho_d = 9.0 \times 10^9/cm^2$, $4\pi r_v \rho_v = 4.85 \times 10^{10}$.

mental result and the predictions of Brailsford and Bullough is not clear at present.

CONCLUSIONS

This experiment measured directly the effect of a factor of 100 in flux on the final observed damage microstructure for the case for all other irradiation parameters remaining the same. Adequate agreement was not obtained in comparison with two of the theoretical predictions of the effect of flux on swelling. The implication of this is not at present clear.

REFERENCES

1. C. Cawthorne and E. J. Fulton, Nature 216, 575 (1967).
2. S. F. Pugh, M. A. Loretto, and D. I. Norris (editors), Voids Formed by Irradiation of Reactor Materials, Reading University, March 1971, British Nuclear Energy Society.
3. J. W. Corbett and L. C. Ianniello (editors), Radiation-Induced Voids in Metals, Albany, New York, June 1971, AEC Symposium Series (CONF-710601).
4. R. Bullough and R. C. Perrin, in Ref. 2, p. 79.
5. A. D. Brailsford and R. Bullough, J. Nucl. Mat. 44, 121 (1972).
6. J. A. Sprague, to be published.
7. Irwin Manning and G. P. Mueller, E-DEP-1, to be published, Computer Physics Communications.
8. G. H. Kinchin and R. S. Pease, Rep. Prog. Phys. 18, 1 (1955).
9. I. M. Torrens and M. T. Robinson, in Ref. 3, p. 739.
10. Ref. 5, p. 130.
11. G. L. Kulcinski, B. Mastel, and H. E. Kissinger, Acta Met. 19, 27 (1971).
12. See D. I. Norris, Rad. Effects 14, 1 (1972), Rad. Effects 15, 1 (1972), a relatively recent two part review article with a comprehensive listing of references.
13. R. Bullough and R. C. Perrin, in Irradiation Effects on Structural Alloys for Nuclear Reactor Applications, ASTM STP 484 (1970), p. 317.
14. Equation (51) in Ref. 5.
15. A. J. E. Foreman, Rad. Effects 14, 175 (1972), Eq. (10).

HEAVY ION-INDUCED VOID FORMATION IN PURE NICKEL*

T. D. Ryan

University of Michigan, Ann Arbor, Mich.   48104

A. Taylor

Argonne National Laboratory, Argonne, Ill.   60439

ABSTRACT

The void structure in pure nickel induced by irradiation
with $^{12}C^+$, $^{20}Ne^+$, $^{35}Cl^+$, and $^{58}Ni^+$ is being investigated as a
function of temperature and dose using a dose rate of 2.5 x
10^{-3} dpa/sec at a preselected depth of 6700 Å. The dose dependence of swelling to 60 dpa, and the temperature dependence at
10 dpa between 500 and 700°C for self-ion irradiations, are in
substantial agreement with previous studies. The peak swelling
was observed in a 600°C irradiation in which the number density
was 2 x $10^{15}/cm^3$, the mean size was 280 Å and the total volume
fraction was 3%. As a function of temperature, the mean void
size increased by 2.5, and the number density decreased an order
of magnitude. Preinjection of He did not influence the total
swelling. Asymmetry in the thickness of the void-denuded layers
across grain boundaries was directly correlated with grain-boundary
motion during the irradiation. Preliminary data for $^{12}C^+$ irradiation at these temperatures indicates that the peak swelling
temperature occurs at least 25°C lower than for self-ion irradiations.

---

*Work performed under the auspices of the U. S. Atomic Energy
Commission.

INTRODUCTION

Several recent radiation-damage studies of pure nickel have shown that void structures similar to those developed during fast-neutron irradiations[1] can be evolved by bombardment with a variety of charged particles.[2] Quantitative correlations between various types of irradiations are still tenuous however, see, for example Norris.[3] To eliminate differences caused by variations in metallurgical structure and composition, it seemed worthwhile to reexamine the swelling behavior of nickel in a set of standardized irradiations using material from a single source. If a systematic dependence of the microstructural development on projectile mass can be established, the correlation between neutron and ion damage for reactor-simulation studies would be strengthened. We are examining the dose and temperature dependence of the microstructure at low swelling values subsequent to irradiation with $^{12}C^+$, $^{20}Ne^+$, $^{35}Cl^+$, and $^{58}Ni^+$ at energies and fluxes such that the depth of the calculated peak damage and the magnitude of the damage rate are held constant. The peak damage depth corresponds to a nickel self-ion irradiation using the 4 MV accelerator facility at Argonne National Laboratory.

SAMPLE PREPARATION

Grade 1, nickel sheet stock 0.18 mm thick was obtained from the Johnson Matthey Company, Ltd. Spectrographic and gas analysis showed the major impurities listed in Table I.

TABLE I

Principal impurities in nickel sheet stock, wt. %

Carbon	0.011
Oxygen	0.01
Hydrogen	0.0011
Nitrogen	0.006
Tungsten	<0.01
Others (Fe, Mn, Si, Cr, Mo, Pb)	< 0.005 each

Disks 3 mm in diameter were punched from the sheet, flattened, and annealed in vacuo or high-purity argon gas for 1 h at 800°C. The disk specimens were mounted in resin, lapped flat, and vibratorily polished in two stages using grinding slurries of 0.3 and 0.05 μm alumina grit. Finally, 1.5 μm of material was electropolished from each specimen, and those with surfaces flat to within <0.06 μm were selected for bombardment. The ASTM grain size of the sample material was 5, and the initial dislocation density was $5 \times 10^8 cm^{-2}$.

ION IRRADIATIONS

Groups of four or eight specimens were clamped between a temperature homogenizer block of copper or tungsten and a mask containing a rectangular array of 2-mm-diameter holes with a 3-mm-diameter counter bore to a depth of 0.3 mm. Annular tantalum spacers were placed between the mask and the specimen to prevent binding. The sample holders were then heated by a resistance furnace or by electron bombardment to the irradiation temperature in a vacuum of better than 10^{-6} Torr. The holder temperature was measured and controlled by two Chromel-K thermocouples inserted into its side. An infrared pyrometer was used to correct for differences between the block and the sample temperatures - typically 5-10°C because of beam heating. [Selected specimens were injected at temperature to a level of 5 ppm using a beam of helium ions with mean energy swept between 0 and 650 keV.] The heavy-ion beam was defocused over an aperature the area of the 4 or 8 specimens. Beam uniformity to within 10% across the target was obtained by wobbling the beam electrostatically at ∿10 kHz, and continuously checking the spatial uniformity with a vibrating wire beam-profile monitor. The target current was periodically measured by inserting a Faraday cup into the beam. The ion fluence was obtained by integrating the current detected by the profile monitor and normalizing to the shutter current. Upon completion of an irradiation, the samples were cooled such that within 1 min. the temperature was decreased by 50°C.

The ion fluxes and energies were selected to obtain an average damage rate of either 2.5 or 4.0×10^{-3} dpa/sec. over a 1500 Å interval about 6700 Å beneath the surface. This depth corresponds to the point of maximum energy deposition for a nickel self-ion irradiation at the maximum accelerator voltage of 4 MV. The energy deposition profiles for $^{12}C^+$, $^{20}Ne^+$, $^{35}Cl^+$, and $^{58}Ni^+$ were computed using the Brice[4] computer codes RASE 3 and DAMG2.[4]

Based on a mean displacement energy of 40 eV, the results of these computations, for the given displacement rate, are given in Table II.

TABLE II

Calculated heavy-ion parameters for displacement rate

of 2.5 x 10^{-3} dpa/sec at 6700 $\overset{\circ}{\text{A}}$ below surface

Ion	Energy	$\mu a/cm^2$ Current Density	Ion Range Projected Standard Deviation $\overset{\circ}{\text{A}}$
$^{12}C^+$	1.0	2.90	1285
$^{20}Ne^+$	1.5	1.30	1273
$^{35}Cl^+$	2.4	0.61	1325
$^{58}Ni^+$	4.0	0.33	1438

The doses reported are based on the total energy deposited in the transmission electron microscopy (TEM) foil.

MICROSCOPY

Postirradiation analyses consisted of optical microscopy of the irradiated surface and a determination of the void microstructure near the peak damage zone by TEM using a JEOL 200 keV microscope. Characterization of the dislocation structure is still in progress. The foil was sectioned to the desired depth by electropolishing the center of the specimen while the surrounding area was protected with lacquer. With the front surface coated with lacquer, the specimen was penetrated from the back. The foil thickness was determined from pairs of stereomicrographs. The void (internal) size distribution was measured with a Zeiss particle analyser equipped with a circular aperture. At least two grains in each specimen were analyzed.

EXPERIMENTAL RESULTS

Results for the mean void size (number average), void number density, and void volume fraction as a function of dose using $^{58}Ni^+$ ions between 0.5 and 60 dpa for specimens not preinjected with He and irradiated at 525°C are shown in Fig. 1. The

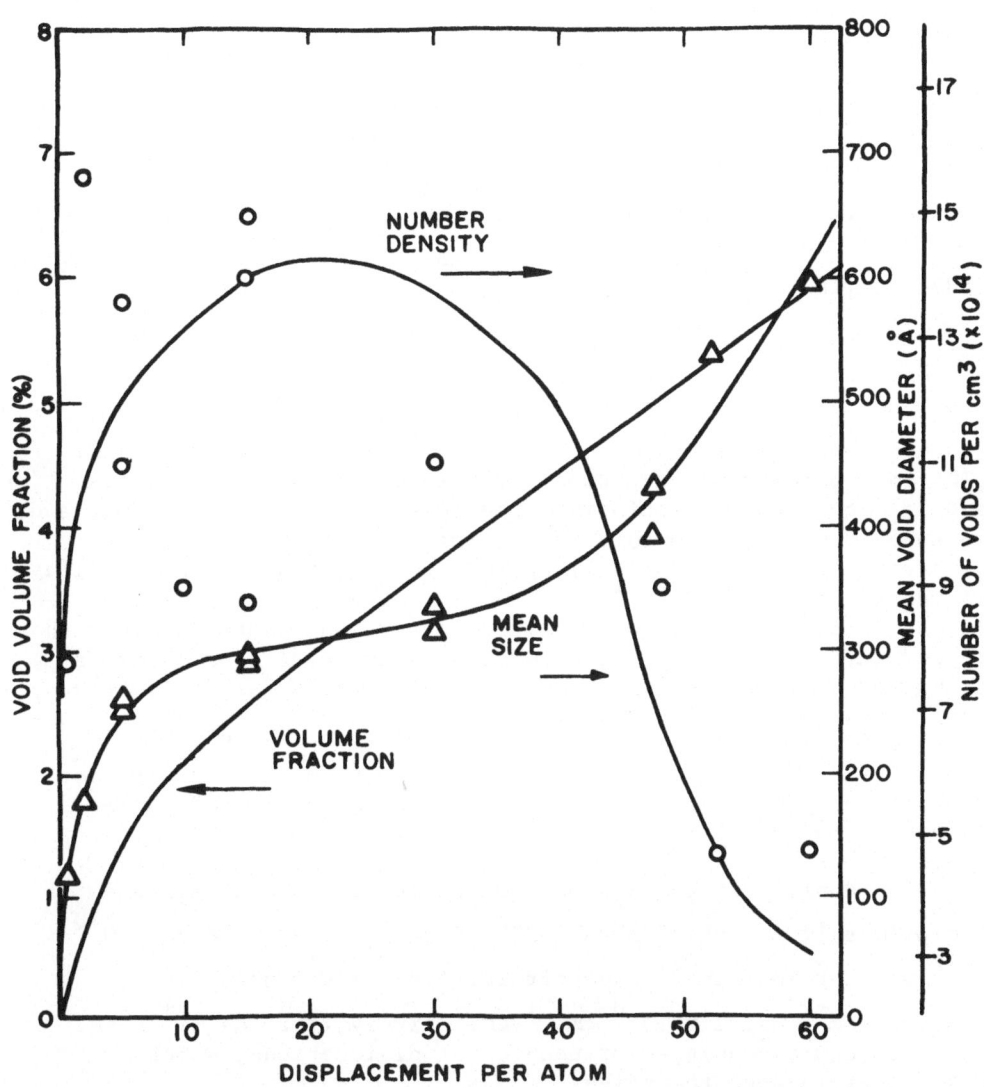

FIGURE 1. Void volume, mean void size, and void number density versus dose for nickel irradiated at 525°C with ^{58}Ni$^+$ ions.

morphology of the voids and the near linear dependence of the
volume fraction on dose are in essential agreement with previous
studies.[5] The evolution of the structure was characterized by a
rapid increase in void number density (void nucleation) to a
maximum $\sim 2 \times 10^{15}$ cm^{-3} followed by a progressive decrease due to
void coalescence. The mean void size increased slowly after the
nulceation phase but increased rapidly as void coalescense occurred.
The swelling within adjacent grains was consistent, although there
were denuded regions associated with twins and grain boundaries.
Preliminary data on the volume swelling of 525°C $^{12}C^+$ irradiations at
2, 10, and 22 dpa were 3.5 times greater than $^{58}Ni^+$ irradiations
under similar conditions.

The temperature dependence of the void volume fraction,
number density, and mean diameter for specimens preinjected with
5 ppm He is shown between 500-700°C in Fig. 2. Some preliminary
data for void volume fraction after $^{12}C^+$ irradiations are also
plotted. Because of variations in the individual irradiations
and in the depth of sectioning, the doses varied from 6-9 dpa, and
we have normalized the void volume fraction to 10 dpa using a
linear relation between swelling and dose. A peak in the swelling
in the vicinity of 600°C is apparently caused by an inflection
in the number density that otherwise decreases monotonically with
temperature. Preliminary carbon irradiation data show a similar
trend with a peak swelling indicated below 575°C, significantly
lower than the self-ion irradiation.

Variations in swelling due to preinjection with He were not
statistically significant; however, an increase in void number
density by a factor of two was indicated.

The density of the dislocation structure was determined
approximately for the 525 and 600°C irradiations to be 2×10^{10}
cm^{-2} and 1×10^{10} cm^{-2} respectively, for the C^+ runs, and
5×10^{10} cm^{-2} and 2×10^{10} cm^{-2} respectively, for the Ni^+ runs.
This structure consisted of tangles of dislocations, which
multiply intersect the voids.

The swelling was essentially independent of grain orientation,
although denuding effects at grain boundaries and twins were observed. Grain-boundary denuding was frequently asymmetrical so
that the grain boundary intersected a normal void distribution on
one side and a wide denuded zone on the other see Fig. 3c. The
region adjacent to the denuded zones frequently contained voids
three times the size of intergranular voids and a high density of

barely resolvable voids. The origin of this structure was correlated with the discontinuous motion of grain boundaries during
the 625°C irradiation, as rendered visible by thermal etching.
This is illustrated in Fig. 3a and b which shows (a) an optical
micrograph of the irradiated surface, and (b) a sketch of the
grain-boundary locations before (dashed line) and after (solid line)
a bombardment of ∿8 dpa. Grain-boundary motion of 20-μm
distances were observed.

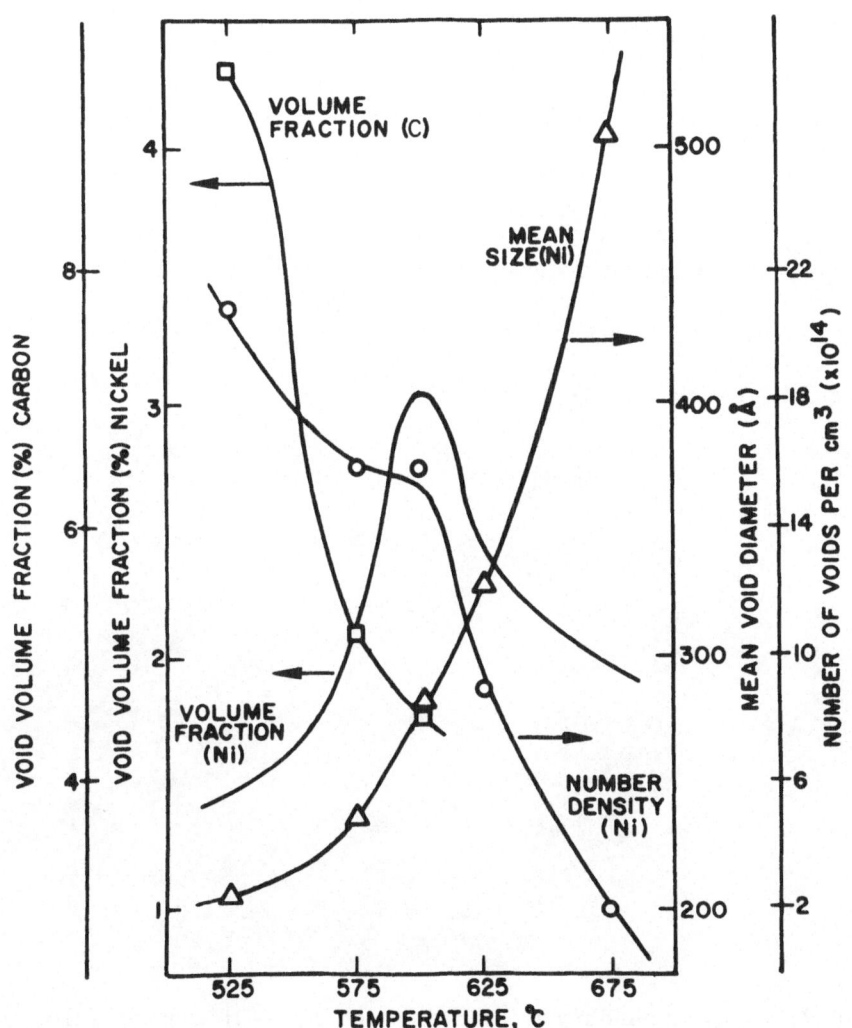

FIGURE 2. Void volume, mean void size, and void number density
for $^{58}Ni^+$ irradiations and preliminary data for C^+ irradiations.

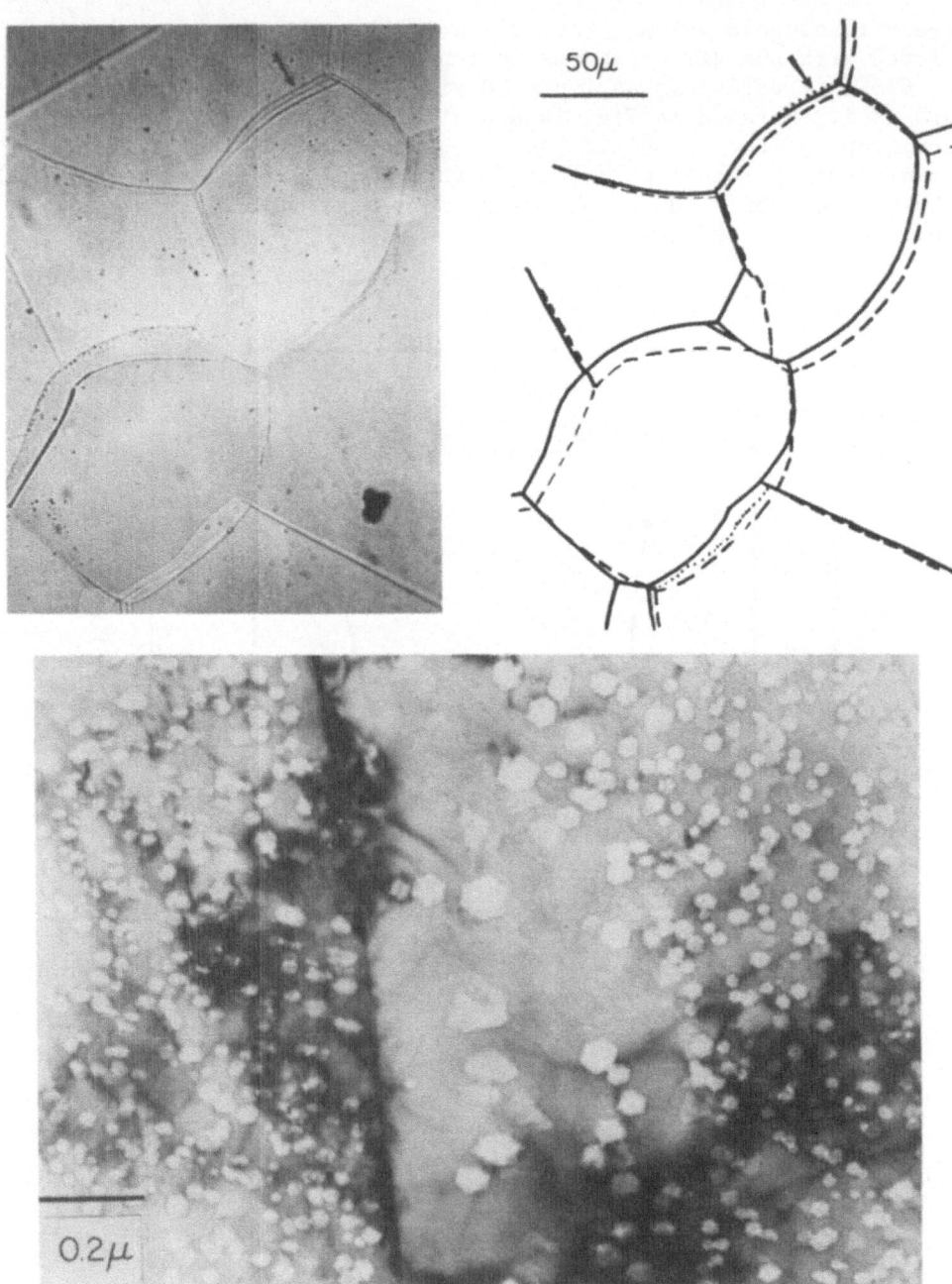

FIGURE 3. Grain boundary motion induced by $^{58}Ni^+$ irradiation at 625°C to 8dpa. (a) Thermally etched surface grains. (b) Schematic showing initial (dashed line) and final position (solid line) of grain boundaries. (c) Electron micrograph of asymmetry in void distribution across grain boundary.

DISCUSSION

Some general conclusions concerning the influence of experimental parameters on void formation in self-ion irradiated pure nickel can be made from a comparison of the present study with three previous investigations.[5,6,7] Although it has been argued that variations in swelling can arise because of dislocation and point-defect depletion effects within $\sim$0.5 μm of the free surface,[8] no systematic variation in swelling over an incident-ion energy range of 0.5 to 10 MeV is apparent. Sputtering has been shown to be negligible with 4-5 MeV ions incident on stainless steel in the void-formation temperature range.[9,10] Segregation of minor constituents to the surface are currently under investigation at ANL. The peak swelling temperature for displacement rates of $\sim 10^{-3}$ dpa/sec lies between 600 and 625°C, increasing with dose rate as theoretically predicted. Inaccuracies in the precise determination of irradiation temperature may mask variations due to materials variables. Within experimental scatter, the dose dependence of the void volume fraction and the void size and number density of the present work at 525°C is in good agreement with that of Kulcinski et al.,[5] by assuming that the irradiation temperature in the latter case was 20° lower than our actual temperature.

The void-number-density curve of Fig. 2 has been smoothed through the low-dose data points. The data between 2 and 10 dpa indicate a factor of two reduction in the number density after the initial rapid void nucleation. Subsequently, the number density increased and then decreased monotonically with dose as the role of void coalescence dominated. A similar behavior may have been observed in $^{12}C^+$ irradiations.[11] We also observed an inflection in the temperature variation of the void number density on the low-temperature side of the swelling peak. Since the mean void diameter is increasing linearly in this region, the volume swelling rises rapidly as the peak is approached.

At 10 dpa, helium preinjection, in 525°C irradiations, resulted in a twofold increase in the void number density and perhaps a slight decrease in swelling. Although helium obviously acts as a nucleation site, the mechanism for nucleation in nonpreinjected samples is indefinite. Preexisting impurities or impurities introduced during sample heat[12] up could explain the observed void number density.

At 525°C the $^{12}C^+$ irradiations show approximately a fourfold increase in void volume fraction compared with the $^{58}Ni^+$ data.

Additionally, an apparent decrease in the $^{12}C^+$ temperature peak
is indicated, but as yet unconfirmed, by data at three temperatures.
The larger swelling data are the result of enhanced void growth.
Since the observed dislocation number densities are similar, this
implies a larger vacancy supersaturation for $^{12}C^+$ irradiations as
opposed to $^{58}Ni^+$ irradiations at the same calculated displacement
rate and/or a significant shift in the point-defect sink efficiencies.

Uniform denuding of voids at grain boundaries and surface
layers, characteristic of a local decrease in point-defect supersaturation, was not observed in the present study. Asymmetry in
the void depletion was caused by irradiation-induced grain-boundary
migration. On the basis of postirradiation grain-boundary etching
and the location and depth of the thermally etched furrows on the
surface, it is believed that the motion of the grain boundary
shown by the arrow in Fig. 3a was discontinuous and irregular in
direction. The large voids close to the grain boundary were
probably formed by coalescence of several smaller voids inter-
secting the grain boundary. These voids were left in the wake
of the boundary. The repopulation of the denuded band with small
voids was also observed. These observations demonstrate that
inhomogeneities in swelling and number density can have their
origin in microstructure effects without hypothesising composi-
tional variations in the materials.

ACKNOWLEDGMENTS

Authors wish to thank B. Kestel and C. Steves for assistance
with experimental work. Useful discussions were conducted with
H. Wiedersich and P. Okamoto.

REFERENCES

1. J. O. Stiegler and E. E. Bloom, Rad. Effects <u>8</u>, 33 (1971).

2. Radiation-Induced Voids in Metals, AEC Symposium Series 26 (1972).

3. D. I. R. Norris, Radiation Effects <u>14</u>, 1 (1972).

4. D. K. Brice, RASE 3 and DAMG 2, Routines for Ion Implantation
    Calculations, Sandia Laboratory Report, SLA-73-0410 (April 1973).

5. G. L. Kulcinski, J. L. Brimhall, and H. E. Kissinger, Ref. 2,
    p. 449.

6. L. LeNaour and N. Azam, Conf. of the Soc. of Electron Microscopy, Nantes, May 1972.

7. J. A. Sprague and F. A. Smidt, NRL Dept. #2555, p. 25.

8. R. S. Nelson, J. A. Hudson, D. J. Masey, T. P. Walters, T. M. Williams, Ref. 2, p. 430.

9. W. G. Johnston, J. H. Rosolowski, A. M. Takalo, and T. Lauritzen, Proc. Sixth ASTM Intern. Symp. on Effects of Radiation on Structural Materials (Los Angeles), 1972.

10. S. G. MacDonald and A. Taylor, Proc. Sixth ASTM Intern. Symp. on Effects of Radiation on Structural Materials (Los Angeles), 1972.

11. J. A. Hudson, D. J. Mazey and R. S. Nelson, J. Nuclear Materials 41, 241, 1971.

12. C. W. N. Silvent, G. Sainfort, J. Nucl. Mat. 46, 353, 1973.

THE TEMPERATURE DEPENDENCE OF IRRADIATION INDUCED VOID SWELLING IN 20% COLD WORKED TYPE 316 STAINLESS STEEL IRRADIATED WITH 5 MeV NICKEL IONS*

K. D. Challenger and T. Lauritzen

Breeder Reactor Dept., General Electric Co.

San Jose, California 95125

ABSTRACT

The temperature dependence of swelling for 20% CW Type 316 SS containing nominally 15 ppm He has been established using 5 MeV nickel ions; swelling was measured using standard transmission electron microscopy techniques. The damage rate of this experiment is roughly 10^3 times faster than fast reactor irradiation resulting in a shift in the peak swelling temperature of +115±25°C.

Swelling is studied in the temperature range of 475 to 725°C after doses causing up to 105 displacements per atom. The technique is experimentally verified as a valid simulation of fast reactor irradiation for temperatures up to 625°C when the appropriate temperature shift is applied.

Swelling in the cold worked steels is less than that previously reported for an annealed steel at all temperatures and doses studied. However, above 625°C thermal recovery of the cold work results in a swelling behavior that gradually approaches that of annealed steel as the temperature and dose are increased. Swelling occurs in the temperature range of 525 to ∿725°C with a maximum swelling at 625°C for both annealed and cold worked steels.

* All work performed under the sole sponsorship of the General Electric Co.

Void distributions in the cold worked steel become increasingly nonuniform as the irradiation temperature is increased; voids are distributed in bands following the original deformation substructure. Voids nucleate easier and grow faster in regions that have undergone recovery during irradiation. The lower swelling in the cold worked steels is brought about by a combination of reduced void size and number density.

INTRODUCTION

Core structural materials in liquid metal cooled fast breeder reactors will be subjected to fast neutron fluences of the order of 3×10^{23} n/cm^2 while operating in the temperature range 300 to 700°C. It has been well established that metals and alloys are susceptible to marked radiation effects under these conditions.

When an energetic particle, such as a fast neutron, enters a metallic lattice it interacts with the lattice atoms in many different ways; the types of interactions germane to this study are those that cause an atom to be displaced from its normal lattice site producing a Frenkel pair. One manifestation of this damage is void formation which results in a volumetric swelling due to the displaced interstitials. Void formation in the 300 series austenitic stainless steels (SS) is now a commonplace occurrence after neutron fluences in excess of about 1×10^{22} n/cm^2. There are, however, several prerequisites for void formation that limit its occurrence to the temperature range of .3 to .55T_m (T_m is the melting point in degrees absolute).

Cold working has been shown to significantly reduce the irradiation induced swelling of austenitic stainless steels at least up to 7×10^{22} n/cm^2 (the highest fluence data available) (1). Based partially on this observation, cold worked Type 316 steel was selected by both the US (FTR and Demo Plant) and the UK (PFR) as the primary core structural material in their respective LMFBR systems.

The US Demo Plant is designed for a total core exposure of about 4×10^{23} n/cm^2. The painfully slow rate at which damage accumulates in the present test reactors ($\sim 6 \times 10^{22}$ n/cm^2 per year) requires approximately six years of normal reactor operation in the US test reactor, EBR-II, to reach this fluence level. Therefore, complete information regarding in-reactor dose and temperature dependence of swelling in this steel will not be available for some time.

The present work was undertaken in an attempt to hasten the availability of this critical design data. Its principal objective was to assess and characterize the high fluence swelling behavior of 20 percent cold worked Type 316 steel by simulation techniques.

TABLE I

Chemical Composition (wt.%) of Target Materials

Heat Numbers

Element	81621 316 SS	(ref. 3) 820467 316 SS
C	0.055	0.05
Mn	1.61	1.86
Si	0.54	0.73
P	0.004	0.018
S	0.006	0.009
Cr	17.40	17.03
Ni	13.60	13.17
Mo	2.29	2.08

TABLE II

Processing History

annealed (ref. 3) 1010°C - 1 hr.(vacuum)-rapid argon quench+age[*]

20% CW   1177°C - 30 sec (dry hydrogen)- rapid cool +
                                          20% CW + age

[*]age = 593°C for 200 hrs. in vacuum

Work at Harwell (2) has shown that damage rates approximately 10^3 times faster than possible in reactor could be attained by heavy ion bombardment and that the resultant material behavior was qualitatively similar to that resulting from fast neutron irradiation. Our own work (3) confirmed that observation and has shown that such techniques can produce quantiative simulation as well. The results presented in this paper were generated by a technique developed by Johnston et al (3) using 5 MeV nickel ions as the bombarding species.

EXPERIMENTAL

The material used for this study was double vacuum melted Type 316 stainless steel supplied by the mill as .012" ($\sim$305µm) thick - 20% cold worked strip. This was reduced to 0.005" ($\sim$125µm) by grinding with 600 grit SiC paper completely submerged in water. The chemical composition of the steel and its processing history are given in Tables I and II. A thermal age at 593°C for 200 hours was performed in an attempt to simulate the purely thermal effects of reactor exposure.

The experimental techniques for the ion bombardment have been previously reported (3). In summary the procedure involves injecting 20 ppm ± 30% helium uniformly to about 25 µm into a 125µm thick slab measuring 2.5x1.25 cm. Disks 3mm in diameter are then spark machined from this slab and a 10 µm deep dimple is electropolished into each disk to remove any possible surface disturbances. The prepared discs thus contain about 15 µm of helium-injected material. This sample is clamped into a molybdenum furnace block, heated to temperature in a vacuum ranging from 3 to $8x10^{-7}$ torr and irradiated with a particle flux of $1.3x10^{13}$ nickel ions per cm^2 per second. The resultant displacement rate is approximately $2x10^{-2}$ dpa-sec^{-1}. The peak damage zone created by the 5 MeV nickel ions is located approximately 9000 Å from the irradiated surface. Thus, it is necessary to remove 8000 Å from the irradiated surface. This is accomplished by placing a small dot of stop-off lacquer on the irradiated surface and electropolishing ($90CH_3COOH-10HClO_4$, 20 volts) for a previously calibrated time. The lacquer is then removed and the height of the step between the irradiated surface and the newly electropolished surface is measured with a diamond-tipped stylus profilometer capable of resolving surface discontinuities as small as 25Å (4). This position is known to within 500 Å. This surface is then masked off and back thinned by a conventional jet polishing technique.

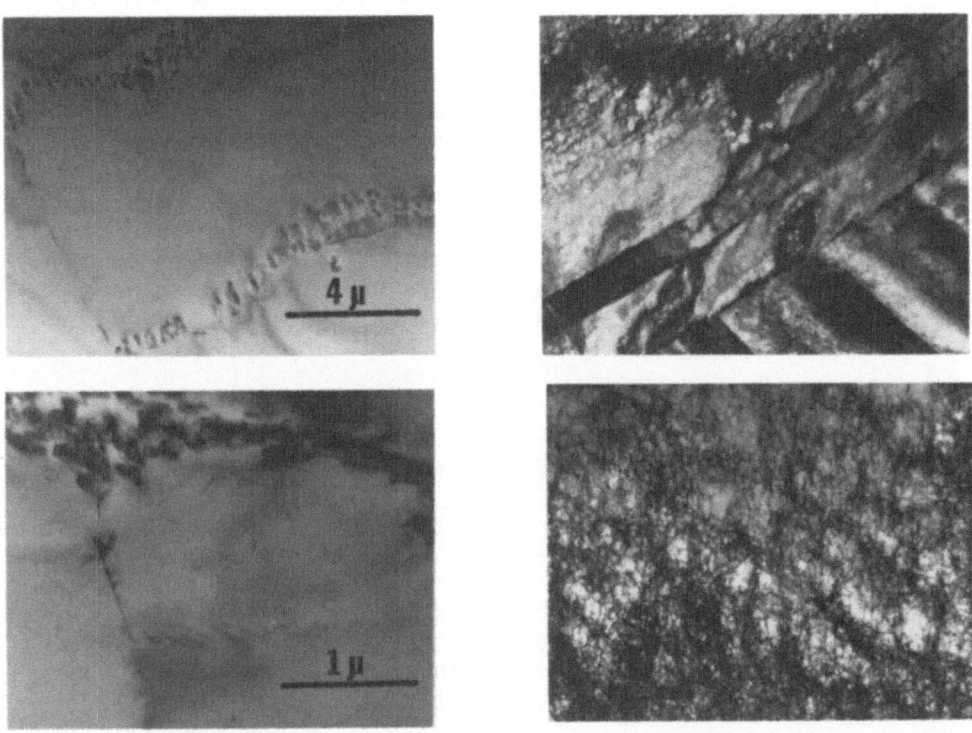

ANNEALED   20% COLD WORKED

FIGURE 1   PRE-IRRADIATION SUBSTRUCTURES

TEM evaluation was performed on a JEM 200 electron microscope operating at 200 KeV. Typical regions were selected and ± 6° stereo pairs recorded. When the swelling (void distribution) was nonuniform and typical regions difficult to select, a low magnification composite photograph of approximately 1/2 of the electron transparent region was constructed and divided into regions of high, low and zero swelling. High magnification stereo pairs representative of each of these regions were taken and the overall swelling calculated from weighted fractions.

The swelling was determined by counting the voids and measuring their diameters. The foil thickness was determined by parallax measurements of the stereo pairs. The swelling reported is $\Delta V/V_o$ (void volume/original volume) while the number density is simply the number of voids per unit volume of swollen material. The average void diameter is the simple arithmetic average.

The displacement damage calculation has also been previously reported (5). The method is consistent with that employed by Argonne workers using Brice's calculation (6). In order to compare this damage with that caused by neutron irradiation we employ the calculation of Doran and Kulcinski (7) that yields a conversion of 77.6 dpa = 1×10^{23} n/cm ^{2} in the core center of EBR-II.

RESULTS AND DISCUSSION

The preirradiation microstructure and substructure of the 20% cold worked steel is shown in Figure 1. The dislocation density was much too high to measure by TEM techniques and thus was assumed to be greater than 5×10^{10} lines/cm^2. The absence of any significant amount of precipitation, either intragranular or intergranular, is quite surprising as the aging treatment at 593°C should have been sufficient to precipitate copious amounts of $M_{23}C_6$ (8). This behavior is not understood at this time. The substructure consists of dislocation tangles, multipoles and microtwins.

The experimental technique employed for this study has been shown to provide a valid simulation of neutron irradiation of cold worked Type 316SS at 575°C. At 625°C the correlation appeared good, but excessive scatter in the reactor swelling data precluded any definitive statement (9). However, very little reactor data are available for this comparison.

In order to make the comparison with reactor data a temperature shift of +115°C (resulting from the higher damage rate) was assumed as this is the magnitude of the shift in the peak swelling temperature reported for annealed Type 316SS irradiated by this same technique (3). We have chosen this empirical approach over a theoretical one although the direction and magnitude of this

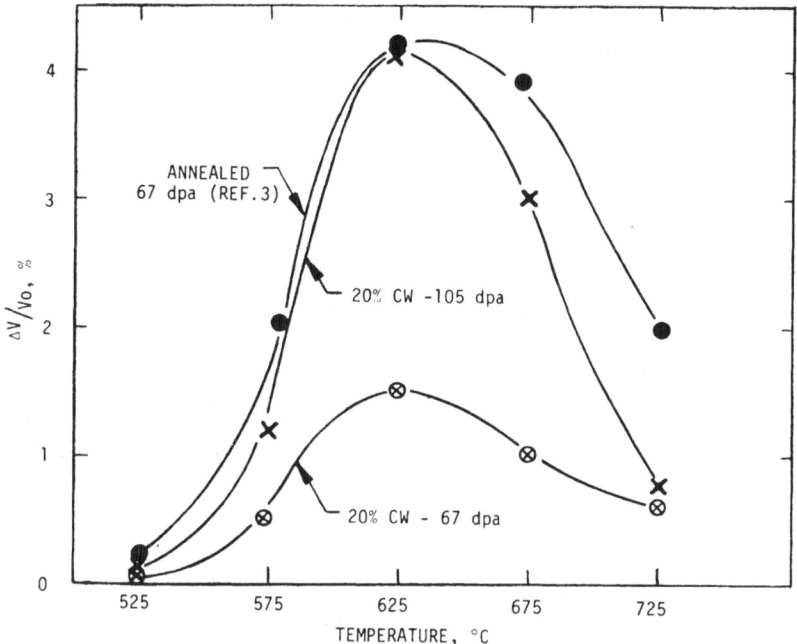

FIGURE 2 - TEMPERATURE DEPENDENCE OF SWELLING FOR TARGET N
(20% CW 316SS) WITH THE SWELLING DATA OF ANNEALED TYPE 316SS
FROM REF. 3 SHOWN FOR COMPARISON

shift is consistent with that theoretically predicted (10). The
temperature dependence of swelling, $\Delta V/V_o$, at nominally 70 and
105 dpa, is shown in Figure 2. The swelling data for annealed
Type 316SS generated by Johnston et al (3) are included for comparison. A bell shaped curve with a lower magnitude of swelling
but similar in shape to that for the annealed steel is observed.

A comparison of the void sizes and void number densities is
presented as a function of temperature in Figures 3 and 4 respectively. The lower swelling in the cold worked steel has been
effected thru the combination of a reduction in void number density and void size. A better understanding of the effect of cold
work on void size can be gained by reference to Figure 5 where
the void size distributions for the annealed and cold worked steel
are compared at different temperatures. A much greater portion of
the void population in the cold worked steel is in the <175Å range
while the maximum void sizes are essentially comparable. However,
as the temperature is increased from 675 to 725°C the void size

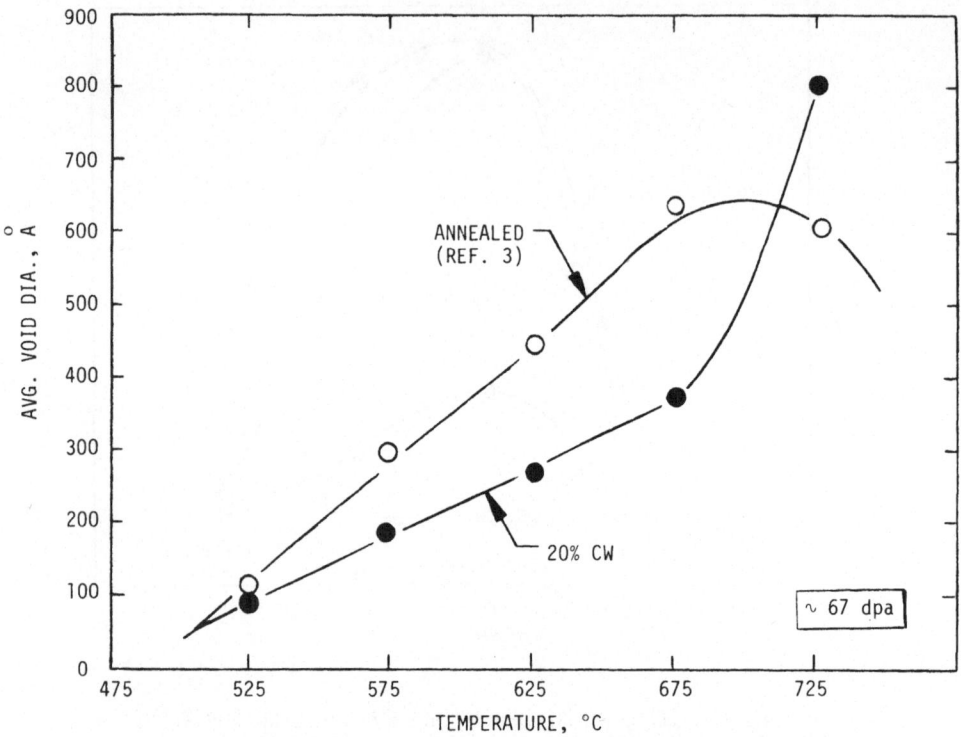

FIGURE 3 - TEMPERATURE DEPENDENCE OF THE AVERAGE VOID DIAMETER

distribution become essentially the same. Thus, at 725°C we are
no longer irradiating a cold worked steel as the material has
completely recovered.

The microstructural distribution of voids as a function of
temperature is shown in Figure 6. Areas containing very large
voids are clearly delineated while other regions contain either
smaller or no voids. The uniformity in void size distribution
decreases with increasing temperature. The regions containing the
larger voids are believed to be regions of initially low dislocation density and/or are regions that have undergone recovery during bombardment. The void size distribution in the cold worked
steel after bombardment at 725°C is very similar to that for an
annealed steel, Figure 5d. Apparently at this temperature the
cold work has recovered and we have irradiated an essentially annealed steel.

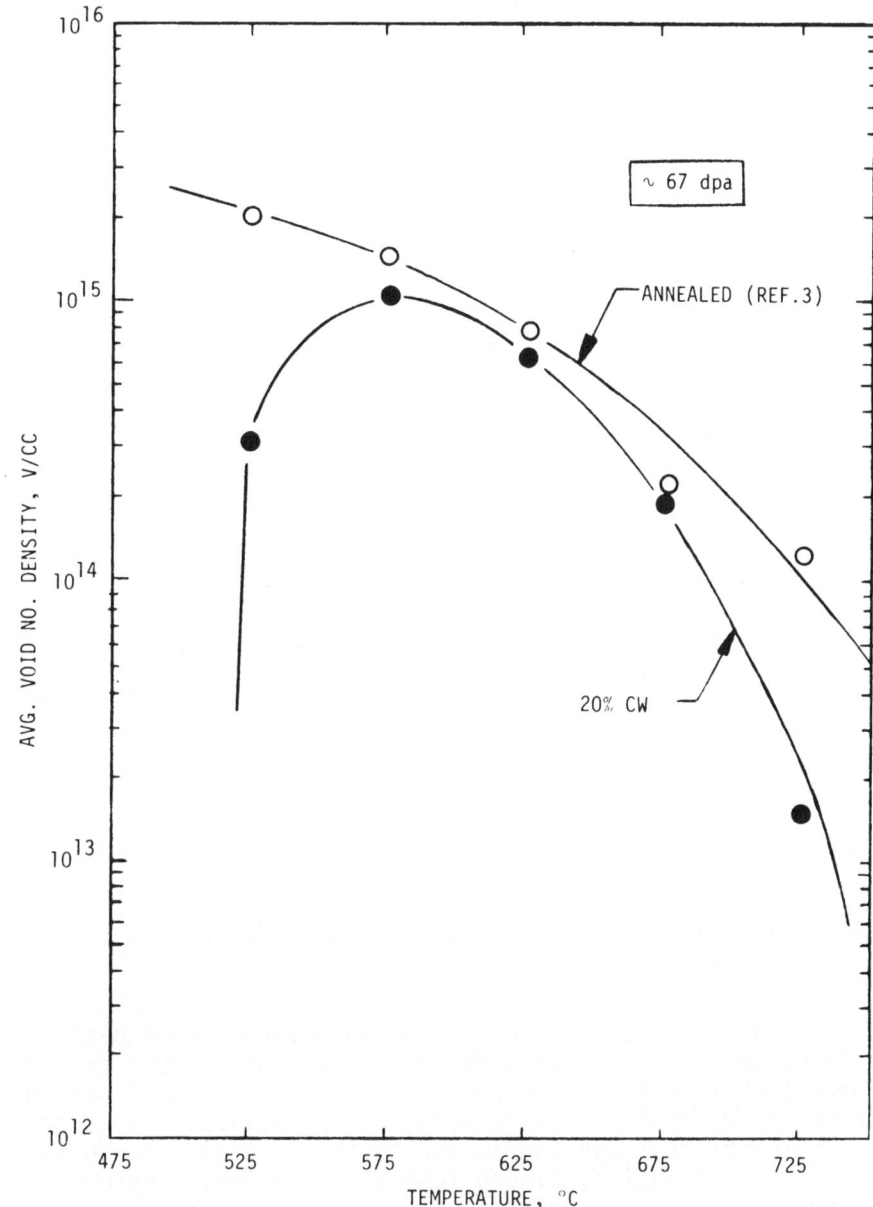

FIGURE 4 - TEMPERATURE DEPENDENCE OF THE AVERAGE VOID NUMBER DENSITY

The high dislocation density of a cold worked steel causes a reduction in the irradiation-induced point defect supersaturation by providing a great number of unsaturable point defect sinks

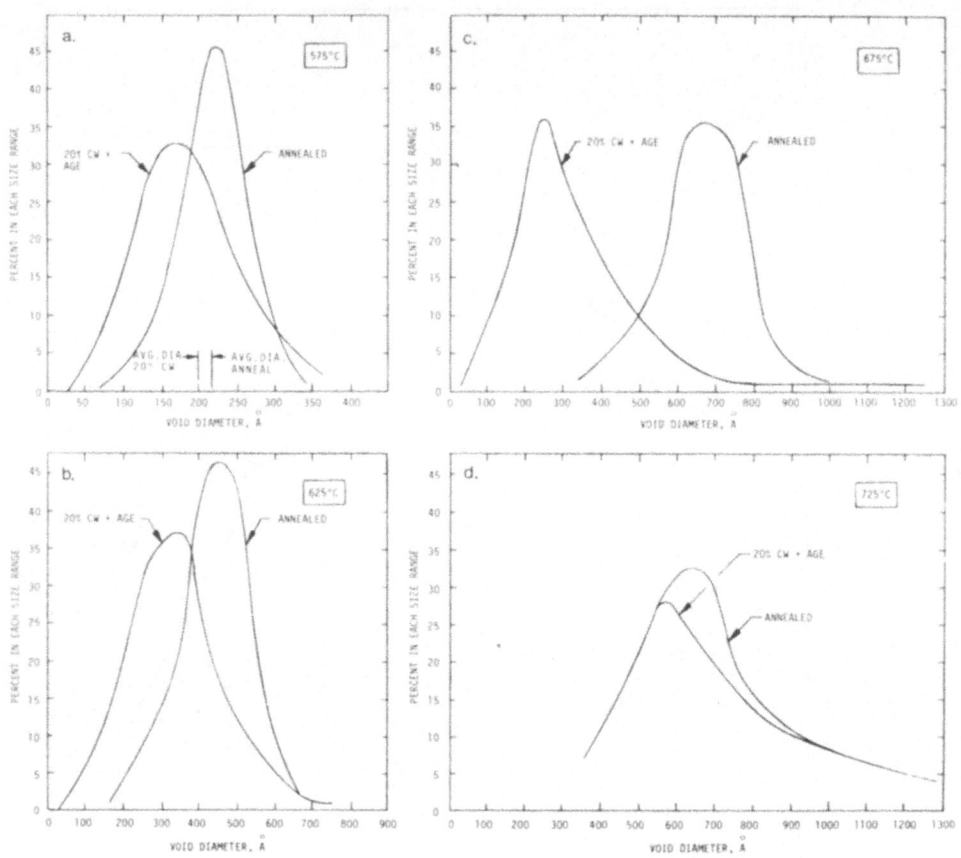

FIGURE 5 - VOID SIZE DISTRIBUTION IN 316 SS AFTER ~ 67 dpa AT
VARIOUS TEMPERATURES

(dislocations). Since void nucleation is strongly dependent on
defect supersaturation, cold working reduces the nucleation rate.
Void growth rate is also expected to be reduced as there are fewer vacancies available for void growth. The maximum void size
is unaffected by cold working as due to the initially nonuniform
dislocation distribution, Figure 1, small very local regions
exist where the CW steel behaves as an annealed steel.

   As previously mentioned, the high damage rate causes an upward shift in the peak swelling temperature. Thus, by necessity,
all irradiations are carried out approximately 115°C higher than
a comparable reactor experiment. When studying an inherently unstable material such as a cold-worked steel one must recognize
the possibility that the behavior observed by ion bombardment may
not be representative of in-reactor behavior due to the higher
temperatures necessarily employed. The fact that the void dis-

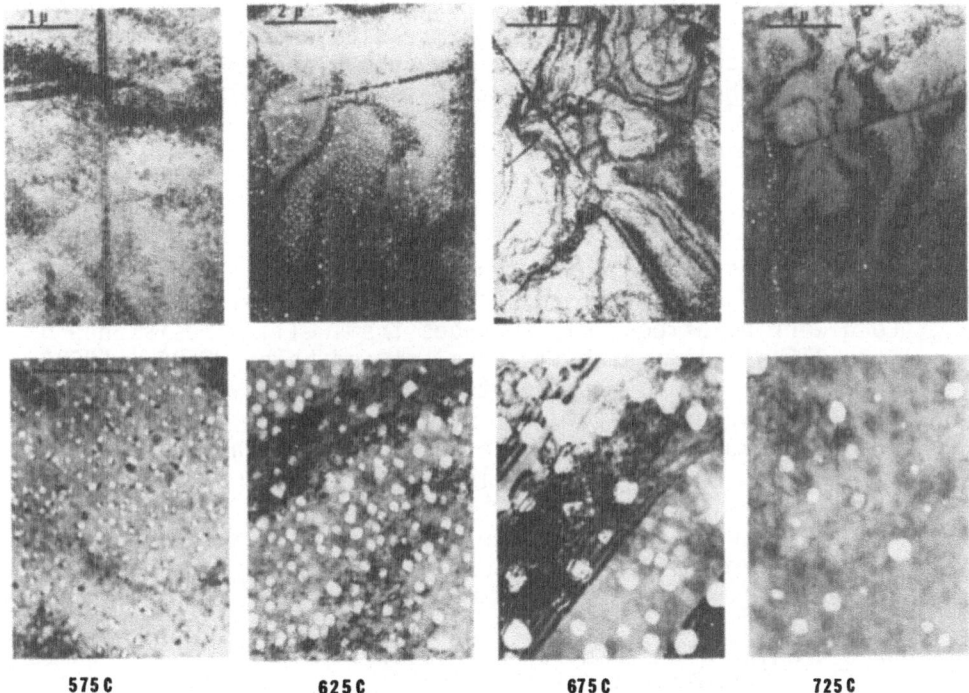

575 C   625 C   675 C   725 C

FIGURE 6   EFFECT OF TEMPERATURE ON THE VOID DISTRIBUTION IN 20%
            CW 316SS AFTER 105 DPA

tribution closely resembles the original deformation substructures
and that the distribution of voids becomes increasingly nonuniform
as the temperature is increased above 625°C supports the previous
statement. However, we must also recognize that although we must
bombard at substantially higher temperatures, the bombardment time
is very short relative to reactor exposure to a comparable damage
level (1 hour as compared to 1 to 2 years in reactor) and allowances
must be made accordingly.

        In an effort to minimize thermal recovery, the samples were
heated to the irradiation temperatures (in less than 10 minutes)
and irradiated within ten minutes after reaching temperature. It
is our judgement that for temperatures below 650°C, the observed
behavior is typical of in-reactor behavior after allowing for the
+115°C temperature shift, while above 650°C, thermal recovery of
the cold work results in a behavior that gradually approaches an
annealed steel.

        The pronounced nonuniformity in the void distribution along
with the observation that the void distribution closely resembles
the original deformation substructure strongly suggests that it is
the dislocation distribution and the thermal stability of this

distribution more than the average dislocation density that are of prime importance in controlling the irradiation induced swelling in a cold worked steel. Any material or processing modifications that would increase the uniformity of the dislocation distribution or the thermal stability of this substructure should enhance the material's swelling resistance.

CONCLUSIONS

1.   A stable dislocation distribution is essential in a cold worked stainless steel to assure swelling resistance.
2.   Cold working effectively enhances the swelling resistance of Type 316 steel at high fluences, at and below 625°C ($\sim$500°C in-reactor) due to a combination of reduced average void size and number density.
3.   Ion bombardment provides an accurate and reliable simulation of in-reactor swelling of cold worked Type 316SS for temperatures below about 650°C.
4.   The peak swelling temperature for both cold worked and annealed Type 316SS is 625°C in this experiment ($\sim$500°C in-reactor).

ACKNOWLEDGEMENTS - The authors are grateful to Drs. W. G. Johnston and J. H. Rosolowski for their assistance in this experiment.

REFERENCES

1.   C. Cawthorne, et al in Voids Formed by Irradiation of Reactor Materials, Symposium Proceedings, Reading University, 1971, p. 35.
2.   R. S. Nelson and D. J. Mazey, Symposium on Radiation Damage in Reactor Materials, IAEA, Vienna, 1969, p. 1971.
3.   W. G. Johnston, J. H. Rosolowski, A. M. Turkalo and T. Lauritzen, to appear in The Effects of Irradiation on Substructure and Mechanical Properties of Metals and Alloys, ASTM STP 529, October 1973.
4.   The instrument used was a Dektak, manufactured by Sloan Industries, Santa Barbara, CA.
5.   W. G. Johnston, J. H. Rosolowski, A. M. Turkalo, and T. Lauritzen, "Nickel Ion Bombardment of Type 304 Stainless Steel: A Comparison with Fast Reactor Swelling Data", to appear in J. Nucl. Mat'l.
6.   D. K. Brice, Rad. Effects, 11, 51 (1971).
7.   D. G. Doran and G. L. Kulcinski, Rad. Effects, 9, 283 (1971).
8.   J. E. Spuiell et al, Met. Trans., 4, 1533 (1973).
9.   K. D. Challenger, PhD Thesis, University of Cincinnati, 1973.
10. R. Bullough, and R. C. Perrin in Voids Formed by Irradiation of Reactor Materials, Symposium Proceedings, Reading University, 1971, p. 79.

AUTHOR INDEX

Abel, F., 377
Adde, R., 27
Alessandrini, E., 169
Alexander, R. B., 365
Ansaldo, E. J., 365, 379

Baglin, J. E. E., 193
Bauer, W., 533
Beeler, J. R., 651
Beeler, M. F., 651
Behrisch, R., 573
Bernas, H., 27, 377, 459
Beyer, G., 361
Biersack, J. P., 307
Bisson, C. L., 423
Blewer, R. S., 557
Borders, J. A., 179
Brown, F., 111
Bruneaux, M., 377
Brusic, V., 169
Buckel, W., 3, 35

Campisano, S. U., 157
Challenger, K. D., 687
Chaumont, J., 27, 377, 459
Choyke, W. J., 87
Chu, W. K., 193
Cohen, C., 377
Cohen, R. L., 361
Crozat, P., 27

Das, S. K., 543
Dearnaley, G., 63, 101, 123
DeBonte, W. J., 147
de Waard, H., 317
Deutch, B. I., 361, 365

Diehl, J., 507
Drentje, S. A., 353

Eckstein, W., 597
Edwards, D. E., 521
Erents, S. K., 585

Feldman, L. C., 317, 365
Fink, D., 307
Fleischer, L. R., 87
Foti, G., 157
Freyhardt, H. C., 47

Gellert, J., 365
Gettings, M., 241
Goh, K. H., 269
Gomez-Giraldez, C., 469
Gotthardt, R., 507
Grasso, F., 157

Harrison, D. E., 427
Hartley, N. E. W., 123
Heim, G., 35
Hertel, B., 469, 507

Jouffrey, B., 459

Kalkman, G. N., 353
Kaminsky, M., 543
Kaufmann, E. N., 379
Kesternich, W., 495
Kornelson, E. V., 521
Kräutle, H., 193
Krien, K., 379
Kulcinski, G. L., 613

SUBJECT INDEX

701